Lecture Notes in Computer Science 15828

Founding Editors

Gerhard Goos
Juris Hartmanis

Editorial Board Members

Elisa Bertino, *Purdue University, West Lafayette, IN, USA*
Wen Gao, *Peking University, Beijing, China*
Bernhard Steffen, *TU Dortmund University, Dortmund, Germany*
Moti Yung, *Columbia University, New York, NY, USA*

The series Lecture Notes in Computer Science (LNCS), including its subseries Lecture Notes in Artificial Intelligence (LNAI) and Lecture Notes in Bioinformatics (LNBI), has established itself as a medium for the publication of new developments in computer science and information technology research, teaching, and education.

LNCS enjoys close cooperation with the computer science R & D community, the series counts many renowned academics among its volume editors and paper authors, and collaborates with prestigious societies. Its mission is to serve this international community by providing an invaluable service, mainly focused on the publication of conference and workshop proceedings and postproceedings. LNCS commenced publication in 1973.

Vincent Chau · Christoph Dürr · Minming Li ·
Pinyan Lu
Editors

Frontiers of Algorithmics

19th International Joint Conference, IJTCS-FAW 2025
Paris, France, June 30 – July 2, 2025
Proceedings

Editors
Vincent Chau
Université Evry Paris-Saclay
Evry Cedex, France

Christoph Dürr
Sorbonne University
Paris, France

Minming Li
Department of Computer Science
City University of Hong Kong
Kowloon Tong, Hong Kong

Pinyan Lu
Shanghai University of Finance
and Economics
Shanghai, China

ISSN 0302-9743 ISSN 1611-3349 (electronic)
Lecture Notes in Computer Science
ISBN 978-981-96-8311-6 ISBN 978-981-96-8312-3 (eBook)
https://doi.org/10.1007/978-981-96-8312-3

© The Editor(s) (if applicable) and The Author(s), under exclusive license
to Springer Nature Singapore Pte Ltd. 2025

This work is subject to copyright. All rights are solely and exclusively licensed by the Publisher, whether the whole or part of the material is concerned, specifically the rights of translation, reprinting, reuse of illustrations, recitation, broadcasting, reproduction on microfilms or in any other physical way, and transmission or information storage and retrieval, electronic adaptation, computer software, or by similar or dissimilar methodology now known or hereafter developed.
The use of general descriptive names, registered names, trademarks, service marks, etc. in this publication does not imply, even in the absence of a specific statement, that such names are exempt from the relevant protective laws and regulations and therefore free for general use.
The publisher, the authors and the editors are safe to assume that the advice and information in this book are believed to be true and accurate at the date of publication. Neither the publisher nor the authors or the editors give a warranty, expressed or implied, with respect to the material contained herein or for any errors or omissions that may have been made. The publisher remains neutral with regard to jurisdictional claims in published maps and institutional affiliations.

This Springer imprint is published by the registered company Springer Nature Singapore Pte Ltd.
The registered company address is: 152 Beach Road, #21-01/04 Gateway East, Singapore 189721, Singapore

If disposing of this product, please recycle the paper.

Preface

We are delighted to present the proceedings of the International Joint Conference on Theoretical Computer Science - Frontier of Algorithmic Wisdom, IJTCS-FAW 2025. This conference was a combination of the 6th International Joint Conference on Theoretical Computer Science (IJTCS) and the 19th International Conference on Frontiers of Algorithmic Wisdom (FAW), which took place in Paris, France, from June 30 to July 2, 2025. The FAW conference originated as the Frontiers in Algorithms Workshop in 2007 in Lanzhou, China, and has been held annually from 2007 to 2024 with published archival proceedings. IJTCS started in 2020 with the aim of attracting presentations on active topics in various tracks related to theoretical computer science.

To support a range of emerging research avenues, IJTCS and FAW collaborated to host an event that fostered the exchange of insights into both recent breakthroughs and enduring contributions in theoretical computer science. The conference featured contributed talks submitted to three tracks in IJTCS-FAW 2025, namely

- Track A: The 19th Conference on Frontiers of Algorithmic Wisdom
- Track B: Blockchain, Distributed Computing, Multi-Agents
- Track C: Game Theory, Algorithmic Game Theory, Machine Learning, Electronic Commerce

The Program Committee, comprising 37 active researchers from the field, reviewed 63 submissions from the three tracks and selected 28 of them as full papers and 2 as short papers. Each submission underwent a rigorous double-blind peer review process and received an average of three reviews.

In addition to the regular talks, IJTCS-FAW 2025 featured keynote talks by Gabrielle Demange (École d'économie de Paris), Xiaoming Sun (Institute of Computing Technology, Chinese Academy of Sciences), and Shanghua Teng (University of Southern California).

The conference also included invited talks in focused tracks on Quantum Computing, Multi-agent Learning/Systems/Games, and a Young Researchers Forum.

We are grateful to Publications Chair Vincent Chau for his editorial work and to the local organizing committee.

June 2025

Christoph Dürr
Minming Li
Pinyan Lu

Organization

Conference Chairs

John E. Hopcroft	Cornell University, USA
Huimin Lin	University of Chinese Academy of Sciences, China

General Chair

Hao Li	LISN, Paris-Saclay University, France

Program Committee Chairs

Christoph Dürr	Sorbonne University, France
Minming Li	City University of Hong Kong, China
Pinyan Lu	Shanghai University of Finance and Economics, China

Track Chairs

The 19th Conference on Frontiers of Algorithmic Wisdom

Thomas Erlebach	Durham University, UK
Jialin Zhang	Institute of Computing Technology, Chinese Academy of Science, China

Blockchain, Distributed Computing, Multi-agents

Ralf Klasing	University of Bordeaux, France
Jiasun Li	George Mason University, USA

Game Theory, Algorithmic Game Theory, Machine Learning, Electronic Commerce

Yukun Cheng	Jiangnan University, China
Johanne Cohen	Paris-Saclay University, France

viii   Organization

Vianney Perchet	ENSAE Paris, France

Local Organization

Evripidis Bampis	Sorbonne University, France
Vincent Chau	Université Évry Paris-Saclay, France
Christoph Dürr	Sorbonne University, France

Advisor Committee

Wen Gao	Peking University, China
Hong Mei	Peking University, China
Pingwen Zhang	CSIAM and Wuhan University, China

Steering Committee

Xiaotie Deng	Peking University, China
Jian Li	Tsinghua University, China
Pinyan Lu	Shanghai University of Finance and Economics, China
Jianwei Huang	Chinese University of Hong Kong, Shenzhen, China
Lijun Zhang	Chinese Academy of Sciences, China

Program Committee

Evripidis Bampis	LIP6 Sorbonne Université, France
Cristina Bazgan	Université Paris Dauphine, France
Xiaohui Bei	Nanyang Technological University, Singapore
Cédric Bentz	Conservatoire national des arts et métiers, France
Gewu Bu	LIMOS, Université Clermont Auvergne, France
Vincent Chau	Université Évry Paris-Saclay, France
Yurong Chen	Inria, École Normale Supérieure, PSL Research University, France
Yukun Cheng	Jiangnan University, China
Johanne Cohen	Université Paris-Saclay, CNRS, LISN, France
Christoph Dürr	Sorbonne University, France
Thomas Erlebach	Durham University, UK

Qilong Feng	Central South University, China
Yiding Feng	Hong Kong University of Science and Technology, China
Ralf Klasing	CNRS and University of Bordeaux, France
Bo Li	Hong Kong Polytechnic University, China
Hao Li	Paris-Saclay University, France
Jiasun Li	George Mason University, USA
Minming Li	City University of Hong Kong, China
Ya-Chun Liang	National Cheng Kung University, Taiwan
Hsiang-Hsuan Liu	Utrecht University, Netherlands
Pinyan Lu	Shanghai University of Finance and Economics, China
Kelin Luo	University at Buffalo, USA
Pan Peng	University of Sheffield, UK
Vianney Perchet	Crest, ENSAE & Criteo AI Lab, France
Qi Qi	Renmin University of China, China
David Saulpic	CNRS, IRIF, France
Sylvain Sené	Aix-Marseille Université, France
Warut Suksompong	National University of Singapore, Singapore
Changjun Wang	Academy of Mathematics and Systems Science, Chinese Academy of Sciences, China
Ye Wang	University of Macau, China
Zihe Wang	Renmin University of China, China
Xiaowei Wu	University of Macau, China
Fang-Yi Yu	George Mason University, USA
Jialin Zhang	Institute of Computing Technology, Chinese Academy of Sciences, China
Jie Zhang	University of Bath, UK
Jinshan Zhang	Zhejiang University, China
Ruilong Zhang	Technical University of Munich, Germany
Dengji Zhao	ShanghaiTech University, China

Additional Reviewers

Faisal Abu-Khzam
Grégoire Beaudoire
Thomas Bellitto
Valentin Bouquet
Xiaolin Bu
Po-An Chen
Shengminjie Chen
Denis Cornaz

Yinhao Dong
Yuejia Dou
Andrew Draganov
Max Dupré La Tour
Romaric Duvignau
Angelo Fanelli
Xiangyu Guo
Weiqiang He

Arnaud Labourel
Stefanos Leonardos
Fu-Hong Liu
Hanbing Liu
Haodong Liu
Yuchao Ma
Christophe Picouleau
Weiran Shen
Jean-Marc Talbot
Nicholas Teh
Ernest van Wijland
Zongqi Wan
Bingzhe Wang

Chenhao Wang
Yuyang Wang
Dimitri Watel
Michalis Xefteris
Xiang Yan
Yu Yokoi
Zhen Zhang
Zhijie Zhang
Muyang Zhao
Yingchao Zhao
Chenghan Zhou
Shengwei Zhou

Contents

Domination in Diameter-Two Graphs and the 2-Club Cluster Vertex
Deletion Parameter 1
  Faisal N. Abu-Khzam and Lucas Isenmann

k-Universality of Regular Languages Revisited 16
  Duncan Adamson, Pamela Fleischmann, Annika Huch, Tore Koß,
  and Florin Manea

Comparing the Hardness of Online Minimization and Maximization
Problems with Predictions 33
  Magnus Berg

Complexity Classes for Online Problems with and Without Predictions 49
  Magnus Berg, Joan Boyar, Lene M. Favrholdt, and Kim S. Larsen

Scheduling with Testing: Competitive Algorithms for Minimizing
the Total Weighted Completion Time in the Adversarial Model 64
  Felix Buld and Andreas S. Schulz

Mixed Graph Covering with Target Constraints 78
  Xujin Chen, Xiyuan Deng, Xiaodong Hu, and Changjun Wang

Multiplication of 0-1 Matrices via Clustering 92
  Jesper Jansson, Mirosław Kowaluk, Andrzej Lingas, and Mia Persson

From MAXCUT to MAXNAESAT: Elegant Proofs and Algorithmic
Advances 103
  Sangram K. Jena and K. Subramani

Exact Algorithms for the Maximum k-Balanced Weighted Biclique Problem ... 118
  Jingyi Liu, Jianxin Wang, Qilong Feng, and Feng Shi

Approximation Algorithms for Individual Preference Facility Location 128
  Shuilian Liu, Yicheng Xu, and Yong Zhang

The Online Power Cover Problem on a Line 142
  Zhonghao Liu, Man Xiao, Xiaofei Liu, and Weidong Li

The Subinterval Cover Problem 152
  Kelin Luo, Chenran Yang, Zonghan Yang, and Yuhao Zhang

Oblivious Robots Under Round Robin: Gathering on Rings 166
   Alfredo Navarra and Francesco Piselli

Finding a Set of Long Common Substrings with Repeats from m Input
Strings 181
   Tiantian Li, Lusheng Wang, and Daming Zhu

An LP-Rounding Based Algorithm for Soft Capacitated Facility Location
Problem with Submodular Penalties 193
   Hanyin Xiao, Jiaming Zhang, Zhikang Zhang, and Weidong Li

Less-Excludable Mechanism for DAOs in Public Good Auctions 207
   Jing Chen and Wentao Zhou

TBDS: Transaction-Based Data Sharing 222
   *Wu Xin, Hongyin Chen, Xiaoqi Dong, Jichen Li, Xiaotie Deng,
   Zhonghai Wu, and Bin Xiao*

Pure Nash Equilibria of Weighted Picking Sequence Protocol is WEF1
for Two Agents 238
   Rufan Bai, Huahua Miao, Xiaowei Wu, Cong Zhang, and Shengwei Zhou

A Comparative Study of Waitlist Mechanisms: Deferral Versus
Pay-Per-Offer 252
   Zhou Chen, Qi Qi, Hao Sun, and Muyang Zhao

Optimal Repurchasing Contract Design for Efficient Utilization
of Computing Resources 264
   Zhengyan Deng, Yusen Zheng, Chenliang Sheng, and Shaowen Qin

Characterizing Strategyproofness Through Score Functions in Voting
Mechanisms 279
   *Felipe V. Furquim, Valentin Dardilhac, Daniel Cordeiro,
   and Johanne Cohen*

Minimizing Blocking Agents for Stable Matching with Partial Approval
Information 293
   Yitian Gao, Jiaxue Li, Junjie Luo, and Yiheng Zhang

The Capacity-Constrained Facility Location Problem with Ordinal
Preferences: Algorithmic and Mechanism Design Perspectives 307
   *Zifan Gong, Alexander Lam, Momcilo Mrkaic, Yachao Yan,
   and Yingchao Zhao*

Regularized Minimax-V Learning for Solving Randomly Terminating
Two-Player Zero-Sum Markov Games 321
  Xinxiang Guo and Yifen Mu

Improved Approximation of Maximin Share Fair Allocation Under
Generalized Assignment Constraints 335
  Bo Li, Md Habibur Rahman Sifat, and Ankang Sun

Optimal Hiring Strategy in Auction-Based Crowdsourcing Systems 343
  Hongtao Liu, Weiran Shen, and Yiheng Shen

Large-Scale Contextual Market Equilibrium Computation Through Deep
Learning 356
  *Yunxuan Ma, Yide Bian, Hao Xu, Weitao Yang, Jingshu Zhao,
  Zhijian Duan, Feng Wang, and Xiaotie Deng*

Fair Value Distribution in Cooperative Committee Election 372
  Ying Qin, Zeyu Ren, Zihe Wang, and Jie Zhang

A Payoff-Based Policy Gradient Method in Stochastic Games
with Long-Run Average Payoffs 385
  Junyue Zhang and Yifen Mu

Mechanism Design for Auctions with Externalities on Budgets 400
  Yusen Zheng, Yukun Cheng, Chenyang Xu, and Xiaotie Deng

Author Index 415

Domination in Diameter-Two Graphs and the 2-Club Cluster Vertex Deletion Parameter

Faisal N. Abu-Khzam[✉] and Lucas Isenmann

Department of Computer Science and Mathematics, Lebanese American University, Beirut, Lebanon
{faisal.abukhzam,lucas.isenmann}@lau.edu.lb

Abstract. The s-club cluster vertex deletion number of a graph, or sccvd, is the minimum number of vertices whose deletion results in a disjoint union of s-clubs, or graphs whose diameter is bounded above by s. We launch a study of several domination problems on diameter-two graphs, or 2-clubs, and study their parameterized complexity with respect to the 2ccvd number as main parameter. We further propose to explore the class of problems that become solvable in sub-exponential time when the running time is independent of some input parameter. Hardness of problems for this class depends on the Exponential-Time Hypothesis. We give examples of problems that are in the proposed class and problems that are hard for it.

Keywords: s-Club Cluster Vertex Deletion · Dominating set · Efficient Dominating Set · Independent Dominating Set · Connected Dominating Set · Roman Domination · Exponential-time hypothesis · Fixed-parameter sub-exponential (FPSUB)

1 Introduction

The parameterized complexity of a problem focuses mainly on the asymptotic running time with respect to specific input parameters, which can provide insights into how the running time can be improved by selecting appropriate parameter(s). This type of algorithmic analysis reveals the potential for more efficient algorithms tailored to instances where some parameters are small, thus enhancing the overall tractability of the problem.

Studying a problem with respect to a variety of possible input parameters has witnessed great increase in attention and there have been several studies in the recent literature [4,8,11,14]. The relation between the various parameters is of interest by itself. For example, the deletion into bounded degree parameter (dbd) is more general than the vertex cover parameter (vc), thus a problem that is fixed-parameter tractable (FPT) with respect to dbd is also FPT with respect to vc. Moreover, if a problem does not admit a polynomial size kernel (modulo some hierarchy condition) with respect to vc, then it is hard to "kernelize" with respect to dbd. In general, given a problem X that can be parameterized by any of two parameters k and t, we say that parameter k is smaller than parameter t if and only if the family of Yes-instances for t is a subset of the family of

Yes-instances for k. As such, it would be more interesting from a complexity standpoint to study the complexity with respect to the smaller parameter.

There are many generalizations of the VERTEX COVER problem that have resulted in smaller parameters. A notable example is the cluster vertex deletion parameter (cvd): number of vertices whose deletion yields a disjoint union of cliques (of course, VERTEX COVER corresponds to the special case where each of the resulting cliques is of size one). Recently, the parameterized complexity of a variety of domination problems was studied with respect to cvd [11]. In this paper, we further refine the cluster deletion parameter by considering the s-club vertex deletion parameter (sccvd): number of vertices whose deletion results in a disjoint union of s-clubs, or graphs of diameter at most s. In particular, we consider the 2-club cluster vertex deletion parameter, or 2ccvd, being the very next parameter in the considered chain of "deletion into s-clubs" parameters for $s = 1, 2 \ldots$

We mainly consider domination problems by studying several variants of the classic dominating set problem, namely: EFFICIENT DOMINATING SET, INDEPENDENT DOMINATING SET and a variety of ROMAN DOMINATION problems. To study parameterization with respect to 2ccvd, we first consider the complexity of each of these problems on graphs of diameter (at most) two. We then obtain the resulting parameterized complexity of the problem with respect to 2ccvd. Finally, we propose a study that can further classify problems from a parameterized complexity viewpoint, by considering problems that admit sub-exponential time algorithms when some input parameter(s) are small enough, or constant.

2 Preliminaries

Basic graph theoretic terminology is used throughout this paper. All the considered graphs are assumed to be undirected, simple and unweighted. The neighborhood of a vertex v of a graph $G = (V, E)$ is the set of vertices adjacent to v in G: $N(v) = N^1(v) = \{w \in V : uw \in E\}$, and its closed neighborhood is $N[v] = N^1[v] = N(v) \cup \{v\}$. This notion can be extended to sets by defining, for $S \subset V$, $N(S) = N^1(S) = \bigcup_{v \in S} N(v)$ and $N[S] = N^1[S] = \bigcup_{v \in S} N[v]$. Furthermore, for $j > 1$, $N^j[v] = N[N^{j-1}(v)]$ and $N^j(v) = N^j[v] \setminus \{v\}$. In other words, $N^j[v]$ is the set of vertices that are within distance j from v.

A graph $G = (V, E)$ is an s-club if for every vertex $v \in V$, $N^s[v] = V$, i.e., every pair of vertices of G are within a distance of s from each other. The subgraph induced by a set $S \subset V$ will be denoted by $G[S]$. A dominating set in $G = (V, E)$ is a set $D \subset V$ such that $N[D] = V$. A set D is said to be an *efficient dominating set* (EDS) if for every $v \in V, |N[v] \cap D| = 1$. It is an *independent dominating set* (IDS) if for every $v \in D, |N[v] \cap D| = 1$ (while $N[D] = V$). A dominating set D is a connected dominating set of G if $G[D]$ is connected. The s-CLUB CLUSTER VERTEX DELETION problem (s-CCVD) is defined as follows:

s-CLUB CLUSTER VERTEX DELETION
Input: A graph $G = (V, E)$ and a non-negative integer k;
Question: Is there a set $S \subset V$ of cardinality at most k such that $G - S$ is a disjoint union of s-clubs?

The s-CCVD problem seems to have been first formulated in [22] where it was shown to be solvable in polynomial-time on trees. The 2-CCVD problem was studied in [17] where it was shown to be NP-hard but fixed-parameter tractable by presenting an $\mathcal{O}^*(3.31^k)$ algorithm. The same was proved in [17] for the edge-deletion version. Further improvements and similar problems were obtained/discussed in a number of recent papers [2,9,24].

We define the *sccvd* of a graph G as the minimum integer k for which s-CCVD has a solution in G. We assume s is a small constant so computing $sccvd(G)$ takes FPT-time. In fact, the s-CCVD problem is trivially FPT when s is a constant being solvable in $\mathcal{O}^*((s+2)^k)$ time: find a path p of length $s+1$ between two vertices that are at distance exactly $s+1$ from each other; then branch by deleting one of the $s+2$ vertices of p in each case.

We recall that a problem is said to be PARA-NP-hard if it is NP-hard for a constant value of the parameter. As mentioned in the introduction, we focus on the 2ccvd parameter in this paper, and we shall first consider variants of the DOMINATING SET problem. We first note that the DOMINATING SET (DS) and CONNECTED DOMINATING SET (CDS) problems are para-NP-hard with respect to sccvd, simply because these problems are W[2]-hard in 2-clubs [18]. Therefore we focus on other variants of DOMINATING SET that have not been studied on diameter-2 graphs. In particular, we consider EFFICIENT DOMINATING SET (EDS), INDEPENDENT DOMINATING SET (IDS), ROMAN DOMINATION (RD), INDEPENDENT ROMAN DOMINATION (IRD) and PERFECT ROMAN DOMINATION (PRD). A summary of the known and proved results is given in the below table (Table 1).

Remark that EDS has a linear time algorithm on diameter-2 graphs because a 2-club has an EDS if and only if it has a vertex connected to all the other vertices. Therefore the algorithm consists only in finding or not a vertex of degree $n-1$ where n is the number of vertices of the graph.

Table 1. Summary of the complexities of domination problems parameterized by 2ccvd.

Problem(s)	On Diameter-2 graphs	Parameterized by 2ccvd
DS, CDS	W[2]-hard [18]	para-NP-hard
IDS, RD, IRD	W[1]-hard (Theorems 3, 4 & 5)	para-NP-hard
EDS	Linear-time	FPT(Theorem 1)
PRD	NP-hard (Theorem 6)	para-NP-hard

3 EFFICIENT DOMINATING SET parameterized by 2ccvd

We prove that EFFICIENT DOMINATING SET parameterized by 2ccvd is FPT by exhibiting a fixed parameter tractable algorithm.

Let G be a graph such that $2ccvd(G) \leq k$. Then there exists a subset $S \subseteq V(G)$ of size k or less such that $G - S$ is a disjoint union of 2-clubs that we denote by

$C_1, \ldots, C_p$. Let $S' \subseteq S$ be such that S' is an EDS of $G[N[S']]$ (meaning that the sets $N[s]$ are disjoint for every $s \in S'$). We define $U = S \setminus N[S']$ the vertices of S not dominated by S'. If there exists $i \in [1,p]$ such that $V(C_i) \subseteq N[S']$, then we can remove C_i. Therefore we suppose that for every $i \in [1,p]$, we have $V(C_i) \not\subseteq N[S']$.

Lemma 1. *A subset D of $V(G)$ is an EDS of G such that $S' = D \cap S$ if and only if there exists $(v_1, \ldots, v_p) \in V(C_1) \times \cdots \times V(C_p)$ such that*

- $D = S' \sqcup_{i \in [1,p]} \{v_i\}$;
- $U = \sqcup_{i \in [1,p]}(N[v_i] \cap U)$;
- $\forall i \in [1,p], V(C_i) \subseteq N[v_i] \cup N[S']$ and $N[v_i] \cap N[S'] = \emptyset$.

Proof. Suppose that D is an EDS of G such that $S' = D \cap S$.

Let $i \in [1,p]$. Remark that for any two vertices v and w in C_i we have $N[v] \cap N[w] \neq \emptyset$ because $d(v,w) \leq 2$. Therefore $|D \cap V(C_i)| \leq 1$. If $D \cap V(C_i) = \emptyset$, then we would have $V(C_i) \subseteq N[S']$ which is excluded. Therefore there exists $v_i \in V(C_i)$ such that $D \cap V(C_i) = \{v_i\}$. We deduce that $D = S' \sqcup_{i \in [1,p]} \{v_i\}$. As D is an EDS, then the sets $N[v_i]$ are disjoint. As $U \cap N[S']$ is by definition empty, then U must be covered by the neighbors of the vertices $v_1, \ldots, v_p$. Thus $U = \sqcup_{i \in [1,p]}(N[v_i] \cap U)$.

Let $i \in [1,p]$. As D is an EDS of G, then $N[v_i]$ and $N[S']$ are disjoint. Furthermore $V(C_i)$ can only be covered by the neighbors of v_i or S'. Thus $V(C_i) \subseteq N[v_i] \cup N[S']$.

Now suppose there exists $(v_1, \ldots, v_p) \in V(C_1) \times \cdots \times V(C_p)$ such that the previous properties are satisfied. Let us prove that D is an EDS of G. Consider i and j in $[1,p]$. Then $N[v_i]$ and $N[v_j]$ does not intersect in the clusters $C_1, \ldots, C_p$ and in U; and they do not intersect $N[S']$. Thus $N[v_i]$ and $N[v_j]$ are disjoint. Therefore the sets $N[v_1], \ldots, N[v_j]$ and $N[S']$ are pairwise disjoint. Moreover, for every $i \in [1,p]$, $V(C_i)$ is covered by v_i and S'. The set U is covered by $v_1, \ldots, v_p$. Hence D is an EDS. □

We now present a dynamic programming method for computing a minimum EDS solution, if it exists.

For every subset $W \subseteq U$ and $j \in [1,p]$ we define $T[W, j]$ as a Boolean which is true if and only if there exists $(v_1, \ldots, v_j) \in V(C_1) \times \cdots \times V(C_j)$ such that

- $W = \sqcup_{i \in [1,j]}(N[v_i] \cap U)$;
- $\forall i \in [1,j], V(C_i) \subseteq N[v_i] \cup N[S']$ and $N[v_i] \cap N[S'] = \emptyset$.

Because of Lemma 1, there exists an EDS, denoted E, of G such that $S' = E \cap S$ if and only if $T[U,p]$ is true. The proof of the following lemma is found in the long version of this paper [1].

Lemma 2. *For every $W \subseteq U$ and $j \in [2,p]$,*

$$T[W, j] = \vee_{v \in V(C_j)_{S'}} T[W - N[v], j-1]$$

where $V(C_j)_{S'} = \{v \in V(C_j) \mid V(C_j) \subseteq N[v_j] \cup N[S'] \wedge N[v_j] \cap N[S'] = \emptyset\}$.

The following algorithm tries to extend a subset S' of S to an EDS of G.

Lemma 3. *The time complexity of Algorithm 1 is in $O(n^3 2^{|U|})$.*

Algorithm 1. CanExtendPartialSolution

1: **procedure** CANEXTENDPARTIALSOLUTION(S, S')
2:     $U \leftarrow S - (N[S'] \cap S)$
3:     **for** $W \subseteq U$ **do**
4:         **for** $j \in [1, p]$ **do**
5:             $T[W, j] \leftarrow$ false
6:     **for** $v_1 \in V(C_1)$ **do**
7:         **if** $V(C_1) \subseteq N[v_1] \cup N[S']$ and $N[v_1] \cap N[S'] = \emptyset$ **then**
8:             $T[N[v_1] \cap U, 1] \leftarrow$ true
9:     **for** $j \in [2, p]$ **do**
10:        **for** $W \subseteq U$ **do**
11:            **for** $v \in V(C_j) \mid V(C_j) \subseteq N[v_j] \cup N[S']$ and $N[v] \cap N[S'] = \emptyset$ **do**
12:                 **if** $T[W - N[v] \cap U, j-1]$ **then**
13:                     $T[W, j] \leftarrow$ true
    return T[U, p]

We now consider the main Algorithm 2 which, given a graph G and a 2-CCVD set S of size at most k, tries to extend every subset S' of S such that S' is an EDS of $N[S']$ to G and returns the size of a minimum EDS of G.

Lemma 4. *The time complexity of Algorithm 2 is $O(n^3 3^k)$ where k is the size of S.*

To conclude:

Theorem 1. EFFICIENT DOMINATING SET *parameterized by 2-CCVD is* FPT.

We can give a lower bound for the complexity of EFFICIENT DOMINATING SET parameterized by 2ccvd thanks to the following Theorem:

Algorithm 2. EDS-2-CCVD fixed parameter tractable algorithm

1: **procedure** EDS(G, S)
2:     $r \leftarrow +\infty$
3:     **for** $S' \subseteq S$ **do**
4:         **if** S' is an EDS of $N[S']$ **then**
5:             $t \leftarrow$ the number of 2-clubs not dominated by S'
6:             **if** CANEXTENDPARTIALSOLUTION(S') and $|S'| + t < r$ **then**
7:                 $r \leftarrow |S'| + t$
    return r

Theorem 2 (Goyal *et al.* [11]). EFFICIENT DOMINATING SET *parameterized by* VERTEX COVER *cannot be solved in $2^{o(k)}$ time unless ETH fails.*

Since any vertex cover is s-club cluster vertex deletion set, we deduce that:

Corollary 1. EFFICIENT DOMINATING SET *parameterized by sccvd cannot be solved in $2^{o(k)}$ time unless ETH fails.*

4 Complexity of INDEPENDENT DOMINATING SET on 2-clubs

The INDEPENDENT DOMINATING SET (IDS) is NP-complete [10]. In order to study its complexity in 2-clubs, we consider the following variant of INDEPENDENT SET:

k-MULTICOLORED INDEPENDENT SET
Input: A graph G and a vertex coloring $c : V(G) \to \{1, 2, ..., k\}$ for G;
Question: Does G have an independent set including vertices of all k colors? That is, are there $v_1, \ldots, v_k \in V(G)$ such that for all $1 \leq i < j \leq k$, $\{v_i, v_j\} \notin E(G)$ and $c(v_i) \neq c(v_j)$?

According to [7], k-MULTICOLORED INDEPENDENT SET is $W[1]$-complete.

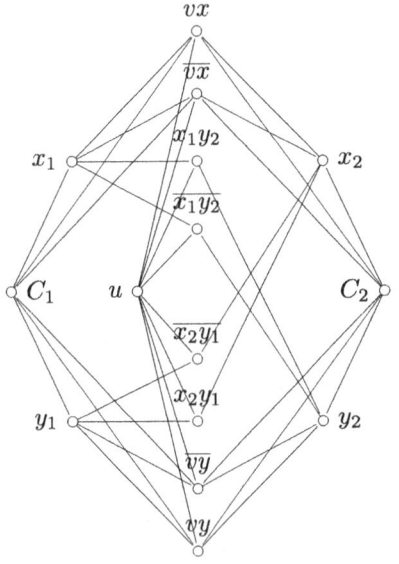

(a) The graph G' obtained from the union of two disjoint vertices colored respectively 1 and 2. The edges between the central vertices are not drawn.

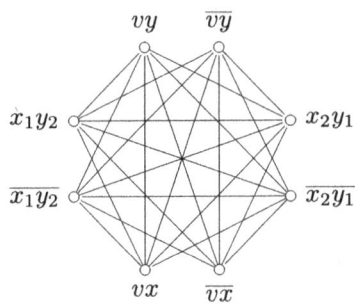

(b) The induced subgraph of the central vertices.

Fig. 1. Example of a graph obtained by the construction from an instance of k-MULTICOLORED INDEPENDENT SET.

Theorem 3. INDEPENDENT DOMINATING SET *is* $W[1]$-*hard on 2-clubs.*

Proof. We use a reduction from k-MULTICOLORED INDEPENDENT SET. Consider a graph G whose vertices are colored with k colors $1, \ldots, k$. For every $i \in [1, k]$, we denote by C_i the subset of vertices colored i. We construct a graph G' from G as follows:

- Add edges between every pair of vertices of the same color (so that C_i becomes a clique in G' for every $i \in [1, k]$);
- For every $i \in [1, k]$, add two vertices x_i and y_i connected to all vertices colored i;
- Add 4 vertices $vx, \overline{vx}, vy$ and $\overline{vy}$, and connect them to all the vertices of G;
- For every i, connect vx and $\overline{vx}$ to x_i, and connect vy and $\overline{vy}$ to y_i;
- For every $i \neq j$, add 2 vertices $x_i y_j$ and $\overline{x_i y_j}$ connected to x_i and y_j;
- Connect the vertices $vx, \overline{vx}, vy, \overline{vy}, x_i y_j, \overline{x_i y_j}$ together except for each vertex and its overlined counterpart;
- Add a vertex u connected to the vertices $vx, \overline{vx}, vy, \overline{vy}, x_i y_j, \overline{x_i y_j}$.

Note that a vertex is the twin of its overlined counterpart if they share the same neighbors. We call *central* vertices each of $vx, \overline{vx}, vy, \overline{vy}, x_i y_j$ and $\overline{x_i y_j}$ (Fig. 1).

We prove that G' is a 2-club by demonstrating in Table 2 a common neighbor between each pair of vertices.

Table 2. Common neighbors of G' for each type of vertices, proving that G' is of diameter 2. Sub-diagonal cases are symmetric.

	C_i	x_i	y_i	vx	vy	$x_i y_j$	u
$C_{i'}$	vx	vx	vy	vx	vy	vx	vx
$x_{i'}$		vx	$x_{i'} y_j$	vx	vx	vx	vx
$y_{i'}$			vy	vy	vy	vy	vy
vx				u	u	u	u
vy					u	u	u
$x_{i'} y_{j'}$						u	u
u							u

Given a colorful independent set X of G, we define the following subset of G' by $X' = X \cup \{u\}$. As X is of size k, X' is of size $k+1$. As X is an independent set of G, then X is still an independent set of G'. Furthermore u is not connected to the original vertices. Therefore X' is an independent set of G'. For every $i \in [1, k]$, every vertex of V_i is dominated by X_i (the vertex of X colored i). Furthermore every central vertex is dominated by u. Thus X' is an IDS of size $k + 1$ of G'.

Assume X' is an IDS of size at most $k + 1$ of G', and assume by contradiction that $u \notin X'$. As u should be dominated by X', there exists a central vertex w in X'. Its twin should also be in X' (it is a general fact that twins are either both in an IDS or they are not in it). Let $i \in [1, k]$ and let us show that $X' \cap (C_i \cup \{x_i, y_i\})$ is not empty. By construction the vertex w cannot be adjacent to both x_i and y_i because the common neighbors of x_i and y_i are the vertices of C_i. We can assume without loss of generality that w is not adjacent to x_i. The vertex x_i cannot be dominated by another central vertex because w is already in X' and is connected to every central vertex except $\overline{w}$. Thus x_i must be dominated by itself or by a vertex of C_i, and $X' \cap (C_i \cup \{x_i, y_i\})$ is not empty for every i. As the subsets $C_i \cup \{x_i, y_i\}$ are disjoint and also disjoint from the central vertices, then $|X'| \geq k + 2$ which contradicts that $X' \leq k + 1$. It follows that $u \in X'$.

Thus no central vertices can be in X'. For each $i \in [1, k]$, the vertices x_i and y_i must be dominated by themselves or by a vertex of C_i. If no vertex of C_i is in X', then x_i and y_i are in X'. By the same counting argument, this is impossible. Thus X' contains a vertex from C_i for every i. Hence there exists a colorful independent set in G. □

Corollary 2. *IDS parameterized by 2ccvd is* **para-NP***-hard.*

5 Roman Domination

The notion of Roman domination was first introduced in [23]. A *Roman dominating function* (RDF) on a graph G is a function $f : V(G) \to \{0, 1, 2\}$ such that for every vertex $v \in f^{-1}(\{0\})$, there exists $u \in N(v)$ such that $f(u) = 2$. The weight $w(f)$ of such a function f is defined as $\sum_{v \in V(G)} f(v)$. There always exists such a function as the constant function 1 is a RDF. We denote by $\gamma_R(G)$ the minimum weight of a RDF in G. This notion is related to dominating sets because to each dominated set X in G, we can define the RDF function f_X by $f_X(v) = 2$ if $v \in X$ and $f_X(v) = 0$ otherwise. Therefore $\gamma_R(G) \leq 2\gamma(G)$. The corresponding minimization problem is formally defined as follows.

k-ROMAN DOMINATION
Input: A graph G, an integer k;
Question: Is there a RDF of weight at most k?

Similarly we recall the notion of *independent Roman dominating function* (IRDF), introduced in [5], over a graph $G = (V, E)$: It is a RDF such that $f^{-1}(\{1, 2\})$ is an independent set. For a survey on INDEPENDENT ROMAN DOMINATION see [21]. We consider the associated minimization problem:

k-INDEPENDENT ROMAN DOMINATION
Input: A graph G, an integer k;
Question: Is there an IRDF of weight at most k?

Another variant is the notion of *perfect Roman dominating function* (PRDF) introduced in [12], over a graph $G = (V, E)$: It is a RDF such that for every $v \in V(G)$ with $f(v) = 0$, there exists exactly one vertex $w \in N(v)$ with $f(w) = 2$. The associated decision problem is W[1]-complete if parameterized by solution size and fixed parameter tractable if parameterized by clique-width [19].

k-PERFECT ROMAN DOMINATION
Input: A graph G, an integer k;
Question: Is there a PRDF of weight at most k?

5.1 Complexity of ROMAN DOMINATION in 2-Clubs

Construction 1. *Let G be an arbitrary graph. We construct the following graph G'. Consider the union of two copies G_1 and G_2 of G. For every $v \in G_1$, connect v to $N[v']$ where v' is the copy of v in G_2. For every $v' \in G_2$, connect v' to $N[v]$ in G_1. Claim: If G is a 2-club then G' is a 2-club.*

Proof. Consider two vertices v and w of G'. If v and w are in the same copy of G, then there exists a common neighbor in this copy (as G is a 2-club). Otherwise, we can suppose without loss of generality that $v \in G_1$ and $w \in G_2$. We denote by v' the copy of v in G_2. As G_2 is a 2-club, v' and w have a common neighbor that we call u. As $N[v] = N[v']$, then u is also a neighbor of v. Thus v and w have a common neighbor. Therefore, G' is a 2-club (Fig. 2). □

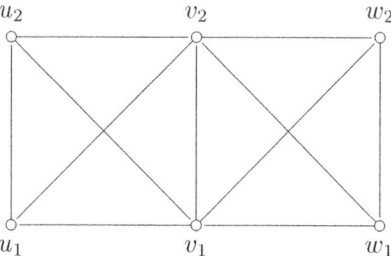

Fig. 2. Example of the duplication construction: The depicted graph is the graph G' obtained from a graph G which is the path (u, v, w).

Theorem 4. ROMAN DOMINATION *is W[2]-hard even restricted to 2-clubs.*

Proof. We use a reduction from DOMINATING SET in 2-clubs, which is $W[2]$-hard [18]. Consider a 2-club G and the 2-club G' obtained from Construction 1. We prove G has a dominating set of size at most k if and only if G' has a RDF of weight at most $2k$.

Suppose that G has a dominating set X of size at most k. We define a function $f : V(G') \to \{0, 1, 2\}$ as follows: for every $v \in V(G)$ we set $f(v_1) = 0$ if $v \notin X$ and $f(v_1) = 2$ otherwise and $f(v_2) = 0$. The weight of f is $2|X| \leq 2k$. Let us prove that f is a RDF. For every $v \in V(G)$, there exists $w \in N[v]$ such that $w \in X$. As $w_1 \in N[v_1]$ and $w_1 \in N[v_2]$ and $w_1 \in V_2$, we deduce that v_1 and v_2 are dominated by w_1 such that $f(w_1) = 2$. Thus f is a RDF of G' of weight at most $2k$.

Suppose G' has a RDF f of weight at most $2k$. We define X as the vertices of $V(G)$ such that $f(v_1) + f(v_2) \geq 2$. We prove X is a dominating set of G. Let $u \in V(G) \notin X$. So $f(v_1) = 0$ or $f(v_2) = 0$. If $f(v_1) = 0$, then $f(v_2) < 2$. In this case, there exists $w \in N(v)$ such that $f(w_1) = 2$ or $f(w_2) = 2$. Therefore $w \in X$ is adjacent to v in G. The case $f(v_2) = 0$ is symmetric. Hence X is a dominating set of G.

Suppose by contradiction that $|X| > k$. Thus the weight of f satisfies:

$$w(f) = \sum_{v \in V(G)} (f(v_1) + f(v_2)) \geq \sum_{v \in X} (f(v_1) + f(v_2)) \geq \sum_{v \in X} 2 \geq 2|X| > 2k \ .$$

This contradicts the fact that f is of weight at most $2k$. We deduce that $|X| \leq k$. □

Corollary 3. ROMAN DOMINATION *parameterized by 2ccvd is* **para-NP-hard**.

5.2 Complexity of INDEPENDENT ROMAN DOMINATION in 2-clubs

By a similar proof to the previous one we obtain the following Theorem.

Theorem 5. INDEPENDENT ROMAN DOMINATION *is* **W[1]-hard in 2-clubs**.

Corollary 4. INDEPENDENT ROMAN DOMINATION *parameterized by 2ccvd is* **para-NP-hard**.

5.3 Complexity of PERFECT ROMAN DOMINATION in 2-clubs

We reduce PERFECT ROMAN DOMINATION from the following problem which has been proved to be NP-complete [15]:

EXACT COVER
Input: A set of elements U, a set S of subsets of U;
Question: Is there an exact cover of U, that is a subset S' of S such that the sets in S' are pairwise disjoint and the union of these sets is U?

Theorem 6. PERFECT ROMAN DOMINATION *is* **NP-complete** *even when restricted to 2-clubs*.

Proof. Let $U = \{u_1, \ldots, u_n\}$ and $S = \{s_1, \ldots, s_k\}$ be an instance of EXACT COVER. We define $w = 2k + 3$ and a graph G as follows (Fig. 3):

- Add one vertex for each set of S;
- Add an independent set U_i with $w + 1$ vertices for every element u_i of U;
- Connect s_i to U_j for every i, j such that $u_j \in s_i$;
- Add independent set S_i with $w + 1$ vertices for each $s_i \in S$ and connect s_i to S_i;
- Add a vertex r_i for every set $s_i \in S$ and connect r_i to s_i and r_i to S_i;
- Add an independent set T_i with $w + 1$ vertices for each set s_i of S and connect r_i, S_i and s_i to T_i;
- Add a vertex a connected to every r_i, S_i, s_i and U_j;
- Add a vertex b connected to every T_i, r_i, S_i, s_i;
- Add a vertex c connected to every T_i and U_j.

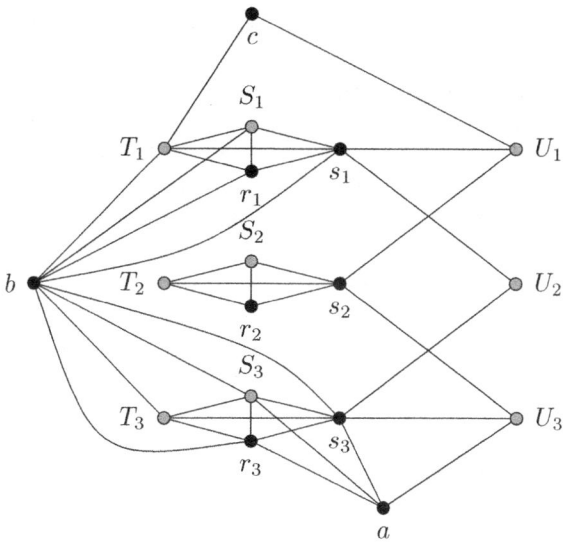

Fig. 3. Example of graph obtained by the construction of Theorem 6 for EXACT COVER instance $(U = \{u_1, u_2, u_3\}, S = \{s_1 = \{u_1, u_2\}, s_2 = \{u_1, u_3\}, s_3 = \{u_2, u_3\}\})$.

Table 3. Common neighbors of G for each type of vertices, proving that G is of diameter 2. Sub-diagonal cases are symmetric.

	U_j	s_i	S_i	r_i	T_i	a	b	c
$U_{j'}$	a	a	a	a	c	a	s_i	c
$s_{i'}$		a	a	a	b	a	b	$u_j \in s_{i'}$
$S_{i'}$			a	a	b	a	b	$T_{i'}$
$r_{i'}$				a	b	a	b	$T_{i'}$
T_i					c	r_i	b	c
a							a s_1	U_1
b								b T_1
c								c

The graph G is of diameter at most 2. We prove this by giving a common neighbor for every pair of vertices of G in Table 3. We now show that (U, S) has a solution of size at most k if and only if G has a PRDF of size at most $w = 2k + 3$. Suppose (U, S) has a solution X. We define the following function $f : V(G) \to \{0, 1, 2\}$. For every $i \in X$, $f(s_i) = 2$. For every $i \notin X$, $f(r_i) = 2$. Furthermore we set $f(a) = f(b) = f(c) = 1$. For every other vertex v, we set $f(v) = 0$. The weight of f is $2k + 3$. We claim that f is a RDF. For every i, each vertex of S_i and T_i is only dominated by s_i and r_i. For every $i \in X$, r_i is only dominated by s_i. For every $i \notin X$, s_i is only dominated by r_i.

For every j, there exists i such that $u_j \in s_i$, thus each vertex of U_j is only dominated by s_i. Hence f is a PRDF.

Suppose G has a PRDF f of weight at most $w = 2k + 3$ and let us prove that (U, S) has an exact cover. Suppose by contradiction that $f(b) = 2$. Then $f(s_i) < 2$, $f(S_i) < 2$, $f(r_i) < 2$ for every i, otherwise one of this vertex v would have value $f(v) = 2$ and then all vertices of T_i would be adjacent to v and to b, so all vertices would have value at least 1. As $|T_i| > w$, it would contradict that f is of weight at most w. In the same way, we deduce that $f(a) < 2$ (otherwise all vertices of S_1 would be adjacent to a and b which have value 2). We also deduce that $f(c) < 2$. Thus for every j, every vertex of U_j is not adjacent to a vertex valued 2 and is therefore of value at least 1, a contradiction. We deduce that $f(b) \leq 1$.

In the same way, we prove that $f(c) \leq 1$. Suppose by contradiction that $f(c) = 2$. Then $f(a) < 2$ (because c and a are connected to every U_j). Furthermore every vertex r_i, S_i, s_i is of value at most 1 because they are all connected to T_i. As S_i must contain a vertex valued 0 (otherwise the weight of f would be greater than $w + 1$), this vertex can only be covered by a 2 situated in T_i. We deduce that every T_i has a vertex mapped to 2. As b is connected to every T_i, it cannot be assigned to 0 and is thus assigned 1. It is not possible that $f(a) = 0$, otherwise there would exist a vertex v in U_j assigned to 2. In this case the weight is at least $f(c) + f(b) + f(a) + f(v) + \sum_i \sum_{v \in T_i} f(v) \geq 2 + 1 + 2 + 2k = 4 + 2k$. This contradicts the upper bound of $w(f)$. Therefore $f(c) \leq 1$.

As $f(b) \leq 1$ and $f(c) \leq 1$ and as every T_i must contain a 0, then for every i, r_i, s_i or a vertex of S_i is assigned to 2. Thus the weight of f is at least $f(a) + f(b) + f(c) + \sum_i f(r_i) + f(s_i) + f(S_i) \geq 3 + 2k$. We deduce that every other vertices are assigned to 0, in particular the vertices of U_j for every j.

We define X as $\{i : f(s_i) = 2\}$. We show that X is an exact cover of U. Let $u_j \in U_j$. As $f(u_j) = 0$ and as u_j is connected to T_i (which are assigned to 0), to a and c (which are assigned to 1) and to some vertices s_i, thus there exists s_i with $f(s_i) = 2$ which is connected to u_j. As f is a perfect roman domination, there exists exactly one vertex s_i with $f(s_i) = 2$ which is connected to u_j. Hence X is an exact cover of U.

Finally, (U, S) has an exact cover if and only if G has a PRDF of weight at most $2k + 3$. We conclude that PERFECT ROMAN DOMINATION is NP-complete in 2-clubs. □

Corollary 5. PERFECT ROMAN DOMINATION-2-CCVD *is para-NP-hard.*

6 Fixed-Parameter Sub-Exponential Algorithms

Despite its W[2]-hardness, the DOMINATING SET problem is solvable in sub-exponential time in 2-clubs. In fact, the minimum size of a dominating set in a 2-club is at most $\mathcal{O}(\sqrt{n \log n})$. This property, which is based on the fact that the neighborhood of any vertex in a 2-club is a dominating set, was used by Mertzios and Spirakis to solve the 3-COLORING problem in sub-exponential time in 2-clubs [20]. To see this: once a dominating set D of size $c\sqrt{n \log n}$ is found, a simple brute-force 3-COLORING algorithm consists of enumerating all the possible $\mathcal{O}(3^{|D|})$ 3-colorings of the elements of D, which restricts the number of colors of each other vertex (in $G - D$) to at most two.

Completing the 3-coloring, or reporting a No-instance, consists of solving an instance of 2-LIST COLORING in polynomial-time via a simple reduction to 2-SAT. More recently, this result was further improved by Debski *et al.* [6] where 3-COLORING and 3-LIST COLORING were shown to be solvable in $2^{\mathcal{O}(n^{\frac{1}{3}} \log^2 n)}$.

Let (G, k) be an instance of 3-COLORING parameterized by 2ccvd, and let S be such that $G - S$ is a disjoint union of 2-clubs, with $|S| \leq k$. In this case, a simple 3-COLORING algorithm consists of enumerating all the $\mathcal{O}(3^k)$ possible colorings of $G[S]$ and then, in each 2-club C of $G - S$ we use the algorithm described in [6] to complete the 3-coloring, if possible. Overall, we obtain the following.

Theorem 7. 3-COLORING *parameterized by 2ccvd is solvable in* $\mathcal{O}(3^k 2^{cn^{\frac{1}{3}} \log^2 n})$ *time.*

A similar result can be obtained for the 3ccvd parameter, due to a sub-exponential time algorithm for 3-LIST COLORING on diameter-3 graphs [6]. In general, and because of the Exponential-Time Hypothesis (ETH), solvability in sub-exponential time when a parameter is fixed can be interesting by itself, as an analogy to solvability in polynomial-time. However, it is not clear yet whether a complexity hierarchy exists that is analogous to the W-hierarchy. On the other hand, one can show that, modulo the ETH, some classical problems are not solvable by fixed-parameter sub-exponential algorithms with respect to 2ccvd. Notable examples include VERTEX COVER and INDEPENDENT SET. To see this, let (G, k) be an arbitrary instance of VERTEX COVER. If the diameter of G is greater than two, we construct an equivalent diameter-two instance $(G', k + 1)$ simply by adding a vertex s that is adjacent to all vertices of G (in G'). The size of the vertex cover increases by 1 simply because G is not a complete graph (being of diameter greater than 2). This shows that solving VERTEX COVER by a sub-exponential algorithm on 2-clubs results in solving it in sub-exponential time in general, which is impossible unless the ETH fails [13]. The same reduction applies to INDEPENDENT SET but with the same value of k.

The above reduction poses more questions. Which problems are solvable via fixed-parameter subexponential algorithms when parameterized by 2ccvd? We showed that the size of a minimum connected dominating set is at most $2\sqrt{n} \log n$ in 2-clubs (proof omitted due to space constraints), which leads to solving CDS in sub-exponential time in 2-clubs. The same is known for DOMINATING SET. Is any of the two problems solvable via a fixed-parameter subexponential algorithm w.r.t. 2ccvd? We believe this is an interesting open problem.

Another interesting scenario where several problems are solvable via fixed-parameter sub-exponential algorithms is when diameter-two is replaced with planarity. This follows for the fact that many NP-hard problems are solvable in sub-exponential time on planar graphs due to the planar separator theorem [16]. The same applies to several other parameters including distance to P_t-free and Broom-free graphs [3].

Motivated by the above, one can define the class of problems solvable via fixed-parameter sub-exponential algorithms, which can possibly be named *Fixed-Parameter Sub-exponential* or **FPSUB**. Obviously **FPT** $\subseteq$ **FPSUB**, but what about higher parameterized complexity classes? We already observed that DOMINATING SET is W[1]-hard w.r.t 2ccvd, so it does not seem possible to place FPSUB in the W-hierarchy. We believe this classification is worth exploring.

References

1. Abu-Khzam, F.N., Isenmann, L.: Domination in diameter-two graphs and the 2-club cluster vertex deletion parameter. CoRR, abs/2408.08418 (2024)
2. Abu-Khzam, F.N., Makarem, N., Shehab, M.: An improved fixed-parameter algorithm for 2-club cluster edge deletion. Theoret. Comput. Sci. **958**, 113864 (2023)
3. Bacsó, G., Lokshtanov, D., Marx, D., Pilipczuk, M., Tuza, Z., Van Leeuwen, E.J.: Subexponential-time algorithms for maximum independent set in Pt-free and broom-free graphs. Algorithmica **81**(2), 421–438 (2019)
4. Banik, A., Kasthurirangan, P.N., Raman, V.: Dominator coloring and cd coloring in almost cluster graphs. In: Morin, P., Suri, S. (eds.) Algorithms and Data Structures, pp. 106–119. Springer, Cham (2023)
5. Cockayne, E.J., Dreyer Jr, P.A., Hedetniemi, S.M., Hedetniemi, S.T.: Roman domination in graphs. Discret. Math. **278**(1–3), 11–22 (2004)
6. Debski, M., Piecyk, M., Rzazewski, P.: Faster 3-coloring of small-diameter graphs. SIAM J. Discret. Math. **36**(3), 2205–2224 (2022)
7. Fellows, M.R., Hermelin, D., Rosamond, F.A., Vialette, S.: On the parameterized complexity of multiple-interval graph problems. Theor. Comput. Sci. **410**(1), 53–61 (2009)
8. Fellows, M.R., Jansen, B., Rosamond, F.: Towards fully multivariate algorithmics: parameter ecology and the deconstruction of computational complexity. Eur. J. Comb. **34**(3), 541–566 (2013)
9. Figiel, A., Himmel, A.-S., Nichterlein, A., Niedermeier, R.: On 2-clubs in graph-based data clustering: theory and algorithm engineering. J. Graph Algorithms Appl. **25**(1), 521–547 (2021)
10. Garey, M.R., Johnson, D.S.: Computers and Intractability, vol. 174. Freeman, San Francisco (1979)
11. Goyal, D., Jacob, A., Kumar, K., Majumdar, D., Raman, V.: Parameterized complexity of dominating set variants in almost cluster and split graphs. CoRR, abs/2405.10556 (2024)
12. Henning, M.A., Klostermeyer, W.F., MacGillivray, G.: Perfect roman domination in trees. Discret. Appl. Math. **236**, 235–245 (2018)
13. Impagliazzo, R., Paturi, R., Zane, F.: Which problems have strongly exponential complexity? J. Comput. Syst. Sci. **63**(4), 512–530 (2001)
14. Jansen, B., Raman, V., Vatshelle, M.: Parameter ecology for feedback vertex set. Tsinghua Sci. Technol. **19**(4), 387–409 (2014)
15. Karp, R.M.: Reducibility among combinatorial problems. In: Miller, R.E., Thatcher, J.W. (eds.) Proceedings of a symposium on the Complexity of Computer Computations, held 20–22 March 1972, at the IBM Thomas J. Watson Research Center, Yorktown Heights, New York, USA, The IBM Research Symposia Series, pp. 85–103. Plenum Press, New York (1972)
16. Lipton, R.J., Tarjan, R.E.: Applications of a planar separator theorem. SIAM J. Comput. **9**(3), 615–627 (1980)
17. Liu, H., Zhang, P., Zhu, D.: On editing graphs into 2-club clusters. In: Snoeyink, J., Lu, P., Su, K., Wang, L. (eds.) AAIM/FAW -2012. LNCS, vol. 7285, pp. 235–246. Springer, Heidelberg (2012). https://doi.org/10.1007/978-3-642-29700-7_22

18. Lokshtanov, D., Misra, N., Philip, G., Ramanujan, M.S., Saurabh, S.: Hardness of r-DOMINATING SET on graphs of diameter (r+1). In: Gutin, G., Szeider, S. (eds.) IPEC 2013. LNCS, vol. 8246, pp. 255–267. Springer, Cham (2013). https://doi.org/10.1007/978-3-319-03898-8_22
19. Mann, K., Fernau, H.: Perfect roman domination: aspects of enumeration and parameterization. In: International Workshop on Combinatorial Algorithms, pp. 354–368. Springer (2024)
20. Mertzios, G.B., Spirakis, P.G.: Algorithms and almost tight results for 3-colorability of small diameter graphs. Algorithmica **74**(1), 385–414 (2016)
21. Padamutham, C., Palagiri, V.: Complexity aspects of variants of independent roman domination in graphs. Bull. Iran. Math. Soc. **47**, 1715–1735 (2021)
22. Schaefer, A.: Exact algorithms for s-club finding and related problems. Ph.D. thesis (2009)
23. Stewart, I.: Defend the roman empire! Sci. Am. **281**(6), 136–138 (1999)
24. Tsur, D.: Algorithms for 2-club cluster deletion problems using automated generation of branching rules. Theoret. Comput. Sci. **984**, 114321 (2024)

k-Universality of Regular Languages Revisited

Duncan Adamson[1](✉)[iD], Pamela Fleischmann[2](✉)[iD], Annika Huch[2](✉)[iD], Tore Koß[3](✉)[iD], and Florin Manea[3](✉)[iD]

[1] School of Computer Science, University of St Andrews, St Andrews, UK
duncan.adamson@st-andrews.ac.uk
[2] Department of Computer Science, Kiel University, Kiel, Germany
{fpa,ahu}@informatik.uni-kiel.de
[3] Department of Computer Science, University of Göttingen, Göttingen, Germany
{tore.koss,florin.manea}@cs.uni-goettingen.de

Abstract. A subsequence of a word w is a word u such that $u = w[i_1]w[i_2]\cdots w[i_k]$, for some set of indices $1 \leq i_1 < i_2 < \cdots < i_k \leq |w|$. A word w is k-*subsequence universal* over an alphabet Σ if every word over Σ up to length k appears in w as a subsequence. In this paper, we revisit the problem k-ESU of deciding, for a given integer k, whether a regular language, given either as nondeterministic finite automaton or as a regular expression, contains a k-universal word. [Adamson et al., ISAAC 2023] showed that this problem is NP-hard, even in the case when $k = 1$, and an FPT algorithm w.r.t. the size of the input alphabet was given. In this paper, we improve the aforementioned algorithmic result and complete the analysis of this problem w.r.t. other parameters. That is, we propose a more efficient FPT algorithm for k-ESU, with respect to the size of the input alphabet, and propose new FPT algorithms for this problem w.r.t. the number of states of the input automaton and the length of the input regular expression. We also discuss corresponding lower bounds. Our results significantly improve the understanding of this problem.

Keywords: String Algorithms · Regular Languages · Subregular Languages · Regular Expressions · Finite Automata · Subsequences · Universality

1 Introduction

Words and subsequences are two fundamental and heavily studied concepts in the field of combinatorics on words. A subsequence of a given word w is a word u that is obtained by deleting some letters of w while preserving the order of letters in w for the non-deleted ones. For instance, the word subregularuniversality has the scattered factors glue, guilty, surreality while versailles is not a scattered factor since the letters do not occur in the same order as in the whole word.

Within theoretical computer science, subsequences were considered mainly in two areas. On the one hand, there are many algorithmic and complexity problems arising from this object, e.g., in problems such as the longest or shortest common subsequence or supersequence problems [1–3,10,12–15,33,36,49,51,52], or matching and analysis problems related to the sets of subsequences occurring in a word [11,17,18,21,24,27,31,40,43,59,61]; see also the survey [44]. On the other hand, subsequences are intensely studied objects in combinatorics on words, formal languages, automata theory and logics, especially with respect to their strong relation to piecewise-testable languages [38,39,57,58], or subword orders and downward closures [29,46,47,62,63]. Such theoretical results found applications in a wide number of fields including bioinformatics [30,56], and modelling concurrency [55]. Still on the applicative side, a new research direction, originating in database theory [9,26,41,42], deals with constrained-subsequences of strings, where the substrings occurring between the positions of the subsequence are subject to regular or length constraints; a series of algorithmic and complexity results related to this setting were obtained [6,19,45,50].

The work of this paper is positioned in between the study of algorithmic properties of subsequences and automata and formal language theory. More precisely, we revisit and extend the results of [5,20,54]. In the respective works, the authors generalized fundamental algorithmic questions related to the subsequences occurring in a single word to the case of subsequences occurring in (finite or infinite) sets of words represented succinctly by automata accepting these sets or by grammars generating them (i.e., formal languages). In general, two main classes of questions were investigated: for some subsequence-related property of words, given a generative or accepting mechanism for a language, one is interested, on the one hand, whether there is at least one word of the language which fulfils the respective property, or, on the other hand, whether all words of the language fulfil the respective property. The main such subsequence-property approached in [5,20,54] is the k-universality of words. Formally, a word over an alphabet Σ is called k-universal if its set of subsequences of length k includes every word of length k over Σ. An extensive literature deals with this concept, mostly from a theoretical perspective [4,11,18,22,23,39,43,53,54]; see [5,18] and the references therein for a detailed discussion of the motivation for the study of k-universality. Coming back to our focus, in [5], the authors investigated two decision problems regarding universality, corresponding to the two directions mentioned above. First is the problem of deciding whether there exists a k-universal word in a given regular language L, specified as a non-deterministic finite automaton, for some given k (for short k-ESU for k existence subsequence universality); this problem was shown to be NP-complete, and an FPT algorithm, with respect to the size of the input alphabet, was given for it. Second is the problem whether all words in L are k-universal (k-USU for k universal subsequence universality); this problem was shown to be solvable in polynomial time. In [20], the authors extended the study of these problems for context-free and context-sensitive languages, specified by grammars generating them; [54] discusses similar problems for strings encoded via straight-line programs. In this context, the problem k-ESU is the harder and less understood of the two aforementioned problems, so we revisit it in this paper.

The research direction proposed in [5,20,54], and revisited here, is not completely new. It originates in the famous result of Higman [32] which states that the downward closure of every language (i.e., the set of all subsequences of the strings of the respective language) is regular; however, an automaton accepting this language is not always computable (e.g., for the class of context-sensitive languages). As downward closures of languages can be seen as structurally-simple yet faithful-enough abstractions of complex (and practically relevant) formal languages, they become useful in practical applications and well studied from a theoretical point of view (see [8,62,63] and the references therein). In our setting, downward closures, the fact that they are regular and it is not hard to compute an automaton for them (since we deal with regular languages only), could also be useful in approaching problems such as k-ESU. But, as discussed in [20], such solutions would be quite inefficient. So, our main aim in revisiting k-ESU is to extend the work of [5,20] on it, and identify new, more efficient solutions, as well as a better understanding of the lower bounds for solving this problem. Moreover, as k-ESU was already shown to be hard for regular languages (even for finite languages), we also consider first this class of languages in this paper, and leave the investigation of more complex classes of languages as future work.

Our Contributions. Our main results provide a very detailed image of the parametrised complexity of the k-ESU problem for regular languages. When the input regular language is given as an NFA, this problem has three parameters: the number n of states of the NFA, the size σ of the input alphabet, and the integer k. We analyse the complexity of k-ESU with respect to all these parameters. As mentioned above, an FPT algorithm w.r.t. σ was known from [5]; we give here a faster algorithm, which runs in linear time for constant alphabets. We also give here the first FPT algorithm for k-ESU w.r.t. n, which runs in linear time for NFAs with a constant number of states. As already 1-ESU is NP-hard (when the input regular language is given as NFA, DFA, or regular expression), we can conclude that, unless P=NP, there is no FPT algorithm, w.r.t. the parameter k, solving k-ESU. When the input regular language is given as a regular expression of length n over an alphabet of size σ, we obtain an FPT algorithm w.r.t. σ; 1-ESU remains NP-complete in this setting. Finally, based on the lower bounds shown here and in [5], we show that our algorithmic results are tight, from a fine-grained complexity perspective. As far as the used techniques are concerned, we extensively use a graph-theoretical tool box, which allows us to gain new combinatorial insights in the structure of the k-ESU problem, and develop efficient methods for solving it. Due to space restrictions, some proofs are omitted and can be found on arxiv.org/abs/2503.18611.

2 Preliminaries

Let $\mathbb{N} = \{1, 2, \ldots\}$ denote the natural numbers and set $\mathbb{N}_0 = \mathbb{N} \cup \{0\}$. Let $[n] = \{1, \ldots, n\}$ and let $[i, n] = \{i, i+1, \ldots, n\}$ for all $i, n \in \mathbb{N}_0$ with $i \leq n$.

An *alphabet* $\Sigma = \{1, 2, \ldots, \sigma\}$ is a finite set of symbols, called *letters* (w.l.o.g., we can assume that the letters are integers). A *word* (also known as a *string*) w is a finite sequence of letters from a given alphabet. The length of a word w, denoted $|w|$ is the number of letters in the word. For $i \in [|w|]$ let $w[i]$ denote the ith letter of w. The set of all finite words over the alphabet Σ, denoted by Σ^*, is the free monoid generated by Σ with concatenation as operation and the neutral element is the empty word ε, i.e., the word of length 0. Given $n \in \mathbb{N}_0$, let Σ^n denote all words in Σ^* exactly of length n and $\Sigma^{\leq n}$ the set of all words of Σ^* of length at most n. Let $\mathrm{alph}(w) = \{\mathtt{a} \in \Sigma \mid \exists i \in [|w|] : w[i] = \mathtt{a}\}$ be the alphabet of w. For $u, w \in \Sigma^*$, u is called a *factor* of w, if $w = xuy$ for some words $x, y \in \Sigma^*$. If $x = \varepsilon$ (resp., $y = \varepsilon$) then u is called a *prefix* (resp., *suffix*) of w. For $1 \leq i \leq j \leq |w|$ define the factor from w's i^{th} letter to the jth letter by $w[i, j] = w[i] \cdots w[j]$. Given a pair of indices $i < j$, we assume $w[j, i] = \varepsilon$. For $w \in \Sigma^*$ and $n \in \mathbb{N}_0$ define inductively $w^0 = \varepsilon$ and $w^n = ww^{n-1}$.

As we are interested in investigating the k-subsequence universality of subregular languages, we first introduce the basic concepts related to subsequences. We then present the definitions for the transformation of these notions to the domain of subregular languages, finite automata and other language models.

Definition 1. *Let $w \in \Sigma^*$ and $n \in \mathbb{N}_0$. A word $u \in \Sigma^*$ is called subsequence of w ($u \in \mathrm{SubSeq}(w)$) if there exist $v_1, \ldots, v_{n+1} \in \Sigma^*$ such that $w = v_1 u[1] v_2 u[2] \cdots v_n u[n] v_{n+1}$. Let $\mathrm{SubSeq}_k(w) = \{u \in \mathrm{SubSeq}(w) \mid |u| = k\}$.*

Subsequences of automatauniversality are auto, tomata, salty, and atom while star, alien are not because their letters occur in the wrong order.

In [11], the authors investigated words which have, for a given $k \in \mathbb{N}_0$, all words from Σ^k as subsequence, namely k-subsequence universal words. Note that this notion is similar to the one of richness introduced and investigated in [38,39]. We stick here to the notion of k-subsequence universality since our focus is subregular languages and thus the well-known notion of the universality of automata and formal languages, i.e., $L(A) = \Sigma^*$ for a given finite automaton A, is close to the one of subsequence universality of words. Further, we recall the *arch factorisation* by Hébrard [31].

Definition 2. *A word $w \in \Sigma^*$ is called k-subsequence universal (w.r.t. Σ), for $k \in \mathbb{N}_0$, if $\mathrm{SubSeq}_k(w) = \Sigma^k$. If the context is clear we call w k-universal. The universality-index $\iota(w)$ is the largest k such that w is k-universal.*

Definition 3. *The arch factorisation of $w \in \Sigma^*$ is defined by $w = \mathrm{ar}_1(w) \cdots \mathrm{ar}_k(w) \, \mathrm{r}(w)$ for $k \in \mathbb{N}_0$ with $\mathrm{ar}_i(w)[|\mathrm{ar}_i(w)|] \notin \mathrm{alph}(\mathrm{ar}_i(w)[1, |\mathrm{ar}_i(w)| - 1])$ for all $i \in [k]$ and $\iota(\mathrm{ar}_i(w)) = 1$, as well as $\mathrm{alph}(\mathrm{r}(w)) \subsetneq \Sigma$. The words $\mathrm{ar}_i(w)$ are the arches and $\mathrm{r}(w)$ is the rest of w.*

Example 1. Consider $w = \mathtt{baaababb} \in \{\mathtt{a}, \mathtt{b}\}^*$. We have $|\mathrm{SubSeq}_3(\mathtt{baaababb})| = |\{\mathtt{aaa}, \mathtt{aab}, \mathtt{aba}, \mathtt{abb}, \mathtt{baa}, \mathtt{bab}, \mathtt{bba}, \mathtt{bbb}\}| = 2^3$. Since $\mathtt{abba} \notin \mathrm{SubSeq}_4(\mathtt{baaababb})$, it follows that $\iota(\mathtt{baaababb}) = 3$. Further, $w = (\mathtt{ba}) \cdot (\mathtt{aab}) \cdot (\mathtt{ab}) \cdot \mathtt{b}$ is the arch factorisation of w where the parentheses denote the three arches and the rest $\mathtt{b}$.

Theorem 1 ([11]). *Let $w \in \Sigma^{\geq k}$ with $\mathrm{alph}(w) = \Sigma$. Then we have $\iota(w) = k$ iff w has exactly k arches.*

We now recall basic notions on finite automata, for further details we refer to [34]. A *non-deterministic finite automaton (NFA)* A is a tuple $(Q, \Sigma, q_0, \delta, F)$ with the finite set of states Q (of cardinality $n \in \mathbb{N}$), an initial state $q_0 \in Q$, the set of final states $F \subseteq Q$, an input alphabet Σ, and a transition function $\delta : Q \times (\Sigma \cup \{\varepsilon\}) \to 2^Q$, where 2^Q is the powerset of Q. If $q_2 \in \delta(q_1, a)$, for some $a \in \Sigma \cup \{\varepsilon\}$, then we have the transition (q_1, a, q_2) in A. If we have $|\delta(q,a)| = 1$ and $\delta(q,\varepsilon) = \emptyset$ for all $q \in Q, a \in \Sigma$, then A is called *deterministic* (DFA). We call a sequence of transitions $\beta = (q, a_1, q_1)(q_1, a_2, q_2)\ldots(q_{\ell-1}, a_\ell, q_\ell)$ an *ℓ-length walk* from any $q \in Q$ to q_ℓ in A; in such a walk, we have $q_i \in Q$ for all $i \in [0, \ell]$, $a_i \in \Sigma \cup \{\varepsilon\}$, and $q_{i+1} \in \delta(q_i, \mathsf{a}_{i+1})$, for all $i \in [0, \ell-1]$. The word $a_1 \cdots a_\ell$ is the *label of β* and is denoted by $\mathrm{label}(\beta)$. A state $q \in Q$ is called *accessible* (respectively, *co-accessible*) in A if there exists a walk connecting q_0 to q (respectively, q to a final state). A walk is called *accepting* if $q_\ell \in F$ holds. The language of A, i.e., the set of words *accepted* by A, is $L(A) = \{w \in \Sigma^* \mid \exists \text{ accepting walk } \beta \text{ in } A : w = \mathrm{label}(\beta)\}$. Note that the class of languages accepted by NFAs is equal to the class of languages accepted by DFAs and it is equal to the class of regular languages. Moreover, for every word $w \in L(A)$ there exists exactly one (in the deterministic case) or a set (in the non-deterministic case) of walk(s) labelled with w.

Further, we define regular expressions (regex, for short) and their languages, e.g. , $L(\mathsf{a}^*\mathsf{b}|\varepsilon) = L(\mathsf{a}^*\mathsf{b}) \cup \{\varepsilon\} = L(\mathsf{a}^*)\{\mathsf{b}\} \cup \{\varepsilon\} = \{\mathsf{a}\}^*\{\mathsf{b}\} \cup \{\varepsilon\}$.

Definition 4. *Let Σ be an alphabet. Then $\emptyset, \varepsilon$, and $\mathsf{a} \in \Sigma$ are regular expressions. Further, for two regex R_1 and R_2, the terms $R_1 R_2$, $(R_1 \mid R_2)$ and $(R_1)^*$ are regex. A regex R that contains no $*$ is called* star-free.

Let R be a regex. Then the semantics of R is inductively defined by the base cases $L(\emptyset) = \emptyset$, $L(\varepsilon) = \{\varepsilon\}$ and $L(\mathsf{a}) = \{\mathsf{a}\}$ for $\mathsf{a} \in \Sigma$. Further, we define $L(R_1 R_2) = L(R_1)L(R_2)$, $L((R_1|R_2)) = L(R_1) \cup L(R_2)$ and $L((R_1)^*) = L(R_1)^*$ for regex R_1, R_2. $L(R)$ is called the language described by the regex R.

Now, we can define the k-subsequence universality of a regular language L. Here, we distinguish whether at least one or all words of L are k-universal w.r.t. Definition 2. Note that we always take the minimal alphabet Σ such that $L \subseteq \Sigma^*$ as a reference when considering the k-universality of words from L, as otherwise, for larger alphabets, the universality is trivially 0.

Definition 5. *Let $k \in \mathbb{N}$ be an integer. A language L is* existence k-subsequence universal *(k-$\exists$-universal) if there exists a k-universal word $w \in L$, and L is* universal k-subsequence universal *(k-$\forall$-universal) if all words $w \in L$ are k-universal.*

If a language L, accepted (represented) by some automaton A (or regular expression R), is k-$\exists$-universal (respectively, k-$\forall$-universal) then we also say that the automaton A (regular expression R) is k-$\exists$-universal (respectively, k-$\forall$-universal). We also say that a *walk π in A is k-universal* if $\mathrm{label}(\pi)$ is k-universal.

For instance, the expression (a*b*c|bc)|a* is a 1-∃-universal regular expression that is not 2-∃-universal.

Starting from Definition 5, two decision problems were introduced in [5].

1. The *existence subsequence universality problem for regular languages (k-ESU)* is to decide for a language, given by an NFA $\mathcal{A}$ recognising it (respectively, by a regex R describing it), and $k \in \mathbb{N}$ whether L is k-∃-universal.
2. The *universal subsequence universality problem for regular languages (k-USU)* is to decide for a language, given by an NFA $\mathcal{A}$ recognising it (regex R describing it), and $k \in \mathbb{N}$ whether L is k-∀-universal.

Here we focus on the first problem, as k-USU can already be solved in polynomial time for all regular languages, given as NFAs or regexes.

Our computational model is the RAM model with word size $\omega \geq \log n$, [16,25] where n is the size of the input; that is, the input size never exceeds 2^ω. We also assume that the given strings are over integer alphabets $\Sigma = \{1, \ldots, \sigma\}$, while input NFAs have state-sets $Q = \{0, \ldots, q\}$, with $\sigma, q \leq n$.

3 Preprocessing: Graph-Theoretic View on NFAs

In the problems discussed in this paper, we consider a regular language L, over the alphabet Σ, with $|\Sigma| = \sigma$, given either as a regular expression or as an NFA. In this section we only consider the case when the language is given as an NFA $A = (Q, \Sigma, q_0, F, \delta)$ with n states and m transitions, and initial state q_0.

We aim to obtain efficient algorithms that, for a given k, check whether L contains a word w of universality $\iota(w) \geq k$. In particular, we will discuss two algorithms, one which is FPT with respect to the parameter n, and one which is FPT w.r.t. the parameter σ. In this section, we describe a series of steps that will be executed as preprocessing for both these algorithms, and introduce a series of definitions used in the description of these algorithms.

Firstly, we note that we can assume w.l.o.g. that all states in A are both accessible and co-accessible and that A has a single final state f (i.e., $F = \{f\}$), which is not the origin of any transition (i.e., $\delta(f, a) = \emptyset$ for all $a \in \Sigma$). Indeed, if we are given an arbitrary NFA A', we can transform it, by standard methods (see, e.g., [34]), in linear time in the total size of A' (that is, number of states plus number of transitions), and obtain an automaton A for which these assumptions hold. Clearly, for A, we also have that $\sigma \leq m \leq n^2\sigma$ (as for each pair of states (q_1, q_2), we can have at most σ transitions from q_1 to q_2, one for each letter, and, for each letter, there should be at least one transition labelled with that letter).

We now make several remarks on NFAs that are useful for our results.

Each NFA A can be seen canonically as a directed graph G_A, whose vertex-set is the set Q of states of A, and there is an edge from state q_1 to state q_2 in the graph G_A if an only if there is a transition from q_1 to q_2 in A. We recall the following lemma, see, e.g. [60].

Lemma 1. *Given an NFA A, we can compute in $O(n + m)$ time the decomposition of the associated graph G_A in strongly connected components $\mathcal{C}_1, \ldots, \mathcal{C}_e$.*

The complexity of this algorithm is determined by the computation of the graph G_A from A, which can be done in $O(n+m)$ time, and then the usage of Tarjan's algorithm [60] for computing the strongly connected components $C_1, \ldots, C_e$ which also requires $O(n+m)$ time. We can also compute, within the same time complexity, the directed acyclic graph $S(A)$ of the strongly connected components of G_A: it has the vertices $\{1, \ldots, e\}$, corresponding to the strongly connected components $C_1, \ldots, C_e$ of G, respectively, and there is an edge from i to j, for $i, j \in [e]$, if and only if there is an edge from a vertex of C_i to some vertex of C_j.

We can assume that each strongly connected component produced in this algorithm is output as a list of vertices, and, for each vertex, the list of the edges leaving that vertex. For a state $q \in Q$, we set $S(q) = i$ if C_i is the strongly connected component of G_A which contains state q.

The decomposition of G_A in strongly connected components canonically corresponds to a decomposition of the NFA A in strongly connected components: for $i \in [e]$, the component C_i of A contains the states corresponding to the vertices of the component C_i of G_A, as well as all the transitions between them. The main difference between the components of A and those of G_A is that in a component of A we might have multiple transitions between two states, while in the respective component of G_A there is at most one edge from one vertex to another. Clearly, we can compute this decomposition of A in strongly connected components in $O(n+m)$ time.

This concludes the description of the preprocessing steps performed at the beginning of our algorithms.

Other graph-theoretic notions can be transferred canonically from G_A to A. Recall that a walk in the NFA A, originating in state q and with target q_n, is a sequence of transitions $\beta = (q, a_1, q_1)(q_1, a_2, q_2) \cdots (q_{n-1}, a_n, q_n)$, where $q_i \in Q$, $a_i \in \Sigma \cup \{\varepsilon\}$ for $i \in [n]$, and $q_i \in \delta(q_{i-1}, a_i)$ for $i \geq 2$, and $q_1 \in \delta(q, a_1)$. The label of β is the word $\text{label}(\beta) = a_1 \cdots a_n$. If $q = q_n$, then β is called a cycle.

It is not hard to see that, for each state q, there is a cycle, denoted in the following c_q, containing q that goes at least once through every transition of the NFA-component $C_{S(q)}$. Indeed, for each transition (q_1, a, q_2), where both q_1 and q_2 are in $C_{S(q)}$, we can follow the walk connecting q to q_1, then the transition (q_1, a, q_2), and then follow the walk from q_2 to q; we then simply take the concatenation of all these cycles, and get c_q. So, in the following, for each state q, we will use c_q to denote a cycle which goes through q and traverses at least once every transition between states contained in $C_{S(q)}$; we emphasise already at this point that this c_q is never used in our algorithmic results, so we do not focus on constructing it efficiently, but, rather, it is important to know that such a cycle exists for each q. Also, note that even if $S(q_1) = S(q_2)$, we do not necessarily have that $c_{q_1} = c_{q_2}$, but we do have that $\text{alph}(\text{label}(c_{q_1})) = \text{alph}(\text{label}(c_{q_2}))$. For states q, q' with $S(q) = S(q')$, let $p_{q,q'}$ denote the shortest prefix of c_q (w.r.t. number of transitions) connecting state q to state q' in the NFA A.

A walk expression is:

- a *single transition* (q_1, a_2, q_2). This denotes a single one-edge walk, namely (q_1, a_2, q_2). The origin of this walk expression is q_1 and the target q_2. Transitions are called atomic walk expressions (atoms, for short).
- an *extended cycle*, written as $(q[c_q]q', b, q'')$, where $q, q', q'' \in Q$, q' is a state of $\mathcal{C}_{S(q)}$, $b \in \Sigma \cup \{\varepsilon\}$, and $q'' \in \delta(q', b)$. This denotes a set of walks, namely those walks $c_q^i p_{q,q'}(q', b, q'')$ obtained by going i times around the cycle c_q (i.e., following i times the transitions of the cycle c_q, starting in q), for some $i \geq 0$, then following the path $p_{q,q'}$, and, finally the transition (q', b, q''), which corresponds to $q'' \in \delta(q', b)$. The origin of this walk expression is q and the target q''. Such walk expressions are also called atomic (or atoms).
- a concatenation $\alpha = \alpha_1 \alpha_2$ of walk expressions, such that the origin of α_2 is the same as the target of α_1. This expression denotes the set containing all walks obtained by concatenating walks denoted by α_1 with walks denoted by α_2. The origin of α is the origin of α_1, the target of α is the target of α_2. These are non atomic walk expressions.

4 FPT-Algorithms for k-ESU

We assume that we are in the setting defined in Sect. 3, and use the notations introduced in that section. We begin with a simple observation.

Lemma 2. *L contains, for each $i \in \mathbb{N}$, a word w_i of universality $\iota(w_i) \geq i$ if and only if there exists a state q of A such that $\mathrm{alph}(\mathrm{label}(c_q)) = \Sigma$.*

It is not hard to see that if L does not contain words or arbitrarily large universality then $\iota(w) \leq n$ for all $w \in L$.

The following lemma follows immediately from the preprocessing described in Sect. 3 and shows that we can test this property efficiently.

Lemma 3. *We can compute in $O(m+n)$ time the sets $V_i \subseteq \Sigma \cup \{\varepsilon\}$ of all labels of transitions of the strongly connected component $\mathcal{C}_i$ of A, for $i \in [e]$. We can then access in $O(1)$ the set $V_{S(q)}$ with $\mathrm{alph}(\mathrm{label}(c_q)) = V_{S(q)}$, for all $q \in Q$.*

Therefore, as a first main consequence of the previous lemma, we can test efficiently the existence in A of a state q such that $\mathrm{alph}(\mathrm{label}(c_q)) = V_{S(q)} = \Sigma$, as in Lemma 2. If such a state exists, then we can trivially answer to k-ESU, for the input NFA A, positively, in polynomial time (in the size of A).

So, let us assume, from now on, that there is no state q with $\mathrm{alph}(\mathrm{label}(c_q)) = \Sigma$. We emphasise that Lemma 3 allows us to compute in polynomial time, for each state $q \in Q$, only the set $V_{S(q)} = \mathrm{alph}(\mathrm{label}(c_q))$; as already mentioned in the previous section, we do not compute effectively the cycle c_q (we just collect for each q the letters which label all the transitions in its strongly connected component). However, in the following, in our combinatorial proofs, we will still make use of the notation c_q, as introduced above. Since there is no transition leaving the final state f of A, $V_{S(f)} = \emptyset$ and c_f is the empty walk.

We will now move on and describe our main results: the first FPT algorithm w.r.t. the parameter n for deciding k-ESU, and a more efficient (compared to the similar algorithm presented in [5]) FPT algorithm w.r.t. parameter σ. In both cases, we assume that L does not contain words of arbitrarily large universality, so we can also assume that $k \leq n$, as explained above.

FPT Algorithm w.r.t. n. To begin with, we show the following theorem:

Theorem 2. *If L contains a word w with $\iota(w) = k$ then there exists a word $u \in L$ with $\iota(u) \geq k$, which is the label of a walk denoted by a walk expression of the form $(q_0[c_{q_0}]q_0', b_1, q_1)(q_1[c_{q_1}]q_1', b_2, q_2) \cdots (q_{h-1}[c_{q_{h-1}}]q_{h-1}', b_h, q_h)$, where: $h \leq n$; q_i' belongs to $\mathcal{C}_{S(q_i)}$ for all $i \in [h-1]$; $S(q_i) \neq S(q_j)$, for all $i \neq j$; and $q_h = f$ is the final state of A.*

Proof. Let us consider a walk $\beta = (q_0, a_1, q_1)(q_1, a_2, q_2) \cdots (q_{r-1}, a_r, q_r)$ with label w, where $q_r = f$. We will rewrite this walk in several steps, in order to derive the result from the statement.

Clearly, the walk expression
$$\beta' = (q_0[c_{q_0}]q_0, a_1, q_1)(q_1[c_{q_1}]q_1, a_2, q_2) \cdots (q_{r-1}[c_{q_{r-1}}]q_{r-1}, a_r, q_r)$$
denotes walks whose labels are words w' for which $\iota(w') \geq \iota(w)$ (as all such words have w as subsequence). Obviously, w is the label of some walk (namely β) denoted by the walk expression β'.

We iterate the following process on β':

- Firstly, find the leftmost atom $(q_1[c_{q_1}]q_1, a, q_1')$ of the walk expression β' such that there exists another atom $(q_2[c_{q_2}]q_2, b, q_2')$ in β', with $S(q_1) = S(q_2)$. Note that, if such states q_1, q_2 exist, then $q_1 \neq f \neq q_2$.
- Secondly, find the rightmost atom $(q_2[c_{q_2}]q_2, b, q_2')$, with $S(q_2) = S(q_1)$. Thus: $\beta' = x(q_1[c_{q_1}]q_1, a, q_1')z(q_2[c_{q_2}]q_2, b, q_2')y$, where neither x nor y contain atoms originating with a state q of $\mathcal{C}_{S(q_1)}$ and z is a walk visiting only states in $\mathcal{C}_{S(q_1)}$. Rewrite β' as $\beta' = x(q_1[c_{q_1}]q_2, b, q_2')y$. Note that no state q of $\mathcal{C}_{S(q_1)}$ may appear in β' as the origin of a walk expression, except q_1. Moreover, if a state g_1 is the origin of some atom of $x(q_1[c_{q_1}]q_2, b, q_2')$, then there is no other atom of β' whose origin is some state g' of $\mathcal{C}_{S(g)}$.

Let $w_0 = w$. Now, we claim that at the end of the ith iteration, for $i \geq 1$, there exists a word w_i which labels one of the walks denoted by the current expression β' (i.e., as computed at the end of the respective iteration) such that $\iota(w_i) \geq \iota(w_{i-1})$. Indeed, we can obtain w_i as follows: we take the prefix of w_{i-1} corresponding to the walk expression x, then we follow transitions of the cycle c_{q_1} at least $2|z| + 1$ times (which is more than enough to cover all the factors of z that had some state q of $\mathcal{C}_{S(q_1)}$ as origin), then we follow the transition (q_2, b, q_2'), and end with the suffix of w_{i-1} corresponding to the expression y.

The above process is finite (in each step we reduce the number of atoms of β'), and shows how each walk β, with label w, is transformed into a walk expression $\beta' = (q_0[c_{q_0}]q_0', b_1, q_1) \cdots (q_{h-1}[c_{q_{h-1}}]q_{h-1}', b_h, q_h)$, where: $h \leq n$; q_i'

belongs to $\mathcal{C}_{S(q_i)}$ for all $i \in [h-1]$; $S(q_i) \neq S(q_j)$, for all $i \neq j$; and $q_h = f$. Moreover, as shown, there exists at least one word u labelling a walk denoted by the expression β', derived at the end of this process, such that $\iota(u) \geq \iota(w)$. □

Note that we will not use Theorem 2 as an algorithmic tool. It just shows that to identify the maximum universality index over the words of L it is enough to consider words w of maximum universality which label walks denoted by walk expressions $(q_0[c_{q_0}]q'_0, b_1, q_1)\cdots(q_{h-1}[c_{q_{h-1}}]q'_{h-1}, b_h, q_h)$, where: $h \leq n$; q'_i belongs to $S(q_i)$ for all $i \in [h-1]$; $S(q_i) \neq S(q_j)$, for all $i \neq j$; and $q_h = f$ is the final state of A. In the following, we show that the search space for the words $w \in L$ with maximum universality index can be restricted further. In particular, our goal is to show that, in order to compute the maximum universality of a word accepted by A along a walk described by expressions as the ones above, it is not important how many times a cycle is traversed, but it is only important to know the states q along this walk and their corresponding sets $V_{S(q)}$.

Before stating our next result, note that the walks denoted by an expression $\beta' = (q_0[c_{q_0}]q'_0, b_1, q_1)(q_1[c_{q_1}]q'_1, b_2, q_2)\cdots(q_{h-1}[c_{q_{h-1}}]q'_{h-1}, b_h, q_h)$ are exactly the walks $\beta = c_{q_0}^{i_0} p_{q_0,q'_0}(q'_0, b_1, q_1)\cdots c_{q_{h-1}}^{i_{h-1}} p_{q_{h-1},q'_{h-1}}(q_{h-1}, b_h, q_h)$, where $i_0, \ldots, i_{h-1}$ are non-negative integers. Our next result analyses certain factors of such walks.

Lemma 4. *Let w be a word with $\iota(w) = k$, which labels a walk*

$$\beta = c_{q_1}^{i_1} p_{q_1,q'_1}(q'_1, b_2, q_2) c_{q_2}^{i_2} p_{q_2,q'_2}(q'_2, b_3, q_3)\cdots c_{q_{t-1}}^{i_{t-1}} p_{q_{t-1},q'_{t-1}}(q'_{t-1}, b_t, q_t) c_{q_t}^{i_t},$$

where q'_i is a state of $\mathcal{C}_{S(q_i)}$ for all $i \in [t]$, and $S(q_i) \neq S(q_j)$ for $i \neq j$.
Then, there exists a word $w' = w'_1 \cdots w'_k$ and integers $s_0, s_1, \ldots, s_k$ such that:

- $1 = s_0 < s_1 < \ldots < s_k = t$
- *For $i \in [k]$, w'_i is 1-universal and labels the walk:*

$$c_{q_{s_{i-1}}} p_{q_{s_{i-1}},q'_{s_{i-1}}}(q'_{s_{i-1}}, b_{s_{i-1}+1}, q_{s_{i-1}+1})\cdots c_{q_{s_i-1}} p_{q_{s_i-1},q'_{s_i-1}}(q'_{s_i-1}, b_{s_i}, q_{s_i}) c_{q_{s_i}}.$$

Proof. As $\iota(w) \geq k$, we have that there exist words $w_1, \ldots, w_k$, with $w = w_1 \cdots w_k$ and $\iota(w_i) = 1$ for all $i \in [k]$. Let β_i be the subwalk of β labelled by w_i. We have $\beta = \beta_1 \beta_2 \cdots \beta_k$. For simplicity, let $q'_t = q_t$, $p_{q_t,q'_t} = \varepsilon$, and $s_0 = 1$.

For $i \in [k]$, let $s_i \geq s_{i-1}$ be the smallest integer such that β_i is a factor (that is, contiguous subwalk) of $c_{q_{s_{i-1}}}^{i_{s_{i-1}}} p_{q_{s_{i-1}},q'_{s_{i-1}}}(q'_{s_{i-1}}, b_{s_{i-1}+1}, q_{s_{i-1}+1})\cdots c_{q_{s_i}}^{i_{s_i}} p_{q_{s_i},q'_{s_i}}$. It is not hard now to see that $s_i > s_{i-1}$ holds, as alph(label($c_{q_{s_i-1}} p_{q_{s_i-1},q'_{s_i-1}}$)) = alph(label($c_{q_{s_i-1}}$)) $\subsetneq \Sigma =$ alph(label(β_i)) = alph(w_i).

Let us now define, for all $i \in [t]$, the walk β'_i as:
$$c_{q_{s_{i-1}}} p_{q_{s_{i-1}},q'_{s_{i-1}}}(q'_{s_{i-1}}, b_{s_{i-1}+1}, q_{s_{i-1}+1})\cdots c_{q_{s_i-1}} p_{q_{s_i-1},q'_{s_i-1}}(q'_{s_i-1}, b_{s_i}, q_{s_i}) c_{q_{s_i}}.$$
Also, let $w'_i =$ label(β'_i). Clearly, alph(w'_i) = alph(w_i) = Σ, as alph($c_{q_{s_i}}^{i_{s_i}} p_{q_{s_i},q'_{s_i}}$) = alph($c_{q_{s_i}} p_{q_{s_i},q'_{s_i}}$) = alph($c_{q_{s_i}}$). The conclusion follows, for $w' = w'_1 \cdots w'_k$. □

Therefore, by Lemma 4, to compute the maximum universality index over the words of L it is enough to identify a word $w' \in L$ of maximum universality that labels some walk obtained by concatenating several subwalks of the form $c_{q_1} p_{q_1, q'_1}(q'_1, b_2, q_2) \cdots c_{q_{h-1}} p_{q_{h-1}, q'_{h-1}}(q'_{h-1}, b_h, q_h) c_{q_h}$, where: $h \leq n$; q'_i belongs to $S(q_i)$ for all $i \in [h-1]$; $S(q_i) \neq S(q_j)$, for all $i \neq j$; and $q_h = f$ is the final state of A. The key observation is that, in fact, to find such a word of maximum universality, it is enough to know the sequence of states $q_1, \ldots, q_h$ as above. We will explain now how this is done efficiently.

Lemma 5. *Given a sequence of $h < n$ states $q_1 \cdots q_h$, such that $S(q_i) \neq S(q_j)$ for all $i \neq j$, we can decide in $O(m + n\sigma\sqrt{n})$ time whether there exist states $g_1, \ldots, g_h$ and $g'_1, \ldots, g'_h$, with $g_1 = q_1$, $S(g_i) = S(g'_i) = S(q_i)$ for all $i \in [h]$, and the walk $\beta = c_{q_1} p_{q_1, g'_1}(g'_1, b_2, g_2) \cdots c_{g_{h-1}} p_{g_{h-1}, g'_{h-1}}(g'_{h-1}, b_h, g_h) c_{g_h} p_{g_h, q_h}$ such that $\mathrm{label}(\beta)$ is 1-universal.*

Proof. Firstly, we can compute in $O(n+m)$ time the sets $V_{q_i, q_{i+1}} \subseteq \Sigma \cup \{\varepsilon\}$ of labels of transitions from states of $\mathcal{C}_{S(q_i)}$ to states of $\mathcal{C}_{S(q_{i+1})}$, for all $i \in [h-1]$. If there exists $i \in [h-1]$ such that $V_{q_i, q_{i+1}} = \emptyset$, then we answer the considered problem negatively. Let us assume, in the following, that this is not the case.

Then, we compute the sets $V_{S(q_i)}$, for all $i \in [h]$, as described in Lemma 3, and the set $V = \bigcup_{i \in h} V_{S(q_i)}$. This takes $O(n+m)$ time. Further, we want to see if there is a way to identify states $g_2, \ldots, g_h$, $g'_1, \ldots, g'_h$, and the letters $b_1, \ldots, b_h$, such that $\beta = c_{q_1} p_{q_1, g'_1}(g'_1, b_2, g_2) \cdots c_{g_{h-1}} p_{g_{h-1}, g'_{h-1}}(g'_{h-1}, b_h, g_h) c_{g_h} p_{g_h, q_h}$ is 1-universal. In other words, $\Sigma \setminus V \subseteq \{b_1, \ldots, b_h\}$.

The first case is when $V = \Sigma$. In that particular case, all walks $\beta = c_{q_1} p_{q_1, g'_1}(g'_1, b_2, g_2) \cdots c_{g_{h-1}} p_{g_{h-1}, g'_{h-1}}(g'_{h-1}, b_h, g_h) c_{g_h} p_{g_h, q_h}$, fulfilling the conditions from the statement, are labelled by 1-universal words, and the considered problem can be answered positively.

The second case is if $\Sigma \subset V$ and $|V| > h$; then, the answer is negative.

Otherwise, we define a bipartite graph G as follows:

- The set of vertices is defined by the following two disjoint sets: on the one side, we have the set $[h-1]$ (so the vertices are integers $i \in [h-1]$) and, on the other side, the set $\Sigma \setminus V$ (so the vertices are letters $b \in \Sigma \setminus V$).
- There exist an edge between vertex $i \in [h-1]$ and vertex $b \in \Sigma \setminus V$ if and only if $b \in V_{q_i, q_{i+1}}$. G contains no other edges.

If $\nu = |\Sigma \setminus V|$, we have $\nu \leq h$, so we can compute in $O(h\nu\sqrt{h+\nu}) \subseteq O(n\sigma\sqrt{n})$ time a maximum-cardinality bipartite matching [35]. If the cardinality of this matching is ν (that is, every letter-vertex of G is covered by one edge of the matching), then we proceed as follows. We first define, for $i \in \{2, \ldots, h\}$, the letter b_i: if $(i-1, b)$ is an edge of the matching, then $b_i = b$; otherwise, b_i is an arbitrary element of V_{q_{i-1}, q_i}. In both cases, we set $g'_{i-1} \in V_{S(q_{i-1})}$ and $g_i \in V_{S(q_i)}$ to be states connected by the letter b_i. It is now immediate that the path $\beta = c_{q_1} p_{q_1, g'_1}(g'_1, b_2, g_2) \cdots c_{g_{h-1}} p_{g_{h-1}, g'_{h-1}}(g'_{h-1}, b_h, g_h) c_{g_h} p_{g_h, q_h}$ is labelled by an 1-universal word: each letter of Σ is either the label of a transition from a cycle c_{q_i} or of a transition between some states g'_{i-1} and g_i. Thus, in this case, we can answer the problem positively. If the cardinality of this matching is strictly smaller than ν, then the problem is answered negatively. In that case, we cannot cover all letters with either transitions of the strongly connected components $\mathcal{C}_{S(q_i)}$, for $i \in [t]$, or the transition between states of these components. □

Note that in Lemma 5 we do not compute the sequence $g_1 \cdots g_h$, we just decide whether there is a walk from q_1 to q_h, going through the strongly connected components of $q_2, \ldots, q_{h-1}$, which is labelled by a 1-universal word.

Now, we can state a result which is the main building block of our algorithm.

Lemma 6. *Given a sequence $q_1 \cdots q_h$ of $h < n$ states with $q_h = f$ and $S(q_i) \neq S(q_j)$, for all $i \neq j$, we can compute in $O(n(m + n\sigma\sqrt{n}))$ time the maximum integer k for which there exists a walk β with origin q_1 and target q_h such that label(β) is k-universal and β goes through the strongly connected components of $q_1, \ldots, q_h$, and through no other strongly connected component.*

The idea of the algorithm introduced in this proof is to iteratively use Lemma 5, within a greedy strategy. We use Lemma 5 to identify the shortest prefix $q_1 \cdots q_i$ of $q_1 \cdots q_h$ for which there is a path from q_1 to q_i, going through the strongly connected components of $q_2, \ldots, q_{i-1}$, which is labelled by a 1-universal word. If such a prefix is found, we repeat this process for $q_i \cdots q_h$. We return the number of times we can successfully execute this process.

We can now state the main result of this section.

Theorem 3. *Given a regular language L, over Σ, with $|\Sigma| = \sigma$, specified as an NFA A with n states, we can solve k-ESU in $O(2^n n(m + n\sigma\sqrt{n}))$ time. That is, in the case when the input is given as an NFA, k-ESU is FPT w.r.t. the number n of states of the input NFA.*

The algorithm of this theorem considers each set of states $\{q_0, f\} \cup \{q_1, \ldots, q_{h-1}\}$ of A, sorts it w.r.t. the topological sorting of $S(A)$ to obtain the sequence $q_0 q_1 \ldots q_{h-1} f$, and then uses Lemma 6 on it, if the prerequisites of that lemma apply. We return the largest universality index computed for such a subsequence.

Note that Theorem 3 shows that k-ESU can be solved by an FPT-algorithm w.r.t. the number of states of the input NFA. Moreover, as $m \in O(n^2\sigma)$, k-ESU can be solved in linear time $O(\sigma)$ for NFAs with $n \in O(1)$.

FPT Algorithm w.r.t. σ. In [5] it was shown that we can solve k-ESU in $O(n^3 poly(\sigma) 2^\sigma)$ time. We improve this result and give a linear time solution for k-ESU in the case of constant alphabets.

Theorem 4. *Given a regular language L, over Σ, with $|\Sigma| = \sigma$, specified as an NFA A with n states, we can solve k-ESU in $O(n + m + m2^\sigma)$ time.*

The algorithm of this theorem uses a dynamic programming approach. If $\mathcal{C}_1 = \mathcal{C}_{S(q_0)}, \ldots, \mathcal{C}_e = \{f\}$ is the topological sorting of $S(A)$, we efficiently compute $D[i][S]$ for $i \in [e]$ and $S \subseteq \Sigma$, the largest universality index of a word with alph(r(w)) = S, which labels a walk starting in q_0, ending in $\mathcal{C}_i$, and going only through the components $\mathcal{C}_1, \ldots, \mathcal{C}_i$. We then return the largest element of $D[e][\cdot]$.

Regular Expressions. We now consider languages represented by regexes. While NFAs and regexes define the same class of languages, the size of a regex required to represent a given language can be exponentially larger than the size of the NFA accepting that language (and vice versa, see, e.g., [28]). In this section, we first show that k-ESU, where the language is specified as a regex of length n, can be solved in $O(n2^\sigma)$ time, using the FPT-algorithm w.r.t. σ described above. Then, we provide a new proof of the NP-completeness of k-ESU when the input language is given as a regex. This result is interesting, as the hardness proof from [5] seems to require, in some cases, a regex exponentially larger than the corresponding NFA to describe the regular language used in the given reduction.

We start with the algorithmic part of this section. We assume from now on that we are given a regular expression of length n. The first result is immediate, by a proof similar to Lemma 2.

Lemma 7. *For regular expression R, if $R = R_1(R_2)^*R_3$ and $\Sigma \subseteq \text{alph}(R_2)$, then $L(R)$ is k-universal, for every $k \geq 1$.*

We can now decide in $O(n)$ time, for a regex R of length n, whether $L(R)$ is k-universal, for every $k \geq 1$. If not, then $\iota(w) \leq n$ for all $w \in L(R)$, as $|w| \leq n$. In this case, we can reduce the case of general regexes to the case of star-free regexes.

Lemma 8. *Given regular expression R, if there does not exist a decomposition $R = U_1(U_2)^*U_3$ with $\Sigma \subseteq \text{alph}(U_2)$, then there exist $k \geq 1$ such that $L(R)$ is not k-universal. Moreover, there exists a star-free regular expression R', which can be computed in linear time $O(|R|)$, such that $\max\{\iota(w) \mid w \in L(R')\} = \max\{\iota(w) \mid w \in L(R)\}$.*

Lemma 8 immediately leads us to the following conclusion.

Theorem 5. *Given a regular language L, over Σ, with $|\Sigma| = \sigma$, specified as a regex R of length n, we can solve k-ESU in $O(n2^\sigma)$ time.*

The idea is to first use Lemma 8 to construct a star-free regular expression R', of size $O(n)$, such that $\max\{\iota(w) \mid w \in L(R')\} = \max\{\iota(w) \mid w \in L(R)\}$. Then we build (as in, e.g., [7]) an NFA for R', of total size $O(n)$, and use Theorem 4 for this NFA.

This result shows that k-ESU can be solved in linear time for regular expressions over constant-size alphabets. Regarding the hardness of k-ESU when the input is given as a regular expression, we can show the following result.

Theorem 6. *k-ESU is NP-complete for languages defined by regular expressions. The problem is already NP-hard for $k = 1$ and star-free regexes.*

This result follows from a reduction that maps 3-SAT-instances with n variables and m clauses to 1-ESU-instances where the input regex has length $O(m)$.

Concluding Remarks. The lower bound of Theorem 6 and its proof allow us to also make several final remarks on the presented FPT algorithms for k-ESU.

We consider first the case when the input regular language is given as an NFA with n states over an alphabet with σ letters. The lower bound derived for k-ESU in [5] and its proof show that, under the Exponential Time Hypothesis [37, 48], there are no $2^{o(\sigma)}poly(n,\sigma)$ time algorithms solving the respective problem. The reduction from the proof of Theorem 6, together with the Sparsification Lemma [37], shows that there are no $2^{o(n)}poly(n,\sigma)$ time algorithms for k-ESU, under ETH. If the input language is given as a regular expression of length n over an alphabet with σ letters, we can once more refer to the reduction from the proof of Theorem 6 and the Sparsification Lemma, and get that there are no $2^{o(n)}poly(n,\sigma)$ time algorithms for k-ESU, in this setting. As such, the algorithmic results we have obtained are tight from a fine-grained complexity perspective.

Acknowledgement. Florin Manea is supported by the German Research Foundation (Deutsche Forschungsgemeinschaft, DFG) in the framework of the Heisenberg Programme – project number 466789228.

References

1. Abboud, A., Backurs, A., Williams, V.V.: Tight hardness results for LCS and other sequence similarity measures. In: IEEE 56th Annual Symposium on Foundations of Computer Science, FOCS 2015, Berkeley, CA, USA, 17–20 October 2015, pp. 59–78 (2015). https://doi.org/10.1109/FOCS.2015.14
2. Abboud, A., Rubinstein, A.: Fast and deterministic constant factor approximation algorithms for LCS imply new circuit lower bounds. In: 9th Innovations in Theoretical Computer Science Conference, ITCS 2018, 11–14 January 2018, Cambridge, MA, USA, pp. 35:1–35:14 (2018). https://doi.org/10.4230/LIPIcs.ITCS.2018.35
3. Abboud, A., Williams, V.V., Weimann, O.: Consequences of faster alignment of sequences. In: Esparza, J., Fraigniaud, P., Husfeldt, T., Koutsoupias, E. (eds.) ICALP 2014. LNCS, vol. 8572, pp. 39–51. Springer, Heidelberg (2014). https://doi.org/10.1007/978-3-662-43948-7_4
4. Adamson, D.: Ranking and unranking k-subsequence universal words. In: Frid, A., Mercaş, R. (eds.) WORDS, pp. 47–59. Springer (2023)
5. Adamson, D., Fleischmann, P., Huch, A., Koß, T., Manea, F., Nowotka, D.: k-universality of regular languages. In: ISAAC 2023. LIPIcs, vol. 283, pp. 4:1–4:21. Schloss Dagstuhl - Leibniz-Zentrum für Informatik (2023)
6. Adamson, D., Kosche, M., Koß, T., Manea, F., Siemer, S.: Longest common subsequence with gap constraints. In: Frid, A., Mercaş, R. (eds.) WORDS, pp. 60–76 (2023)
7. Allauzen, C., Mohri, M.: A unified construction of the glushkov, follow, and antimirov automata. In: Královič, R., Urzyczyn, P. (eds.) MFCS 2006. LNCS, vol. 4162, pp. 110–121. Springer, Heidelberg (2006). https://doi.org/10.1007/11821069_10
8. Anand, A., Zetzsche, G.: Priority downward closures. In: Pérez, G.A., Raskin, J. (eds.) 34th International Conference on Concurrency Theory, CONCUR 2023, 18–23 September 2023, Antwerp, Belgium. LIPIcs, vol. 279, pp. 39:1–39:18. Schloss Dagstuhl - Leibniz-Zentrum für Informatik (2023). https://doi.org/10.4230/LIPICS.CONCUR.2023.39

9. Artikis, A., Margara, A., Ugarte, M., Vansummeren, S., Weidlich, M.: Complex event recognition languages: tutorial. In: DEBS, pp. 7–10 (2017)
10. Baeza-Yates, R.A.: Searching subsequences. Theor. Comput. Sci. **78**(2), 363–376 (1991)
11. Barker, L., Fleischmann, P., Harwardt, K., Manea, F., Nowotka, D.: Scattered factor-universality of words. In: DLT, pp. 14–28. Springer (2020)
12. Bergroth, L., Hakonen, H., Raita, T.: A survey of longest common subsequence algorithms. In: SPIRE, pp. 39–48. IEEE Computer Society (2000)
13. Bringmann, K., Chaudhury, B.R.: Sketching, streaming, and fine-grained complexity of (weighted) LCS. In: Ganguly, S., Pandya, P. (eds.) 38th IARCS Annual Conference on Foundations of Software Technology and Theoretical Computer Science (FSTTCS 2018). LIPIcs, vol. 122, pp. 40:1–40:16 (2018). https://doi.org/10.4230/LIPIcs.FSTTCS.2018.40
14. Bringmann, K., Künnemann, M.: Multivariate fine-grained complexity of longest common subsequence. In: Proceedings of SODA 2018, pp. 1216–1235 (2018)
15. Chvatal, V., Sankoff, D.: Longest common subsequences of two random sequences. J. Appl. Probab. **12**(2), 306–315 (1975)
16. Crochemore, M., Hancart, C., Lecroq, T.: Algorithms on strings. Cambridge University Press (2007)
17. Crochemore, M., Melichar, B., Troníček, Z.: Directed acyclic subsequence graph - overview. J. Discrete Algorithms **1**(3–4), 255–280 (2003)
18. Day, J., Fleischmann, P., Kosche, M., Koß, T., Manea, F., Siemer, S.: The edit distance to k-subsequence universality. In: STACS, vol. 187, pp. 25:1–25:19 (2021)
19. Day, J.D., Kosche, M., Manea, F., Schmid, M.L.: Subsequences with gap constraints: complexity bounds for matching and analysis problems. In: Bae, S.W., Park, H. (eds.) 33rd International Symposium on Algorithms and Computation, ISAAC 2022, 19–21 December 2022, Seoul, Korea. LIPIcs, vol. 248, pp. 64:1–64:18. Schloss Dagstuhl - Leibniz-Zentrum für Informatik (2022). https://doi.org/10.4230/LIPICS.ISAAC.2022.64
20. Fazekas, S.Z., Koß, T., Manea, F., Mercaş, R., Specht, T.: Subsequence matching and analysis problems for formal languages (2024). https://arxiv.org/abs/2410.07992
21. Fleischer, L., Kufleitner, M.: Testing simon's congruence. In: MFCS. Schloss Dagstuhl-Leibniz-Zentrum fuer Informatik (2018)
22. Fleischmann, P., Germann, S., Nowotka, D.: Scattered factor universality–the power of the remainder. preprint arXiv:2104.09063 (published at RuFiDim) (2021)
23. Fleischmann, P., Höfer, J., Huch, A., Nowotka, D.: α-β-factorization and the binary case of simon's congruence. In: FCT. LNCS, vol. 14292, pp. 190–204. Springer (2023)
24. Fleischmann, P., Kim, S., Koß, T., Manea, F., Nowotka, D., Siemer, S., Wiedenhöft, M.: Matching patterns with variables under simon's congruence. In: RP 2023. LNCS, vol. 14235, pp. 155–170. Springer (2023)
25. Fredman, M.L., Willard, D.E.: BLASTING through the information theoretic barrier with FUSION TREES. In: Ortiz, H. (ed.) Proceedings of the 22nd Annual ACM Symposium on Theory of Computing, 13–17 May 1990, Baltimore, Maryland, USA, pp. 1–7. ACM (1990). https://doi.org/10.1145/100216.100217
26. Frochaux, A., Kleest-Meißner, S.: Puzzling over subsequence-query extensions: disjunction and generalised gaps. In: AMW 2023. CEUR Workshop Proceedings, vol. 3409. CEUR-WS.org (2023)
27. Gawrychowski, P., Kosche, M., Koß, T., Manea, F., Siemer, S.: Efficiently testing simon's congruence. In: STACS, vol. 187, pp. 34:1–34:18 (2021)

28. Gruber, H., Holzer, M.: From finite automata to regular expressions and back - a summary on descriptional complexity. Int. J. Found. Comput. Sci. **26**(8), 1009–1040 (2015). https://doi.org/10.1142/S0129054115400110
29. Halfon, S., Schnoebelen, P., Zetzsche, G.: Decidability, complexity, and expressiveness of first-order logic over the subword ordering. In: LICS, pp. 1–12. IEEE (2017)
30. Han, R., Wang, S., Gao, X.: Novel algorithms for efficient subsequence searching and mapping in nanopore raw signals towards targeted sequencing. Bioinformatics **36**(5), 1333–1343 (2020)
31. Hebrard, J.J.: An algorithm for distinguishing efficiently bit-strings by their subsequences. TCS **82**(1), 35–49 (1991)
32. Higman, G.: Ordering by divisibility in abstract algebras. Proc. Lond. Math. Soc. **3**(1), 326–336 (1952)
33. Hirschberg, D.S.: Algorithms for the longest common subsequence problem. J. ACM **24**(4), 664–675 (1977)
34. Hopcroft, J.E., Ullman, J.D.: Introduction to Automata Theory, Languages and Computation. Addison-Wesley (1979)
35. Hopcroft, J.E., Karp, R.M.: An $n^{5/2}$ algorithm for maximum matchings in bipartite graphs. SIAM J. Comput. **2**(4), 225–231 (1973). https://doi.org/10.1137/0202019
36. Hunt, J.W., Szymanski, T.G.: A fast algorithm for computing longest subsequences. Commun. ACM **20**(5), 350–353 (1977)
37. Impagliazzo, R., Paturi, R., Zane, F.: Which problems have strongly exponential complexity? J. Comput. Syst. Sci. **63**(4), 512–530 (2001). https://doi.org/10.1006/JCSS.2001.1774
38. Karandikar, P., Kufleitner, M., Schnoebelen, P.: On the index of Simon's congruence for piecewise testability. Inf. Process. Lett. **115**(4), 515–519 (2015)
39. Karandikar, P., Schnoebelen, P.: The height of piecewise-testable languages with applications in logical complexity. In: CSL (2016)
40. Kim, S., Ko, S., Han, Y.: Simon's congruence pattern matching. In: ISAAC. LIPIcs, vol. 248, pp. 60:1–60:17. Schloss Dagstuhl - Leibniz-Zentrum für Informatik (2022)
41. Kleest-Meißner, S., Sattler, R., Schmid, M.L., Schweikardt, N., Weidlich, M.: Discovering event queries from traces: laying foundations for subsequence-queries with wildcards and gap-size constraints. In: ICDT. LIPIcs, vol. 220, pp. 18:1–18:21 (2022)
42. Kleest-Meißner, S., Sattler, R., Schmid, M.L., Schweikardt, N., Weidlich, M.: Discovering multi-dimensional subsequence queries from traces - from theory to practice. In: BTW. LNI, vol. P-331, pp. 511–533 (2023)
43. Kosche, M., Koß, T., Manea, F., Siemer, S.: Absent subsequences in words. In: RP, pp. 115–131. Springer (2021)
44. Kosche, M., Koß, T., Manea, F., Siemer, S.: Combinatorial algorithms for subsequence matching: a survey. In: Bordihn, H., Horváth, G., Vaszil, G. (eds.) NCMA (2022)
45. Kosche, M., Koß, T., Manea, F., Pak, V.: Subsequences in bounded ranges: matching and analysis problems. In: Lin, A.W., Zetzsche, G., Potapov, I. (eds.) RP 2022. LNCS, vol. 13608, pp. 140–159. Springer, Cham (2022). https://doi.org/10.1007/978-3-031-19135-0_10
46. Kuske, D.: The subtrace order and counting first-order logic. In: CSR. LNCS, vol. 12159, pp. 289–302. Springer (2020)
47. Kuske, D., Zetzsche, G.: Languages ordered by the subword order. In: FOSSACS. LNCS, vol. 11425, pp. 348–364. Springer (2019)

48. Lokshtanov, D., Marx, D., Saurabh, S.: Lower bounds based on the exponential time hypothesis. Bull. EATCS **105**, 41–72 (2011). http://eatcs.org/beatcs/index.php/beatcs/article/view/92
49. Maier, D.: The complexity of some problems on subsequences and supersequences. J. ACM **25**(2), 322–336 (1978)
50. Manea, F., Richardsen, J., Schmid, M.L.: Subsequences with generalised gap constraints: upper and lower complexity bounds. In: Inenaga, S., Puglisi, S.J. (eds.) 35th Annual Symposium on Combinatorial Pattern Matching, CPM 2024, 25–27 June 2024, Fukuoka, Japan. LIPIcs, vol. 296, pp. 22:1–22:17. Schloss Dagstuhl - Leibniz-Zentrum für Informatik (2024). https://doi.org/10.4230/LIPICS.CPM.2024.22
51. Masek, W.J., Paterson, M.: A faster algorithm computing string edit distances. J. Comput. Syst. Sci. **20**(1), 18–31 (1980). https://doi.org/10.1016/0022-0000(80)90002-1
52. Nakatsu, N., Kambayashi, Y., Yajima, S.: A longest common subsequence algorithm suitable for similar text strings. Acta Informatica **18**, 171–179 (1982)
53. Schnoebelen, P., Karandikar, P.: The height of piecewise-testable languages and the complexity of the logic of subwords. Log. Methods Comput. Sci. **15** (2019)
54. Schnoebelen, P., Veron, J.: On arch factorization and subword universality for words and compressed words. In: WORDS. LNCS, vol. 13899, pp. 274–287. Springer (2023)
55. Shaw, A.C.: Software descriptions with flow expressions. IEEE Trans. Software Eng. **3**, 242–254 (1978)
56. Shikder, R., Thulasiraman, P., Irani, P., Hu, P.: An openmp-based tool for finding longest common subsequence in bioinformatics. BMC. Res. Notes **12**, 1–6 (2019)
57. Simon, I.: Hierarchies of events with dot-depth one. Ph.D. thesis, University of Waterloo, Department of Applied Analysis and Computer Science (1972)
58. Simon, I.: Piecewise testable events. In: Automata Theory and Formal Languages, 2nd GI Conference. LNCS, vol. 33, pp. 214–222. Springer (1975)
59. Simon, I.: Words distinguished by their subwords. WORDS **27**, 6–13 (2003)
60. Tarjan, R.E.: Depth-first search and linear graph algorithms. SIAM J. Comput. **1**(2), 146–160 (1972). https://doi.org/10.1137/0201010
61. Troníček, Z.: Common subsequence automaton. In: CIAA, pp. 270–275 (2003)
62. Zetzsche, G.: The complexity of downward closure comparisons. In: ICALP, vol. 55, pp. 123:1–123:14 (2016)
63. Zetzsche, G.: Separability by piecewise testable languages and downward closures beyond subwords. In: LICS 2018, pp. 929–938. ACM (2018)

Comparing the Hardness of Online Minimization and Maximization Problems with Predictions

Magnus Berg(✉)

University of Southern Denmark, Odense, Denmark
magbp@imada.sdu.dk
https://imada.sdu.dk/u/magbp

Abstract. We build on the work of Berg, Boyar, Favrholdt, and Larsen, who developed a complexity theory for online problems with and without predictions (IJTCS-FAW 2025) where they define a hierarchy of complexity classes that classifies online problems based on the competitiveness of best possible deterministic online algorithms for each problem. Their work focused on online minimization problems and we continue their work by considering online maximization problems.

First, we compare the competitiveness of the base online minimization problem from Berg, Boyar, Favrholdt, and Larsen, Asymmetric String Guessing, to the competitiveness of Online Bounded Degree Independent Set. Formally, we show that there exists algorithms of any given competitiveness for Asymmetric String Guessing if and only if there exists algorithms of the same competitiveness for Online Bounded Degree Independent Set, while respecting that the competitiveness of algorithms is measured differently for minimization and maximization problems. Moreover, we give several hardness preserving reductions between different online maximization problems, which imply new membership, hardness, and completeness results for the complexity classes. Finally, we show new positive and negative algorithmic results for (among others) Online Bounded Degree Independent Set, Online Interval Scheduling, Online Set Packing, and Online Bounded Degree Clique.

Keywords: Online Algorithms with Predictions · Complexity Theory · Minimization vs Maximization · Independent Set

1 Introduction

Recently, Berg, Boyar, Favrholdt, and Larsen introduced a complexity theory for online problems with (and without) predictions [5], where they classify several hard online minimization problems based on the competitiveness of best possible

---

Supported in part by the Independent Research Fund Denmark, Natural Sciences, grant DFF-4283-00079B and in part by the Innovation Fund Denmark, grant 9142-00001B, Digital Research Centre Denmark, project P40: Online Algorithms with Predictions.

online algorithms for each problem. In [5], the authors only considers online minimization problems, and in this work, we prove several membership, hardness, and completeness results of online maximization problems for the complexity classes. Towards this goal, we directly compare the competitiveness of online minimization problems to the competitiveness of online maximization problems, while respecting that the competitiveness of algorithms is measured differently for minimization and maximization problems.

An *online problem* is an optimization problem, where the input is divided into a sequence of *requests* that are given one-by-one. When an online algorithm receives a request, it must make an irrevocable decision about the request before receiving the next request. We analyze the performance of algorithms using competitive analysis [7,23,28]; the most standard framework for online algorithms. Competitive analysis is a framework for worst-case guarantees, where we measure the quality of an algorithm by comparing its performance over all possible input sequences to the performance of an optimal offline algorithm.

In recent years, there has been an increasing interest in algorithms with predictions [1], not least in the context of online algorithms [24,27]. Here, algorithms are endowed with a predictor to help them create their solution to a given instance. In the perfect world, online algorithms perform optimally when given perfect predictions and no worse than the best known purely online algorithm when the predictions are erroneous or adversarial. This ideal case is often not realizable and instead we determine an error measure through which we evaluate the quality of the predictions, and then we express the competitiveness of the algorithms as a function of the error.

The recently developed complexity theory for online problems with predictions [5] allows for classifying online problems based on their *hardness*, which is given by the competitiveness of best possible deterministic online algorithms for each problem. Formally, for any $t \in \mathbb{Z}^+ \cup \{\infty\}$ and any pair of error measures (η_0, η_1), the complexity class $\mathcal{C}^t_{\eta_0,\eta_1}$ is defined as the closure of $(1,t)$-*Asymmetric String Guessing with Unknown History and Predictions* (ASG_t), under the as-hard-as relation (see Definition 2) with respect to (η_0, η_1) [5]. We give a brief review of the central components of the complexity classes in Sect. 2.1.

1.1 Our Contribution

Our contribution revolves around proving several membership, hardness, and completeness results for the complexity classes from [5], focusing on online maximization problems. Our first effort in this direction (see Sect. 3) is a hardness comparison between the base (minimization) problem of $\mathcal{C}^t_{\eta_0,\eta_1}$, ASG_t, and *Online t-Bounded Degree Independent Set with Predictions* (IS_t). We show that IS_t is exactly as hard as ASG_t with respect to the canonical pair of error measures (μ_0, μ_1) (see Eq. (3)), in the sense that there exists an algorithm of any given performance guarantee for ASG_t if and only if there exists an algorithm of the exact same performance guarantee for IS_t, while respecting that the performance guarantee is measured differently for online minimization problems and online maximization problems (see Definition 1). This relation implies that IS_t is

$\mathcal{C}^t_{\mu_0,\mu_1}$-complete. Since the dual problem of IS_t, *Online t-Bounded Degree Vertex Cover with Predictions* (VC_t), is also $\mathcal{C}^t_{\mu_0,\mu_1}$-complete [5], the above hardness comparison between ASG_t and IS_t also imply that IS_t is exactly as hard as VC_t with respect to (μ_0, μ_1).

Secondly, in Sect. 4, we compare the hardness of IS_t to the hardness of five other online maximization problems, which proves a number of membership, hardness and completeness results for $\mathcal{C}^t_{\mu_0,\mu_1}$. Finally, we use these new membership, hardness, and completeness results for $\mathcal{C}^t_{\mu_0,\mu_1}$ to extend positive algorithmic results from [5] to all new members and complete problems, and strong negative results from [2] to all complete and hard problems.

We extend the *hardness graph* from [5] to illustrate our results (see Fig. 1). Omitted details and proofs are contained in the full paper [4].

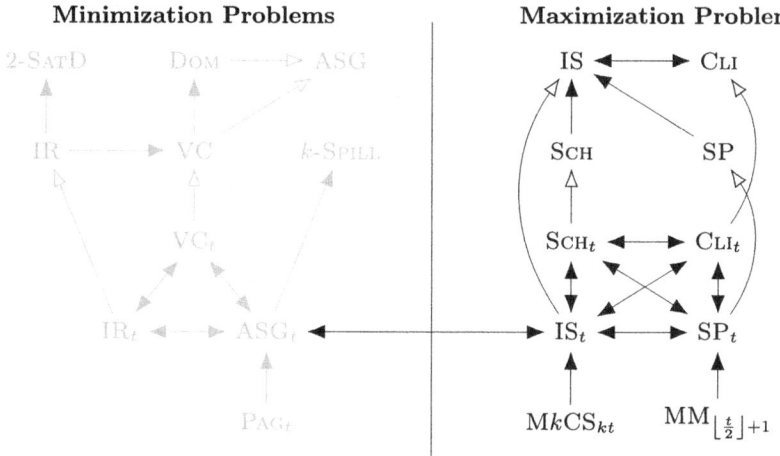

Fig. 1. An (incomplete) representation of the hardness graph for $t \in \mathbb{Z}^+$ with respect to (μ_0, μ_1). The vertices are online problems, and there is an arc $P \to Q$ when Q is as hard as P (see Definition 2). If the arrowhead of an arc $P \to Q$ is only outlined then $Q \to P$ cannot exist. All problems and arcs in gray are from [5]. The definitions of the remaining problems and the existence of the remaining arcs are in Sects. 3 and 4. All arcs in the transitive closure also exist but are omitted for simplicity. Hence, if there is a directed (P,Q)-path in the hardness graph, then $P \to Q$ also exists (see Lemma 1).

2 Preliminaries

We consider online problems, P, where algorithms must output a bit as its answer to each request. Following [5], we think of an instance of P as a triple $I = (x, \hat{x}, r)$, where $x, \hat{x} \in \{0,1\}^n$ are bitstrings and $r = \langle r_1, r_2, \ldots, r_n \rangle$ is a sequence of requests for P. We let $\mathcal{I}_P$ be the set of all instances for P, and we let OPT_P be a fixed optimal offline algorithm for P. Given an instance, $I = (x, \hat{x}, r)$,

x must be an encoding of OPT_P's solution and $\hat{x}$ is a prediction of x. When an algorithm for P, ALG_P, receives a request, r_i, it also receives the prediction $\hat{x}_i$ which predicts the optimal output to r_i, x_i. The contents of the requests is defined for each problem separately. If P is a maximization problem we let $\text{OPT}_P^p(I)$ and $\text{ALG}_P^p(I)$ be the *profit* of OPT_P and ALG_P, respectively, and if P is a minimization problem we let $\text{OPT}_P^c(I)$ and $\text{ALG}_P^c(I)$ be the *cost* of OPT_P and ALG_P, respectively. Finally, we let $\mathcal{A}_P$ be the set of deterministic online algorithms for P.

We analyze the performance of algorithms using competitive analysis. An online algorithm, ALG_P, for an online problem, P, is called α-*competitive* if there exists a constant $\kappa \in \mathbb{R}$, called the *additive constant*, such that for all instances $I \in \mathcal{I}_P$, we have that $\text{OPT}_P^p(I) \leq \alpha \cdot \text{ALG}_P^p(I) + \kappa$, if P is a maximization problem, or $\text{ALG}_P^c(I) \leq \alpha \cdot \text{OPT}_P^c(I) + \kappa$, if P is a minimization problem. We extend the definition of competitiveness to take into account the quality of the predictions. Following previous work on online algorithms with binary predictions [2,5], we measure the error through a pair of error measures (η_0, η_1) where η_b, for $b \in \{0,1\}$, is a function of the incorrectly predicted requests where the prediction is b. For any instance, I, we require that $0 \leq \eta_b(I) < \infty$.

Definition 1. *Let P be an online problem with binary predictions, let (η_0, η_1) be a pair of error measures, and let $\alpha, \beta, \gamma \colon \mathcal{I}_P \to [0, \infty)$ be any maps. If P is a maximization problem, then $\text{ALG}_P \in \mathcal{A}_P$ is called (α, β, γ)-competitive with respect to (η_0, η_1) if there exists $\kappa \in \mathbb{R}$ such that for all $I \in \mathcal{I}_P$,*

$$\text{OPT}_P^p(I) \leq \alpha \cdot \text{ALG}_P^p(I) + \beta \cdot \eta_0(I) + \gamma \cdot \eta_1(I) + \kappa. \tag{1}$$

Similarly, if P is a minimization problem, then $\text{ALG}_P \in \mathcal{A}_P$ is called (α, β, γ)-competitive with respect to (η_0, η_1) if there exists $\kappa \in \mathbb{R}$ such that for all $I \in \mathcal{I}_P$,

$$\text{ALG}_P^c(I) \leq \alpha \cdot \text{OPT}_P^c(I) + \beta \cdot \eta_0(I) + \gamma \cdot \eta_1(I) + \kappa, \tag{2}$$

If (η_0, η_1) is clear from the context, we say that ALG_P is (α, β, γ)-competitive.

Following [5], we suppress α's, β's, and γ's dependency on I. Moreover, observe that any α-competitive algorithm without predictions for P gives rise to an $(\alpha, 0, 0)$-competitive algorithm with predictions for P with respect to any pair of error measures, and vice versa. All results in this paper hold with respect to the canonical pair of error measures [2,5], (μ_0, μ_1), given by

$$\mu_0(I) = \sum_{i=1}^n x_i \cdot (1 - \hat{x}_i) \quad \text{and} \quad \mu_1(I) = \sum_{i=1}^n (1 - x_i) \cdot \hat{x}_i, \tag{3}$$

and the pair of error measures (Z_0, Z_1) given by $Z_0(I) = Z_1(I) = 0$. In words, for $b \in \{0,1\}$, μ_b counts the number of incorrect predictions in I, where the prediction is b, and Z_b is the zero measure. The reason that (Z_0, Z_1) is interesting is that it models the purely online case, in the sense that an (α, β, γ)-competitive algorithm with respect to (Z_0, Z_1) is α-competitive as a purely online algorithm.

2.1 Measuring Hardness and the Complexity Classes $\mathcal{C}^t_{\eta_0,\eta_1}$

We recall the central components of the complexity framework from [5]. The aim is to classify online problems based on their hardness. Given an online problem, P, the *hardness of* P is given by the competitiveness of best possible deterministic online algorithms for P, where an algorithm is called *best possible* if it is Pareto-optimal. An (α, β, γ)-competitive algorithm for P is called *Pareto-optimal*, if, for any $\varepsilon > 0$, there is no $(\alpha - \varepsilon, \beta, \gamma)$-, $(\alpha, \beta - \varepsilon, \gamma)$-, or $(\alpha, \beta, \gamma - \varepsilon)$-competitive algorithm for P. Formally, the hardness of online problems is compared using the as-hard-as relation:

Definition 2 ([5]). *Let P and Q be online problems with binary predictions and let (η_0, η_1) be a pair of error measures. We say that Q is (at least) as hard as P with respect to (η_0, η_1), if the existence of an (α, β, γ)-competitive Pareto-optimal algorithm for Q implies the existence of an (α, β, γ)-competitive algorithm for P.*

Lemma 1 ([5]). *The as-hard-as relation is reflexive and transitive.*

For any $t \in \mathbb{Z}^+ \cup \{\infty\}$, the basis of the complexity classes, $\mathcal{C}^t_{\eta_0,\eta_1}$, is $(1,t)$-*Asymmetric String Guessing with Unknown History and Predictions* (ASG_t). Given a bitstring, x, the task of an algorithm for ASG_t is to guess the contents of x. A request for ASG_t, r_i, is a prompt for guessing the next bit, x_i. Together with r_i, algorithms for ASG_t receive a prediction, $\hat{x}_i$, that predicts x_i. When the algorithm has given a guess, y_i, for each bit, x_i, in x, the full contents of x is revealed to the algorithm. For any instance $I = (x, \hat{x}, r) \in \mathcal{I}_{\text{ASG}_t}$, the cost of an algorithm, $\text{ALG} \in \mathcal{A}_{\text{ASG}_t}$, is given by

$$\text{ALG}^c(I) = \sum_{i=1}^{n} (y_i + t \cdot x_i \cdot (1 - y_i)), \quad (4)$$

and the goal is to minimize the cost.

For any $t \in \mathbb{Z}^+ \cup \{\infty\}$ and any pair of error measures (η_0, η_1), the complexity class $\mathcal{C}^t_{\eta_0,\eta_1}$ is defined as the closure of ASG_t under the as-hard-as relation. Therefore, for an online problem, P,

- $P \in \mathcal{C}^t_{\eta_0,\eta_1}$ if ASG_t is as hard as P,
- P is $\mathcal{C}^t_{\eta_0,\eta_1}$-hard if P is as hard as ASG_t, and
- P is $\mathcal{C}^t_{\eta_0,\eta_1}$-complete if $P \in \mathcal{C}^t_{\eta_0,\eta_1}$ and P is $\mathcal{C}^t_{\eta_0,\eta_1}$-hard.

Remark 1. For any two $\mathcal{C}^t_{\eta_0,\eta_1}$-complete problems, P and Q, there exists an (α, β, γ)-competitive algorithm for P with respect to (η_0, η_1) if and only if there exists an (α, β, γ)-competitive algorithm for Q with respect to (η_0, η_1).

3 Online Independent Set vs. ASG_t

In this section, we give our main result which is a comparison of the hardness of an online variant of Independent Set and the hardness of ASG_t, the base problem of $\mathcal{C}^t_{\mu_0,\mu_1}$, while respecting that hardness is measured differently for minimization problems and maximization problems (see Definition 1).

Independent Set is one of Karp's original 21 NP-complete problems [22], and it has been studied in many variants through the years [9,10,13,17,18]. Given a graph, $G = (V, E)$, the goal of an algorithm for Independent Set is to select a largest possible subset of vertices $V' \subseteq V$ such that for all $u, v \in V'$, the edge (u, v) is not in E. We consider an online vertex-arrival variant of Independent Set called *Online t-Bounded Degree Independent Set with Predictions* (IS_t), where all input graphs have bounded degree t. In the *vertex-arrival* model [3,6,17], each request, r_i, contains a vertex, $v_i \in V$, and a collection of edges $E_i = \{(v_i, v_j) \mid j < i\} \subseteq E$ to previously revealed vertices. Given a request, r_i, an algorithm for IS_t outputs $y_i = 0$ to add v_i to V', and $\{v_i \mid x_i = 0\}$ is an optimal independent set. The profit of $\text{ALG} \in \mathcal{A}_{\text{IS}_t}$ on $I \in \mathcal{I}_{\text{IS}_t}$ is

$$\text{ALG}^p(I) = \begin{cases} \sum_{i=1}^n (1 - y_i), & \text{if ALG outputs an independent set,} \\ -\infty, & \text{otherwise,} \end{cases} \quad (5)$$

and the goal is to maximize the profit. For $t = \infty$, we abbreviate IS_∞ by IS.

In Lemma 2, we show that ASG_t is as hard as IS_t with respect to (μ_0, μ_1). To this end, we recall the following result from [5]:

Theorem 1 ([5]). *For any $t \in \mathbb{Z}^+$ and any $\alpha, \beta \in \mathbb{R}^+$ with $\alpha \geqslant 1$ and $\alpha + \beta \geqslant t$, there exist an $(\alpha, \beta, 1)$-competitive algorithm for ASG_t with respect to (μ_0, μ_1).*

Remark 2. Let ALG be an (α, β, γ)-competitive Pareto-optimal algorithm for ASG_t with respect to (μ_0, μ_1). Then, $\gamma \leqslant 1$.

Lemma 2. ASG_t *is as hard as* IS_t *with respect to* (μ_0, μ_1) *for any* $t \in \mathbb{Z}^+$.

Proof. First, we outline the main structure of the proof. We define two maps $\rho_\text{A} \colon \mathcal{A}_{\text{ASG}_t} \to \mathcal{A}_{\text{IS}_t}$ and $\rho_\text{R} \colon \mathcal{A}_{\text{ASG}_t} \times \mathcal{I}_{\text{IS}_t} \to \mathcal{I}_{\text{ASG}_t}$, with the following purposes. Given an algorithm $\text{ALG}_{\text{ASG}_t} \in \mathcal{A}_{\text{ASG}_t}$, the purpose of ρ_A is to define an algorithm for IS_t, $\text{ALG}_{\text{IS}_t} = \rho_\text{A}(\text{ALG}_{\text{ASG}_t})$, that performs as well as $\text{ALG}_{\text{ASG}_t}$. A central part of ALG_{IS_t} is that it gives requests to $\text{ALG}_{\text{ASG}_t}$ the answers to which ALG_{IS_t} uses to create its own solution. Given an instance $I \in \mathcal{I}_{\text{IS}_t}$, we use ρ_R to keep track of which requests ALG_{IS_t} gives to $\text{ALG}_{\text{ASG}_t}$ during the processing of I. Thus, ρ_R is mainly a tool for analysis that makes it easier to compare the solutions produced by ALG_{IS_t} and OPT_{IS_t} to the solutions produced by $\text{ALG}_{\text{ASG}_t}$ and $\text{OPT}_{\text{ASG}_t}$.

After defining ρ_A and ρ_R, we show that if $\text{ALG} \in \mathcal{A}_{\text{ASG}_t}$ is an (α, β, γ)-competitive Pareto-optimal algorithm for ASG_t with respect to (μ_0, μ_1), then $\rho_\text{A}(\text{ALG})$ is an (α, β, γ)-competitive for IS_t with respect to (μ_0, μ_1), which implies that ASG_t is as hard as IS_t. Observe by Definition 1, that (α,β,γ)-competitiveness is defined differently for maximization and minimization problems. We therefore ensure that our reduction respects this difference, such that when an algorithm is (α,β,γ)-competitive for IS_t, it satisfies Eq. (1), and when an algorithm is (α,β,γ)-competitive for ASG_t, it satisfies Eq. (2).

Defining ρ_A and ρ_R: Let $\text{ALG}_{\text{ASG}_t} \in \mathcal{A}_{\text{ASG}_t}$, let $I = (x,\hat{x},r) \in \mathcal{I}_{\text{IS}_t}$, and let G be the underlying graph of I. Then, we define $\text{ALG}_{\text{IS}_t} = \rho_\text{A}(\text{ALG}_{\text{ASG}_t})$ and $I' = (x',\hat{x}',r') = \rho_\text{R}(\text{ALG}_{\text{ASG}_t},I)$ as follows.

Whenever ALG_{IS_t} receives a vertex v_i with prediction $\hat{x}_i$ as part of processing I, it determines the *level of* v_i, denoted $\ell(v_i)$. If there exists a vertex v_j, for some $j < i$, with $y_j = 0$ and $(v_i, v_j) \in E_i$, then ALG_{IS_t} places v_i on level 2, and rejects v_i (i.e. outputs $y_i = 1$). Otherwise, ALG_{IS_t} places v_i on level 1, and asks $\text{ALG}_{\text{ASG}_t}$ to guess the next the next bit given $\hat{x}_i$. Then, ALG_{IS_t} outputs the same bit as $\text{ALG}_{\text{ASG}_t}$.

Let $\{v_{i_1}, v_{i_2}, \ldots, v_{i_k}\}$ be the vertices on level 1 when ALG_{IS_t} does not receive any more requests. Observe that for each $j = 1, 2, \ldots, k$, ALG_{IS_t} had the option of accepting v_{i_j}, and therefore it made $\text{ALG}_{\text{ASG}_t}$ to guess a bit, the answer to which determined whether or not ALG_{IS_t} accepted v_{i_j}.

Having defined ALG_{IS_t}, it only remains to define x' and $\hat{x}'$ for I'. We let x'_{i_j} and $\hat{x}'_{i_j}$ be the true and predicted bit of the request given to $\text{ALG}_{\text{ASG}_t}$ when ALG_{IS_t} received the vertex v_{i_j}. For all $j = 1, 2, \ldots, k$ we let $\hat{x}'_{i_j} = \hat{x}_{i_j}$ and

$$x'_{i_j} = \begin{cases} 1 - x_{i_j}, & \text{if } x_{i_j} = y_{i_j} = 1 \\ x_{i_j}, & \text{otherwise.} \end{cases} \quad (6)$$

Bounding the Error of I': By definition of I', the number of requests for ASG_t may be fewer than the number of requests for IS_t, and so $\mu_b(I') \leqslant \mu_b(I)$, for $b \in \{0,1\}$, before taking into account new prediction errors that may be introduced in (6). In particular, if $x_{i_j} = y_{i_j} = 1$ then we set $x'_{i_j} = 0$, and so if $\hat{x}_{i_j} = 0$ then we change an incorrect prediction into a correct prediction, and if $\hat{x}_{i_j} = 1$ we change a correct prediction into an incorrect prediction. Therefore,

$$\mu_0(I') \leqslant \mu_0(I) - \sum_{j=1}^{k} x_{i_j} \cdot y_{i_j} \cdot (1 - \hat{x}_{i_j}) \leqslant \mu_0(I), \quad (7)$$

$$\mu_1(I') \leqslant \mu_1(I) + \sum_{j=1}^{k} x_{i_j} \cdot y_{i_j} \cdot \hat{x}_{i_j}. \quad (8)$$

Structure of the Analysis: To show that ASG_t is as hard as IS_t, it remains to show that if $\text{ALG}_{\text{ASG}_t} \in \mathcal{A}_{\text{ASG}_t}$ is an (α,β,γ)-competitive Pareto-optimal algorithm for ASG_t, then $\rho_\text{A}(\text{ALG}_{\text{ASG}_t})$ is (α,β,γ)-competitive for IS_t. To this end, observe by Remark 2 that any (α,β,γ)-competitive Pareto-optimal algorithm for ASG_t satisfies that $\gamma \leqslant 1$.

For any (α, β, γ)-competitive Pareto-optimal algorithm, $\text{ALG}_{\text{ASG}_t} \in \mathcal{A}_{\text{ASG}_t}$, and any $I \in \mathcal{I}_{\text{IS}_t}$ with $\text{ALG}_{\text{IS}_t} = \rho_A(\text{ALG}_{\text{ASG}_t})$ and $I' = \rho_R(\text{ALG}_{\text{ASG}_t}, I)$, we show that

$$\alpha \cdot \text{OPT}^c_{\text{ASG}_t}(I') + \text{OPT}^p_{\text{IS}_t}(I)$$
$$\leqslant \alpha \cdot \text{ALG}^p_{\text{IS}_t}(I) + \text{ALG}^c_{\text{ASG}_t}(I') - \sum_{j=1}^{k} x_{i_j} \cdot y_{i_j} \cdot \hat{x}_{i_j}. \tag{9}$$

Before verifying Eq. (9), we argue that the definition of ρ_A and ρ_R together with Eq. (9) implies that ALG_{IS_t} is (α, β, γ)-competitive.

Since $\text{ALG}_{\text{ASG}_t}$ is (α, β, γ)-competitive with additive constant κ (see Definition 1), then, for any $I' \in \mathcal{I}_{\text{ASG}_t}$, we have that

$$-\alpha \cdot \text{OPT}^c_{\text{ASG}_t}(I') \leqslant -\text{ALG}^c_{\text{ASG}_t}(I') + \beta \cdot \mu_0(I') + \gamma \cdot \mu_1(I') + \kappa.$$

Hence, by Eq. (9) and the (α, β, γ)-competitiveness of $\text{ALG}_{\text{ASG}_t}$,

$$\text{OPT}^p_{\text{IS}_t}(I) \leqslant \alpha \cdot \text{ALG}^p_{\text{IS}_t}(I) + \beta \cdot \mu_0(I') + \gamma \cdot \mu_1(I') + \kappa - \sum_{j=1}^{k} x_{i_j} \cdot y_{i_j} \cdot \hat{x}_{i_j}.$$

Now, by Eqs. (7) and (8), and since ALG is Pareto-optimal so $\gamma \leqslant 1$,

$$\text{OPT}^p_{\text{IS}_t}(I) \leqslant \alpha \cdot \text{ALG}^p_{\text{IS}_t}(I) + \beta \cdot \mu_0(I) + \gamma \cdot \mu_1(I) + \kappa.$$

Hence, verifying that Eq. (9) is true for any Pareto-optimal algorithm for ASG_t and any instance of IS_t, we get that ASG_t is as hard as IS_t.

Verifying Eq. (9): First, observe that ALG_{IS_t} can only gain profit from the vertices on level 1. Therefore, $\text{ALG}^p_{\text{IS}_t}(I) = \sum_{j=1}^{k}(1 - y_{i_j})$. Hence, using Eqs. (4) and (5), we rewrite Eq. (9) as

$$\underbrace{\alpha \cdot \sum_{j=1}^{k} x'_{i_j} + \sum_{i=1}^{n}(1 - x_i)}_{\text{LHS}}$$
$$\leqslant \underbrace{\alpha \cdot \sum_{j=1}^{k}(1 - y_{i_j}) + \sum_{j=1}^{k}\left(y_{i_j} + t \cdot x'_{i_j} \cdot (1 - y_{i_j})\right) - \sum_{j=1}^{k} x_{i_j} \cdot y_{i_j} \cdot \hat{x}_{i_j}}_{\text{RHS}}. \tag{10}$$

We show that Eq. (10) is satisfied at each step during the processing of ALG_{IS_t}. Clearly, when no requests have been given to ALG_{IS_t}, Eq. (10) is satisfied as both LHS and RHS evaluate to 0.

Next, assume that Eq. (10) is satisfied after $i - 1$ vertices has been revealed to ALG_{IS_t}, and suppose that the i'th vertex, v_i, has just been revealed. We split the rest of this argument into six subcases:

(a) *Case $\ell(v_i) = 1$ and $x_i = y_i = 1$*: Since $\ell(v_i) = 1$, there exists $j \in \{1, 2, \ldots, k\}$ such that $i = i_j$. Further, as $x_{i_j} = y_{i_j} = 1$ then $x'_{i_j} = 1 - x_{i_j} = 0$. Therefore, LHS does not increase, and since $y_{i_j} = 1$ RHS increases by 1 when $\hat{x}_{i_j} = 0$ and by 0 otherwise. Hence, the inequality in Eq. (10) remains satisfied.

(b) *Case $\ell(v_i) = 1$, $x_i = 1$ and $y_i = 0$*: Since $\ell(v_i) = 1$, there exists some $j \in \{1, 2, \ldots, k\}$ such that $i = i_j$. Further, since $x_{i_j} \neq y_{i_j}$, $x'_{i_j} = x_{i_j} = 1$. Hence, LHS increases by α. Further, since $y_{i_j} = 0$ and $x'_{i_j} = 1$, RHS increases by $\alpha + t$ and so the inequality in Eq. (10) remains satisfied. Observe that RHS increase by t more than LHS. We will use this observation later.

(c) *Case $\ell(v_i) = 1$ and $x_i = y_i = 0$*: Since $\ell(v_i) = 1$, there exist $j \in \{1, 2, \ldots, k\}$ such that $i = i_j$. Since $x_{i_j} = 0$, then $x'_{i_j} = 0$. Therefore, LHS increases by 1. Since $y_{i_j} = 0$ and $x'_{i_j} = 0$, RHS increases by α. Hence, the inequality in Eq. (10) remains satisfied.

(d) *Case $\ell(v_i) = 1$, $x_i = 0$ and $y_i = 1$*: Since $\ell(v_i) = 1$, there exists some $j \in \{1, 2, \ldots, k\}$ such that $i = i_j$. Since $x_{i_j} = 0$ then $x'_{i_j} = 0$. Therefore, LHS increases by 1. Since $y_{i_j} = 1$, RHS increases by 1, and so the inequality in Eq. (10) remains satisfied.

(e) *Case $\ell(v_i) = 2$ and $x_i = 1$*: Since $\ell(v_i) = 2$ then $y_i = 1$, and for all $j \in \{1, 2, \ldots, k\}$, $i_j \neq i$. Since $x_i = 1$, neither LHS nor RHS increases by anything and therefore the inequality in Eq. (10) remains satisfied.

(f) *Case $\ell(v_i) = 2$ and $x_i = 0$*: Since $\ell(v_i) = 2$ then $y_i = 1$, and for all $j \in \{1, 2, \ldots, k\}$, $i_j \neq i$. Since $x_i = 0$, then LHS increase by 1, and since there does not exist j such that $i_j = i$, then RHS is unchanged.

In this case, there exist a smallest $j \in \{1, 2, \ldots, k\}$ with $i_j < i$ such that $(v_{i_j}, v_i) \in E$ and $y_{i_j} = 0$, as otherwise $\ell(v_i)$ would have been 1. Since $x_i = 0$ and $(v_{i_j}, v_i) \in E$, then $x_{i_j} = 1$ as otherwise x encodes an infeasible solution. Since $y_{i_j} = 0$ and $x_{i_j} = 1$, we were in case (b) when v_{i_j} was revealed, which left RHS t larger than LHS. Therefore, we can assign the profit of 1 gained by OPT_{IS_t} on the i'th request to the cost incurred by $\text{ALG}_{\text{ASG}_t}$ on the i_j'th request. Since $d_G(v_{i_j}) \leq t$, v_{i_j} has at most t neighbours in G and so we do this accounting at most t times for v_{i_j}, corresponding to the amount that the RHS is larger than the LHS after accounting for v_{i_j}. Hence, the inequality in Eq. (10) remains satisfied.

This verifies Eq. (10) and thus Eq. (9), which finishes the proof. □

Next, in Lemma 4, we build a reduction in the other direction. This reduction is inspired by the reduction template from [5]. For this reduction, we need an observation about the competitiveness of Pareto-optimal algorithms for IS_t.

Lemma 3. *Let* ALG *be an (α, β, γ)-competitive Pareto-optimal algorithm for* IS_t *with respect to (μ_0, μ_1). Then, $\alpha + \beta \leq t$.*

Lemma 4. IS_t *is as hard as* ASG_t *with respect to (μ_0, μ_1) for all $t \in \mathbb{Z}^+$.*

Proof. Similarly to the proof of Lemma 2, we define two maps $\rho_A \colon \mathcal{A}_{\mathrm{IS}_t} \to \mathcal{A}_{\mathrm{ASG}_t}$ and $\rho_R \colon \mathcal{A}_{\mathrm{IS}_t} \times \mathcal{I}_{\mathrm{ASG}_t} \to \mathcal{I}_{\mathrm{IS}_t}$ which allow us to prove that for any (α, β, γ)-competitive Pareto-optimal algorithm, $\mathrm{ALG}_{\mathrm{IS}_t} \in \mathcal{A}_{\mathrm{IS}_t}$, for IS_t with respect to (μ_0, μ_1), we have that $\rho_A(\mathrm{ALG}_{\mathrm{IS}_t})$ is an (α, β, γ)-competitive algorithm for ASG_t with respect to (μ_0, μ_1).

Definition of ρ_A and ρ_R: Let $\mathrm{ALG}_{\mathrm{IS}_t} \in \mathcal{A}_{\mathrm{IS}_t}$ and $I = (x, \hat{x}, r) \in \mathcal{I}_{\mathrm{ASG}_t}$. Then, we define $\mathrm{ALG}_{\mathrm{ASG}_t} = \rho_A(\mathrm{ALG}_{\mathrm{IS}_t})$ and $I' = (x', \hat{x}', r') = \rho_R(\mathrm{ALG}_{\mathrm{IS}_t}, I)$ as follows.

When $\mathrm{ALG}_{\mathrm{ASG}_t}$ is asked to guess the bit x_i given the prediction $\hat{x}_i$, it gives a new request containing an isolated vertex, a *challenge request*, v'_i, with prediction $\hat{x}'_i = \hat{x}_i$ to $\mathrm{ALG}_{\mathrm{IS}_t}$. Then, $\mathrm{ALG}_{\mathrm{ASG}_t}$ outputs the same guess of x_i as $\mathrm{ALG}_{\mathrm{IS}_t}$ outputs on v'_i.

When $\mathrm{ALG}_{\mathrm{ASG}_t}$ receives no more requests, and therefore learns the true contents of x, it determines x', by setting

$$x'_i = \begin{cases} 1 - x_i, & \text{if } x_i = y_i = 0, \\ x_i, & \text{otherwise,} \end{cases} \tag{11}$$

where x'_i is the true bit of v'_i. By construction, there is no challenge request with $x'_i = y_i = 0$.

After computing x', $\mathrm{ALG}_{\mathrm{ASG}_t}$ adds a block of requests for each challenge request given to $\mathrm{ALG}_{\mathrm{IS}_t}$. We let $B_{v'_i}$ be the block of requests associated to the challenge request, v'_i. The blocks are defined as follows.

(a) *Case $x'_i = 1$ and $y_i = 0$:* Request t new vertices, $v'_{i,j}$, for $j = 1, 2, \ldots, t$, where $v'_{i,j}$ is connected to v'_i with prediction $\hat{x}'_{i,j} = 0$. Clearly, $\mathrm{OPT}_{\mathrm{IS}_t}$ accepts all $v'_{i,j}$, and $\mathrm{ALG}_{\mathrm{IS}_t}$ must reject all $v'_{i,j}$ to avoid creating an infeasible solution. Hence, $x'_{i,j} = 0$ and $y_{i,j} = 1$ for all $j = 1, 2, \ldots, t$.

(b) *Case $x'_i = 0$ and $y_i = 1$:* This block is empty. Since $x'_i = 0$, and v'_i is isolated, it is optimal to accept v'_i into the independent set.

(c) *Case $x'_i = y_i = 1$:* Request a new vertex, $v'_{i,1}$, that is connected to v_i with $x'_{i,1} = \hat{x}'_{i,1} = 0$. Observe that v'_i is necessary to ensure that x' encodes an optimal solution. $\mathrm{ALG}_{\mathrm{IS}_t}$ may either accept or reject $v'_{i,1}$.

Bounding the Error of I': By Eq. (11), the true bits of some challenge requests for IS_t are different from the true bits of the corresponding requests for ASG_t, while the predicted bits of the same requests coincide. In particular, if $x_i = y_i = 0$, then $x'_i = 1$. Hence, if $x_i = y_i = 0$ and $\hat{x}'_i = \hat{x}_i = 1$ then we change an incorrect prediction into a correct prediction, and if $x_i = y_i = 0$ and $\hat{x}'_i = \hat{x}_i = 0$ then we change a correct prediction into an incorrect prediction. In the following, we bound the error of I' as a function of the error of I:

$$\mu_0(I') = \mu_0(I) + \sum_{i=1}^{n}(1 - x_i) \cdot (1 - y_i) \cdot (1 - \hat{x}_i) \tag{12}$$

$$\mu_1(I') = \mu_1(I) - \sum_{i=1}^{n}(1 - x_i) \cdot (1 - y_i) \cdot \hat{x}_i \leqslant \mu_1(I). \tag{13}$$

Structure of the Analysis: Observe that each request, r_i, in I induces a subgraph H'_i in the underlying graph of I'. By construction, H'_i contains the vertex v'_i and the vertices and edges from $B_{v'_i}$. We let $\text{ALG}^p_{\text{IS}_t}(H'_i)$ and $\text{OPT}^p_{\text{IS}_t}(H'_i)$ be the profit of ALG_{IS_t} and OPT_{IS_t} on the requests in I' corresponding to H'_i. Further, observe that for any $i, j \in \{1, 2, \ldots, n\}$ with $i \neq j$, H'_i and H'_j are not connected by any edges, and so $\text{ALG}^p(H'_i \cup H'_j) = \text{ALG}^p(H'_i) + \text{ALG}^p(H'_j)$, for any $\text{ALG} \in \mathcal{A}_{\text{IS}_t}$ (including OPT_{IS_t}).

Next, we explain the structure of the rest of the proof. We show that there exists a function $d \colon \{0,1\} \times \{0,1\} \to \mathbb{Z}^+$ such that for all $x_i, y_i \in \{0, 1\}$,

$$y_i + t \cdot x_i \cdot (1 - y_i) \leq d(x_i, y_i) - \alpha \cdot \text{ALG}^p_{\text{IS}_t}(H'_i) - \beta \cdot (1 - x_i) \cdot (1 - y_i) \cdot (1 - \hat{x}_i) \tag{C1}$$

and

$$d(x_i, y_i) - \text{OPT}^p_{\text{IS}_t}(H'_i) \leq \alpha \cdot x_i. \tag{C2}$$

Before proving the existence of d, we argue that the existence of d implies that if ALG_{IS_t} is an (α, β, γ)-competitive for IS_t with respect to (μ_0, μ_1), then $\text{ALG}_{\text{ASG}_t} = \rho_A(\text{ALG}_{\text{IS}_t})$ is an (α, β, γ)-competitive for ASG_t with respect to (μ_0, μ_1). To this end, assume that ALG_{IS_t} is (α, β, γ)-competitive, let κ be the additive constant of ALG_{IS_t}, let $I \in \mathcal{I}_{\text{ASG}_t}$, and let $I' = \rho_R(\text{ALG}_{\text{IS}_t}, I)$. Then, by Eqs. (4) and (C1), we have that

$$\text{ALG}^c_{\text{ASG}_t}(I) = \sum_{i=1}^{n} (y_i + t \cdot x_i \cdot (1 - y_i))$$

$$\leq \sum_{i=1}^{n} (d(x_i, y_i) - \alpha \cdot \text{ALG}^p_{\text{IS}_t}(H'_i) - \beta \cdot (1 - x_i) \cdot (1 - y_i) \cdot (1 - \hat{x}_i)).$$

Further, by Eq. (12) and since $\sum_{i=1}^{n} \text{ALG}^p_{\text{IS}_t}(H'_i) = \text{ALG}^p_{\text{IS}_t}(I')$, we get that

$$\text{ALG}^c_{\text{ASG}_t}(I) \leq \left(\sum_{i=1}^{n} d(x_i, y_i) \right) - \alpha \cdot \text{ALG}^p_{\text{IS}_t}(I') - \beta \cdot (\mu_0(I') - \mu_0(I)).$$

Using the (α, β, γ)-competitiveness of ALG_{IS_t} (see Definition 1), we get that

$$\text{ALG}^c_{\text{ASG}_t}(I) \leq \left(\sum_{i=1}^{n} d(x_i, y_i) \right) - \text{OPT}^p_{\text{IS}_t}(I') + \beta \cdot \mu_0(I) + \gamma \cdot \mu_1(I') + \kappa.$$

Now, by Eq. (13) and since $\text{OPT}^p_{\text{IS}_t}(I') = \sum_{i=1}^{n} \text{OPT}^p_{\text{IS}_t}(H'_i)$, we have

$$\text{ALG}^c_{\text{ASG}_t}(I) \leq \left(\sum_{i=1}^{n} d(x_i, y_i) - \text{OPT}^p_{\text{IS}_t}(H'_i) \right) + \beta \cdot \mu_0(I) + \gamma \cdot \mu_1(I) + \kappa.$$

Finally, by Eqs. (C2) and (4), we get that

$$\text{ALG}^c_{\text{ASG}_t}(I) \leqslant \alpha \cdot \text{OPT}^p_{\text{IS}_t}(I) + \beta \cdot \mu_0(I) + \gamma \cdot \mu_1(I) + \kappa.$$

Hence, if ALG_{IS_t} is an (α, β, γ)-competitive algorithm with respect to (μ_0, μ_1), then the existence of d is sufficient to show that $\text{ALG}_{\text{ASG}_t}$ is an (α, β, γ)-competitive algorithm for ASG_t with respect to (μ_0, μ_1).

Recall from Definition 2 that we only have to preserve the competitiveness of Pareto-optimal algorithms for IS_t in order to conclude that IS_t is as hard as ASG_t. Observe, by Lemma 3, that any (α, β, γ)-competitive Pareto-optimal algorithm for IS_t satisfies that $\alpha + \beta \leqslant t$. Hence, proving the existence of a function d satisfying (C1) and (C2) whenever $\alpha + \beta \leqslant t$ implies that IS_t is as hard as ASG_t.

The Existence of d: We show that the function

$$d(x_i, y_i) = |H'_i| - (1 - x_i) \cdot (1 - y_i) + (\alpha - 1) \cdot x_i \tag{14}$$

where $|H'_i|$ is the number of vertices in H'_i, satisfies Eqs. (C1) and (C2) whenever $\alpha + \beta \leqslant t$. We consider four cases:

Case $x_i = 1$ and $y_i = 0$: In this case $x'_i = 1$ and so H_i contains v'_i and the t vertices in the block $B_{v'_i}$ (see (a)). Therefore, $|H'_i| = t + 1$, $\text{OPT}_{\text{IS}_t}(H'_i) = t$, and $\text{ALG}_{\text{IS}_t}(H'_i) = 1$, and so $d(1, 0) = t + 1 - 0 + \alpha - 1 = t + \alpha$. Hence, Eq. (C1) is satisfied as $t \leqslant t + \alpha - \alpha - 0 = t$, and Eq. (C2) is satisfied as $t + \alpha - t \leqslant \alpha$.

Case $x_i = 0$ and $y_i = 1$: In this case $x'_i = 0$ and so H'_i only contains v'_i as $B_{v'_i}$ is empty (see (b)). Therefore, $|H'_i| = 1$. $\text{OPT}_{\text{IS}_t}(H'_i) = 1$, and $\text{ALG}_{\text{IS}_t}(H'_i) = 0$, and so $d(0, 1) = 1 - 0 + 0 = 1$. Hence, Eq. (C1) is satisfied as $1 \leqslant 1 - 0 - 0$, and Eq. (C2) is satisfied as $1 - 1 \leqslant 0$.

Case $x_i = y_i = 1$: In this case, $x'_i = 1$ and so H'_i contains the two vertices v'_i and $v'_{i,1}$ (see (c)). Therefore, $|H'_i| = 2$, $\text{OPT}_{\text{IS}_t}(H'_i) = 1$, and $\text{ALG}_{\text{IS}_t}(H'_i) \in \{0, 1\}$. Observe that $\text{ALG}_{\text{IS}_t}(H'_i) \in \{0, 1\}$ as ALG_{IS_t} may either accept or reject $v'_{i,1}$. Hence, $d(1, 1) = 2 - 0 + \alpha - 1 = \alpha + 1$, and so Eq. (C1) is satisfied as

$$1 \leqslant \alpha + 1 - \alpha - 0 = 1 \ (\text{ALG}_{\text{IS}_t} \text{ accepts } v_{i,1})$$
$$1 \leqslant \alpha + 1 - 0 - 0 = 1 + \alpha \ (\text{ALG}_{\text{IS}_t} \text{ rejects } v_{i,1})$$

and (C2) is satisfied as: $\alpha + 1 - 1 \leqslant \alpha$.

Case $x_i = y_i = 0$: In this case $x'_i = 1$ (see Eq. (11)), and so H'_i contains the vertex v'_i and the t vertices in the block $B_{v'_i}$ (see @refblock:10). Therefore, $|H'_i| = t + 1$, $\text{OPT}_{\text{IS}_t}(H'_i) = t$, and $\text{ALG}_{\text{IS}_t}(H'_i) = 1$, and so $d(0, 0) = t + 1 - 1 + 0 = t$. Hence Eq. (C1) is satisfied as

$$0 \leqslant t - \alpha - 0 (\text{if } \hat{x}_i = 1)$$
$$0 \leqslant t - \alpha - \beta (\text{if } \hat{x}_i = 0)$$

Observe that both inequalities are satisfied as $\alpha + \beta \leqslant t$. Further, (C2) is satisfied as $t - t \leqslant 0$.

Hence, the function d proposed in Eq. (14) satisfies Eq. (C1) and (C2) for all $x_i, y_i \in \{0,1\}$, whenever $\alpha + \beta \leq t$. □

We are now ready to state the main result of this section, which proves the first $\mathcal{C}_{\mu_0,\mu_1}^t$-completeness result of a maximization problem:

Theorem 2. IS_t is $\mathcal{C}_{\mu_0,\mu_1}^t$-complete for all $t \in \mathbb{Z}^+$.

The main motivation for comparing the hardness of IS_t to the hardness of ASG_t is that it extends to a hardness comparison between IS_t and *Online t-Bounded Degree Vertex Cover with Predictions* (VC_t), the dual problem of IS_t. In particular, by [5], it is known that VC_t is $\mathcal{C}_{\mu_0,\mu_1}^t$-complete, and so, by Remark 1, there exists an (α, β, γ)-competitive algorithm for VC_t if and only if there exists an (α, β, γ)-competitive algorithm for IS_t[1]. This is rather surprising as, in the context of the related field of approximation algorithms, it is well-known that (Bounded Degree) Vertex Cover is much easier to approximate than (Bounded Degree) Independent Set [12,19].

4 Relative Hardness of Maximization Problems

In the full paper [4], we consider five other online maximization problems, the hardness of which we compare to IS_t. In particular, we consider *Online t-Bounded Set-Arrival Set Packing with Predictions* (SP_t), *Online t-Bounded Overlap Interval Scheduling with Predictions* (SCH_t), *Online t-Bounded Clique with Predictions* (CLI_t), *Online t-Bounded Maximum k-Colorable Subgraph with Predictions* ($\text{M}k\text{CS}_t$), and *Online t-Bounded Maximum Matching with Predictions* (MM_t), all of which have been studied before in several different variations [8,9,11,14–16,19–21,25,26]. For all these problems except $\text{M}k\text{CS}_t$, we also consider their unbounded variants, abbreviated SP, SCH, CLI, and MM, respectively. The goal of an algorithm for any of the above problems is to create a largest possible solution, i.e. accept as many requests as possible. For SP_t, finite sets arrive one-by-one, and an algorithm must irrevocably accept or reject these sets upon arrival, while ensuring that no two accepted sets intersect. For SCH_t, intervals arrive one-by-one, and an algorithm must irrevocably accept or reject each interval upon arrival, while ensuring that no two overlapping intervals intersect. CLI_t is a vertex-arrival problem, and algorithms must accept a largest possible set of vertices that form a clique. $\text{M}k\text{CS}_t$ is a vertex-arrival problem, and algorithms must irrevocably accept or reject vertices upon arrival while ensuring that the subgraph induced by the accepted vertices is k-colorable. Finally, MM_t is an edge-arrival problem meaning that edges arrive one-by-one together with their endpoints. Algorithms for MM_t must accept as many edges as possible while ensuring that no two accepted edges share an endpoint. The next theorem summarizes our main results.

---

[1] As mentioned in the preliminaries, this result also holds with respect to (Z_0, Z_1), and thus in the purely online setting.

Theorem 3. *Let $t, k \in \mathbb{Z}^+$ and let $\tilde{t} = \lfloor \frac{t}{2} \rfloor$. Then,*

- IS_t, SCH_t, SP_t, *and* CLI_t *are* $\mathcal{C}^t_{\mu_0,\mu_1}$- *and* $\mathcal{C}^t_{Z_0,Z_1}$-*complete*,
- $\text{M}k\text{CS}_{kt}, \text{MM}_{\tilde{t}}$ *are members of* $\mathcal{C}^t_{\mu_0,\mu_1}$ *and* $\mathcal{C}^t_{Z_0,Z_1}$, *and*
- $\text{SCH}, \text{IS}, \text{CLI}, \text{SP}$ *are* $\mathcal{C}^t_{\mu_0,\mu_1}$- *and* $\mathcal{C}^t_{Z_0,Z_1}$-*hard*.

From [5], we know that proving membership, hardness, and completeness results for $\mathcal{C}^t_{\mu_0,\mu_1}$ yields several positive and negative algorithmic results:

Corollary 1. *Let $t \in \mathbb{Z}^+ \cup \{\infty\}$, let $P \in \{\text{IS}_t, \text{SCH}_t, \text{SP}_t, \text{CLI}_t\}$, and let ALG be an (α, β, γ)-competitive algorithm for P with respect to (μ_0, μ_1). Then,*

(i) $\alpha + \beta \geqslant t$,
(ii) $\alpha + (t-1) \cdot \gamma \geqslant t$, and

Corollary 2. *Let $k, t \in \mathbb{Z}^+$ and $P \in \{\text{IS}_t, \text{SP}_t, \text{SCH}_t, \text{CLI}_t, \text{M}k\text{CS}_{kt}, \text{MM}_{\tilde{t}}\}$, where $\tilde{t} = \lfloor \frac{t}{2} \rfloor$. Then, for all $\alpha, \beta \in \mathbb{R}^+$ with $\alpha \geqslant 1$ and $\alpha + \beta \geqslant t$, there exists an $(\alpha, \beta, 1)$-competitive and a $(t, 0, 0)$-competitive algorithm for P. If $P \in \{\text{IS}_t, \text{SP}_t, \text{SCH}_t, \text{CLI}_t\}$, these algorithms are Pareto-optimal.*

References

1. Algorithms with predictions. https://algorithms-with-predictions.github.io/. Accessed 4 Aug 2025
2. Antoniadis, A., et al.: Paging with succinct predictions. In: 40th International Conference on Machine Learning (ICML), vol. 202, pp. 952–968. PMLR (2023). https://proceedings.mlr.press/v202/antoniadis23a.html
3. Antoniadis, A., Broersma, H., Meng, Y.: Online graph coloring with predictions. In: 8th International Symposium on Combinatorial Optimization (ISCO). Lecture Notes in Computer Science, vol. 14594, pp. 289–302. Springer (2024). https://doi.org/10.1007/978-3-031-60924-4_22
4. Berg, M.: Comparing the hardness of online minimization and maximization problems with predictions (2024). https://doi.org/10.48550/arXiv.2409.12694. arXiv:2409.12694
5. Berg, M., Boyar, J., Favrholdt, L.M., Larsen, K.S.: Complexity classes for online problems with and without predictions. In: International Joint Conference on Theoretical Computer Science - Frontier of Algorithmic Wisdom (IJTCS-FAW). Lecture Notes in Computer Science, Springer (2025, accepted for publication). arXiv:2406.18265
6. Bianchi, M.P., Böckenhauer, H., Brülisauer, T., Komm, D., Palano, B.: Online minimum spanning tree with advice. Int. J. Found. Comput. Sci. **29**(4), 505–527 (2018). https://doi.org/10.1142/S0129054118410034
7. Borodin, A., El-Yaniv, R.: Online Computation and Competitive Analysis. Cambridge University Press (1998)
8. Boyar, J., Favrholdt, L.M., Kamali, S., Larsen, K.S.: Online interval scheduling with predictions. In: 18th International Algorithms and Data Structures Symposium (WADS). Lecture Notes in Computer Science, vol. 14079, pp. 193–207. Springer (2023). https://doi.org/10.1007/978-3-031-38906-1_14

9. Boyar, J., Favrholdt, L.M., Kudahl, C., Mikkelsen, J.W.: The advice complexity of a class of hard online problems. Theor. Comput. Syst. **61**, 1128–1177 (2017). https://doi.org/10.1007/s00224-016-9688-y
10. Braverman, V., Dharangutte, P., Shah, V., Wang, C.: Learning-augmented maximum independent set. In: Approximation, Randomization, and Combinatorial Optimization. Algorithms and Techniques (APPROX/RANDOM). Leibniz International Proceedings in Informatics (LIPIcs), vol. 317, pp. 24:1–24:18. Schloss Dagstuhl – Leibniz-Zentrum für Informatik (2024). https://doi.org/10.4230/LIPIcs.APPROX/RANDOM.2024.24
11. Chrobak, M., Dürr, C., Fabijan, A., Nilsson, B.J.: Online clique clustering. Algorithmica **82**(4), 938–965 (2019). https://doi.org/10.1007/s00453-019-00625-1
12. Cormen, T.H., Leiserson, C.E., Rivest, R.L., Stein, C.: Introduction to Algorithms, 4th edn. The MIT Press (2022). https://mitpress.mit.edu/9780262046305/introduction-to-algorithms/
13. Dobrev, S., Královič, R., Královič, R.: Independent set with advice: the impact of graph knowledge. In: Erlebach, T., Persiano, G. (eds.) WAOA 2012. LNCS, vol. 7846, pp. 2–15. Springer, Heidelberg (2013). https://doi.org/10.1007/978-3-642-38016-7_2
14. Emek, Y., Halldórsson, M.M., Mansour, Y., Patt-Shamir, B., Radhakrishnan, J., Rawitz, D.: Online set packing. SIAM J. Comput. **41**(4), 728–746 (2012). https://doi.org/10.1137/110820774
15. Gagnon, F., Hertz, A., Montagné, R.: Online algorithms for the maximum k-colorable subgraph problem. Comput. Oper. Res. **91**, 209–224 (2018). https://doi.org/10.1016/j.cor.2017.10.003
16. Halldórsson, M.M.: Approximations of weighted independent set and hereditary subset problems. J. Graph Algorithms Appl. **4**(1), 1–16 (2000). https://doi.org/10.7155/JGAA.00020
17. Halldórsson, M.M., Iwama, K., Miyazaki, S., Taketomi, S.: Online independent sets. Theoret. Comput. Sci. **289**(2), 953–962 (2002). https://doi.org/10.1016/S0304-3975(01)00411-X
18. Halldórsson, M.M., Radhakrishnan, J.: Greed is good: approximating independent sets in sparse and bounded-degree graphs. Algorithmica **18**, 145–163 (1997). https://doi.org/10.1007/BF02523693
19. Håstad, J.: Clique is hard to approximate within n^{1-}. Acta Math. **182**(1), 105–142 (1999). https://doi.org/10.1007/BF02392825
20. Huang, Z., Kang, N., Tang, Z.G., Wu, X., Zhang, Y., Zhu, X.: Fully online matching. J. ACM **67**(3), 17:1–17:25 (2020). https://doi.org/10.1145/3390890
21. Karavasilis, C.: Interval selection with binary predictions (2025). https://doi.org/10.48550/arXiv.2502.10314. arXiv:2502.10314
22. Karp, R.M.: Reducibility among combinatorial problems. In: Proceedings of a Symposium on the Complexity of Computer Computations, pp. 85–103. Plenum Press (1972). https://doi.org/10.1007/978-1-4684-2001-2_9
23. Komm, D.: An Introduction to Online Computation: Determinism, Randomization, Advice. Texts in Theoretical Computer Science. Springer, Switzerland (2016). https://doi.org/10.1007/978-3-319-42749-2
24. Kumar, R., Purohit, M., Svitkina, Z.: Improving online algorithms via ml predictions. In: 32nd International Conference on Neural Information Processing Systems (NIPS), pp. 9684–9693. Curran Associates, Inc. (2018). https://dl.acm.org/doi/10.5555/3327546.3327635

25. Lipton, R.J., Tomkins, A.: Online interval scheduling. In: 5th Annual ACM-SIAM Symposium on Discrete algorithms (SODA), pp. 302–311. SIAM (1994). https://dl.acm.org/doi/10.5555/314464.314506
26. Lo, A.: Cliques in graph with bounded minimum degree. Comb. Probab. Comput. **21**(3), 457–482 (2012). https://doi.org/10.1017/S0963548311000745
27. Lykouris, T., Vassilvitskii, S.: Competitive caching with machine learned advice. J. ACM **68**(4), 24:1–24:25 (2021). https://doi.org/10.1145/3447579
28. Sleator, D.D., Tarjan, R.E.: Amortized efficiency of list update and paging rules. Commun. ACM **28**(2), 202–208 (1985). https://doi.org/10.1145/2786.2793

Complexity Classes for Online Problems with and Without Predictions

Magnus Berg⬤, Joan Boyar⬤, Lene M. Favrholdt⬤, and Kim S. Larsen(✉)⬤

University of Southern Denmark, Odense, Denmark
{magbp,joan,lenem,kslarsen}@imada.sdu.dk
https://imada.sdu.dk/u/magbp, https://imada.sdu.dk/u/joan,
https://imada.sdu.dk/u/lenem, https://imada.sdu.dk/u/kslarsen
https://imada.sdu.dk/u/magbp , https://imada.sdu.dk/u/joan ,
https://imada.sdu.dk/u/lenem , https://imada.sdu.dk/u/kslarsen

Abstract. With the developments in machine learning, there has been a surge in interest and results focused on algorithms utilizing predictions, not least in online algorithms where most new results incorporate the prediction aspect for concrete online problems. While the structural computational hardness of problems with regards to time and space is quite well developed, not much is known about online problems where time and space resources are typically not in focus. Some information-theoretical insights were gained when researchers considered online algorithms with oracle advice, but predictions of uncertain quality is a very different matter.

We initiate the development of a complexity theory for online problems with predictions, focusing on binary predictions for minimization problems. Based on the most generic hard online problem type, string guessing, we define a family of hierarchies of complexity classes (indexed by pairs of error measures) and develop notions of reductions, class membership, hardness, and completeness. Our framework contains all the tools one expects to find when working with complexity, and we illustrate our tools by analyzing problems with different characteristics. In addition, we show that known lower bounds for paging with predictions apply directly to all hard problems for each class in the hierarchy based on the canonical pair of error measures.

Our work also implies corresponding complexity classes for classic online problems without predictions, with the corresponding complete problems.

1 Introduction

In computational complexity theory, one aims at classifying computational problems based on their hardness, by relating them via hardness-preserving map-

---

Supported in part by the Independent Research Fund Denmark, Natural Sciences, grants DFF-0135-00018B and DFF-4283-00079B and in part by the Innovation Fund Denmark, grant 9142-00001B, Digital Research Centre Denmark, project P40: Online Algorithms with Predictions.

ⓒ The Author(s), under exclusive license to Springer Nature Singapore Pte Ltd. 2025
V. Chau et al. (Eds.): IJTCS-FAW 2025, LNCS 15828, pp. 49–63, 2025.
https://doi.org/10.1007/978-981-96-8312-3_4

pings, referred to as reductions. Most commonly seen is time and space complexity, where problems are classified based on how much time or space is needed to solve the problem. Our primary aim is to classify online minimization problems with predictions based on the *competitiveness* of best possible deterministic online algorithms for each problem. As a starting point, we consider minimization problems with binary predictions. Our framework has recently been extended to maximization problems [4].

An *online problem* is an optimization problem where the input is revealed to an online algorithm in a piece-wise fashion in the form of *requests*. When a request arrives, an online algorithm must make an irrevocable decision about the request before the next request arrives. When comparing the quality of online algorithms, we use the standard *competitive analysis* framework [29] (see [9,23]). An algorithm is c-competitive if, asymptotically over all possible input sequences, its cost is at most a factor c times the cost of an optimal offline algorithm.

With the increased availability and improved quality of predictions from machine learning software, efforts to utilize predictions in online algorithms have increased dramatically [1]. Typically, one studies the competitiveness of online algorithms that have access to additional information about the instance through (unreliable) predictions. Ideally, such algorithms should perform perfectly when the predictions are error-free (the competitiveness in this case is called the *consistency*), and perform as well as the best purely online algorithm when the predictions are erroneous (*robustness*). There is also a desire that an algorithm's competitiveness degrades gracefully from the consistency to the robustness as the predictions get worse (often referred to as *smoothness*). To establish smoothness, it is necessary to have some measure of how wrong a prediction is. Thus, results of this type are based on some *error measure*.

The complexity of algorithms with predictions has also been considered in a different context, dynamic graph problems [19]. However, Henzinger et al. study the *time* complexity of dynamic data structures, whereas we create complexity classes where the hardness is based on *competitiveness*.

The basis for our complexity classes is *asymmetric string guessing* [11,25,26], a generic hard online problem, where each request is simply a prompt for the algorithm to guess a bit. String guessing has played a fundamental rôle in what is often referred to as advice complexity [7,15,17,20], where online algorithms have access to oracle-produced information about the instance which, in our context, can be considered infallible predictions. Specifically, we use *Online $(1,t)$-Asymmetric String Guessing with Unknown History and Predictions* (ASG_t), which will be our base family of complete problems, establishing a strict hierarchy based on the parameter, t. The *cost* of processing an input is the number of guesses of 1 plus t times the number of wrong guesses of 0.

We define complexity classes, $\mathcal{C}^t_{\eta_0,\eta_1}$, parameterized by $t \in \mathbb{Z}^+ \cup \{\infty\}$ and a pair of error measures, (η_0, η_1). A problem, P, is $\mathcal{C}^t_{\eta_0,\eta_1}$-*hard*, if it is as hard as ASG_t, and it is a *member of* $\mathcal{C}^t_{\eta_0,\eta_1}$, if ASG_t is as hard as P. If both are true, P is $\mathcal{C}^t_{\eta_0,\eta_1}$-*complete*. The as-hard-as relation is transitive, so our framework provides all the usual tools: if a subproblem of some problem is hard, the problem itself is hard, one can reduce from the most convenient complete problem, etc. Thus, working with our complexity classes is similar to working with, e.g., NP, MAX-SNP [27], the W-hierarchy [16], and APX [3], in that hardness results are obtained by proving the existence of special types of reductions that preserve properties related to hardness. However, we obtain performance bounds independent of any conjectures.

Deriving lower bounds on the competitiveness of algorithms based on the hardness of string guessing has been considered before [6,8,17], with different objectives. The closest related work is in [11], where one of the base problems we use in this paper, $(1,\infty)$-Asymmetric String Guessing with Unknown History, was used as the base problem for the complexity class AOC; AOC-complete problems are hard online problems with advice. Note that despite the similarities, working with advice and predictions are quite different matters. In advice complexity, the competitive ratio is a function of the number of advice bits available. Working with predictions, competitiveness is a function of the quality, not the quantity, of information about the input. Thus, results cannot be translated between AOC and the complexity classes of this paper. Moreover, AOC is only one complexity class, not a hierarchy.

Proving that a problem is $\mathcal{C}^t_{\eta_0,\eta_1}$-hard suggests that, when using binary predictions, one cannot solve it better than blindly trusting the predictions, giving a rather poor result. Hence, proving that a problem is hard serves as an argument for needing a richer prediction scheme for the problem, or possibly a more accurate way of measuring prediction error.

Our main contribution is the framework enabling a complexity theory for online algorithms with predictions. Using this framework, we prove hardness and class membership results for several problems, including showing completeness of Online t-Bounded Degree Vertex Cover (VC_t) for $\mathcal{C}^t_{\eta_0,\eta_1}$. Thus, VC_t, or any other complete problem, could be used as the basis for the complexity classes instead of ASG_t. However, we follow the tradition from advice complexity and use a string guessing problem, ASG_t, as its lack of structure offers simpler proofs. We illustrate the relative hardness of the problems we investigate in Fig. 1.

Worth noting is that by choosing the appropriate pair of error measures, our set-up immediately gives the same hardness results for purely online problems, that is, for algorithms without predictions.

Omitted proofs and other details can be found in the full paper [5].

2 Preliminaries

We consider online problems, where algorithms must make an irrevocable decision for each request, by outputting a bit.[1] For any problem, P, we let $\mathcal{I}_P$ be the collection of instances of P, we let $\mathcal{A}_P$ be the set of online algorithms for P, and we let OPT_P be a fixed optimal algorithm for P. An instance of P is a triple $I = (x, \hat{x}, r)$, consisting of two bitstrings $x, \hat{x} \in \{0,1\}^n$, and a sequence of

---
[1] The only exception to this is Paging with Discard Predictions, where the output is a page (see Sect. 6).

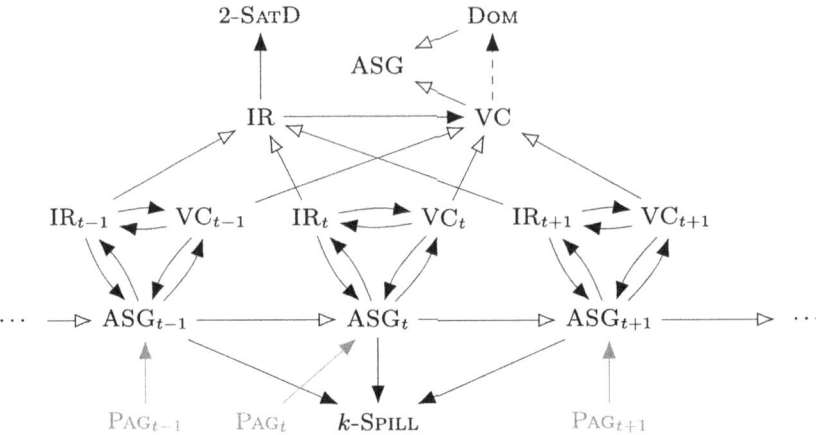

Fig. 1. A hardness graph based on our complexity hierarchy. The problems shown are defined in the beginning of Sects. 3, 5.1–5.3, and 6. Given two problems P and Q, we let $P \to Q$ indicate that Q is as hard as P (see Definition 1). If the arrowhead is only outlined, P is not as hard as Q. If the arrow is dashed, P is as hard as Q in the weak sense (see Definition 4). We leave out most arrows that can be derived by transitivity. The gray arrows hold with respect to the pair of error measures (μ_0, μ_1) (see Sect. 2) and the remaining arrows hold with respect to all pairs of *insertion monotone* error measures (see Sect. 5).

requests $r = \langle r_1, r_2, \ldots, r_n \rangle$ for P. The bitstring x is an encoding of OPT_P's solution, and $\hat{x}$ is a prediction of x. When an algorithm, ALG, receives the request r_i, it also receives the prediction $\hat{x}_i$, to aid its decision, y_i, for r_i. What information is contained in each request, r_i, and the meaning of the bits x_i, $\hat{x}_i$, and y_i, will be specified for each problem.

For graph problems, we consider the *vertex-arrival* model, where each request contains a vertex, v, and the edges connecting v to previously arrived vertices.

An algorithm, ALG, for an online minimization problem without predictions, P, is c-*competitive* if there exists a constant $\kappa \in \mathbb{R}^+$, called the *additive constant*, such that for all instances $I \in \mathcal{I}_P$, $\text{ALG}(I) \leqslant c \cdot \text{OPT}(I) + \kappa$.[2] We extend the definition of competitiveness to online minimization problems with binary predictions, based on [2]. Here, the competitiveness of an algorithm is written as a function of two error measures[3], η_0 and η_1, where η_b is a function of the bits

---

[2] Our formulation is analogous to the definition in [9,23]. In the full paper [5], we allow for κ to be "sublinear in OPT", which is similar to the asymptotic competitive ratio, where we require that $\limsup_{|I| \to \infty} \frac{\text{ALG}(I)}{\text{OPT}(I)} \leqslant c$.

[3] We work with separate error measures for predicted 0s and 1s to allow for more detailed results. Our reductions and structural results would also work if we used only one error measure. For instance, Theorem 2 implies that FTP is $(1,1)$-competitive with respect to $\text{Ham}_{t-1,1}(x, \hat{x}) = (t-1) \cdot \mu_0(x, \hat{x}) + \mu_1(x, \hat{x})$. However, combining the two error measures into one, we lose some detail such as the tradeoffs between α, β, and γ given in Theorem 9.

incorrectly predicted to be b. We assume that $0 \leqslant \eta_b(I) < \infty$, for all instances I. If there exist three maps $\alpha, \beta, \gamma \colon \mathcal{I}_P \to [0, \infty)$ and a constant $\kappa \in \mathbb{R}$ such that for all $I \in \mathcal{I}_P$,

$$\mathrm{ALG}(I) \leqslant \alpha \cdot \mathrm{OPT}(I) + \beta \cdot \eta_0(I) + \gamma \cdot \eta_1(I) + \kappa,$$

then ALG is (α, β, γ)-*competitive with respect to* (η_0, η_1). When $\kappa \leqslant 0$ then ALG is *strictly* (α, β, γ)-*competitive*. Observe that the definition of (α, β, γ)-competitiveness does not require α, β, and γ to be constants, though it is desirable that they are, especially α which is the consistency. In particular, α, β, and γ are allowed to be functions of the instance, $I \in \mathcal{I}_P$, for example of n. In our notation, α, β, and γ's dependency on I is, however, kept implicit. Further, observe that any α-competitive algorithm without predictions is $(\alpha, 0, 0)$-competitive with respect to any pair of error measures (η_0, η_1).

We define a pair of error measures equivalent to the pair of error measures in [2], where μ_b is the number of wrong predictions of $b \in \{0, 1\}$. For any instance $I = (x, \hat{x}, r)$, we let

$$\mu_0(I) = \sum_{i=1}^{n} x_i \cdot (1 - \hat{x}_i) \text{ and } \mu_1(I) = \sum_{i=1}^{n} (1 - x_i) \cdot \hat{x}_i.$$

3 Asymmetric String Guessing: A Collection of Hard Problems

Given a $t \in \mathbb{Z}^+ \cup \{\infty\}$ and an input, $I = (x, \hat{x}, r)$, for *Online* $(1, t)$-*Asymmetric String Guessing with Unknown History and Predictions* (ASG$_t$), each request, r_i, is a prompt for guessing the bit x_i. Together with r_i, an algorithm, ALG, receives a prediction, $\hat{x}_i$, for x_i. The cost of ALG on I is given by

$$\mathrm{ALG}(I) = \sum_{i=1}^{n} \left(y_i + t \cdot x_i \cdot (1 - y_i) \right),$$

where y_i is ALG's i'th guess. When $t = \infty$, the problem corresponds to the string guessing problem considered in [11], and we call it ASG.

For all $t \in \mathbb{Z}^+ \cup \{\infty\}$, we may consider ASG$_t$ as a purely online problem by omitting $\hat{x}$. We briefly state the main results on the competitiveness of online algorithms for ASG$_t$, with and without predictions. We note that, in Theorem 1, Item (iii) is a direct consequence of Item (ii).

Observation 1 ([11]). *For any algorithm,* ALG, *for* ASG *without predictions, there is no function, f, such that* ALG *is $f(n)$-competitive.*

Theorem 1. *Let $t \in \mathbb{Z}^+$ and let $0 < \varepsilon < 1$. Then, for* ASG$_t$, *the following hold.*

(i) The algorithm that always guesses 0 is t-competitive.
(ii) There is no $(t - \varepsilon)$-competitive deterministic algorithm without predictions.

(iii) For any pair of error measures (η_0, η_1), there is no $(t - \varepsilon, 0, 0)$-competitive deterministic algorithm with predictions with respect to (η_0, η_1).

An obvious algorithm for ASG_t is *Follow-the-Predictions* (FTP), which always sets its guess, y_i, to $\hat{x}_i$.

Theorem 2. *For any $t \in \mathbb{Z}^+$ and any $\alpha, \beta \in \mathbb{R}^+$ such that $\alpha \geqslant 1$ and $\alpha + \beta \geqslant t$, FTP is strictly $(\alpha, \beta, 1)$-competitive for ASG_t with respect to (μ_0, μ_1).*

4 A Hierarchy of Complexity Classes

We formally introduce the complexity classes, prove that they form a strict hierarchy, and show fundamental structural properties of the complexity classes.

4.1 Relative Hardness and Reductions

Labeling one problem, Q, as hard as another problem, P, usually means that no algorithm for Q can be better than a best possible algorithm for P. However, there is no total order on the (α, β, γ)-triples capturing competitiveness, so we call an algorithm "best possible" if it is Pareto-optimal. An (α, β, γ)-competitive algorithm is *Pareto-optimal* for P with respect to (η_0, η_1) if, for any $\varepsilon > 0$, there is no $(\alpha - \varepsilon, \beta, \gamma)$-, $(\alpha, \beta - \varepsilon, \gamma)$-, or $(\alpha, \beta, \gamma - \varepsilon)$-competitive algorithm for P. Before defining the as-hard-as relation, we classify all Pareto-optimal algorithms for ASG_t with respect to (μ_0, μ_1):

Theorem 3. *Let ALG be an (α, β, γ)-competitive algorithm for ASG_t with respect to (μ_0, μ_1). Then, ALG is Pareto-optimal if and only if*

(i) $\alpha = t$ *and* $\beta = \gamma = 0$ *or*
(ii) $\alpha + \beta = t$ *and* $\gamma = 1$.

Thus, by Theorem 2, FTP is Pareto-optimal.

Definition 1. *Let P and Q be online problems with binary predictions and let (η_0, η_1) be any error measures pair.[4] We say that Q is (at least) as hard as P with respect to (η_0, η_1), if the existence of an (α, β, γ)-competitive Pareto-optimal algorithm for Q implies the existence of an (α, β, γ)-competitive algorithm for P.*

Trivially, the as-hard-as relation is both reflexive and transitive.

As a tool for proving hardness, we introduce the notion of reductions. A reduction from a problem, P, to another problem, Q, consists of a mapping of instances of P to instances of Q and a mapping from algorithms for Q to algorithms for P, with the requirement that (α, β, γ)-competitive Pareto-optimal algorithms for Q map to (α, β, γ)-competitive algorithms for P. In this paper, we use a stronger type of reduction which we call strict online reductions.

---

[4] For more generality, in the full paper [5], we have two pairs of error measures; one for P and one for Q.

Definition 2. Let P and Q be online minimization problems with binary predictions, and let (η_0, η_1) be any pair of error measures.[5] Let $\rho = (\rho_A, \rho_R)$ be a tuple consisting of two maps, $\rho_A \colon \mathcal{A}_Q \to \mathcal{A}_P$ and $\rho_R \colon \mathcal{A}_Q \times \mathcal{I}_P \to \mathcal{I}_Q$. Then, ρ is called a strict online reduction from P to Q with respect to (η_0, η_1), if there exists a constant $k_A \in \mathbb{R}$, called the reduction constant of ρ, such that for each $I_P \in \mathcal{I}_P$ and each $\mathrm{ALG}_Q \in \mathcal{A}_Q$, letting $\mathrm{ALG}_P = \rho_A(\mathrm{ALG}_Q)$ and $I_Q = \rho_R(\mathrm{ALG}_Q, I_P)$,

(O1) $\mathrm{ALG}_P(I_P) \leqslant \mathrm{ALG}_Q(I_Q) + k_A$,
(O2) $\mathrm{OPT}_Q(I_Q) \leqslant \mathrm{OPT}_P(I_P)$,
(O3) $\eta_0(I_Q) \leqslant \eta_0(I_P)$, and
(O4) $\eta_1(I_Q) \leqslant \eta_1(I_P)$,

Abusing notation slightly, we write $\rho \colon P \xrightarrow{\mathrm{red}} Q$.

Lemma 1. Let $\rho \colon P \xrightarrow{\mathrm{red}} Q$ be a strict online reduction, then Q is as hard as P.

Throughout, we refer to strict online reductions, simply as reductions.

4.2 Defining the Complexity Classes

For any pair, (η_0, η_1), of error measures and any $t \in \mathbb{Z}^+ \cup \{\infty\}$, we define the complexity class $\mathcal{C}^t_{\eta_0, \eta_1}$ as the set of online problems with binary predictions that are no harder than ASG_t with respect to (η_0, η_1):

Definition 3. For each $t \in \mathbb{Z}^+ \cup \{\infty\}$ and each pair of error measures, (η_0, η_1), the complexity class $\mathcal{C}^t_{\eta_0, \eta_1}$ is the closure of ASG_t under the as-hard-as relation with respect to (η_0, η_1). Hence, for an online problem, P,

- $P \in \mathcal{C}^t_{\eta_0, \eta_1}$, if ASG_t is as hard as P,
- P is $\mathcal{C}^t_{\eta_0, \eta_1}$-hard, if P is as hard as ASG_t, and
- P is $\mathcal{C}^t_{\eta_0, \eta_1}$-complete, if $P \in \mathcal{C}^t_{\eta_0, \eta_1}$ and P is $\mathcal{C}^t_{\eta_0, \eta_1}$-hard.

Thus, if P and Q are $\mathcal{C}^t_{\eta_0, \eta_1}$-complete problems, then there exists an (α, β, γ)-competitive algorithm for P if and only if there exists an (α, β, γ)-competitive algorithm for Q.

Since the as-hard-as relation is reflexive, ASG_t is $\mathcal{C}^t_{\eta_0, \eta_1}$-complete for any pair of error measures, (η_0, η_1), and any t. Further, due to transitivity:

Lemma 2. Let $t \in \mathbb{Z}^+ \cup \{\infty\}$ and let (η_0, η_1) be any pair of error measures.

- If $P \in \mathcal{C}^t_{\eta_0, \eta_1}$, and P is as hard as Q, then $Q \in \mathcal{C}^t_{\eta_0, \eta_1}$.
- If P is $\mathcal{C}^t_{\eta_0, \eta_1}$-hard, and Q is as hard as P, then Q is $\mathcal{C}^t_{\eta_0, \eta_1}$-hard.

This lemma implies results concerning special cases of a problem:

---

[5] In the full paper [5], we have two pairs of error measures; one for P and one for Q.

Corollary 1. *Let $t \in \mathbb{Z}^+ \cup \{\infty\}$ and let (η_0, η_1) be any pair of error measures. Let P and P_{sub} be online minimization problems such that $\mathcal{I}_{P_{\text{sub}}} \subseteq \mathcal{I}_P$. If $P \in \mathcal{C}^t_{\eta_0, \eta_1}$, then $P_{\text{sub}} \in \mathcal{C}^t_{\eta_0, \eta_1}$, and if P_{sub} is $\mathcal{C}^t_{\eta_0, \eta_1}$-hard, then P is $\mathcal{C}^t_{\eta_0, \eta_1}$-hard.*

Observe that Lemma 2 implies that any $\mathcal{C}^t_{\eta_0,\eta_1}$-complete problem can be used as the base problem when defining $\mathcal{C}^t_{\eta_0,\eta_1}$. For instance, by Lemmas 4 and 5, the better known Online t-Bounded Degree Vertex Cover with Predictions is $\mathcal{C}^t_{\eta_0,\eta_1}$-complete with respect to a wide range of pairs of error measures, and may therefore be used as the basis of these complexity classes instead of ASG_t. Also, after establishing the first $\mathcal{C}^t_{\eta_0,\eta_1}$-hard problems, we may reduce from any $\mathcal{C}^t_{\eta_0,\eta_1}$-hard problem to prove hardness. Finally, a problem Q is $\mathcal{C}^t_{\eta_0,\eta_1}$-hard, if and only if Q is as hard as P, for all $P \in \mathcal{C}^t_{\eta_0,\eta_1}$.

Lemma 3. *For any $t \in \mathbb{Z}^+$ and any pair of error measures, (η_0, η_1), ASG_{t+1} is as hard as ASG_t, and ASG_t is not as hard as ASG_{t+1}.*

Theorem 4. *For any pair of error measures (η_0, η_1), we have a strict hierarchy of complexity classes:*

$$\mathcal{C}^1_{\eta_0,\eta_1} \subsetneq \mathcal{C}^2_{\eta_0,\eta_1} \subsetneq \mathcal{C}^3_{\eta_0,\eta_1} \subsetneq \cdots \subsetneq \mathcal{C}_{\eta_0,\eta_1}.$$

4.3 Purely Online Algorithms

Observe that our complexity theory extends to a complexity theory for purely online algorithms. Using the error measures (Z_0, Z_1), given by $Z_0(I) = Z_1(I) = 0$, for any $I = (x, \hat{x}, r)$, any (α, β, γ)-competitive algorithm, ALG, for an online minimization problem P, satisfies

$$\text{ALG}(I) \leqslant \alpha \cdot \text{OPT}(I) + \beta \cdot Z_0(I) + \gamma \cdot Z_1(I) + k_A = \alpha \cdot \text{OPT}(I) + k_A,$$

for all instances $I \in \mathcal{I}_P$, and so ALG is an α-competitive purely online algorithm for P. Hence, we obtain a similar complexity theory for purely online algorithms, since all reductions in the (full) paper are valid with respect to (Z_0, Z_1).

Observation 2. *All of our results involving problems and complexity classes also hold for the same problems and classes without predictions.*

Observation 3. *In Sect. 6 we discuss a strategy for proving general lower bounds for all $\mathcal{C}^t_{\eta_0,\eta_1}$-hard problems. Using the same strategy, the lower bound from Theorem 1(ii) on the competitive ratio of any purely online algorithm for ASG_t extends to a lower bound on the competitiveness of any purely online algorithm for any $\mathcal{C}^t_{Z_0,Z_1}$-hard problems.*

5 A List of $\mathcal{C}^t_{\eta_0,\eta_1}$-Hard Problems

In this section, we provide a list of $\mathcal{C}^t_{\eta_0,\eta_1}$-hard problems. For some of the problems, we prove $\mathcal{C}^t_{\eta_0,\eta_1}$-hardness using a *reduction template*. We briefly explain how to prove $\mathcal{C}^t_{\eta_0,\eta_1}$-hardness of a problem P using the reduction template. For any $(x, \hat{x}, r) \in \mathcal{I}_{\text{ASG}_t}$ and any $\text{ALG}_P \in \mathcal{A}_P$, the reduction works in two stages:

- For each request, r_i, of r, give a *challenge request*, c_i, Read ALG$_P$'s output, y_i', for c_i and output $y_i = y_i'$. The instance of P is constructed such that no information about the true bit, x_i', of c_i can be inferred from the information about the instance obtained so far. This continues until the end of r is reached and the true contents of x are learned.
- After this, a sequence of *blocks* is created, with a block, $B(x_i, y_i')$, for each request, r_i, in r. The block $B(x_i, y_i')$ consists of a number of requests ensuring that (O1)–(O4) from Definition 2 are satisfied.

Reductions created using this template are valid strict online reductions with respect to all pairs of *insertion monotone* error measures. An error measure, η, is called insertion monotone if the insertion of a finite number of correctly predicted bits into the instance does not increase the error. Clearly, μ_b and Z_b are insertion monotone for all $b \in \{0,1\}$. We provide a non-exhaustive list of pairs of insertion monotone error measures in the full paper [5].

5.1 Vertex Cover

Given a graph, $G = (V, E)$, a vertex cover for G is a subset $V' \subseteq V$ such that each edge in E has at least one endpoint in V'. The cost of a vertex cover is the number of vertices it contains. We consider *Online t-Bounded Degree Vertex Cover with Predictions* (VC$_t$); an online version of Vertex Cover where the maximum degree of the graph is at most t. VC$_t$ is a vertex-arrival problem. Together with each request, r_i, algorithms receive a prediction, $\hat{x}_i$, for x_i that predicts whether or not v_i is in a given optimal vertex cover. When an algorithm, ALG, receives a vertex, it outputs $y_i = 1$ to accept this vertex into its vertex cover and $y_i = 0$ to reject it. Thus, the cost of ALG on an instance, $I = (x, \hat{x}, r) \in \mathcal{I}_{\text{VC}_t}$, is

$$\text{ALG}(I) = \begin{cases} \sum_{i=1}^n y_i, & \text{if ALG's output defines a vertex cover,} \\ \infty, & \text{otherwise,} \end{cases}$$

and $\text{OPT}(I) = \sum_{i=1}^n x_i$. The standard unbounded Online Vertex-Arrival Vertex Cover with Predictions is also considered, and is abbreviated VC.

Other work on Vertex Cover includes the following. In the purely online case, there exists a t-competitive algorithm for t-Bounded Degree Vertex Cover, but a $(t-\varepsilon)$-competitive online algorithm cannot exist, for any $\varepsilon > 0$ [13]. In the offline setting, Vertex Cover is MAX-SNP-hard [27], APX-complete [14,28], and NP-complete [18]. Lastly, Online Vertex-Arrival Vertex Cover is AOC-complete [11].

Lemma 4. *For any $t \in \mathbb{Z}^+$ and any pair of insertion monotone error measures (η_0, η_1), VC$_t$ is $\mathcal{C}_{\eta_0,\eta_1}^t$-hard.*

Proof. We give a strict online reduction $\rho \colon \text{ASG}_t \xrightarrow{\text{red}} \text{VC}_t$, using the reduction template described in the beginning of Sect. 5 (see Fig. 2 for an example):

Consider any $I = (x, \hat{x}, r) \in \mathcal{I}_{\text{ASG}_t}$ and any ALG$' \in \mathcal{A}_{\text{VC}_t}$ and let $n = |x|$. Each challenge request, c_i, is an isolated vertex, v_i, with $x_i' = x_i$ and $\hat{x}_i' = \hat{x}_i$. Let y_i' be the output of ALG$'$ on c_i. The ith block, $B(x_i, y_i')$, is constructed as follows, with all true and predicted bits equal to 0.

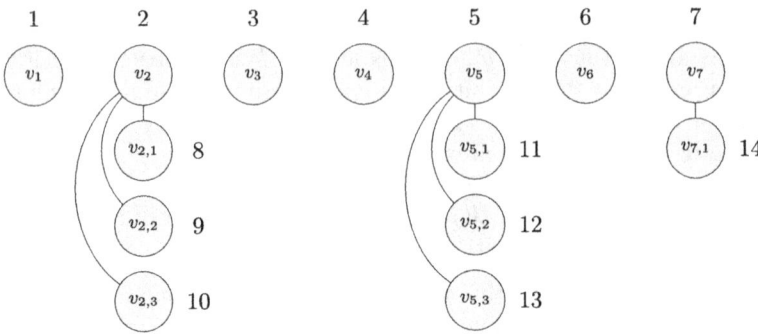

Fig. 2. Depicting the reduction graph for $t = 3$, $x = 0100101$, and $y' = 1011001(\_ \_ \_)(\_ \_ \_)(\_)$. The first seven bits of y' are the VC_t algorithm's response to the challenge requests. The algorithm's response to the requests of the three nonempty blocks are simply shown as '$\_$', since they do not influence the definition of the graph. The numbers next to the vertices indicate the order in which the vertices are revealed.

- If $x_i = 0$, then $B(x_i, y_i')$ is empty. Thus, no optimal solution will contain v_i.
- If $x_i = y_i' = 1$, then $B(x_i, y_i')$ contains one request to a vertex, $v_{i,1}$, connected to v_i, ensuring that there is an optimal solution containing v_i.
- If $x_i = 1$ and $y_i' = 0$, then $B(x_i, y_i')$ contains requests to t vertices, $v_{i,j}$, $j = 1, 2, \ldots, t$, each connected to v_i, giving ALG' a cost of t and ensuring that there is an optimal solution containing v_i.

Note that x' encodes an optimal solution for the VC_t-instance. Further, recall that, following the template, the algorithm, ALG, for ASG_t produces the same output as ALG' does on the challenge requests, i.e., $y_i = y_i'$, $1 \leqslant i \leqslant n$. Thus, it is not hard to check that (O1)–(O4) are all satisfied, with $k_A = 0$. □

Lemma 5. *For any $t \in \mathbb{Z}^+$ and any error measure pair (η_0, η_1), $\text{VC}_t \in \mathcal{C}_{\eta_0,\eta_1}^t$.*

Sketch of Proof. We define a strict online reduction, $\rho \colon \text{VC}_t \xrightarrow{\text{red}} \text{ASG}_t$, with reduction constant $k_A = 0$ as follows. Consider any $I = (x, \hat{x}, r) \in \mathcal{I}_{\text{VC}_t}$ and any $\text{ALG}' \in \mathcal{A}_{\text{ASG}_t}$. We give an instance, $I' = (x', \hat{x}', r') \in \mathcal{I}_{\text{ASG}_t}$, and an algorithm, ALG, for handling I using the output of ALG'. For each request, r_i, in r, give $\hat{x}_i' = \hat{x}_i$ to ALG' and let y_i' be the output of ALG'. If v_i has a neighbor among $v_1, \ldots, v_{i-1}$ which is not in the vertex cover constructed so far, ALG outputs $y_i = 1$. Otherwise, it outputs $y_i = y_i'$. After the last request of I, compute an optimal solution, x, for I and present $x' = x$ to ALG', in order to finish the instance I'. Now, it only remains to check (O1)–(O4). □

Our results about VC_t and VC are summarized in Theorem 5. Note that the results on VC_t follow directly from Lemmas 4 and 5.

Theorem 5. *For any $t \in \mathbb{Z}^+$ and any pair of insertion monotone error measures, (η_0, η_1), VC_t is $\mathcal{C}_{\eta_0,\eta_1}^t$-complete, VC is $\mathcal{C}_{\eta_0,\eta_1}^t$-hard, and $\text{VC} \notin \mathcal{C}_{\eta_0,\eta_1}^t$.*

Moreover, for any $t \in \mathbb{Z}^+$ and any pair of error measures, (η_0, η_1), $\text{VC}_t \in \mathcal{C}_{\eta_0,\eta_1}^t$, VC is not $\mathcal{C}_{\eta_0,\eta_1}$-hard, and $\text{VC} \in \mathcal{C}_{\eta_0,\eta_1}$.

5.2 k-Minimum-Spill

Given a graph $G = (V, E)$, the objective of k-Minimum-Spill is to determine a smallest possible subset $V_1 \subseteq V$ such that the subgraph of G induced by $V \setminus V_1$ is k-colorable. We consider *Online k-Minimum-Spill with Predictions* (k-SPILL), which is a vertex-arrival problem. Together with each request, r_i, algorithms receive a prediction, $\hat{x}_i$, predicting the optimal output to r_i, x_i. When an algorithm, ALG, receives a vertex, it outputs $y_i = 1$ to mark this vertex as a spill, and thus place it in V_1. Given an instance, $I \in \mathcal{I}_{k\text{-SPILL}}$,

$$\text{ALG}(I) = \begin{cases} \sum_{i=1}^{n} y_i, & \text{if the graph induced by } V \setminus V_1 \text{ is } k\text{-colorable,} \\ \infty, & \text{otherwise.} \end{cases}$$

and $\text{OPT}(I) = \sum_{i=1}^{n} x_i$.

As discussed below, 1-SPILL is equivalent to VC.

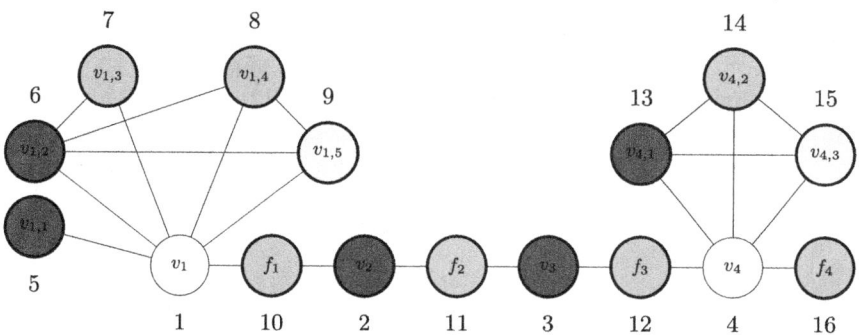

Fig. 3. Depicting the reduction graph for $t = k = 3$, $x = 1001$, and $y' = 0011(10101\_)(\_)(\_)(\_\_\_\_)$. The first four bits of y' are the k-SPILL algorithm's response to the four challenge requests, and the bits in parentheses correspond to its responses to the four blocks. Bits that do not influence the definition of the graph are simply shown as '_'. The numbers next to the vertices indicate the order in which they are revealed. For later reference, we show an optimal coloring of the graph.

Theorem 6. *For all $k, t \in \mathbb{Z}^+$ and all pairs of insertion monotone error measures (η_0, η_1), k-SPILL is $\mathcal{C}_{\eta_0,\eta_1}^t$-hard.*

Sketch of Proof. 1-SPILL is equivalent to VC, and so 1-SPILL is $\mathcal{C}_{\eta_0,\eta_1}^t$-hard by Theorem 5. For k-SPILL with $k \geqslant 2$, we use the reduction template to construct a reduction, $\rho \colon \text{ASG}_t \xrightarrow{\text{red}} k\text{-SPILL}$, with respect to (η_0, η_1) (see Fig. 3 for an example). To this end, consider any $I = (x, \hat{x}, r) \in \mathcal{I}_{\text{ASG}_t}$ and any $\text{ALG}' \in \mathcal{A}_{k\text{-SPILL}}$. Let $n = |x|$ and let $I' = (x', \hat{x}', r')$ be the k-SPILL-instance created in the reduction.

For each x_i, the corresponding challenge request, c_i, for k-SPILL is an isolated vertex, v_i, with $x'_i = x_i$ and $\hat{x}'_i = \hat{x}_i$. Let y'_i be the response of ALG' on c_i. Recall that if $y'_i = 1$, then ALG' adds v_i to V_1.

For each x_i, the corresponding block, $B(x_i, y'_i)$ is defined as follows. All true and predicted bits are 0. The last vertex, f_i, of the block is called a *final* vertex. If $i < n$, f_i is connected to v_i and v_{i+1}, and if $i = n$, f_i is connected to $v_i = v_n$.

- If $x_i = 0$, then $B(x_i, y'_i)$ has a single request containing the final vertex, f_i.
- If $x_i = y'_i = 1$, then $B(x_i, y'_i)$ contains $k+1$ requests:
  - For $j = 1, 2, \ldots, k$, a vertex, $v_{i,j}$, connected to v_i and to $v_{i,l}$, $1 \leqslant l < j$.
  - The final vertex, f_i.
- If $x_i = 1$ and $y'_i = 0$, then $B(x_i, y'_i)$ contains the following requests:
  (a) Let $j = 1$. Until ALG' has added t new vertices to V_1 by outputting 1:
    - A vertex, $v_{i,j}$, connected to v_i and to each vertex in $\bigcup_{l=1}^{j-1} v_{i,l} \setminus V_1$. If ALG' has added more than k vertices to $V \setminus V_1$ in this block, only include the edge $(v_{i,l}, v_{i,j})$ to the first k vertices in $\bigcup_{l=1}^{j-1} v_{i,l} \setminus V_1$ and go to (c) (this happens only if ALG' creates an infeasible solution). Otherwise, increment j.
  (b) If $\{v_i\} \cup \bigcup_j \{v_{i,j}\}$ does not contain a $(k+1)$-clique:
    - For $j = 1, 2, \ldots, k$, a vertex, $v_{i,j}$, connected to v_i and to $v_{i,l}$, $l < j$.
  (c) The final vertex, f_i.

Now, it only remains to check that I' is a valid instance of k-SPILL, i.e., that x' is an optimal solution, and that (O1)–(O4) from Definition 2 are satisfied. □

5.3 Other Problems

We state our remaining results on membership, hardness, and completeness without proof. We consider *Online Dominating Set with Predictions* (DOM), *Online t-Bounded Overlap Interval Rejection with Predictions* (IR_t), and *Online Minimum 2-SAT Deletion with Predictions* (2-SATD). These problems have been considered without predictions in many variants, including in [10,12,13,21,22,24,27]. We also consider an unbounded version Interval Rejection (IR). DOM is a vertex-arrival problem, the most standard model for online graph problems. Hence, upon receiving a vertex, an algorithm has to include or exclude a vertex from its dominating set. For IR_t and IR all requests are intervals, and an algorithm must output a set of intervals to remove, such that no two remaining intervals intersect. The goal is to minimize the number of removed intervals. For IR_t, all intervals intersect at most t other intervals. Finally, 2-SATD is a variable arrival problem. Each request consists of a variable and all clauses that contain the newly arrived variable and a previously arrived variable. When a variable arrives, the algorithm has to assign it a truth value. Its objective is to minimize the number of clauses the algorithm must delete to make φ satisfied.

For describing our results on these problems, we introduce a slightly weaker notion of hardness:

Definition 4. *Let P and Q be online problems with binary predictions. If any (α, β, γ)-competitive Pareto-optimal algorithm for P, with $\alpha \in o(\text{OPT}_P)$, implies the existence of an (α, β, γ)-competitive algorithm for Q, then P is said to be as hard as Q in the weak sense. If Q is complete for some complexity class, $\mathcal{C}$, then P is said to be* weakly $\mathcal{C}$-hard.

For the next Theorem, recall that both (Z_0, Z_1) and (μ_0, μ_1) are pairs of insertion monotone error measures.

Theorem 7. *For all $t \in \mathbb{Z}^+$ and insertion monotone error measures, (η_0, η_1),*

- *IR_t is $\mathcal{C}^t_{\eta_0, \eta_1}$-complete.*
- *IR and 2-SATD are $\mathcal{C}^t_{\eta_0, \eta_1}$-hard and not members of $\mathcal{C}^t_{\eta_0, \eta_1}$.*
- *DOM is weakly $\mathcal{C}^t_{\eta_0, \eta_1}$-hard and not a member of $\mathcal{C}^t_{\eta_0, \eta_1}$.*
- *IR and DOM are members of $\mathcal{C}_{\eta_0, \eta_1}$ and not $\mathcal{C}_{\eta_0, \eta_1}$-hard.*

6 Establishing Upper and Lower Bounds for Problems Related to $\mathcal{C}^t_{\mu_0, \mu_1}$

Discard Predictions (PAG_k), which is the standard paging problem with a binary prediction scheme, with respect to (μ_0, μ_1). In the full paper [5], we show that PAG_t is a member of $\mathcal{C}^t_{\mu_0, \mu_1}$, for any $t \in \mathbb{Z}^+$ (Theorem 8), which means that all lower bounds for PAG_t are valid for any $\mathcal{C}^t_{\mu_0, \mu_1}$-hard problem, resulting in Theorem 9.

Theorem 8. *For all $t \in \mathbb{Z}^+$, $\text{PAG}_t \in \mathcal{C}^t_{\mu_0, \mu_1}$.*

Theorem 9. *Let $t \in \mathbb{Z}^+$ and let P be any $\mathcal{C}^t_{\mu_0, \mu_1}$-hard problem. Then, for any (α, β, γ)-competitive algorithm for P with respect to (μ_0, μ_1),*

(i) $\alpha + \beta \geq t$,
(ii) $\alpha + (t-1) \cdot \gamma \geq t$.

Finally, we re-prove existing positive results for PAG_t (see Remark 3.2 in [2]) using that $\text{PAG}_t \in \mathcal{C}^t_{\mu_0, \mu_1}$:

Theorem 10. *For any $\alpha, \beta \in \mathbb{R}$ with $\alpha \geq 1$ and $\alpha + \beta \geq t$, there exists an $(\alpha, \beta, 1)$-competitive algorithm for PAG_t with respect to (μ_0, μ_1).*

7 Concluding Remarks and Future Work

Note that our definition of relative hardness also applies to maximization problems. In a very recent follow-up to the arXiv version of our paper, online maximization problems with binary predictions are considered and several maximization problems are shown to be members, hard and complete for $\mathcal{C}^t_{\mu_0, \mu_1}$ [4].

Possible directions for future work include considering randomization and changing the hardness measure from competitiveness to, e.g., relative worst order or random order.

References

1. Algorithms with predictions. https://algorithms-with-predictions.github.io/. Accessed 28 Mar 2025
2. Antoniadis, A., et al.: Paging with succinct predictions. In: 40th International Conference on Machine Learning (ICML), vol. 202, pp. 952–968. PMLR (2023)
3. Ausiello, G., Protasi, M., Marchetti-Spaccamela, A., Gambosi, G., Crescenzi, P., Kann, V.: , Complexity and Approximation: Combinatorial Optimization Problems and Their Approximability Properties. Springer (1999)
4. Berg, M.: Comparing the hardness of online minimization and maximization problems with predictions. In: International Joint Conference on Theoretical Computer Science - Frontier of Algorithmic Wisdom (IJTCS-FAW). Lecture Notes in Computer Science. Springer (2025). Accepted for publication. arXiv:2409.12694
5. Berg, M., Boyar, J., Favrholdt, L.M., Larsen, K.S.: Complexity classes for online problems with and without predictions (2024). arXiv:2406.18265
6. Böckenhauer, H.-J., Hromkovič, J., Komm, D., Krug, S., Smula, J., Sprock, A.: The string guessing problem as a method to prove lower bounds on the advice complexity. Theoret. Comput. Sci. **554**, 95–108 (2014). https://doi.org/10.1016/j.tcs.2014.06.006
7. Böckenhauer, H.-J., Komm, D., Královič, R., Královič, R., Mömke, T.: On the advice complexity of online problems. In: Dong, Y., Du, D.-Z., Ibarra, O. (eds.) ISAAC 2009. LNCS, vol. 5878, pp. 331–340. Springer, Heidelberg (2009). https://doi.org/10.1007/978-3-642-10631-6_35
8. Borodin, A., Boyar, J., Larsen, K.S., Pankratov, D.: Advice complexity of priority algorithms. Theory Comput. Syst. **64**, 593–625 (2020). https://doi.org/10.1007/s00224-019-09955-7
9. Borodin, A., El-Yaniv, R.: Online Computation and Competitive Analysis. Cambridge University Press (1998)
10. Boyar, J., Eidenbenz, S.J., Favrholdt, L.M., Kotrbčík, M., Larsen, K.S.: Online dominating set. Algorithmica **81**(5), 1938–1964 (2019). https://doi.org/10.1007/s00453-018-0519-1
11. Boyar, J., Favrholdt, L.M., Kudahl, C., Mikkelsen, J.W.: The advice complexity of a class of hard online problems. Theory Comput. Syst. **61**, 1128–1177 (2017). https://doi.org/10.1007/s00224-016-9688-y
12. Chlebík, M., Chlebíkova, J.: On approxmiation hardness of the minimum 2SAT-deletion problem. Discret. Appl. Math. **155**(2), 172–179 (2007). https://doi.org/10.1016/j.dam.2006.04.039
13. Demange, M., Paschos, V.T.: On-line vertex-covering. Theoret. Comput. Sci. **332**(1–3), 83–108 (2005). https://doi.org/10.1016/j.tcs.2004.08.015
14. Dinur, I., Safra, S.: On the hardness of approximating minimum vertex cover. Ann. Math. **162**(1), 439–485 (2005). https://doi.org/10.4007/annals.2005.162.439
15. Dobrev, S., Královic, R., Pardubská, D.: Measuring the problem-relevant information in input. RAIRO Theor. Inform. Appl. **43**(3), 585–613 (2009). https://doi.org/10.1051/ita/2009012
16. Downey, R.G., Fellows, M.R.: Parameterized Complexity. Springer (1999). https://doi.org/10.1007/978-1-4612-0515-9
17. Emek, Y., Fraigniaud, P., Korman, A., Rosén, A.: Online computation with advice. Theoret. Comput. Sci. **412**(24), 2642–2656 (2011). https://doi.org/10.1016/j.tcs.2010.08.007

18. Garey, M.R., Johnson, D.S.: Computers and Intractability; A Guide to the Theory of NP-Completeness. W. H. Freeman and Co. (1990)
19. Henzinger, M., Saha, B., Seybold, M.P., Ye, C.: On the complexity of algorithms with predictions for dynamic graph problems. In: 15th Innovations in Theoretical Computer Science Conference (ITCS). Leibniz International Proceedings in Informatics (LIPIcs), vol. 287, pp. 62:1–62:25. Schloss Dagstuhl — Leibniz-Zentrum für Informatik (2024). https://doi.org/10.4230/LIPIcs.ITCS.2024.62
20. Hromkovič, J., Královič, R., Královič, R.: Information complexity of online problems. In: Hliněný, P., Kučera, A. (eds.) MFCS 2010. LNCS, vol. 6281, pp. 24–36. Springer, Heidelberg (2010). https://doi.org/10.1007/978-3-642-15155-2_3
21. Karp, R.M.: Reducibility among combinatorial problems. In: Proceedings of a Symposium on the Complexity of Computer Computations, pp. 85–103. Plenum Press (1972). https://doi.org/10.1007/978-1-4684-2001-2_9
22. Kleinberg, J., Tardos, É.: Algorithm Design. Addison-Wesley Longman Publishing Co., Inc. (2005)
23. Komm, D.: An Introduction to Online Computation: Determinism, Randomization, Advice. Springer (2016). https://doi.org/10.1007/978-3-319-42749-2
24. Mahajan, M., Raman, V.: Parameterizing above guaranteed values: maxsat and maxcut. J. Algorithms **31**(2), 335–354 (1999). https://doi.org/10.1006/jagm.1998.0996
25. Mikkelsen, J.W.: Randomization can be as helpful as a glimpse of the future in online computation (2015). arXiv:1511.05886
26. Mikkelsen, J.W.: Randomization can be as helpful as a glimpse of the future in online computation. In: 43rd International Colloquium on Automata, Languages, and Programming (ICALP). Leibniz International Proceedings in Informatics (LIPIcs), vol. 55, pp. 39:1–39:14. Schloss Dagstuhl – Leibniz-Zentrum für Informatik (2016). https://doi.org/10.4230/LIPIcs.ICALP.2016.39
27. Papadimitriou, C.H., Yannakakis, M.: Optimization, approximation, and complexity classes. J. Comput. Syst. Sci. **43**(3), 425–440 (1991). https://doi.org/10.1016/0022-0000(91)90023-X
28. Savage, C.: Depth-first search and the vertex cover problem. Inf. Process. Lett. **14**(5), 233–235 (1982). https://doi.org/10.1016/0020-0190(82)90022-9
29. Sleator, D.D., Tarjan, R.E.: Amortized efficiency of list update and paging rules. Commun. ACM **28**(2), 202–208 (1985). https://doi.org/10.1145/2786.2793

Scheduling with Testing: Competitive Algorithms for Minimizing the Total Weighted Completion Time in the Adversarial Model

Felix Buld[(✉)] and Andreas S. Schulz

Operations Research, School of Management and School of Computation, Information and Technology, TU Munich, Arcisstr. 21, 80333 Munich, Germany
{felix.buld,andreas.s.schulz}@tum.de

Abstract. We establish the first theoretical results for scheduling with testing on a single machine and on identical parallel machines to minimize the total *weighted* completion time in the adversarial model. We present a deterministic algorithm with a competitive ratio of 2.3166 for single-machine scheduling and show that a randomized variant has a competitive ratio of 2.1523. These algorithms, combined with list scheduling, yield competitive ratios of 2.7763 and 2.5110 for identical parallel machine scheduling.

Keywords: Testing · Explorable Uncertainty · Online Algorithms

1 Introduction

An emerging research branch in scheduling and beyond considers *explorable uncertainty*, where uncertainty in data can be reduced through actions such as conducting tests. However, these actions typically come at a cost, such as money or time. The problem considered here, *scheduling with testing*, belongs to this category. We are given jobs that must be assigned to machines, servers, or personnel over time. The processing times of jobs are initially unknown but can be determined through additional time-consuming individual tests. Therefore, one has to balance the benefit of additional certainty against the cost incurred from exploration. This framework is motivated by various real-world applications, see, e.g., [1,7,12] for comprehensive overviews. One prominent example is *medical care*, where patients are tested to determine their needed treatment. For instance, [20] considers emergency situations following mass casualty events, where limited medical personnel handle both the testing and the treatment. Another example is *fault diagnosis* in *maintenance*, where tests are conducted to explore the duration of repairs. In a further application, *code optimizers* can be used in advance to potentially reduce the runtime of computer programs.

1.1 Problem Description

We examine the following adversarial scheduling with testing problem that is a mixture of both online and robust optimization frameworks and was introduced in [7]. We are given n jobs to be scheduled on a single machine. Each job j has a weight $w_j > 0$ expressing its relative importance, which is known in advance. There are two possible processing options per job. The first is to execute the job untested in a safe mode, requiring u_j time. The second option is to test a job, taking $t_j > 0$ time, which reveals the actual processing time $p_j \in [0, u_j]$, which is initially unknown. The job can then be executed taking p_j time at any later point. The completion time of a job j in a schedule is denoted as C_j. The objective is to minimize the total weighted completion time, $\sum_j w_j C_j$. It is important to note that at any point in time the machine can test or process at most one job at a time, i.e., testing requires the same capacity as processing.

The classic counterpart–without testing–is the scheduling problem known as $1 \mid \mid \sum w_j C_j$, which considers n jobs with known weights w_j and processing times p_j. This problem is optimally solved by *Smith's rule*, aka the *Weighted Shortest Processing Time First* (WSPT) rule [19]. That is, in an optimal sequence jobs appear in order of non-decreasing ratios p_j/w_j. If the processing times p_j were fully known in the setting with testing, an optimal solution would therefore be obtained by simply testing all jobs with $t_j + p_j \leq u_j$ and processing the jobs in non-decreasing order of $1/w_j \cdot \min\{t_j + p_j, u_j\}$, ensuring that the execution of a tested job directly follows its test. However, in the model considered here, a job's processing time p_j is only revealed after testing. The goal is to devise a (randomized) γ-competitive algorithm whose outcome–for any problem instance–has an (expected) objective value at most γ times the objective value of an optimal solution that would be achievable with full information.

1.2 Related Work

The problem with $w_j = 1$ for all jobs j has been studied under three different settings: preemptive, test-preemptive, and non-preemptive, as discussed in [1]. Accordingly, the (testing and) processing of a job may be interrupted at any point in time and resumed later, or only between testing and processing, or not at all. The most commonly considered setting is the test-preemptive one that we will always consider, unless stated otherwise. This work introduces the first competitive algorithms for this scheduling problem with arbitrarily given weights $w_j > 0$.

Table 1. Research on scheduling with testing to minimize the total (weighted) completion time in the adversarial model.

Model	Objective	References	Testing Time	Hidden Information
Adversarial	$\sum_j C_j$	[7]	Uniform	$p_j \in [0, u_j]$
		[1,9,14,15]	Job-dependent	
	$\sum_j w_j C_j$	[21]	Uniform	$p_j \in [0, u]$
		This work	Job-dependent	$p_j \in [0, u_j]$

Scheduling with testing has so far also been explored in [2–6,10,12,13,20], albeit in various settings. Table 1 provides an overview of research on the model from Subsect. 1.1. Early research on the single-machine variant focused on uniform testing times [7], but this work builds on the more general framework including general job-dependent testing times, as in [1,9,14,15]. Table 2 summarizes the achieved bounds for this model on a single machine, where the *Golden Ratio* is denoted by $\varphi := (1+\sqrt{5})/2 \approx 1.618$. Objectives studied include the makespan and the total (weighted) completion time.

Table 2. Bounds for adversarial single-machine scheduling with testing.

Objective	Type of Algorithm	Lower Bound for $t_j = 1$	Upper Bound for $t_j = 1$	Upper Bound for t_j arbitrary
Makespan	Deterministic	φ [7]	φ [7]	φ [1]
	Randomized	4/3 [7]	4/3 [7]	4/3 [1]
Total Completion Time	Deterministic	1.8546 [7]	2 [7]	2.3166 [14]
	Randomized	1.6257 [7]	1.7453 [7]	2.1523 [14]

Minimizing the total completion time on identical parallel machines has been considered in [9,15]. The currently best-known upper bounds for the unweighted case are 3.2361 for a deterministic algorithm and 2.8307 for a randomized algorithm, established in [9]. Recently, [15] presented ideas for an improved upper bound of 2.7763 for a deterministic algorithm. The work of [21] made the first attempt to address minimizing the total *weighted* completion time on a single machine in the adversarial model. However, all outlined strategies have unbounded competitive ratios. Our main contributions in this paper are:

A deterministic 2.3166-competitive and a randomized 2.1523-competitive algorithm for minimizing $\sum_j w_j C_j$ on a single machine.

A deterministic 2.7763-competitive and a randomized 2.5110-competitive algorithm for minimizing $\sum_j w_j C_j$ on identical parallel machines.

1.3 Paper Outline

This paper initiates research on scheduling with testing in the adversarial model and under the total *weighted* completion time objective. In Sect. 2, we demonstrate how the GOLDEN ROUND ROBIN RULE from [1] can be naturally extended to account for job-dependent weights. We show that its test-preemptive version corresponds to the algorithm of [14] for a certain choice of parameters. In Sect. 3 we extend the best-known deterministic and randomized versions of [14] to incorporate job-dependent weights. With this approach, the currently best-known competitive ratios are preserved in the weighted setting. In Sect. 4, we extend the single-machine algorithms to identical parallel machines. The resulting deterministic performance guarantee matches that of the best-known deterministic algorithm for uniform weights. The competitive ratio of the randomized algorithm even improves upon that of the best randomized algorithm known for the case of uniform weights.

2 From Preemptive to Test-Preemptive Scheduling

In scheduling with testing, two kinds of decisions are required: which jobs to test and in which order to schedule the tasks. The standard approach in the adversarial model has been to decouple those decisions. The decision of which jobs to test is solely based on thresholds, and the ordering decisions depend on the outcomes of tests. We follow this regime as well. In Subsect. 2.1, we (re)state the main results for testing and makespan minimization. In Subsect. 2.2 we present a *weighted* version of the (α, β)-PCP algorithm from [14]. In Subsect. 2.3, we introduce a *weighted* version of the GOLDEN ROUND ROBIN RULE from [1] and link it to the previous algorithm by making it test-preemptive.

2.1 Preliminaries: Testing a Single Job and Makespan Minimization

The performance of algorithms for scheduling with testing is analyzed using competitive analysis, i.e., by comparing them to optimal offline algorithms. The best possible algorithm for minimizing the makespan on a single machine boils down to making optimal testing decisions for single jobs. The basic rule is to test a job if and only if its ratio $r_j := u_j/t_j$ is at least a chosen threshold α fixed for all jobs in advance and then scheduling jobs feasibly and without interruption.

The optimal processing time of job j is given by $p_j^* := \min\{t_j + p_j, u_j\}$ and the algorithmic processing time is denoted by p_j^A, which is either $t_j + p_j$ or u_j, depending on whether the algorithm decides to test or not.

Lemma 1 ([1,7]). *If job j is tested, then $p_j^A \leq (1 + 1/r_j)\,p_j^* \leq (1 + 1/\alpha)\,p_j^*$. If j is not tested, then $p_j^A \leq r_j p_j^* \leq \alpha p_j^*$. Testing job j if and only if $r_j \geq \varphi$ leads to a φ-competitive algorithm for minimizing the makespan. This is best possible in the deterministic case. The best possible randomized algorithm tests a job j with probability $(r_j^2 - r_j)/(r_j^2 - r_j + 1)$, if $r_j > 1$, and achieves a competitive ratio of $4/3$.*

The following algorithms for minimizing the total weighted completion time contain testing decisions using thresholds or probability functions as in Lemma 1.

2.2 An Algorithm for Test-Preemptive Scheduling

In this subsection, we introduce the WEIGHTED (α, β)-PCP algorithm, which is a generalization of the (α, β)-PCP algorithm of [14] for the unweighted case, see Algorithm 1. The abbreviation PCP was introduced in [14] and stands for *Prioritizing-Certain-Processing* (times) (not to be confused with Probabilistically Checkable Proofs). First, the testing decision is made independently for each job according to a threshold α, as in Subsect. 2.1. Secondly, we keep a priority list, containing currently available tasks. These are sorted according to priority values, which are $\beta t_j/w_j$ for the testing part, u_j/w_j if a job is executed without testing, and $(t_j + p_j)/w_j$ for the execution of a job j that was tested. In each iteration, the task with the smallest priority value, proportionally weighted, is scheduled next. After completion of a testing task of a job, its processing time is revealed and a corresponding tested execution task is added. The intuition behind the priority values in the algorithm (for $\beta = 1$) is that the jobs are completed in WSPT-order w.r.t. the algorithmic processing times p_j^A.

Algorithm 1: WEIGHTED (α, β)-PCP

Input: Jobs $\mathcal{J}$, a threshold $\alpha \geq 1$, a delay-factor $\beta > 0$.
Output: A feasible test-preemptive schedule on a single machine.

Jobs to be tested: $\mathcal{J}_T := \{j \in \mathcal{J} \mid \alpha t_j \leq u_j\}$; Jobs left untested $\mathcal{J}_U := \mathcal{J} \setminus \mathcal{J}_T$;
$L :=$ Empty priority list with priority values σ_j;
Add tasks for jobs from $\mathcal{J}_T$ with $\sigma_j = \beta t_j/w_j$, from $\mathcal{J}_U$ with $\sigma_j = u_j/w_j$ to L;

while *there is a task in L* **do**
  Let $j_{\min}$ be the job index of a task in L with smallest σ_j;
  Process $j_{\min}$ and remove it from L;
  if *the task was a testing task* **then**
    Reinsert a task into L with $\sigma_{j_{\min}} \leftarrow (t_{j_{\min}} + p_{j_{\min}})/w_{j_{\min}}$.

Theorem 1. *The* WEIGHTED $(\varphi, 1)$-PCP *algorithm is 2φ-competitive.*

In Subsect. 2.3, we present a proof of Theorem 1 based on the analysis of the WEIGHTED GOLDEN ROUND ROBIN RULE for the related preemptive setting.

2.3 WEIGHTED GOLDEN ROUND ROBIN for Preemptive Scheduling

Consider the GOLDEN ROUND ROBIN RULE from [1]. It tests job j if its ratio r_j is at least $\alpha = \varphi$ and performs the usual ROUND ROBIN RULE [17], but on the algorithmic processing times, cycling through all unfinished tasks and processing each for an equal amount of time until all are completed. Using the WEIGHTED ROUND ROBIN RULE (WR^3) [11] on the algorithmic processing times as a subroutine, in which all jobs are processed simultaneously in proportion to their weights, leads to the WEIGHTED GOLDEN ROUND ROBIN RULE as described in Algorithm 2.

It was already pointed out in [5,7,9,14] that algorithms generating preemptive schedules can be made test-preemptive without increasing the objective value. We explicitly make Algorithm 2 test-preemptive to connect the research of [1] and [14] and to show Theorem 1. For the analysis of Algorithms 1 and 2, we define the following relations. For known p_j, it is optimal to process the jobs in order of non-decreasing p_j^*/w_j. Consider a fixed optimal order. We write $k <_o j$ if job k appears in this order before job j. In addition, we fix the order of jobs in which Algorithm 1 (with $\alpha = \varphi$ and $\beta = 1$) completes them. For this order, we write $k <_A j$, if job k finishes before job j. Note that the jobs are completed in order of non-decreasing ratios p_j^A/w_j.

Algorithm 2: WEIGHTED GOLDEN ROUND ROBIN RULE (WGR³)
Input: Jobs $\mathcal{J}$.
Output: A feasible preemptive schedule on a single machine.
Jobs to be tested: $\mathcal{J}_T := \{j \in \mathcal{J} \mid \varphi t_j \leq u_j\}$; Jobs left untested $\mathcal{J}_U := \mathcal{J} \setminus \mathcal{J}_T$;
$L :=$ Set of tasks (weight w_j and time t_j for job $j \in \mathcal{J}_T$ and w_j, u_j for $j \in \mathcal{J}_U$).
while *there is a task in L* **do**
  Process the task of job k at rate $w_k/\sum_{j \in L} w_j$ for each $k \in L$ until a task finishes; Let $j_{\min}$ be its job index; Remove it from L;
  if *the task of job $j_{\min}$ was a testing task* **then**
  | Reinsert a task into L with weight $w_{j_{\min}}$ and processing time $p_{j_{\min}}$.

Lemma 2 (cf. Full version, [11,18]). *Let $C_j^{WGR^3}$ be the completion time of job j in Algorithm 2. Consider two jobs $j \neq k$. We have that $C_j^{WGR^3} \leq C_k^{WGR^3}$ if and only if $p_j^A/w_j \leq p_k^A/w_k$. If j is completed and k is not completed yet, the remaining processing time of k at time $C_j^{WGR^3}$ is $p_k^A - w_k/w_j \cdot p_j^A \geq 0$.*

(Partly) processing a job k before another job j is finished contributes to the completion time of the latter and, therefore, to the total cost $\sum_j w_j C_j$. We define $c(k, j)$ as the *contribution* of job k to the completion time of job j, i.e., the amount of time spent on k before j is finished. The total cost therefore can also be expressed as $\sum_j \sum_k w_j \cdot c(k, j)$. We denote the optimal value (i.e., full information) by OPT.

Lemma 3. *Algorithm 2 is 2φ-competitive. This bound is tight.*

Proof. Consider the (preemptive) schedule constructed by Algorithm 2. As in [1], but generalized to job-dependent weights, we obtain for a job j:

$$\sum_k w_j \cdot c(k,j) = \sum_{k \leq_A j} w_j \cdot p_k^A + \sum_{k >_A j} w_j \cdot \frac{w_k}{w_j} \cdot p_j^A = \sum_{k \leq_A j} w_j \cdot p_k^A + \sum_{k >_A j} w_k \cdot p_j^A .$$

Summing up the contributions, the total cost of Algorithm 2 can be bounded by

$$\sum_{j=1}^n w_j C_j^{WGR^3} = \sum_j \sum_k w_j \cdot c(k,j) \leq 2 \cdot \sum_j \sum_{k \leq_A j} w_j \cdot p_k^A$$
$$\leq 2 \cdot \sum_j \sum_{k \leq_o j} w_j \cdot p_k^A \leq 2\varphi \cdot \sum_j \sum_{k \leq_o j} w_j \cdot p_k^* = 2\varphi \cdot OPT .$$

The first inequality is obtained by rearranging the terms as for the proof of 2-competitiveness of the WR³, cf. the full version or [11,18]. Because the order is w.r.t. p_j^A, a standard exchange argument as in [19] shows the second inequality. Finally, the transition from p_j^A to p_j^* yields another factor of φ according to Lemma 1. Tightness of the bound is achieved by the example given in [1]. □

We show now that the WEIGHTED $(\varphi, 1)$-PCP algorithm (i.e., Algorithm 1) is the test-preemptive version of WGR³ (i.e., Algorithm 2).

Proof (of Theorem 1). Using Lemma 2, we obtain that both Algorithm 1 with $\alpha = \varphi, \beta = 1$ and Algorithm 2 complete tasks in the same order. Especially, jobs complete in order of non-decreasing ratios p_j^A/w_j in both algorithms. With C_j^A denoting the completion time of job j in the first algorithm, we have:

$$C_j^A \leq \sum_{k \leq_A j} p_k^A + \sum_{\substack{k >_A j \\ k \text{ is tested} \\ w_j \cdot t_k \leq w_k \cdot p_j^A}} t_k \leq \sum_{k \leq_A j} p_k^A + \sum_{k >_A j} \frac{w_k}{w_j} \cdot p_j^A = C_j^{\text{WGR}^3}. \quad (1)$$

Multiplying by w_j, summing up over j and using Lemma 3 we can conclude that Algorithm 1 with $\alpha = \varphi$ and $\beta = 1$ is also 2φ-competitive. □

No tight example for this bound of the WEIGHTED $(\varphi, 1)$-PCP algorithm is known. The example that shows the tightness of the bound for Algorithm 2 only gives a lower bound of $2\varphi - 1$. The competitive ratio of 2φ can also be established differently following the analysis of [9] or of [14], see Theorem 2, as well.

3 Delaying Testing and a Refined Analysis

In Subsect. 3.1, we extend the analysis of [14] for the (α, β)-PCP algorithm to incorporate weights (cf. Algorithm 1) and give a lower bound of 2 for any choice of parameters. In Subsect. 3.2, we show how the results from [14] for the best-known randomized algorithm generalize to the weighted setting.

3.1 Deterministic Algorithms for $\sum_j w_j C_j$

Table 3 gives an overview over the following results. The algorithms (α, β)-SORT from [1] and (α, β)-PCP from [14] are generalized to the broader setting of arbitrary job weights by scaling the priority values. Note that the only difference between those two algorithms is the priority value of a job that is reinserted. The lower bound of 2 for the first one stems from [1], while the proof for the second one is new. The upper bounds on the competitive ratios are shown by [14] for the unweighted case. We prove that they also hold for the more general case.

Since we can rewrite the objective function as

$$\sum_{j=1}^n w_j \cdot C_j^A = \sum_{j=1}^n \left(w_j \cdot c(j,j) + \left(\sum_{k <_o j} (w_j \cdot c(k,j) + w_k \cdot c(j,k)) \right) \right),$$

we can apply and generalize the idea due to [14] to bound $w_j \cdot c(k,j) + w_k \cdot c(j,k)$ against its value in an optimal solution. Lemma 4 generalizes the analysis of [14] for (α, β)-PCP to job-dependent weights. Their results for (α, β)-SORT generalize analogously to job-dependent weights, cf. the full version of this paper.

Table 3. Overview of results for the competitive ratios of the algorithms.

Algorithm	Testing decision	Priority values	Lower bound	Upper bound
WEIGHTED (α, β)-SORT	Test job iff $\alpha t_j \leq u_j$	$\frac{1}{w_j}\{\beta t_j, u_j, p_j\}$	2 [1]	$1 + \sqrt{2} \leq 2.4143$ for $(\sqrt{2}, \sqrt{2})$
WEIGHTED (α, β)-PCP	Test job iff $\alpha t_j \leq u_j$	$\frac{1}{w_j}\{\beta t_j, u_j, t_j + p_j\}$	2	2.3166 for $(\varphi, 2.3166)$

Lemma 4. *Consider jobs $j \neq k$, and parameters $\alpha \geq 1$ and $\beta > 0$. If the WEIGHTED (α, β)-PCP algorithm does not test job k, it holds*

$$w_j \cdot c(k,j) + w_k \cdot c(j,k) \leq w_j \cdot \left(1 + \frac{1}{\beta}\right) u_k \leq w_j \cdot \alpha \cdot \left(1 + \frac{1}{\beta}\right) \cdot p_k^* .$$

If the WEIGHTED (α, β)-PCP algorithm does test job k, it holds

$$w_j \cdot c(k,j) + w_k \cdot c(j,k) \leq w_j \cdot \max\left\{\beta \cdot t_k, 2t_k + p_k, (t_k + p_k)\left(1 + \frac{1}{\beta}\right)\right\}$$

$$\leq w_j \cdot \max\left\{\beta, 2, 1 + \frac{2}{\alpha}, \left(1 + \frac{1}{\alpha}\right)\left(1 + \frac{1}{\beta}\right)\right\} \cdot p_k^* .$$

Proof. The proof naturally extends the proof of [14] to job-dependent weights. We consider a comparison tree based on the actions of the algorithm. Similar reasoning has also been used in [1,5]. Assume that $k <_o j$ for a given pair of jobs k, j. In the optimal solution, we have mutual contributions of $c^*(k,j) = p_k^*$ and $c^*(j,k) = 0$. In each case, we find an upper bound on the ratio between the weighted sum of mutual contributions $w_j \cdot c(k,j) + w_k \cdot c(j,k)$ in the algorithm and $w_j \cdot p_k^*$ in the optimum, depending on α and β. We extend the notation from [5] and denote the testing task by τ_j, the execution task by π_j, and the untested execution task of job j by v_j. An ordered list, such as $[v_k, \tau_j, \pi_j]$ describes the testing and ordering decisions of the algorithm for jobs k, j compactly.

First, we consider the cases in which the algorithm does not test job k, see Table 4. Overall, it follows that $w_j \cdot c(k,j) + w_k \cdot c(j,k) \leq w_j \cdot (1 + 1/\beta) u_k$ in this case. Bounding $u_k \leq \alpha p_k^*$ by Lemma 1 yields the stated bound.

Secondly, we consider the cases when the algorithm tests job k, see Table 5. Overall, we obtain $w_j \cdot c(k,j) + w_k \cdot c(j,k) \leq w_j \cdot \max\{\beta \cdot t_k, 2t_k + p_k, (1 + 1/\beta)(t_k + p_k)\}$ in these cases, as stated above. We further bound the expressions with respect to the optimal running time p_k^*. First, we have $\beta t_k \leq \beta p_k^*$, since $p_k^* = \min\{t_k + p_k, u_k\}$ and $t_k \leq u_k$. Secondly, if $p_k^* = t_k + p_k$, we have $2t_k + p_k \leq 2t_k + 2p_k = 2p_k^*$. If, however, $p_k^* = u_k$, we have $2t_k + p_k = t_k + t_k + p_k \leq p_k^*/\alpha + (1 + 1/\alpha) \cdot p_k^* = (1 + 2/\alpha) p_k^*$ by Lemma 1 and $t_k \leq u_k/\alpha$ since k was tested. Lastly, $(1 + 1/\beta)(t_k + p_k) \leq (1 + 1/\beta) \cdot (1 + 1/\alpha) p_k^*$ by Lemma 1. These calculations prove the second inequality in the statement of the Lemma. □

Table 4. Case $p_k^A = u_k$

	Order of Tasks	$w_j \cdot c(k,j) + w_k \cdot c(j,k)$		Used Inequality
$p_j^A = u_j$	$[v_k, v_j]$	$w_j \cdot u_k$		
	$[v_j, v_k]$	$w_k \cdot u_j \leq w_j \cdot u_k$		$\frac{u_j}{w_j} \leq \frac{u_k}{w_k}$
$p_j^A = t_j + p_j$	$[v_k, \tau_j, \pi_j]$	$w_j \cdot u_k$		
	$[\tau_j, v_k, \pi_j]$	$w_j \cdot u_k + w_k \cdot t_j \leq \left(1 + \frac{1}{\beta}\right) w_j \cdot u_k$		$\frac{\beta \cdot t_j}{w_j} \leq \frac{u_k}{w_k}$
	$[\tau_j, \pi_j, v_k]$	$w_k \cdot (t_j + p_j) \leq w_j \cdot u_k$		$\frac{t_j + p_j}{w_j} \leq \frac{u_k}{w_k}$

Table 5. Case $p_k^A = t_k + p_k$

	Order of Tasks	$w_j \cdot c(k,j) + w_k \cdot c(j,k)$	Used Inequality
$p_j^A = u_j$	$[\tau_k, \pi_k, v_j]$	$w_j \cdot (t_k + p_k)$	
	$[\tau_k, v_j, \pi_k]$	$w_j \cdot t_k + w_k \cdot u_j \leq w_j \cdot (2t_k + p_k)$	$\frac{u_j}{w_j} \leq \frac{t_k + p_k}{w_k}$
	$[v_j, \tau_k, \pi_k]$	$w_k \cdot u_j \leq \beta \cdot w_j \cdot t_k$	$\frac{u_j}{w_j} \leq \frac{\beta \cdot t_k}{w_k}$
$p_j^A = t_j + p_j$	$[\tau_k, \pi_k, \tau_j, \pi_j]$	$w_j \cdot (t_k + p_k)$	
	$[\tau_k, \tau_j, \pi_k, \pi_j]$	$w_j \cdot (t_k + p_k) + w_k \cdot t_j \leq w_j \cdot \left(1 + \frac{1}{\beta}\right)(t_k + p_k)$	$\frac{\beta \cdot t_j}{w_j} \leq \frac{t_k + p_k}{w_k}$
	$[\tau_k, \tau_j, \pi_j, \pi_k]$	$w_j \cdot t_k + w_k \cdot (t_j + p_j) \leq w_j \cdot (2t_k + p_k)$	$\frac{t_j + p_j}{w_j} \leq \frac{t_k + p_k}{w_k}$
	$[\tau_j, \tau_k, \pi_k, \pi_j]$	$w_j \cdot (t_k + p_k) + w_k \cdot t_j \leq w_j \cdot (2t_k + p_k)$	$\frac{\beta \cdot t_j}{w_j} \leq \frac{\beta \cdot t_k}{w_k}$
	$[\tau_j, \tau_k, \pi_j, \pi_k]$	$w_j \cdot t_k + w_k \cdot (t_j + p_j) \leq w_j \cdot (2t_k + p_k)$	$\frac{t_j + p_j}{w_j} \leq \frac{t_k + p_k}{w_k}$
	$[\tau_j, \pi_j, \tau_k, \pi_k]$	$w_k \cdot (t_j + p_j) \leq \beta \cdot w_j \cdot t_k$	$\frac{t_j + p_j}{w_j} \leq \frac{\beta \cdot t_k}{w_k}$

Intuitively, two different trade-offs in terms of uncertainty are incorporated in the bounds of Lemma 4. The parameter α balances the risks of testing jobs that turn out to be long and not testing jobs that are actually short, cf. Lemma 1. The parameter β balances the risk of delaying many jobs when one tests jobs that turn out to be long early and delaying tests of actually short jobs by too much. The bounds from Lemma 4 will now be applied to derive an upper bound on the competitive ratio of the WEIGHTED (α, β)-PCP as in [14].

Theorem 2. *The competitive ratio of the* WEIGHTED (α, β)-PCP *algorithm with $\alpha \geq 1$ and $\beta > 0$ is at most*

$$Q(\alpha, \beta) := \max\left\{\alpha\left(1 + \frac{1}{\beta}\right), \beta, 2, 1 + \frac{2}{\alpha}, \left(1 + \frac{1}{\alpha}\right)\left(1 + \frac{1}{\beta}\right)\right\}.$$

The optimal choice of the parameters to minimize this term is $\alpha = \varphi$ and $\beta = \left(\varphi + \sqrt{\varphi(\varphi + 4)}\right)/2$, resulting in a competitive ratio of at most $\beta \leq 2.3166$.

Proof.

$$\sum_{j=1}^{n} w_j \cdot C_j^A = \sum_{j=1}^{n} \left(w_j \cdot c(j,j) + \sum_{k <_o j} \left(w_j \cdot c(k,j) + w_k \cdot c(j,k) \right) \right)$$

$$\leq \sum_{j=1}^{n} \left(w_j \cdot p_j^A + \sum_{k <_o j} \left(w_j \cdot Q(\alpha, \beta) \cdot p_k^* \right) \right) \quad (2)$$

$$\leq Q(\alpha, \beta) \cdot \sum_{j=1}^{n} \sum_{k \leq_o j} w_j \cdot p_k^* = Q(\alpha, \beta) \cdot OPT$$

The first inequality follows from Lemma 4, the second one holds, since $p_j^A \leq \max\{\alpha, 1 + 1/\alpha\} \cdot p_j^* \leq Q(\alpha, \beta) \cdot p_j^*$. Minimizing this upper bound over α and β yields $\alpha = \varphi$ and $\beta = \left(\varphi + \sqrt{\varphi(\varphi+4)}\right)/2$ as optimal values, cf. [14]. □

Lemma 5 (cf. [1] and Full Version). *Both algorithms* WEIGHTED (α, β)-SORT *and* WEIGHTED (α, β)-PCP *cannot be better than 2-competitive for any combination of $\alpha \geq 1$ and $\beta \geq 0$ for scheduling with testing on a single machine to minimize the weighted total completion time.*

For the choice of parameters α and β as in Theorem 2, a lower bound of $2\varphi - 1 > 2.2360$ as for $(\varphi, 1)$-PCP can be constructed, cf. the full version of this paper.

3.2 A Randomized Algorithm for $\sum_j w_j C_j$

An idea previously used when designing randomized algorithms for scheduling with testing, is to randomize the testing decisions in the beginning and then use a deterministic algorithm for the choice of the next task to be scheduled, see [1,9,14]. One typically uses a non-decreasing function $f : \mathbb{R}_{>0} \to [0,1]$ that maps the ratio $r_j = u_j/t_j$ to a probability $f(r_j)$ for testing job j. The randomized algorithm from [14] for the unweighted case uses the following function:

$$f(x) := \begin{cases} 0, & x < 1, \\ \frac{x^2 - x}{x^2 - \frac{4}{3}x + 1}, & 1 \leq x \leq 3, \\ 1, & x \geq 3. \end{cases} \quad (3)$$

They give a worst-case analysis that provides such a function based on a general β and set $\beta = 2$ to achieve the best possible bound within their analysis. For better readability, we only use the optimized version here to show how it can be generalized to the weighted case. We define the algorithm WEIGHTED (f, β)-PCP, as the one that tests job j with probability $f(r_j)$ and then always schedules the task with minimal priority value (either $1/w_j \cdot \beta \cdot t_j$ or $1/w_j \cdot u_j$ or $1/w_j \cdot (t_j + p_j)$ depending on the task's type) among the available tasks.

Theorem 3. *The randomized algorithm* WEIGHTED $(f, 2)$-PCP *achieves a competitive ratio of at most $3(7+3\sqrt{6})/20 \leq 2.1523$.*

Proof. Using the randomized algorithm WEIGHTED $(f,2)$-PCP and following the proof ideas and structure for the unweighted case [14], the expected weighted total completion time can be bounded by

$$E\left[\sum_j w_j \cdot C_j^A\right] = \sum_j \left(E\left[w_j \cdot c(j,j)\right] + E\left[\sum_{k<_o j} w_j \cdot c(k,j) + w_k \cdot c(j,k)\right]\right)$$

$$\leq \sum_j w_j \cdot \sum_{k \leq_o j} \underbrace{3/2 \cdot u_k \left(1 - f(r_k)\right) + \max\left\{2t_k + p_k, 3/2\left(t_k + p_k\right)\right\} f(r_k)}_{=:h(r_k)},$$

where the inequality follows from Lemma 4 for $\beta = 2$ for the second expectation. The first expectation $E[w_j \cdot c(j,j)]$ is clearly also bounded by $w_j \cdot h(r_j)$. Overall, $h(r_k)$ can be bounded based on a case distinction on r_k and on $p_k^* \in \{u_k, t_k + p_k\}$ as in [14]. This yields $h(r_k) \leq 3(7+3\sqrt{6})/20 \cdot p_k^* \leq 2.1523 \cdot p_k^*$. □

4 Parallel Machines

Combining ideas from single-machine algorithms with list scheduling yields algorithms for scheduling on identical parallel machines. One essentially keeps the same task order as in the single-machine case (i.e., that of the WEIGHTED (α, β)-PCP algorithm) and schedules the next available task once a machine idles, albeit respecting that a tested execution task is only available after the completion of a testing task of the same job, similar as in [9,15] for minimizing the total completion time. It is important to note that a tested execution task is placed at the relative position in the priority list it would have had in the single-machine case and is not simply reinserted with its priority value. The analysis is similar to proving a performance ratio of $3/2$ for WSPT on identical parallel machines in the offline setting based on the lower bounds from [8], cf. [16]. For the setting considered here, we obtain a deterministic algorithm with a competitive ratio of 2.7763, replicating the performance guarantees in [15] where the unweighted case was studied. Furthermore, we obtain a randomized 2.5110-competitive algorithm beating the so far best-known bound for randomized algorithms on parallel machines of 2.8307 for the unweighted case from [9].

Lemma 6 (cf. Full Version). *The completion time C_j of a job j in the schedule constructed by the list scheduling approach can be bounded by:*

$$C_j \leq \frac{1}{m} \sum_{k \neq j} c(k,j) + p_j^A, \tag{4}$$

where $c(k,j)$ is the contribution of job k to the completion time of job j in the single-machine schedule constructed by the WEIGHTED (α, β)-PCP algorithm.

The idea is that list scheduling puts a task on the least loaded machine or a tested execution task immediately after the completion of its testing task. If another task appearing later in the single-machine order than the execution task of job j is scheduled on another machine, while its testing task is ongoing, one can show that the execution of j directly follows its testing. Then, the bound holds since one only needs to consider the contributions of tasks before the testing of j.

Since we do not have the concrete structure of the optimal solution to the underlying offline problem $P||\sum w_j C_j$ at hand, we use the following lower bound of [8] to derive an upper bound on the competitive ratio:

$$OPT_m \geq \max\left\{\sum_j w_j p_j^*, \frac{1}{m}OPT_1 + \left(\frac{1}{2} - \frac{1}{2m}\right)\sum_j w_j p_j^*\right\}, \quad (5)$$

where OPT_m is the optimal objective value on m machines and OPT_1 on a single machine. Equations (4) and (5) were similarly used for the unweighted case in [9,15] and in many other scheduling contexts before.

Lemma 7. WEIGHTED (α, β)-PCP *with parameters as in Theorem 2 combined with list scheduling yields a competitive ratio of at most* $2.7763 - 0.4597/m$.

Proof.

$$\sum_j w_j C_j \overset{(4)}{\leq} \frac{1}{m}\sum_j w_j \sum_{k \neq j} c(k, j) + \sum_j w_j p_j^A = \frac{1}{m}ALG_1 + \left(1 - \frac{1}{m}\right)\sum_j w_j p_j^A$$

$$\overset{(2),2.1}{\leq} \frac{1}{m}Q(\alpha, \beta)OPT_1 + \left(1 - \frac{1}{m}\right)\varphi \sum_j w_j p_j^*$$

$$= Q(\alpha, \beta)\left(\frac{1}{m}OPT_1 + \left(\frac{1}{2} - \frac{1}{2m}\right)\sum_j w_j p_j^*\right)$$

$$+ \left(\left(1 - \frac{1}{m}\right)\varphi - Q(\alpha, \beta)\left(\frac{1}{2} - \frac{1}{2m}\right)\right)\sum_j w_j p_j^*$$

$$\overset{(5)}{\leq} OPT_m\left(Q(\alpha, \beta)\left(\frac{1}{2} + \frac{1}{2m}\right) + \left(1 - \frac{1}{m}\right)\varphi\right)$$

For $m = 1$, we obtain the single-machine bound $Q(\alpha, \beta)$. The bound is monotonically increasing in m, and converges to $\frac{1}{2}Q(\alpha, \beta) + \varphi \leq 2.7763$ for $m \to \infty$. □

Following essentially the same steps, we can use this analysis to beat the best-known bound for randomized algorithms on parallel machines of 2.8307 from [9] for the unweighted case, even for job-dependent weights.

Theorem 4. WEIGHTED $(f, 2)$-PCP *combined with list scheduling has a competitive ratio of at most* $(7 - 1/m)(7+3\sqrt{6})/40 \leq 2.5110 - 0.3587/m$.

Proof. We follow the proof of Lemma 7, but replace $\sum_j w_j C_j$ by $E\left[\sum_j w_j C_j\right]$ and p_j^A by $E\left[p_j^A\right]$ and make use of the linearity of expectation. Using the upper bound for a single machine $3(7+3\sqrt{6})/20$ from Theorem 3 and calculating $E\left[p_j^A\right] \leq p_j^* \cdot (7+3\sqrt{6})/10$ for f as above similar to the proof of Theorem 3, we obtain $E[\sum_j w_j C_j] \leq OPT_m \frac{(7+3\sqrt{6})}{10}(\frac{3}{2}(\frac{1}{2} + \frac{1}{2m}) + (1 - \frac{1}{m})) \leq OPT_m \frac{(7+3\sqrt{6})}{40}(7 - \frac{1}{m})$. □

5 Conclusion

In this paper, we give the first algorithms with constant competitive ratios for scheduling with testing on a single machine to minimize the total *weighted* completion time objective. These algorithms are natural extensions of algorithms that had earlier been proposed for the unweighted case [14]. We present upper bounds of 2.3166 and 2.1523 on a single machine and of 2.7763 and 2.511 on identical parallel machines for the competitive ratio for a deterministic and a randomized algorithm, respectively.

Acknowledgments. Felix Buld was funded by the Deutsche Forschungsgemeinschaft (DFG, German Research Foundation) – GRK 2201/2 – Projektnummer 277991500. The authors are grateful to Simon Gmeiner for suggesting WGR^3. In addition, the authors thank the anonymous reviewers of IJTCS-FAW 2025 for their valuable feedback and suggestions.

Disclosure of Interests. The authors have no competing interests to declare that are relevant to the content of this article.

References

1. Albers, S., Eckl, A.: Explorable uncertainty in scheduling with non-uniform testing times. In: Kaklamanis, C., Levin, A. (eds.) WAOA 2020. LNCS, vol. 12806, pp. 127–142. Springer, Cham (2021). https://doi.org/10.1007/978-3-030-80879-2_9
2. Albers, S., Eckl, A.: Scheduling with testing on multiple identical parallel machines. In: Lubiw, A., Salavatipour, M., He, M. (eds.) WADS 2021. LNCS, vol. 12808, pp. 29–42. Springer, Cham (2021). https://doi.org/10.1007/978-3-030-83508-8_3
3. Bampis, E., Dogeas, K., Kononov, A., Lucarelli, G., Pascual, F.: Speed scaling with explorable uncertainty. In: Proceedings of the 33rd ACM Symposium on Parallelism in Algorithms and Architectures (SPAA 2021), pp. 83–93. Association for Computing Machinery, New York (2021). https://doi.org/10.1145/3409964.3461812
4. Damerius, C., Kling, P., Li, M., Xu, C., Zhang, R.: Scheduling with a limited testing budget: tight results for the offline and oblivious settings. In: Gørtz, I.L., Farach-Colton, M., Puglisi, S.J., Herman, G. (eds.) 31st Annual European Symposium on Algorithms (ESA 2023). Leibniz International Proceedings in Informatics (LIPIcs), vol. 274, pp. 38:1–38:15. Schloss Dagstuhl – Leibniz-Zentrum für Informatik, Dagstuhl, Germany (2023). https://doi.org/10.4230/LIPIcs.ESA.2023.38
5. Dogeas, K., Erlebach, T., Liang, Y.C.: Scheduling with obligatory tests. In: Chan, T., Fischer, J., Iacono, J., Herman, G. (eds.) 32nd Annual European Symposium on Algorithms (ESA 2024). Leibniz International Proceedings in Informatics (LIPIcs), vol. 308, pp. 48:1–48:14. Schloss Dagstuhl – Leibniz-Zentrum für Informatik, Dagstuhl, Germany (2024). https://doi.org/10.4230/LIPIcs.ESA.2024.48
6. Dufossé, F., Dürr, C., Nadal, N., Trystram, D., Vásquez, Ó.C.: Scheduling with a processing time oracle. Appl. Math. Model. **104**, 701–720 (2022). https://doi.org/10.1016/j.apm.2021.12.020
7. Dürr, C., Erlebach, T., Megow, N., Meißner, J.: An adversarial model for scheduling with testing. Algorithmica **82**(12), 3630–3675 (2020). https://doi.org/10.1007/s00453-020-00742-2

8. Eastman, W.L., Even, S., Isaacs, I.M.: Bounds for the optimal scheduling of n Jobs on m processors. Manage. Sci. **11**(2), 268–279 (1964). https://doi.org/10.1287/mnsc.11.2.268
9. Gong, M., Chen, Z.Z., Hayashi, K.: Approximation algorithms for multiprocessor scheduling with testing to minimize the total job completion time. Algorithmica **86**(5), 1400–1427 (2024). https://doi.org/10.1007/s00453-023-01198-w
10. Gong, M., Goebel, R., Lin, G., Miyano, E.: Improved approximation algorithms for non-preemptive multiprocessor scheduling with testing. J. Comb. Optim. **44**(1), 877–893 (2022). https://doi.org/10.1007/s10878-022-00865-y
11. Kim, J.H., Chwa, K.Y.: Non-clairvoyant scheduling for weighted flow time. Inf. Process. Lett. **87**(1), 31–37 (2003). https://doi.org/10.1016/S0020-0190(03)00231-X
12. Levi, R., Magnanti, T., Shaposhnik, Y.: Scheduling with testing. Manage. Sci. **65**(2), 776–793 (2019). https://doi.org/10.1287/mnsc.2017.2973
13. Levi, R., Magnanti, T., Shaposhnik, Y.: Scheduling with testing of heterogeneous jobs. Manage. Sci. **70**(5), 2934–2953 (2024). https://doi.org/10.1287/mnsc.2023.4833
14. Liu, A.H.H., Liu, F.H., Wong, P.W., Zhang, X.O.: The power of amortization on scheduling with explorable uncertainty. In: Byrka, J., Wiese, A. (eds.) WAOA 2023. LNCS, vol. 14297, pp. 90–103. Springer, Cham (2023). https://doi.org/10.1007/978-3-031-49815-2_7
15. Liu, A.H.H., Liu, F.H., Wong, P.W., Zhang, X.O.: Amortization helps for scheduling with explorable uncertainty, even on parallel machines. In: Proceedings of the 16th Workshop on Models and Algorithms for Planning and Scheduling Problems (MAPSP 2024), pp. 259–262 (2024)
16. Möhring, R.H., Schulz, A.S., Uetz, M.: Approximation in stochastic scheduling: the power of LP-based priority policies. J. ACM **46**(6), 924–942 (1999). https://doi.org/10.1145/331524.331530
17. Motwani, R., Phillips, S., Torng, E.: Nonclairvoyant scheduling. Theoret. Comput. Sci. **130**(1), 17–47 (1994). https://doi.org/10.1016/0304-3975(94)90151-1
18. Pinedo, M.L.: Scheduling: Theory, Algorithms, and Systems, 6th edn. Springer, Cham (2022). https://doi.org/10.1007/978-3-031-05921-6
19. Smith, W.E.: Various optimizers for single-stage production. Naval Res. Logist. Q. **3**(1–2), 59–66 (1956). https://doi.org/10.1002/nav.3800030106
20. Sun, Z., Argon, N.T., Ziya, S.: Patient triage and prioritization under austere conditions. Manage. Sci. **64**(10), 4471–4489 (2018). https://doi.org/10.1287/mnsc.2017.2855
21. Zhang, X.: Scheduling with explorable uncertainty for minimising the total weighted completion time. Master's thesis, Utrecht University (2023)

Mixed Graph Covering with Target Constraints

Xujin Chen[1,2], Xiyuan Deng[1,2(✉)], Xiaodong Hu[1,2], and Changjun Wang[1]

[1] SKLMS, Academy of Mathematics and Systems Science, Chinese Academy of Sciences, Beijing 100190, China
{xchen,dengxiyuan,xdhu,wcj}@amss.ac.cn
[2] School of Mathematical Sciences, University of Chinese Academy of Sciences, Beijing 100049, China

Abstract. The *target-constrained mixed graph covering* (TMGC) problem considers a graph where edges and vertices are each assigned a cost and a weight. The goal is to select a minimum-cost subset of vertices and edges subject to the covering-target constraint that its covered weight (i.e., the total weight of the selected vertices, selected edges, and edges incident to the selected vertices) meets or exceeds a given threshold. This problem models real-world scenarios, like optimizing the removal of facilities (vertices) and roads (edges) in a network while ensuring the value of the remaining network (including the value of remaining facilities and their connecting roads) remains below a specified limit. From a theoretical perspective, this TMGC model extends the weighted partial vertex cover problem in two significant ways: it incorporates covering weights for both edges and vertices, and it allows a direct selection of edges alongside vertices to satisfy the covering target. Despite this increased complexity and generality compared to (the partial version of) the classic vertex cover problem, we develop a 2-approximation primal-dual algorithm for TMGC, whose ratio 2 matches the known lower bound for the simpler vertex cover problem.

Keywords: Approximation algorithms · Primal-dual algorithms · Partial Vertex cover

1 Introduction

The *vertex cover* (VC) problem is a classic combinatorial optimization problem that has been extensively studied over the years [8,9]. Given a graph $G = (V, E)$ with n vertices and m edges, the problem is to find a smallest subset $S \subseteq V$ that covers all edges in E, where an edge is *covered* by S if at least one of its end vertices belongs to S. This problem is known to be NP-hard [5], Moreover, if the unique games conjecture (UGC) is true, no polynomial-time algorithm can approximate the minimum vertex cover better than a factor of $2-\epsilon$ for any $\epsilon > 0$ unless P = NP [6]. The *partial vertex cover* (PVC) problem, a natural variant of the classic VC problem, requires covering only at least a specified number of edges instead of all edges in the graph. This added flexibility introduces new challenges compared to the standard VC problem. The PVC problem arises in various practical scenarios, such as network design, data clustering, and power grid management, where partial coverage often suffices to meet operational requirements. Bshouty and Burroughs [2] were the first to propose a 2-approximation algorithm for PVC using LP-rounding. Hochbaum [4] presented a 2-approximation algorithm that runs in $O(nm \log \frac{n^2}{m} \log n)$ time based on Lagrangian relaxation. Subsequently, Bar-Yehuda [1] employed the local-ratio technique to achieve an $O(n^2)$ running time. Gandhi et al. [3] developed a primal-dual algorithm with a running time of $O(n(n \log n + m))$. Finally, building on the work of Gandhi et al. [3], Mestre [7] further improved the running time to $O(n \log n + m)$.

The PVC model can be generalized in two directions. The first is the problem of *partial cover with general weights*, where both vertices and edges have weights. The goal is to select a minimum number of vertices such that the total weight of the selected vertices and their incident edges meets or exceeds a given threshold. The second is the problem of *partial cover with fine selection*, where vertices and edges have costs, and the objective is to find a minimum-cost set of vertices and edges that covers at least a specified number of edges. In this generalization, a vertex covers all its incident edges, while an edge covers only itself. This model extends the PVC problem by allowing the selection of edges. While both generalizations represent theoretically motivated optimization problems with practical applications, to the best of our knowledge, they appear to be largely unexplored in the literature. Given their relationship to PVC, they inherit a lower bound of 2 on their approximation ratios. In this paper, we design a 2-approximation algorithm for it.

Interestingly, the above two generalizations from PVC can be unified into the framework of the *target-constrained mixed graph covering* (TMGC) problem. In TMGC, each edge and vertex has both a cost and a weight. The objective is to select a minimum-cost subset of vertices and edges such that the total weight of selected vertices, selected edges, and edges incident to selected vertices reaches a given threshold. The TMGC has a variety of practical applications, such as modeling network security scenarios. For example, consider a network of servers and fiber optics, where servers handle data processing and storage, and fiber optics serve as the high-speed communication medium connecting them. Both the servers and fiber optics have associated costs and values. Now, imagine a hacker aiming to attack this network by disabling some servers and fiber optics to reduce the system's total value below a certain threshold (equivalently, achieving a reduction of at least the given threshold). The hacker's optimization task is equivalent to solving the TMGC problem, where the goal is to minimize the attack costs while meeting the target reduction in system value.

Our Contribution. In this paper, we design a 2-approximation algorithm (Algorithm 1) for the TMGC problem that runs in a subquadratic time.

We begin by formulating the TMGC problem as an integer linear program (ILP), which serves as the foundation for our primal-dual algorithm. While some variables in our ILP formulation may appear redundant compared to the linear program for PVC in [7], they are essential for the success of the primal-dual approach. In the TMGC problem, both vertices and edges can be selected and contribute weights toward meeting the target threshold. This requires a more careful strategy for increasing the variables in the dual program and for selecting the appropriate vertices and edges. Our method achieves the best possible (under UGC) approximation ratio of 2 in $O((n+m)\log(n+m))$ time (Theorems 1 and 2).

Furthermore, we show that for the special case of TMGC where the selection cost of each edge is infinite (i.e., only vertices can be selected), the running time of our algorithm is reduced to $O(n \log n + m)$. This extends the main result of [7] for PVC to a more general setting of partial cover with general weights (Remark 3).

Paper Organization. The remainder of this paper is structured as follows. Section 2 formally introduces the TMGC model and presents an integer linear program formulation of the problem. Section 3 presents a primal-dual approximation algorithm for TMGC, along with the proofs of its 2-approximation ratio and subquadratic time complexity. Finally, Sect. 4 discusses key technical insights and future research directions.

2 Model

Given an undirected graph $G = (V, E)$ with vertex set V and edge set E, each *element* (vertex or edge) in $\mu \in V \cup E$ is associated with a nonnegative cost c_μ and a nonnegative weight w_μ. A vertex *covers* itself and its incident edges, while an edge *covers* only itself. For any subset $M \subseteq V \cup E$, its cost $c(M)$ and weight $w(M)$ are defined as the sum of the costs and weights of its elements, respectively. An element is *covered* by M if it is covered by some member of M. So, the set (of elements) covered by M consists of the elements in M, along with the edges incident to vertices in M. The weight covered by M equals the weight of the set it covers. Given a nonnegative target weight $D \in \mathbb{R}_+$, we call M a *mixed cover* (or simply a *cover*) if it satisfies the *target constraint* that the weight covered by M is at least D. The *target-constrained mixed graph covering* (TMGC) problem is to find a mixed cover M with the minimum cost.

To formulate the problem, we introduce two sets of binary variables: x_μ and p_μ for all $\mu \in V \cup E$, to indicate whether μ is selected as a member of the cover and whether μ is covered by the cover, respectively. Specifically, $x_\mu = 1$ if and only if μ is selected, and $p_\mu = 1$ if and only if μ is *not* covered. To facilitate linear program construction and analysis, we transform the target constraint equivalently: instead of requiring the covered weight to be at least the lower bound D, we require that the uncovered weight in total does not exceed the upper threshold $T = w(V \cup E) - D$. For convenience, we use $(G, c, w \,|\, T)$ or simply

$(G\,|\,T)$ to denote the instance of the TMGC problem. The TMGC problem can be written as follows.

$$\begin{aligned}
\min \quad & \sum_{\mu \in V \cup E} c_\mu x_\mu \\
\text{s.t.} \quad & p_v + x_v \geq 1 && \forall v \in V, \\
& p_e + x_u + x_v + x_e \geq 1 && \forall e = \{u,v\} \in E, \quad \text{(ILP)} \\
& \sum_{\mu \in V \cup E} w_\mu p_\mu \leq T, \\
& x_\mu, p_\mu \in \{0,1\} && \forall \mu \in V \cup E.
\end{aligned}$$

We relax the integer constraint to $x_\mu \geq 0$, $p_\mu \geq 0$ for $\mu \in V \cup E$. Then, the relaxation has the following program as its dual, where $\delta(v) = \{e \in E \,|\, e \cap v \neq \emptyset\}$ denotes the set edges incident with vertex $v \in V$,

$$\begin{aligned}
\max \quad & \sum_{\mu \in V \cup E} y_\mu - Tz \\
\text{s.t.} \quad & y_v + \sum_{e \in \delta(v)} y_e \leq c_v && \forall v \in V, \quad \text{(DP)} \\
& y_e \leq c_e && \forall e \in E, \\
& 0 \leq y_\mu \leq w_\mu z && \forall \mu \in V \cup E.
\end{aligned}$$

Although introducing variables p_v for each vertex $v \in V$ in the primal program may seem redundant, we will discuss in Sect. 4 that these variables are necessary for solving the problem.

3 Efficient Approximation

We present a primal-dual algorithm for the TMGC problem and prove its correctness in Sect. 3.1. Its 2-approximation guarantee (Theorem 1) and $O((|V| + |E|) \log(|V| + |E|))$ running time (Theorem 2) are verified in Sects. 3.2 and 3.3, respectively. The result implies an immediate extension of the main result of [7] for PVC, a 2-approximation in $O((|V| \log |V| + |E|))$ time, to the problem of partial cover with general weights (Remark 3).

For ease of exposition, given any $R \subseteq V$ and $K \subseteq E$, we often represent $R \cup K$ as the *ordered* pair (R, K), and denote its cost $c(R \cup K)$ and weight $w(R \cup K)$ by $c(R, K)$ and $w(R, K)$, respectively. Let $E(R)$ stand for the set of edges fully covered by R, i.e., those with both end vertices in R. Let $E[R] = \cup_{v \in R} \delta(v)$ denote the set of edges covered by R, i.e., those with at least one end vertex in R. Furthermore, we use $\tilde{w}(R, K) = \sum_{v \in R} w_v + \sum_{e \in E(R) \cup K} w_e$ to represent the total weight of elements fully covered by $R \cup K$, where an element is considered *fully covered* by $R \cup K$ if it belongs to $R \cup K$, or if it is an edge and both its end vertices belong to R.

3.1 Algorithm

The TMGC becomes trivial when $D = 0$, as the optimal solution is simply the empty set. Thus, we assume $D > 0$, or equivalently $T < w(V \cup E)$, throughout the remainder of this paper. Our algorithm for the TMGC problem constructs a series of covers, called *candidate covers*, and adds them to a set $\mathcal{M}$. The algorithm then selects the minimum-cost cover from $\mathcal{M}$ as its output. The pseudo-code is presented in Algorithm 1 below.

Algorithm 1:

Input: graph $G = (V, E)$, cost $c \in \mathbb{R}_+^{V \cup E}$, weight $w \in \mathbb{R}_+^{V \cup E}$, threshold $T \in [0, w(T \cup E))$

Output: a cover for $(G \mid T)$

1 initialize all sets C, J, R, K, and $\mathcal{M}$ to be $\emptyset$;
2 $\mathcal{M} \leftarrow \emptyset$;
3 **for** *each* $\mu \in (V \setminus R) \cup (E \setminus K)$ *s.t.* $\{\mu\}$ *forms a cover* **do**
4   $\quad \mathcal{M} \leftarrow \mathcal{M} \cup \{\{\mu\}\}$;
5   $\quad c_\mu \leftarrow +\infty$;
6   $\quad$ **if** $\mu \in V$ **then** $R \leftarrow R \cup \{\mu\}$ **else** $K \leftarrow K \cup \{\mu\}$;
7 **end**
8 $F \leftarrow E$;
9 $z \leftarrow 0$; $y_\mu \leftarrow 0$ for all $\mu \in V \cup E$;
10 **while** $\tilde{w}(R, K) \leq T$ **do**
11   $\quad$ raise z at the rate of 1, raise y_μ at the rate of w_μ for $\mu \in (V \setminus C) \cup F$ simultaneously until some element in $(V \setminus C) \cup F$ becomes tentatively selectable;
12   $\quad$ take $\mu \in (V \setminus C) \cup F$ to be one of the tentatively selectable elements;
13   $\quad$ **if** $\mu \in V \setminus C$ **then**
14   $\quad\quad$ $C \leftarrow C \cup \{\mu\}$; $J \leftarrow J \setminus \delta(\mu)$; $F \leftarrow F \setminus \delta(\mu)$;
15   $\quad$ **else**
16   $\quad\quad$ $J \leftarrow J \cup \{\mu\}$; $F \leftarrow F \setminus \{\mu\}$
17   $\quad$ **end**
18   $\quad$ **for** *each* $\mu \in (V \cup E) \setminus (C \cup J \cup R \cup K)$ *s.t.* $\{\mu\} \cup (C \cup J)$ *forms a cover* **do**
19   $\quad\quad$ $c_\mu \leftarrow +\infty$;
20   $\quad\quad$ **if** $\mu \in V$ **then**
21   $\quad\quad\quad$ $\mathcal{M} \leftarrow \mathcal{M} \cup \{(C \cup \{\mu\}, J \setminus \delta(\mu))\}$; $R \leftarrow R \cup \{\mu\}$;
22   $\quad\quad$ **else**
23   $\quad\quad\quad$ $\mathcal{M} \leftarrow \mathcal{M} \cup \{(C, J \cup \{\mu\})\}$; $K \leftarrow K \cup \{\mu\}$;
24   $\quad\quad$ **end**
25   $\quad$ **end**
26 **end**
27 **return** *output a minimum-cost cover in* $\mathcal{M}$.

In the process of constructing candidate covers, the algorithm iteratively expands a vertex set C and updates an edge set J, maintaining the following invariants: (1) No edge covered by C is in J; (2) $C \cup J$ is a subset of some cover in $\mathcal{M}$; and (3) The set of elements covered by $C \cup J$ never shrinks throughout the algorithm's execution.

To facilitate the description, we assign a *pending* status to each element (vertex or edge) initially. Once a pending element is added to $C \cup J$, its status changes to *tentatively selected*. Associated with the tentatively selected elements (i.e., those in $C \cup J$) are disallowed elements and the corresponding candidate covers. A pending element μ becomes *disallowed* when $\{\mu\} \cup (C \cup J)$ forms a cover, which is immediately added to $\mathcal{M}$ as a candidate (we call it μ's *associated candidate cover*). The algorithm uses sets R and K to store disallowed vertices and disallowed edges, respectively. The sets C, $\mathcal{M}$, R, and K grow throughout the algorithm. We employ a cost-resetting mechanism (see Steps 5 and 19) to ensure that a disallowed element will never become tentatively selected (i.e., is disallowed to be added to $C \cup J$).

Importantly, the algorithm ultimately outputs a single cover from $\mathcal{M}$, called the final cover, which consists of some tentatively selected elements and one pending element. For the disallowed elements not included in the final cover, the elements they fully cover are very likely not covered by the final cover. Because, before the algorithm terminates, it is impossible to determine which, or even if any, of the current disallowed elements will be included in the final cover, we use $\tilde{w}(R, K)$ as an estimated lower bound for the weight of elements not covered by the final cover. We expand the set of elements covered by $C \cup J$, and correspondingly expand R and K, only if this estimated lower bound does not exceed threshold T (i.e., the condition $\tilde{w}(R, K) \leq T$ in Step 10 is met).

To determine which elements become tentatively selected (i.e., are newly added to $C \cup J$) and which become disallowed (i.e., are newly added to $R \cup K$), we first employ a dual-ascending process to improve the quality of the feasible solution of the dual program (DP). Initially, all dual variables in (DP) are set to 0. At this point, no elements are tentatively selected (both C and J are empty), and all elements are either pending or disallowed. The disallowed elements and their corresponding candidate covers are identified in Steps 3 – 7. We now check if the weight fully covered by disallowed elements exceeds T. If it does, the algorithm (skips the while-loop, and) selects the minimum-cost cover from the current candidate set $\mathcal{M}$ and terminates. Otherwise, we have room to expand C, R, and K, as well as $\mathcal{M}$, via the while-loop (Steps 10 – 26), where J might not be expanded due to the removal of edges incident to the vertex newly added to C (see Step 14). The while-loop consists of two subroutines:

(1) We first execute the *primal-dual subroutine* (Steps 11 – 17): We increase the dual variables in this way to keep the inequality $y_\mu \leq w_\mu z$ holds with equality: increase z at a rate of 1, and increase dual variables y_μ at a rate of w_μ for all elements μ not covered by $C \cup J$ (i.e., all $\mu \notin C \cup E[C] \cup J$). We increase the dual variables proportionally (which is comparable to the uniform approach in [7]), until some element μ becomes tentatively selectable. An element μ is *tentatively selectable* if
  – either μ is a vertex and $y_\mu + \sum_{e \in \delta(\mu)} y_e \leq c_\mu$ in (DP) becomes tight,
  – or μ is an edge and $y_\mu \leq c_\mu$ in (DP) becomes tight.

Here, an inequality becomes *tight* if it holds with equality. We then tentatively select an arbitrary one (*only one*) such selectable element by adding it to $C \cup J$, and removing from J any edge it covers if the element is a vertex.

(2) Second, with the updated $C \cup J$, we execute the *candidate-set expansion subroutine* (the for-loop at Steps 18 – 25): We identify and add more disallowed elements to $R \cup K$ and more candidate covers (associated with these elements) to $\mathcal{M}$.

We then return to Step 16 to check the condition $\tilde{w}(R, K) \leq T$. If satisfied, we iterate with a new round of the primal-dual and candidate-set expansion subroutines. Otherwise, the algorithm terminates and returns the best candidate cover in $\mathcal{M}$.

In the pseudo-code of Algorithm 1, we use F to hold all the edges not covered by the current $C \cup J$. So, the set $(V \setminus C) \cup F$ in Step 11 for raising dual variables is the set of elements not covered by $C \cup J$.

It is worth noting that once an element μ is disallowed, it will never be tentatively selectable, as we reset c_μ to be $+\infty$ as soon as μ is added to $R \cup K$. It follows that $C \cap R = \emptyset$ and $J \cap K = \emptyset$ throughout the algorithm.

Lemma 1. *Throughout Algorithm 1, the set $C \cup J$ is not a cover of $(G \,|\, T)$.*

Proof. We prove this by induction on the algorithm. The initial $C \cup J$, which is empty, does satisfy the assertion. Observe that the two for-loops (Steps 3 –7 and Steps 18 – 25) in the algorithm for exhaustively identifying new disallowed elements (as well as candidate covers) can guarantee that adding any single pending element (i.e., any single element outside $C \cup J \cup R \cup K$) to $C \cup J$ does not make a cover of $(G \,|\, T)$. Therefore, the expansion of $C \cup J$ by μ conducted in Steps 13 – 17 does not violate the invariant that $C \cup J$ is not a cover. □

Note that once a vertex becomes tentatively selected, it remains tentatively selected forever. However, an edge can revert from tentatively selected to pending after being removed from J. Nevertheless, once an edge e changes from tentatively selected back to pending, it remains permanently pending. This is because e is removed from J when it becomes covered by a tentatively selected vertex. Since that vertex will remain in C thereafter, e will continue to be covered by C. According to the algorithm's principle for tentatively selecting elements (only elements not covered by $C \cup J$ can be added to $C \cup J$), e cannot become tentatively selected again. Furthermore, this pending edge e cannot become disallowed either, because: a disallowed element must cover some elements that $C \cup J$ cannot cover (note that the previous lemma tells us: $C \cup J$ is not a cover, while the algorithm requires that adding a disallowed element to $C \cup J$ would form a cover). Edge e can only cover itself, which is already covered by C. In summary, the discussion leads to the following observation.

Remark 1. Under any circumstances, an element can be added to $C \cup J$ (i.e., become tentatively selected) at most once.

Lemma 2. *Algorithm 1 solves the target-constrained mixed graph covering problem correctly.*

Proof. To establish correctness, given that $\mathcal{M}$ is constructed to either be empty or contain several covers of (G, T), we need only show that the algorithm terminates with a nonempty $\mathcal{M}$. If the algorithm exits the while-loop, the total weight fully covered by $R \cup K$ exceeds T. This implies that $R \cup K$, and consequently $\mathcal{M}$, are nonempty, as desired.

Suppose, for contradiction, that the algorithm does not exit the while-loop. This implies a state is reached where $\tilde{w}(R, K) \leq T$ and no more elements can be added to $C \cup J \cup R \cup K$. In other words, neither the primal-dual subroutine nor the candidate-set expansion subroutine can be executed. By Lemma 1, the weight not covered by $C \cup J$ exceeds T. Thus, there exists an element μ that is neither covered by $C \cup J$ nor fully covered by $R \cup K$. This means μ belongs to both $(V \setminus C) \cup F$ and $(V \cup E) \setminus (R \cup K)$, implying $c_\mu < \infty$. Consequently, the primal-dual subroutine (Steps 11 – 17) can be executed again, adding one more element to $C \cup J$, which yields a contradiction. □

3.2 Approximation Ratio

In this subsection, we prove that Algorithm 1 approximates the minimum-cost mixed cover with a ratio of 2. Let OPT denote any fixed optimal solution of the TMGC problem on $(G, c, w \,|\, T)$.

Lemma 3. $(R \cup K) \cap \text{OPT} \neq \emptyset$ *holds at the end of Algorithm 1.*

Proof. If no element in $R \cup K$ belongs to OPT, then the optimal solution covers none of the elements that are fully covered by $R \cup S$. Therefore, the uncovered weight of OPT must be at least $\tilde{w}(R, K)$. Since OPT is a cover, this uncovered weight and therefore $\tilde{w}(R, K)$ cannot exceed T. However, the algorithm terminates (by Lemma 2) precisely when $\tilde{w}(R, K) > T$, leading to a contradiction. □

By Lemma 3, let h be the first element of the optimal solution OPT that becomes disallowed during the execution of Algorithm 1. Let M denote the candidate cover associated with h. We will establish the 2-approximation guarantee by proving that the cost of M is at most twice the optimal cost by considering two cases: h is an edge (see Lemma 4), or h is a vertex (see Lemma 5).

In the remainder of this subsection, let (C, J) denote the set of tentatively selected elements at the moment h becomes disallowed and (R, K) be the set of disallowed elements immediately before h becomes disallowed.

Remark 2. If $C \cup J = \emptyset$, then $c(M) \leq c(\text{OPT})$

Proof. If $C \cup J = \emptyset$, then $\{h\}$ is a candidate cover added to $\mathcal{M}$, from which we deduce that $c(M) \leq c_h \leq c(\text{OPT})$. □

For ease of algorithm analysis, we call an element *active* if it is not covered by the current set of tentatively selected elements during the execution of Algorithm 1.

Lemma 4. *If $h \in E$, then $c(M) \leq 2 \cdot c(\text{OPT})$.*

Proof. To prove the inequality, we first strengthen the primal program (ILP) to (ILP'), by adding two requirements: first, h must be selected in the cover, and second, no disallowed elements in $R \cup K$ can be selected in the cover, which is implemented by assigning a cost of $+\infty$ to the elements in $R \cup K$.

$$\begin{aligned}
\min \quad & \sum_{\mu \in V \cup E} c'_\mu x_\mu \\
\text{s.t.} \quad & p_v + x_v \geq 1 && \forall v \in V, \\
& p_e + x_u + x_v + x_e \geq 1 && \forall e = \{u,v\} \in E, \\
& x_h \geq 1, \\
& \sum_{\mu \in V \cup E} w_\mu p_\mu \leq T, \\
& x_\mu, p_\mu \in \{0,1\} && \forall \mu \in V \cup E,
\end{aligned} \quad \text{(ILP')}$$

where $c'_\mu = c_\mu$ if $\mu \notin R \cup K$, and $c'_\mu = +\infty$ otherwise.

It is necessary to argue that the program (ILP') has the same optimal objective value as (ILP), which is $c(\text{OPT})$. By our choices of h and (R, K), no element in $R \cup K$ belongs to OPT, implying $c'(\text{OPT}) = c(\text{OPT})$. Since OPT remains feasible in (ILP'), the optimal objective value of (ILP') cannot exceed $c'(\text{OPT}) = c(\text{OPT})$. Conversely, as (ILP') has an additional constraint and element costs no smaller than those in (ILP), its optimal objective value cannot be smaller than that of (ILP). Therefore, the optimal objective value of (ILP') equals $c(\text{OPT})$, as claimed.

The relaxation of the strengthened program (ILP') has the following dual:

$$\begin{aligned}
\max \quad & \sum_{\mu \in V \cup E} y_\mu - Tz + \gamma \\
\text{s.t.} \quad & y_v + \sum_{e \in \delta(v)} y_e \leq c'_v && \forall v \in V, \\
& y_e \leq c'_e && \forall e \in E, \\
& y_h + \gamma \leq c'_h, \\
& 0 \leq y_\mu \leq w_\mu z && \forall \mu \in V \cup E.
\end{aligned} \quad \text{(DP')}$$

Let the values of dual variables z, y_μ (for all $\mu \in V \cup E$) be those produced by Algorithm 1 at the time immediately before h becomes disallowed. Setting $\gamma = c'_h - y_h = c_h - y_h$ makes (y, z, γ) a feasible solution to (DP'). By the weak duality, we obtain

$$c(\text{OPT}) \geq \sum_{\mu \in V \cup E} y_\mu - Tz + \gamma.$$

Recall from Algorithm 1 that a vertex v belongs to C only if $c_v = y_v + \sum_{e \in \delta(v)} y_e$, and an edge e belongs to J only if $c_e = y_e$. It follows from $\gamma = c_h - y_h$ that the cost of $M = (C, J \cup \{h\})$ is

$$c(M) = \sum_{v \in C} \left(y_v + \sum_{e \in \delta(v)} y_e \right) + \sum_{e \in J} y_e + y_h + \gamma. \tag{1}$$

By Remark 2, we may assume nonempty $C \cup J$, and consider the last element added to $C \cup J$, which we denote as r. Let F denote the set of the active edges at the time immediately before the addition of r to $C \cup J$. We then analyze two cases based on whether $r \in C$ or $r \in J$.

Case 1: r is a vertex in C. In the case, since (C, J) is not a cover by Lemma 1, we have $\tilde{w}(V \setminus C, E \setminus J) = \sum_{v \in V \setminus C} w_v + \sum_{e \in F} w_e - \sum_{e \in \delta(r) \cap F} w_e > T$ and

$$-\sum_{e \in F} y_e + \sum_{e \in \delta(r) \cap F} y_e < \sum_{v \in V \setminus C} y_v - Tz.$$

Also, the fact that $(C \setminus \{r\}, J \cup \{h\})$ is not a cover implies $\tilde{w}((V \setminus C) \cup \{r\}, E \setminus J \setminus \{h\}) = \sum_{v \in (V \setminus C) \cup \{r\}} w_v + \sum_{e \in F} w_e - w_h > T$ and

$$-\sum_{e \in F} y_e + y_h < \sum_{v \in (V \setminus C) \cup \{r\}} y_v - Tz.$$

Therefore, because $J \cap F = \emptyset$ and the dual variable y_e of each edge $e \in F \cap \delta(r)$ appears in the summation (1) at precisely once, we have

$$c(M) \leq \sum_{v \in C} y_v + 2 \left(\sum_{e \in E} y_e - \sum_{e \in F} y_e \right) + \sum_{e \in F \cap \delta(r)} y_e + y_h + \gamma$$

$$\leq \sum_{v \in C} y_v + 2 \left(\sum_{e \in E} y_e - Tz \right) + \sum_{v \in V \setminus C} y_v + \sum_{v \in (V \setminus C) \cup \{r\}} y_v + \gamma$$

$$\leq 2 \cdot c(\text{OPT}).$$

Case 2: r is an edge in J. Similarly, since (C, J) is not a cover, we have $-\sum_{e \in F} y_e + y_r < \sum_{v \in V \setminus C} y_v - Tz$. Because $(C, (J \setminus \{r\}) \cup \{h\})$ is not a cover, we have $-\sum_{e \in F} y_e + y_h < \sum_{v \in V \setminus C} y_v - Tz$. Thus, the lemma is established by

$$c(M) \leq \sum_{v \in C} y_v + 2 \left(\sum_{e \in E} y_e - \sum_{e \in F} y_e \right) + y_h + y_r + \gamma$$

$$\leq \sum_{v \in C} y_v + 2 \left(\sum_{e \in E} y_e - Tz \right) + 2 \sum_{v \in V \setminus C} y_v + \gamma$$

$$\leq 2 \cdot c(\text{OPT}),$$

where the first inequality follows from the fact that $J \cap F = \{r\}$, and each edge covered by C is outside $F \cup J$, and its dual variable appears at most twice in the summation given by (1). □

When h is a vertex, similar to the proof of Lemma 4, we strengthen (ILP) by imposing the additional requirement that h must be included in the cover and assigning a cost of $+\infty$ to the elements in $R \cup K$. The strengthened primal program retains the same form as (ILP'), except for the membership of h. However, the corresponding dual program in this setting differs from (DP'). By leveraging these updated primal and dual programs, we establish similar lower and upper bounds for the costs of OPT and M, which lead to the following lemma. The detailed proof is provided in the full version of the paper.

Lemma 5. *If $h \in V$, then $c(M) \leq 2 \cdot c(\text{OPT})$.*

Combining Lemmas 4 and 5, we conclude this subsection with the following 2-approximation guarantee.

Theorem 1. *Algorithm 1 achieves an approximation ratio of 2 for the target-constrained mixed graph covering problem.*

3.3 Time Complexity

In this subsection, we verify the subquadratic time complexity of Algorithm 1. Suppose that the input graph of the TMGC problem has exactly n vertices and m edges.

Theorem 2. *Algorithm 1 runs in $O((n + m) \log(n + m))$ time.*

Proof. In view that the dual variable z (with initial value 0) always grows its value at a rate 1 during Algorithm 1, We naturally refer to z's value as the current *time* in the execution of the algorithm. We implement the algorithm by maintaining two arrays whose members are labeled by $\mu \in V \cup E$. The value of the μ-labeled member in the first array, denoted $k_1(\mu)$, records the time at which μ becomes tentatively selected, while the value of the μ-labeled member in the second array, denoted $k_2(\mu)$, records the total weight of active elements covered by μ. Initially, $k_1(\mu) \leftarrow \frac{c_\mu}{\sum_{e \in \delta(\mu)} w_e + w_\mu}$ for $\mu \in V$ and $k_1(\mu) \leftarrow c_\mu/w_\mu$ for $\mu \in E$; and $k_2(\mu) \leftarrow \sum_{e \in \delta(\mu)} w_e + w_\mu$ for $\mu \in V$ and $k_2(\mu) \leftarrow w_\mu$ for $\mu \in E$. Let g be the leftover weight uncovered by $C \cup J$. Initially, $g \leftarrow \tilde{w}(V, E)$. At the entrance of the candidate-set expansion subroutine (either of the for-loops in Algorithm 1), we fetch an element μ with a maximum k_2 value. If $g - k_2(\mu) > T$, then we skip the for-loop, and proceed to the primal-dual subroutine (Steps 11 – 17). Otherwise, we make μ disallowed by removing the μ-labeled members from both arrays and adding μ to (R, K). We then repeat the disallowing process again and again for the next μ until $\tilde{w}(R, K) > T$ or $g - k_2(\mu) > T$. In the primal-dual subroutine, we fetch an element μ with the minimum k_1 value, reset $g \leftarrow g - k_2(\mu)$, add μ to (C, J), update J to be $J \setminus \delta(\mu)$ if μ is a vertex, and remove the μ-labeled member from both arrays.

- If μ is a vertex, we then further remove all members with labels in $\delta(\mu)$ from both arrays, and for every neighbor $u \in V \setminus (C \cup R)$ of μ, we increase $k_1(u)$ to $z + (k_1(u) - z)\frac{k_2(u)}{k_2(u) - w_{u\mu}}$, decrease $k_2(u)$ to $k_2(u) - w_{u\mu}$.
- Alternatively, if μ is an edge, then, for every end vertex u of μ, we increase $k_1(u)$ to $z + (k_1(u) - z)\frac{k_2(u)}{k_2(u) - w_\mu}$, and decrease $k_2(u)$ to $k_2(u) - w_\mu$.

To find the cheapest candidate cover, we keep a record of the currently best candidate cover. When an element h becomes disallowed, we check if h's associated candidate cover is cheaper than the current cheapest. If it is, we update the current cheapest to h's associated cover.

Clearly, the running time of the algorithm is dominated by the array-related update operations. We can perform at most $2n + 2m$ member removals, $4m$ array value modifications, and m updates of the edge set J. Using Fibonacci heaps, all these operations take $O((n + m) \log(n + m))$ time. □

Recalling the introduction presented in Sect. 1, the TMGC can be viewed as a combination of two problems: partial cover with general weights (PCGW) and partial cover with fine selection. By assuming sufficiently large selection costs for the edges, the TMGC simplifies to focus solely on vertex selection, which corresponds to PCGW. Formally, Given an undirected graph $G = (V, E)$ with vertex costs $c \in \mathbb{R}_+^V$ and element weights $w \in \mathbb{R}_+^{V \cup E}$, the objective of PCGW is to find a vertex subset $S \subseteq V$ with minimum $c(S)$, subject to the covering constraint that the total weight of the vertices and edges covered by S is at least a given threshold. The following remark gives an extension of the main result of [7] for PVC to the PCGW problem.

Remark 3. When applying Algorithm 1 to the special case PCGW of the TMGC, we keep track of the two arrays, whose members are only labeled by vertices, and perform at most $2n$ vertex removals and $2m$ value modifications. Using Fibonacci heaps, the implementation of the algorithm outputs a 2-approximation solution of the PCGW problem in $O(n \log n + m)$ time.

4 Conclusion

We conclude the paper with a technical remark on our primal-dual approach, followed by a brief discussion on future research directions.

Our primal-dual algorithm builds upon (ILP) and the dual (DP) of its relaxation. While the variables p_v for vertices v in (ILP) may appear superfluous and potentially complicate matters, their presence is crucial to our approach.

Necessity of Redundant Expressions. To understand the necessity and benefits of these variables, setting $T' = T - w(V) = w(E) - D$, let us examine the following IP formulation without the p_v variables, which is equivalent to (ILP),

$$\min \sum_{\mu \in V \cup E} c_\mu x_\mu$$
$$\text{s.t.} \quad p_e + x_u + x_v + x_e \geq 1 \quad \forall e = \{u,v\} \in E, \quad \text{(ILP1)}$$
$$- \sum_{v \in V} w_v x_v + \sum_{e \in E} w_e p_e \leq T',$$
$$x_v, x_e, p_e \in \{0,1\} \quad \forall v \in V, \forall e \in E.$$

Its LP relaxation has the following dual program

$$\max \sum_{e \in E} y_e - T'z$$
$$\text{s.t.} \quad w_v z + \sum_{e \in \delta(v)} y_e \leq c_v \quad \forall v \in V, \quad \text{(DP1)}$$
$$y_e \leq c_e \quad \forall e \in E,$$
$$0 \leq y_e \leq w_e z \quad \forall e \in E,$$

which presents a challenge due to the conflicting influences on the variable y_e. The first set of constraints favors larger y_e values when z is smaller. However, the third set of constraints dictates that y_e can only increase as z increases. This inherent tension complicates the analysis of the dual problem (DP1). In contrast, the dual program (DP) avoids this dilemma, as the dual variable z appears in only one set of constraints.

Future Research. Several promising directions exist for future research. One avenue is extending the TMGC problem from graphs to hypergraphs. Another direction involves incorporating multiple covering constraints. Exploring our algorithm's feasibility and adaptability to other covering problem variants would also be valuable. These variants include scenarios where selecting an edge covers both of its incident vertices, or where covering an edge requires selecting both end vertices. Additionally, we could explore adding structural constraints on the selected vertex subset, such as requiring it to induce a connected subgraph.

Acknowledgments. Xujin Chen was supported by the Chinese Academy of Sciences [Grant No. XDA27000000 and ZDBS-LY-7008] and National Natural Science Foundation of China [Grant No. 12331014]. Xiyuan Deng was supported by the National Natural Science Foundation of China [Grant No. 12331014]. Changjun Wang was supported by the National Natural Science Foundation of China [Grant No. 72192804] and the Beijing Natural Science Foundation [Grant No. Z220001].

Disclosure of Interests. The authors have no competing interests to declare that are relevant to the content of this article.

References

1. Bar-Yehuda, R.: Using homogeneous weights for approximating the partial cover problem. J. Algorithms **39**(2), 137–144 (2001)
2. Bshouty, N., Burroughs, L.: Massaging a linear programming solution to give a 2-approximation for a generalization of the vertex cover problem. In: Proceedings of the 15th Annual Symposium on, Theoretical Aspects of Computer Science, pp. 298–308. Springer Berlin Heidelberg, Berlin, Heidelberg (1998)
3. Gandhi, R., Khuller, S., Srinivasan, A.: Approximation algorithms for partial covering problems. J. Algorithms **53**(1), 55–84 (2004)
4. Hochbaum, D.S.: The t-vertex cover problem: extending the half integrality framework with budget constraints. In: Proceedings of the 1st International Workshop on Approximation Algorithms for Combinatorial Optimization, pp. 111–122. Springer Berlin Heidelberg (1998)
5. Karp, R.: Reducibility among Combinatorial Problems. Springer, US, Boston, MA (1972)
6. Khot, S., Regev, O.: Vertex cover might be hard to approximate to within 2-ϵ. J. Comput. Syst. Sci. **74**(3), 335–349 (2008)
7. Mestre, J.: A primal-dual approximation algorithm for partial vertex cover: making educated guesses. Algorithmica **55**(1), 227–239 (2009)
8. Vazirani, V.V.: Approximation Algorithms. Springer Berlin, Heidelberg (2003)
9. Williamson, D., Shmoys, D.: The Design of Approximation Algorithms. Cambridge University Press (2011)

Multiplication of 0-1 Matrices via Clustering

Jesper Jansson[1], Mirosław Kowaluk[2], Andrzej Lingas[3]([✉]), and Mia Persson[4]

[1] Graduate School of Informatics, Kyoto University, Kyoto, Japan
jj@i.kyoto-u.ac.jp
[2] Institute of Informatics, University of Warsaw, Warsaw, Poland
kowaluk@mimuw.edu.pl
[3] Department of Computer Science, Lund University, Lund, Sweden
Andrzej.Lingas@cs.lth.se
[4] Department of Computer Science and Media Technology, Malmö University,
Malmö, Sweden
Mia.Persson@mau.se

Abstract. We study applications of clustering (in particular the k-center clustering problem) in the design of efficient and practical deterministic algorithms for computing an approximate and the exact arithmetic matrix product of two 0-1 rectangular matrices A and B with clustered rows or columns, respectively. Let λ_A and λ_B denote the minimum maximum radius of a cluster in an ℓ-center clustering of the rows of A and in a k-center clustering of the columns of B, respectively. In particular, when A and B are square matrices of size $n \times n$, we obtain the following results.

1. A simple deterministic algorithm that approximates each entry of the arithmetic matrix product of A and B within an additive error of at most $2\lambda_A$ in $O(n^2\ell)$ time or at most $2\lambda_B$ in $O(n^2 k)$ time.
2. A simple deterministic preprocessing of the matrices A and B in $O(n^2\ell)$ time or $O(n^2 k)$ time after which every query asking for the exact value of an arbitrary entry of the arithmetic matrix product of A and B can be answered in $O(\lambda_A)$ time or $O(\lambda_B)$ time, respectively.
3. A simple deterministic algorithm for the exact arithmetic matrix product of A and B running in time $O(n^2(\ell + k + \min\{\lambda_A, \lambda_B\}))$.

Keywords: arithmetic matrix multiplication · clustering · Hamming space · minimum spanning tree

1 Introduction

The arithmetic matrix product of two 0-1 matrices is closely related to the Boolean one of the corresponding Boolean matrices. For square $n \times n$ matrices, both can be computed in $O(n^{2.372})$ time [1,23]. Both are basic tools in science and engineering. Unfortunately, no truly subcubic practical algorithms for any of them are known. Therefore, many researchers studied the complexity of these products for special input matrices, e.g., sparse or structured matrices [3,4,11, 15,21,24], providing faster and often more practical algorithms.

The method of multiplying matrices with clustered rows or columns, proposed for Boolean matrix product in [4] and subsequently generalized in [11,15] and used in [2], relies on the construction of an approximate spanning tree of the rows of the first input matrix or the columns of the second input matrix in a Hamming space. Then, each column or each row of the product matrix is computed with the help of a traversal of the tree in time proportional to the total Hamming cost of the tree up to a logarithmic factor. Simply, the next entry in a column or a row in the product matrix can be obtained from the previous one in time roughly proportional to the Hamming distance between the consecutive (in the tree traversal) corresponding rows or columns of the first or the second input matrix, respectively. Thus, in case the entire tree cost is substantially subquadratic in n, the total running time of this method becomes substantially subcubic provided that a good approximation of a minimum spanning tree of the rows of the first input matrix or the columns of the second one can be constructed in substantially subcubic time. As for simplicity and practicality, a weak point of this method is that in order to construct such an approximation relatively quickly, it employs a randomized dimension reduction.

In case of the arithmetic matrix product of 0-1 matrices, in some cases, a faster approximate arithmetic matrix multiplication can be more useful [8,21]. Among other things, it can enable to identify largest entries in the product matrix and it can be also used to provide a fast estimation of the number of: the so called witnesses for the Boolean product of two Boolean matrices [14], triangles in a graph, or more generally, subgraphs isomorphic to a small pattern graph [12] etc. There is a number of results on approximate arithmetic matrix multiplication, where the quality of approximation is expressed in terms of the Frobenius matrix norm $\|\ \|_F$ (i.e., the square root of the sum of the squares of the entries of the matrix) [8,21].

Cohen and Lewis [8] and Drineas *et al.* [9] used random sampling to approximate arithmetic matrix product. Their papers provide an approximation D of the matrix product AB of two $n \times n$ matrices A and B such that $\|AB - D\|_F = O(\|AB\|_F/\sqrt{c})$, for a parameter $c > 1$ (see also [21]). The approximation algorithm in [9] runs in $O(n^2 c)$ time. Drineas et al. [9] also derived bounds on the entrywise differences between the exact matrix product and its approximation. Unfortunately, the best of these bounds is $\Omega(M^2 n/\sqrt{c})$, where M is the maximum value of an entry in A and B. By using a sketch technique, Sarlós [22] obtained the same Frobenius norm guarantees, also in $O(n^2 c)$ time. However, he derived stronger individual upper bounds on the additive error of each entry D_{ij} of the approximation matrix D. They are of the form $O(\|A_{i*}\|_2 \|B_{*j}\|_2 / \sqrt{c})$, where A_{i*} and B_{*j} stands for the i-th row of A and the j-th column of B, respectively, that hold with high probability. More recently, Pagh [21] presented a randomized approximation $\tilde{O}(n(n+c))$-time algorithm for the arithmetic product of $n \times n$ matrices A and B such that each entry of the approximate matrix product differs at most by $\|AB\|_F/\sqrt{c}$ from the correct one. His algorithm first compresses the matrix product to a product of two polynomials and then uses the fast Fourier transform to multiply the polynomials. Subsequently, Kutzkov [20] developed analogous deterministic algorithms employing different techniques. For approximation results related to sparse arithmetic matrix products, see [19,21].

1.1 Our Contributions

In this paper, we exploit the possibility of applying the classic simple 2-approximation algorithm for the k center clustering problem [16] in order to derive efficient and practical deterministic algorithms for computing an approximate and the exact arithmetic matrix product of two 0-1 rectangular matrices A and B with clustered rows or columns, respectively.

The k-center clustering problem in a Hamming space $\{0,1\}^d$ is for a set P of n points in $\{0,1\}^d$ to find a set T of k points in $\{0,1\}^d$ that minimize $\max_{v \in P} \min_{u \in T} \mathrm{ham}(v, u)$, where $\mathrm{ham}(v, u)$ stands for the Hamming distance between v and u, (the number of coordinate positions they differ from each other). Each center in T induces a cluster consisting of all points in P for which it is the nearest center.

Let λ_A and λ_B denote the minimum maximum radius of a cluster in an ℓ-center clustering of the rows of A or in a k-center clustering of the columns of B, respectively. Assuming that A and B are of sizes $p \times q$ and $q \times r$, respectively, we obtain the following results.

1. A simple deterministic algorithm that approximates each entry of the arithmetic matrix product of A and B within an additive error of at most $2\lambda_A$ in $O(pq\ell + pr)$ time if $p \geq r$ or at most $2\lambda_B$ in $O(qrk + pr)$ time if $p \leq r$.
2. A simple deterministic preprocessing of the matrices A and B in $O(pq\ell + pr)$ time if $p \geq r$ or $O(qrk + pr)$ time if $p \leq r$ after which every query asking for the exact value of an arbitrary entry of the arithmetic matrix product of A and B can be answered in $O(\lambda_A)$ time if $p \geq r$ or $O(\lambda_B)$ time if $p \leq r$.
3. A simple deterministic algorithm for the exact arithmetic matrix product of A and B running in time $O(pq\ell + rqk + \min\{pr\lambda_A + rq\ell, pr\lambda_B + pqk\})$.

1.2 Techniques

All our main results rely on the classical, simple 2-approximation algorithm for the k-center clustering problem (farthest-point clustering) due to Gonzalez [16] (see also Fact 1). Two of them rely also on the idea of updating the inner product of two vectors a and b in $\{0,1\}^q$ over the Boolean or an arithmetic semi-ring to that of two vectors a' and b' in $\{0,1\}^q$, where $a = a'$ or $b = b'$, in time roughly proportional to $\mathrm{ham}(a, a') + \mathrm{ham}(b, b')$. The idea has been used in [4,11,15]. As in the aforementioned papers, we combine it with a traversal of an approximate minimum spanning tree of the rows or columns of an input matrix in the Hamming space $\{0,1\}^q$, where q is the length of the rows or columns (see also Lemma 6).

1.3 Paper Organization

The next section contains basic definitions. Section 3 presents our approximation algorithm for the arithmetic product of two 0-1 matrices and the preprocessing enabling efficient answers to queries asking for the value of an arbitrary entry of the arithmetic product matrix. Section 4 is devoted to our algorithm for the exact arithmetic matrix product of two 0-1 matrices. We conclude with a short discussion on possible extensions of our results.

2 Preliminaries

For a positive integer r, $[r]$ stands for the set of positive integers not exceeding r.

The transpose of a matrix D is denoted by $D^\top$. If the entries of D are in $\{0,1\}$ then D is a 0-1 matrix.

The *Hamming distance* between two points a, b (vectors) in $\{0,1\}^d$ is the number of the coordinates in which the two points differ. Alternatively, it can be defined as the distance between a and b in the L_1 metric over $\{0,1\}^d$. It is denoted by $\operatorname{ham}(a,b)$.

The k-center clustering problem in a Hamming space $\{0,1\}^d$ is as follows: given a set P of n points in $\{0,1\}^d$, find a set T of k points in $\{0,1\}^d$ that minimize $\max_{v \in P} \min_{u \in T} \operatorname{ham}(v,u)$.

The minimum-diameter k-clustering problem in a Hamming space $\{0,1\}^d$ is as follows: given a set P of n points in $\{0,1\}^d$, find a partition of P into k subsets $P_1, P_2, \ldots, P_k$ that minimize $\max_{i \in [k]} \max_{v,u \in P_i} \operatorname{ham}(v,u)$. Note that the k-center clustering problem could be also termed as the minimum-radius k-clustering problem. It is known to be NP-hard and even NP-hard to approximate within $2-\epsilon$ for any constant $\epsilon > 0$ [10,13].

Fact 1. *[16] Let P be a set of n points in $\{0,1\}^d$, and let $k \in [n]$. There is a simple deterministic 2-approximation algorithm for the k-center clustering and minimum-diameter k-clustering problems running in $O(ndk)$ time.*

3 An Approximate Arithmetic Matrix Product of 0-1 Matrices

Our approximation algorithm for the arithmetic matrix product of two 0-1 matrices is specified by the following procedure.

procedure $APPROXMMCLUS(A, B, \ell)$
Input: Two 0-1 matrices A and B of sizes $p \times q$ and $q \times r$, respectively, where $p \geq r$, and a positive integer ℓ not exceeding p.
Output: A $p \times r$ matrix D, where for $1 \leq i \leq p$ and $1 \leq j \leq r$, D_{ij} is an approximation of the inner product C_{ij} of the i-th row A_{i*} of A and the j-th column B_{*j} of B.

1. Determine an approximate ℓ-center clustering of the rows of the matrix A in $\{0,1\}^q$. For each row A_{i*} of A, set $cen_\ell(A_{i*})$ to the center of the cluster in the ℓ-clustering to which A_{i*} belongs.
2. Form the $\ell \times q$ matrix A', where the i'-th row is the i'-th center in the approximate ℓ-center clustering of the rows of A.
3. Compute the arithmetic $\ell \times r$ matrix product C' of A' and B.
4. For $1 \le i \le p$ and $1 \le j \le r$, set D_{ij} to $C'_{i'j}$, where the i'-th row $A'_{i'*}$ of A' is $cen_\ell(A_{i*})$.

For a 0-1 $p \times q$ matrix A, let $\lambda(A, \ell, row)$ be the minimum, over all ℓ-center clusterings of the rows of A in the Hamming space $\{0,1\}^q$, of the maximum Hamming distance between a center of a cluster and a member of the cluster. Similarly, for a 0-1 $q \times r$ matrix B, let $\lambda(B, k, col)$ be the minimum, over all k-center clusterings of the columns of B in the Hamming space $\{0,1\}^q$, of the maximum Hamming distance between a center of a cluster and a member of the cluster.

Lemma 1. *Suppose a 2-approximation algorithm for the ℓ-center clustering is used in $APPROXMMCLUS(A,B,\ell)$ and C stands for the arithmetic product of A and B. Then, for $1 \le i \le p$ and $1 \le j \le r$, $|C_{ij} - D_{ij}| \le 2\lambda(A, \ell, row)$.*

Proof. Recall that $p \ge r$ is assumed in the input to $APPROXMMCLUS(A,B,\ell)$. For $1 \le i \le p$ and $1 \le j \le r$, D_{ij} is the inner product of $cen_\ell(A_{i*})$, where $ham(A_{i*}, cen_\ell(A_{i*})) \le 2\lambda(A, \ell, row)$, with B_{*j}. Hence, C_{ij}, which is the inner product of A_{i*} with B_{*j}, can differ at most by $2\lambda(A, \ell, row)$ from D_{ij}. □

By $T(s, q, t)$, we shall denote the worst-case time taken by the multiplication of two 0-1 matrices of sizes $s \times q$ and $q \times t$, respectively.

Lemma 2. *$APPROXMMCLUS(A,B,\ell)$ can be implemented in $O(pq\ell + pr + T(\ell, q, r))$ time.*

Proof. Recall that $p \ge r$. Step 1, which includes the assignment of the closest center to each row of A, can be done in $O(pq\ell)$ time by using Fact 1, i.e., the classic algorithm of Gonzalez [16]. Step 2 takes $O(\ell q)$ time, which is $O(T(\ell, q, r))$ time. Finally, Step 3 takes $T(\ell, q, r)$ time while Step 4 can be done in $O(pr)$ time. Thus, the overall time is $O(pq\ell + pr + T(\ell, q, r))$. □

We can use the straightforward $O(sqt)$-time algorithm for the multiplication of two matrices of sizes $s \times q$ and $q \times t$, respectively. Since $T(\ell, q, r) = O(\ell q\, r) = O(\ell qp)$ if $p \ge r$, Lemmata 1 and 2 yield the first part (1) of our first main result (Theorem 1 below), for $p \ge r$. Its second part (2) for $p \le r$ follows from the first part by $(AB)^\top = B^\top A^\top$. Note that then the number of rows in $B^\top$, which is r, is not less than the number of columns in $A^\top$, which is p. Simply, we run $APPROXMMCLUS(B^\top, A^\top, k)$ in order to compute an approximation of the transpose of the arithmetic matrix product of A and B. Note also that a k-clustering of columns of B is equivalent to a k-clustering of the rows of $B^\top$ and that $\lambda(B^\top, k, row) = \lambda(B, k, col)$.

Theorem 1. Let A and B be two 0-1 matrices of sizes $p \times q$ and $q \times r$, respectively. There is a simple deterministic algorithm which provides an approximation of all entries of the arithmetic matrix product of A and B within an additive error of at most:

1. $2\lambda(A, \ell, row)$ in time $O(pq\ell + pr)$ if $p \geq r$,
2. $2\lambda(B, k, col)$ in time $O(rqk + pr)$ if $p \leq r$.

We slightly extend $APPROXMMCLUS(A, B, \ell)$ in order to obtain a preprocessing for answering queries about single entries of the arithmetic matrix product of A and B.

procedure $PREPROCMMCLUS(A, B, \ell)$
Input: Two 0-1 matrices A and B of sizes $p \times q$ and $q \times r$, respectively, where $p \geq r$, and a positive integer ℓ not exceeding p.
Output: The $p \times r$ matrix D returned by $APPROXMMCLUS(A, B, \ell)$, and for $1 \leq i \leq p$, the set of coordinate indices $ind(A, i)$ on which A_{i*} differs from its cluster center.

1. Run $APPROXMMCLUS(A, B, \ell)$.
2. For $1 \leq i \leq p$, determine the set $ind(A, i)$ of coordinate indices on which A_{i*} differs from $cen_\ell(A_{i*})$.

Lemma 3. $PREPROCMMCLUS(A, B, \ell)$ can be implemented in $O(pq\ell + pr + T(\ell, q, r))$ time.

Proof. Recall that $p \geq r$. Step 1 can be done in $O(pq\ell + pr + T(\ell, q, r))$ time by Lemma 2. Step 2 can be easily implemented in $O(pq)$ time. □

Our procedure for answering a query about a single entry of the matrix product of A and B is as follows.

procedure $QUERYMMCLUS(A, B, \ell, i, j)$
Input: The preprocessing done by $PREPROCMMCLUS(A, B, \ell)$ for 0-1 matrices A and B of sizes $p \times q$ and $q \times r$, respectively, where $p \geq r$, $\ell \in [p]$, and two query indices $i \in [p]$ and $j \in [r]$.
Output: The inner product C_{ij} of the i-th row A_{i*} of A and the j-th column B_{*j} of B.

1. Set C_{ij} to the entry D_{ij} of the matrix D computed by $APPROXMMCLUS(A, B, \ell)$ in $PREPROCMMCLUS(A, B, \ell, k)$.
2. For $m \in ind(A, i)$ do
   (a) If the m-th coordinate of the center assigned to A_{i*} is 0 and $B_{mj} = 1$ then $C_{ij} \leftarrow C_{ij} + 1$.
   (b) If the m-th coordinate of the center assigned to A_{i*} is 1 and B_{mj} is also 1 then $C_{ij} \leftarrow C_{ij} - 1$.

Lemma 4. $QUERYMMCLUS(A, B, \ell, i, j)$ is correct, i.e., the final value of C_{ij} is the inner product of the i-th row A_{i*} of A and the j-th column B_{*j} of B.

Proof. C_{ij} is initially set to D_{ij}, which is the inner product of the center assigned to A_{i*} and B_{*j}. Then, C_{ij} is appropriately corrected by increasing or decreasing with 1 for each coordinate index $m \in ind(A, i)$ which contributes 1 to the inner product of A_{i*} and B_{*j} and 0 to the inner product of the center of A_{i*} and B_{*j} or *vice versa*. □

Lemma 5. *QUERYMMCLUS(A, B, ℓ, k, i, j) takes $O(\lambda(A, \ell, row))$ time.*

Proof. Recall that $2\lambda(A, \ell, row)$ is an upper bound on the maximum Hamming distance between a row of A and its center in the ℓ-center clustering computed by $APPROXMMCLUS(A, B, \ell)$ in $PREPROCMMCLUS(A, B, \ell)$. Recall also that $p \geq r$. Step 1 takes $O(1)$ time. Since the m-th coordinate in the centers can be accessed in the matrix A' computed by $APPROXMMCLUS(A, B, \ell)$, each of the two substeps in the block of the loop in Step 2 can be done in $O(1)$ time. Finally, since $|ind(A, j)| \leq 2\lambda(A, \ell, row)$, the block is iterated at most $2\lambda(A, \ell, row)$ times. Consequently, the whole Step 2 takes $O(\lambda(A, \ell, row))$ time. □

By putting Lemmata 3, 4, and 5 together, and using the straightforward $O(sqt)$-time algorithm to multiply matrices of size $s \times q$ and $q \times t$, we obtain our next main result for $p \geq r$. The case $p \leq r$ reduces to the case $p \geq r$ by $(AB)^\top = B^\top A^\top$. Recall that then the number of rows in $B^\top$, which is r, is not less than the number of columns in $A^\top$, which is p. Also, we have $\lambda(B^\top, k, row) = \lambda(B, k, col)$. We simply run $PREPROCMMCLUS(B^\top, A^\top, k)$ and $QUERYMMCLUS(B^\top, A^\top, k, j, i)$ instead.

Theorem 2. *Let A and B be two 0-1 matrices of sizes $p \times q$ and $q \times r$, respectively. Given parameters $\ell \in [p]$ and $k \in [r]$, the matrices can be preprocessed by a simple deterministic algorithm in $O(pq\ell + pr)$ time if $p \geq r$ or $O(rqk + pr)$ time if $p \leq r$ such that a query asking for the exact value of a single entry C_{ij} of the arithmetic matrix product C of A and B can be answered in $O(\lambda(A, \ell, row))$ time if $p \geq r$ or $O(\lambda(B, k, col))$ time if $p \leq r$.*

4 The Exact Arithmetic Matrix Product of 0-1 Matrices

Theorem 2 yields the following corollary.

Corollary 1. *Let A and B be two 0-1 matrices of sizes $p \times q$ and $q \times r$, respectively. Given parameters $\ell \in [p]$ and $k \in [r]$, the arithmetic matrix product of A and B can be computed by a simple deterministic algorithm in $O(pq\ell + pr\lambda(A, \ell, row))$ time if $p \geq r$ or $O(rqk + pr\lambda(B, k, col))$ time if $p \leq r$.*

There is however a slightly better way of obtaining a simple deterministic algorithm for the arithmetic matrix product of two 0-1 matrices via ℓ-center clustering of the rows of the first matrix or k-center clustering of the columns of the second matrix. The idea is to use the aforementioned technique of traversing an approximate minimum spanning tree of the rows of the first matrix or the columns of the second matrix in an appropriate Hamming space in order to compute a row or column of the product matrix [4,11,15]. The technique easily generalizes to 0-1 rectangular matrices. We shall use the following procedure and lemma in the spirit of [4,11,15].

procedure $MMST(A, B, T)$
Input: Two matrices A and B of sizes $p \times q$ and $q \times r$, respectively, and a spanning tree T of the rows of A in the Hamming space $\{0, 1\}^q$.
Output: The arithmetic matrix product C of A and B.

1. Construct a traversal (i.e., a non-necessarily simple path visiting all vertices) U of T.
2. For any pair A_{m*}, A_{i*}, where the latter row follows the former in the traversal U, compute the set $diff(m, i)$ of indices $h \in [q]$ where $A_{ih} \neq A_{mh}$.
3. For $j = 1, \ldots, r$, iterate the following steps:
   (a) Compute C_{sj} where A_{s*} is the row of A from which the traversal U of T starts.
   (b) While following U, iterate the following steps:
      i. Set m, i to the indices of the previously traversed row of A and the currently traversed row of A, respectively.
      ii. Set C_{ij} to C_{mj}.
      iii. For each $h \in diff(m, i)$, if $A_{ih} B_{hj} = 1$ then set C_{ij} to $C_{ij} + 1$ and if $A_{mh} B_{hj} = 1$ then set C_{ij} to $C_{ij} - 1$.

Define the Hamming cost $\text{ham}(S)$ of a spanning tree S of a point set $P \subset \{0, 1\}^d$ by $\text{ham}(S) = \sum_{(v,u) \in S} \text{ham}(v, u)$.

Lemma 6. *Let A and B be two 0-1 matrices of sizes $p \times q$ and $q \times r$, respectively. Given a spanning tree T_A of the rows of A and a spanning tree T_B of the columns of B in the Hamming space $\{0, 1\}^q$, the arithmetic matrix product of A and B can be computed in time $O(pq + qr + pr + \min\{r \times \text{ham}(T_A), p \times \text{ham}(T_B)\})$.*

Proof. First, we shall prove that $MMST(A, B, T_A)$ computes the arithmetic matrix product of A and B in time $O(pq + qr + r \times \text{ham}(T_A))$. The correctness of the procedure $MMST$ follows from the correctness of the updates of C_{ij} in the block of the inner loop, i.e., in Step 3(b). Step 1 of $MMST(A, B, T_A)$ can be done in $O(p)$ time while Step 2 requires $O(pq)$ time. The first step in the block under the outer loop, i.e., computing C_{sj} in Step 3(a), takes $O(q)$ time. The crucial observation is that the second step in this block, i.e., Step 3(b), requires $O(p + \text{ham}(T_A))$ time. Simply, the substeps (i), (ii) take $O(1)$ time while the substep (iii) requires $O(|diff(m, i)| + 1)$ time. Since the block is iterated r times, the whole outer loop, i.e., Step 3, requires $O(qr + pr + r\text{ham}(T_A))$ time. Thus, $MMST(A, B, T_A)$ can be implemented in time $O(pq + qr + rp + r \times \text{ham}(T_A))$.

Similarly, we can run $MMST(B^\top, A^\top, T_B)$ to obtain the transpose of the arithmetic matrix product of A and B. So, to obtain the lemma, we can alternate the steps of $MMST(A, B, T_A)$ and $MMST(B^\top, A^\top, T_B)$, and stop whenever any of the calls is completed. □

Theorem 3. *Let A and B be two 0-1 matrices of sizes $p \times q$ and $q \times r$, respectively. Given parameters $\ell \in [p]$ and $k \in [r]$, the arithmetic matrix product of A and B can be computed by a simple deterministic algorithm in time $O(pq\ell + rqk + \min\{pr\lambda(A, \ell, row) + rq\ell, \ pr\lambda(B, k, col) + pqk\})$.*

Proof. We determine an ℓ-center clustering of the rows of A in $\{0,1\}^q$ of maximum cluster radius not exceeding $2\lambda(A, \ell, row)$ in $O(pq\ell)$ time by employing Fact 1. Similarly, we construct a k-center clustering of the columns of B in $\{0,1\}^q$ of maximum cluster radius not exceeding $2\lambda(B, k, col)$ in $O(rqk)$ time. The centers in both aforementioned clusterings are some rows of A and some columns of B, respectively, by the specification of the method in [16]. Hence, the ℓ-center clustering gives rise to a spanning tree T_A of the rows of A with all members of a cluster being pendants of their cluster center and the centers connected by a path of length $\ell - 1$. The Hamming cost of T_A is at most $(p-\ell)2\lambda(A, \ell, row) + (\ell-1)q$. Similarly, we obtain a spanning tree T_B of the columns of B having the Hamming cost not exceeding $(r-k)2\lambda(B, k, col) + (k-1)q$. The theorem follows from Lemma 6 by straightforward calculations. □

5 Extensions

The rows or columns in the input 0-1 matrices can be very long. Also, a large number of clusters might be needed in order to obtain a low upper bound on their radius. Among other things, for these reasons, we have picked Gonzalez's classical algorithm for the k-center clustering problem [16] as a basic tool in our approach to the arithmetic matrix product of two 0-1 matrices with clustered rows or columns. The running time of his algorithm is linear not only in the number of input points but also in their dimension, and in the parameter k. Importantly, it is very simple and provides a solution within 2 of the optimum. For instance, there exist faster (in terms of n and k) 2-approximation algorithms for k-center clustering with hidden exponential dependence on the dimension in their running time, see [10, 17].

One could easily generalize our main results by replacing Gonzalez's algorithm with a simple and efficient approximation algorithm for the more general problem of k-center clustering with outliers [6]. In the latter problem, a given number z of input points could be discarded as outliers when trying to minimize the maximum cluster radius. Unfortunately, the algorithms for this more general problem tend to be more complicated and the focus seems to be the approximation ratio achievable in polynomial time (e.g., 3 in [6] and 2 in [18]) not the time complexity.

There are many other variants of clustering than k-center clustering, and plenty of methods have been developed for them in the literature. In fact, in the design of efficient algorithms for the exact arithmetic matrix product of 0-1 matrices with clustered rows or columns, using the k-median clustering could seem more natural. The objective in the latter problem is to minimize the sum of distances between the input points and their nearest centers. Unfortunately, no simple deterministic $O(1)$-approximation algorithms for the latter problem that are efficient in case the dimension and k parameters are large seem be to available [5,7].

Our approximate and exact algorithms for the matrix product of 0-1 matrices as well as the preprocessing of the matrices can be categorized as supervised since they assume that the user has some knowledge on the input matrices and can choose reasonable values of the parameters ℓ and k guaranteeing relatively low overall time complexity. Otherwise, one could try the ℓ-center and k-center clustering subroutines for a number of combinations of different values of ℓ and k in order to pick the combination yielding the lowest upper bound on the overall time complexity of the algorithm or preprocessing.

Acknowledgments. J.J. was partially supported by KAKENHI grant 24K22294.

References

1. Alman, J., Duan, R., Vassilevska Williams, V., Xu, Y., Xu, Z., Zhou, R.: More asymmetry yields faster matrix multiplication. In: Proceedings of the Annual ACM-SIAM Symposium on Discrete Algorithms (SODA 2025), pp. 2005–2039. ACM-SIAM (2025)
2. Alves, J., Moustafa, S., Benkner, S., Francisco, A.: Accelerating graph neural networks with a novel matrix compression format (2024). https://doi.org/10.48550/arXiv.2409.02208
3. Anand, E., van den Brand, J., McCarthy, R.: The structural complexity of matrix-vector multiplication. arXiv:2502.21240 (2025)
4. Björklund, A., Lingas, A.: Fast boolean matrix multiplication for highly clustered data. In: Dehne, F., Sack, J.-R., Tamassia, R. (eds.) WADS 2001. LNCS, vol. 2125, pp. 258–263. Springer, Heidelberg (2001). https://doi.org/10.1007/3-540-44634-6_24
5. Charikar, M., Guha, S., Tardos, E., Shmoys, D.: A constant-factor approximation algorithm for the k-median problem. In: Proceedings of the 31st Annual ACM Symposium on Theory of Computing (STOC 1999), pp. 1–10. ACM (1999)
6. Charikar, M., Khuller, S., Mount, D., Narasimhan, G.: Algorithms for facility location problems with outliers. In: Proceedings of the 12th Annual ACM-SIAM Symposium on Discrete Algorithms (SODA 2001), pp. 642–651. ACM-SIAM (2001)
7. Chen, K.: On coresets for k-median and k-means clustering in metric and euclidean spaces and their applications. SIAM J. Comput. **39**, 923–947 (2009)
8. Cohen, E., Lewis, D.D.: Approximating matrix multiplication for pattern recognition tasks. J. Algorithms **30**(2), 211–252 (1999)
9. Drineas, P., Kannan, R., Mahoney, M.: Fast monte carlo algorithms for matrices I: approximating matrix multiplication. SIAM J. Comput. **36**(1), 132–157 (2006)
10. Feder, T., Greene, D.: Optimal algorithms for approximate clustering. In: Proceedings of ACM Symposium on Theory of Computing (STOC 1988), pp. 434–444. ACM (1988)
11. Floderus, P., Jansson, J., Levcopoulos, C., Lingas, A., Sledneu, D.: 3D rectangulations and geometric matrix multiplication. Algorithmica **80**(1), 136–154 (2018)
12. Floderus, P., Kowaluk, M., Lingas, A., Lundell, E.: Detecting and counting small pattern graphs. SIAM J. Discret. Math. **29**(3), 1322–1339 (2015)

13. Gąsieniec, L., Jansson, J., Lingas, A.: Approximation algorithms for hamming clustering problems. J. Discrete Algorithms **2**(2), 289–301 (2004)
14. Gąsieniec, L., Kowaluk, M., Lingas, A.: Faster multi-witnesses for Boolean matrix multiplication. Inf. Process. Lett. **109**(4), 242–247 (2009)
15. Gąsieniec, L., Lingas, A.: An improved bound on boolean matrix multiplication for highly clustered data. In: Dehne, F., Sack, J.-R., Smid, M. (eds.) WADS 2003. LNCS, vol. 2748, pp. 329–339. Springer, Heidelberg (2003). https://doi.org/10.1007/978-3-540-45078-8_29
16. Gonzalez, T.: Clustering to minimize the maximum intercluster distance. Theor. Comput. Scirnce **38**, 293–306 (1985)
17. Har-Peled, S., Mendel, M.: Fast construction of nets in low-dimensional metrics and their applications. SIAM J. Comput. **35**(5), 1148–1184 (2006)
18. Harris, D., Pensyl, T., Srinivasan, A., Trinh, K.: A lottery model for center-type problems with outliers. In: Proceedings of the International Workshop on Approximation Algorithms for Combinatorial Optimization Problems (APPROX 2017), pp. 10:1–10:19. Schloss Dagstuhl - Leibniz-Zentrum für Informatik (2017)
19. Iven, M., Spencer, C.: A note on compressed sensing and the complexity of matrix multiplication. Inf. Process. Lett. **109**(10), 468–471 (2009)
20. Kurzkov, K.: Deterministic algorithms for skewed matrix products. In: Proceedings of the International Symposium on Theoretical Aspects of Computer Science (STACS 2013), pp. 466–477. Schloss Dagstuhl- Leibniz-Zentrum fuer Informatik, vol. 20 (2013)
21. Pagh, R.: Compressed matrix multiplication. ACM Trans. Comput. Theory (TOCT) **5**(3), 1–17 (2013)
22. Sarlós, T.: Improved approximation algorithms for large matrices via random projections. In: Proceedings of the IEEE Symposium on Foundations of Computer Science (FOCS 2006), pp. 143–152. IEEE Computer Society (2006)
23. Vassilevska Williams, V., Xu, Y., Xu, Z., Zhou, R.: New bounds for matrix multiplication: from alpha to omega. In: Proceedings of the Annual ACM-SIAM Symposium on Discrete Algorithms (SODA 2024). ACM-SIAM (2024)
24. Yuster, R., Zwick, U.: Fast sparse matrix multiplication. ACM Trans. Algorithms **1**, 2–13 (2005)

From MAXCUT to MAXNAESAT: Elegant Proofs and Algorithmic Advances

Sangram K. Jena[1] and K. Subramani[2(✉)]

[1] Department of Computer Science, University of Alaska Fairbanks, Fairbanks, AK, USA
sangram063@gmail.com
[2] LDCSEE, West Virginia University, Morgantown, WV, USA
k.subramani@mail.wvu.edu

Abstract. In this paper, we investigate the approximation and parameterized complexities of MAXNAESAT variants. We begin by presenting a simple yet rigorous proof establishing the **APX-completeness** of the MAXNAE2SAT problem. Notably, **APX-completeness** holds even when the repetition factor of each variable is bounded by 3, i.e., each variable appears in at most three clauses in the MAXNAE2SAT instance. Our **APX-completeness** proof is a strict reduction that directly establishes a new inapproximability bound for the MAXNAE2SAT problem. The decision version of MAXNAE2SAT remains **NP-complete** when the repetition factor of each variable is bounded by 3, mirroring the **NP-completeness** of MAXCUT in cubic graphs. We further establish a tight computational dichotomy by proving that the MAXNAE2SAT problem is solvable in linear time when the repetition factor of each variable is bounded by 2. Finally, we present a **fixed-parameter tractable** algorithm for MAXNAE2SAT instances where the repetition factor of each variable is bounded by 3.

1 Introduction

This paper focuses on a variant of the SAT problem known as the Not-All-Equal satisfiability (NAESAT) problem. In the NAESAT problem, we are given a CNF formula Φ, where each clause contains at least two literals. An assignment is considered *nae-satisfying* if it satisfies all the clauses of Φ such that at least one literal in each clause is assigned **false**. The central question is: Does there exist a nae-satisfying assignment for Φ, i.e., a satisfying assignment in which every clause contains at least one literal set to **false**? In this paper, we study a natural optimization variant of NAESAT, where the objective is to maximize the number of nae-satisfied clauses. This problem is known as MAX-NAESAT. As expected, the **NP-completeness** of NAESAT directly implies the **NP-hardness** of MAXNAESAT.

In the case of **NP-hard** problems, one line of attack is to study subclasses that may be amenable to efficient approaches. At the very least, we can establish boundaries between the "easy" and "hard" cases of the problem [15]. Such boundaries are particularly important when we consider design paradigms such as backdoor design [14], wherein the presence of feasible subclasses is exploited to design fixed-parameter algorithms. We focus on two types of restrictions, viz., clause width and repetition factor (see Sect. 2) and their combinations. From the computational complexity perspective, we draw the exact line between easy and hard cases of the MAXNAE2SAT problem. More specifically, we show that the MAXNAE2SAT problem is in **P** when the repetition factor of each variable is bounded by two. Previous research has established that the MAXNAE2SAT problem is **NP-hard** when the repetition factor of each variable is bounded by 3. We establish a new inapproximability bound for the MAXNAE2SAT problem. We also consider the parameterized version of the MAXNAE2SAT problem when the repetition factor of each variable is bounded by three and design an **FPT** algorithm. Here, the parameter is the number of nae-satisfied clauses in the optimal solution.

2 Statement of Problems

Let $X = \{x_1, x_2, \ldots, x_n\}$ denote a collection of n Boolean variables. A clause C is a disjunction of literals over the variable set X. For instance, $C = (x_1 \vee x_2 \vee x_3)$. A formula is said to be in Conjunctive Normal Form (CNF) if it is the conjunction of clauses. For instance, $(\bar{x}_1 \vee x_2) \wedge (x_1 \vee \bar{x}_2 \vee x_3) \wedge (x_2 \vee \bar{x}_3 \vee \bar{x}_4)$ is a CNF formula with three clauses over the variable set $X = \{x_1, x_2, x_3, x_4\}$. A CNF formula is said to be in kCNF form if the width of each clause is exactly k; in other words, each clause consists of exactly k literals.

Definition 1. *SAT: Given a CNF formula Φ, does there exist an assignment that satisfies each clause, i.e., sets at least one literal to* **true** *in each clause?*

Definition 2. *NAESAT: Given a CNF formula Φ, does there exist an assignment that nae-satisfies each clause, i.e., sets at least one literal to* **true** *and at least one literal to* **false** *in each clause?*

Note that unit clauses cannot be nae-satisfied. Only clauses of size at least two can be nae-satisfied. We denote the decision and optimization versions of the MAXNAEkSAT problem as MAXNAEkSAT$_D$ and MAXNAEkSAT$_O$, respectively.

Definition 3. *MAXNAEkSAT$_D$: Given a kCNF formula Φ with m clauses over n variables and an integer $m' \leq m$, does there exist an assignment to the variables of Φ that nae-satisfies at least m' clauses?*

Definition 4. *MAXNAEkSAT$_O$: Given a kCNF formula Φ with m clauses over n variables, find an assignment for the variables of Φ that maximizes the number of nae-satisfied clauses in Φ.*

Definition 5. *Repetition factor of a variable: Given a CNF formula Φ with m clauses over n variables, the repetition factor of a variable in Φ is defined as the number of clauses in which the variable or its complement appears.*

We denote the decision and optimization versions of the MAXNAEkSAT problem, where the repetition factor of each variable is bounded by s, as MAXNAEkSAT$_D(n,s)$ and MAXNAEkSAT$_O(n,s)$ respectively.

Definition 6. *MAXNAEkSAT$_D(n,s)$: Given a kCNF formula Φ with m clauses over n variables, with the repetition factor of each variable bounded by s and an integer $m' \leq m$, does there exist an assignment to the variables of Φ that nae-satisfies at least m' clauses?*

Definition 7. *MAXNAEkSAT$_O(n,s)$: Given a kCNF formula Φ with m clauses over n variables, with the repetition factor of each variable bounded by s, find an assignment for the variables of Φ that maximizes the number of nae-satisfied clauses in Φ.*

3 Related Work

This section briefly discusses some of the related work in the literature concerning NAESAT variants. The NAESAT problem is an important problem in theoretical computer science due to its applicability in modeling a wide range of graph-theoretic properties [15, 17]. There exist linear time algorithms for NAESAT variants in the literature [17]. The NAE3SAT problem is a starting point for proving the **NP-completeness** of several graph coloring variants [6]. Several approximation algorithms have been proposed for the MAXNAE-SAT problem. Zhang et al. [21] proposed a 0.7499-approximation algorithm for the MAXNAESAT problem. They also considered the nae-satisfiable instances of the MAXNAESAT problem, i.e., all the clauses can be nae-satisfied in an optimal assignment. In that case, they proved that the performance ratio of their proposed algorithm is 0.8097. A simple randomized algorithm for the NAE2SAT problem can be found in [19]. Moreover, on a nae-satisfiable instance involving n variables, they claimed that the probability that their algorithm does not find a nae-satisfying assignment is at most $\frac{1}{24}$. They also proved that a restricted variant of NAE2SAT is **L-complete**. Xian et al. [20] proposed a $\frac{2^{k-1}}{2^{k-1}-1}$-approximation algorithm for the MAXNAEkSAT problem using the local search algorithm for any $k \geq 2$. They also proved that the MAXNAEkSAT problem cannot be approximated within the factor $\frac{2^{k-1}}{2^{k-1}-1}$ in polynomial time unless **P=NP**. Observe that the MAXNAE2SAT problem has not been extensively studied in the literature, although the MAXNAE3SAT problem has been. The MAXNAE3SAT problem also admits a 0.87-approximation algorithm [10], while the **APX-completeness** of the problem is discussed in [16]. The approximation complexities of SAT and NAESAT variants on bounded-degree instances has been explored in [1]. Exact exponential-time algorithms for these problems are presented in [13]. Additional related results can be found in [4, 7]. The MAXNAEkSAT problem is **NP-hard** for $k \geq 3$. Indeed, the problem stays **NP-hard** when the clauses contain only positive literals [6].

The MAXCUT problem is a special case of the MAXNAE2SAT problem in which all the literals are positive. There exists a 0.87856-approximation algorithm for the MAXCUT problem using semi-definite programming relaxations [8]. The above approximation bound holds for the MAXNAE2SAT problem. An inapproximability bound for the MAXCUT problem can be found in [11]. Several results for the MAXCUT and MAXNAESAT problems can be found in [3]. Note that the MAXCUT problem is inapproximable within a factor of 0.941 [9], whereas we prove that the MAXNAE2SAT problem cannot be approximated within a factor of $\frac{11}{12}$ (≈ 0.92) unless **P=NP**. It is also worth noting that the reduction in [2] is a randomized reduction, whereas our reduction is a deterministic **L**-reduction.

4 Approximation Complexity

In this section, we discuss an **L**-reduction [12] from the MAXNAE3SAT$_O$ problem to the MAXNAE2SAT$_O$ problem. Note that the MAXNAE3SAT$_O$ problem is known to be **APX-complete** under **L**-reductions [16]. It therefore follows that the MAXNAE2SAT$_O$ problem is **APX-complete** under **L**-reductions.

Now, we briefly describe the nature of **L**-reductions: Let P and Q denote two **NP** optimization problems. Let $OPT_P(x)$ and $OPT_Q(y)$ denote the optimal solutions (values) for instance x of P, and instance y of Q respectively. Let $VAL_P(x)$ and $VAL_Q(y)$ denote the values of the objective function for arbitrary instances x and y of problems P and Q respectively.

We say P is **L**-reducible to Q (written $P \leq_L Q$) if:

1. There exists a polynomial time computable function f, such that if x is an instance of P, then $f(x)$ is an instance of Q.
2. There exists a polynomial time computable function g, such that if y is a solution to $f(x)$, then $g(y)$ is a solution to x.
3. There exists a constant α, such that for any instance x of P:
   $OPT_Q(f(x)) \leq \alpha \cdot OPT_P(x)$.
4. There exists a constant β, such that for any instance x of P with image $f(x)$ and solution y to $f(x)$: $|VAL_P(g(y)) - OPT_P(x)| \leq \beta \cdot |VAL_Q(y) - OPT_Q(f(x))|$.

Lemma 1. *The MAXNAE2SAT$_O$ problem is* **APX-hard**.

Proof. • f: The function f is defined as follows: It takes a 3CNF formula Φ and constructs a 2CNF formula $R(\Phi)$. Corresponding to each clause $C_i = (x_1 \vee x_2 \vee x_3)$ in Φ, f adds three clauses $(x_1 \vee x_2)$, $(x_2 \vee x_3)$, and $(x_1 \vee x_3)$ to $R(\Phi)$. If the number of clauses in Φ is m, then from the above construction, the number of clauses in $R(\Phi)$ is $3 \cdot m$.

Let $\mathcal{A}$ be an assignment for the MAXNAE3SAT$_O$ problem instance Φ that nae-satisfies k clauses. We argue that the same assignment $\mathcal{A}$ nae-satisfies $k' = 2 \cdot k$ clauses in $R(\Phi)$.

Consider a single clause C_i from Φ. In the construction from Φ to $R(\Phi)$, for a single clause $C_i = (x_1 \vee x_2 \vee x_3)$ in Φ, we add three clauses $(x_1 \vee x_2)$, $(x_1 \vee x_3)$, and $(x_2 \vee x_3)$ to $R(\Phi)$. If there is an assignment that nae-satisfies clause C_i in Φ, then with that assignment, exactly two of the three corresponding clauses of C_i will be nae-satisfied in $R(\Phi)$. If clause C_i is not nae-satisfied, then none of the three corresponding clauses will be nae-satisfied.

We can extend this argument to conclude that, if there exists an assignment $\mathcal{A}$ that nae-satisfies k clauses in Φ, then $\mathcal{A}$ nae-satisfies at least $k' = 2 \cdot k$ clauses in $R(\Phi)$.

- α: Let k_{opt} and k'_{opt} denote the maximum number of nae-satisfiable clauses in Φ and $R(\Phi)$ respectively. It is clear that $k'_{opt} = 2 \cdot k_{opt}$ and hence, $k'_{opt} \leq 2 \cdot k_{opt}$. Here, $\alpha = 2$.

- g: The function g is defined as follows: g is the identity function which considers the same assignment of the variables in $R(\Phi)$ to Φ. Consider any assignment to the triplet associated with clause C_i. This assignment either nae-satisfies two of the three clauses in the triplet or none at all. If it nae-satisfies two of the three clauses, then it must nae-satisfy C_i. If it fails to nae-satisfy any of the three clauses in the triplet, then it fails to nae-satisfy C_i as well. We can extend this argument to conclude that any assignment that nae-satisfies k' clauses in $R(\Phi)$ must nae-satisfy at least $k = \frac{k'}{2}$ clauses in Φ.

- β: Now, let k' be the number of clauses nae-satisfied in $R(\Phi)$ and k be the number of corresponding clauses nae-satisfied in Φ. Let k'_{opt} and k_{opt} be the maximum number of clauses nae-satisfied in $R(\Phi)$ and Φ respectively. Then $k' - k'_{opt} \geq 2 \cdot k - 2 \cdot k_{opt} = 2 \cdot (k - k_{opt})$. Here, $\beta = 2$

This gives an **L**-reduction from MAXNAE3SAT$_O$ to MAXNAE2SAT$_O$. □

Theorem 1. *The MAXNAE2SAT$_O$ problem is* **APX-complete**.

Proof. The MAXNAE2SAT$_O$ problem is in **APX** [8] and **APX-hard** (from Lemma 1). Therefore, the MAXNAE2SAT$_O$ problem is **APX-complete**. □

5 Inapproximability

This section establishes inapproximability bounds for the MAXNAE2SAT$_O$ problem. Using the following lemma, we can establish a lower bound for the approximability of the MAXNAE2SAT$_O$ problem.

Lemma 2. *[16] Suppose a problem Π L-reduces to another problem Π' (where each of Π and Π' is maximization problem), and there is a polynomial time approximation algorithm for Π' with approximation ratio $(1 - \epsilon)$. Then, there is a polynomial time approximation algorithm for Π with approximation ratio $(1 - \alpha \cdot \beta \cdot \epsilon)$.*

We have established the **APX-completeness** of the MAXNAE2SAT$_O$ problem via an **L**-reduction from the MAXNAE3SAT$_O$ problem (see Sect. 4). It is known that no polynomial-time approximation algorithm can achieve a ratio better than $\frac{11}{12}$ for the MAXNAE3SAT$_O$ problem unless **P**=**NP** [22]. By applying Lemma 2 and using the inapproximability bound of $\frac{11}{12}$ along with the parameters $\alpha = \beta = 2$, we derive an inapproximability bound of $\frac{47}{48}$ for the MAXNAE2SAT$_O$ problem. Suppose there exists a $(1-0.02) \approx \frac{47}{48}$-approximation algorithm for MAXNAE2SAT$_O$. Then, by Lemma 2, this implies a $(1-\alpha \cdot \beta \cdot \epsilon) = (1 - 2 \cdot 2 \cdot 0.02) \approx \frac{11}{12}$-approximation algorithm for MAXNAE3SAT$_O$, which contradicts the known inapproximability result unless **P**=**NP** [22]. We now show that the inapproximability bound for MAXNAE2SAT$_O$ can be further strengthened through an approximation-preserving reduction.

Lemma 3. *If there exists an α-approximation algorithm for the MAXNAE2SAT$_O$ problem, then there exists an α-approximation algorithm for the MAXNAE3SAT$_O$ problem.*

Proof. Let $\mathcal{A}$ denote an α-approximation algorithm for the MAXNAE2SAT$_O$ problem. Let Φ denote an instance of the MAXNAE3SAT$_O$ problem and OPT_Φ denote the cardinality of the maximum size subset of Φ that can be nae-satisfied considering all possible assignments for Φ. Transform Φ into a MAXNAE2SAT$_O$ instance $R(\Phi)$ as discussed in Sect. 4. The cardinality of the maximum size subset of $R(\Phi)$ that is nae-satisfied is clearly $2 \cdot OPT_\Phi$. Let $R(\Phi)$ be provided to $\mathcal{A}$ as input. The output of $\mathcal{A}$ has a cardinality of at least $2 \cdot \alpha \cdot OPT_\Phi$. As argued in Sect. 4, any assignment either nae-satisfies two of the three clauses on the triplet in $R(\Phi)$ corresponding to a clause in Φ or none of them. Thus, the output of $\mathcal{A}$ corresponds to at least $\alpha \cdot OPT_\Phi$ clauses in Φ, i.e., at least $\alpha \cdot OPT_\Phi$ clauses in Φ can be nae-satisfied. Therefore, it follows that an α-approximation algorithm exists for the MAXNAE3SAT$_O$ problem. □

Observe that the above reduction is not only an approximation preserving reduction but also a *strict* reduction.

Corollary 1. *The MAXNAE2SAT$_O$ problem does not have an algorithm with an approximation ratio better than $\frac{11}{12}$ unless* **P = NP**.

Proof. The inapproximability bound of MAXNAE3SAT$_O$ problem is $\frac{11}{12}$ [22]. As per Lemma 3, the inapproximability bound of MAXNAE2SAT$_O$ is also $\frac{11}{12}$. □

The following corollary is immediate.

Corollary 2. *If there exists an inapproximability bound for the MAXNAE3SAT$_O$ problem predicated on* **P $\neq$ NP**, *then the same bound holds for the MAXNAE2SAT$_O$ problem as well.*

6 Linear Time Algorithm

In this section, we present a linear-time algorithm for solving the MAXNAE2SAT$_O(n, 2)$ problem. Let Φ be a 2CNF formula representing an instance of the MAXNAE2SAT$_O(n, 2)$ problem. By definition, the repetition factor of each variable in Φ is bounded by two.

We categorize the variables in Φ into three disjoint sets: O, S, and N. The set O contains variables that appear only once in Φ. The set S consists of variables that appear exactly twice in the same form (either both positive or both negative). The set N includes variables that occur twice but in different forms, i.e., one occurrence is a pure literal, while the other is its complement. Observe that these sets are mutually exclusive and comprehensive.

We now introduce the notion of an *independent clause set* with respect to a CNF formula:

Definition 8. Independent clause set: *Given a CNF formula Φ with m clauses over n variables, a set of clauses $I \subseteq \Phi$ is called an independent clause set if no variable that appears in a clause of I appears in any clause $\Phi \setminus I$.*

Example 1. In the formula $\Phi = (x_1 \vee x_2) \wedge (x_7 \vee \bar{x}_8 \vee x_9) \wedge (\bar{x}_2 \vee x_3 \vee \bar{x}_4) \wedge (x_5 \vee \bar{x}_7) \wedge (\bar{x}_1 \vee \bar{x}_2 \vee x_3) \wedge (x_5 \vee x_6 \vee x_7)$, an independent clause set in Φ is $I = (x_1 \vee x_2), (\bar{x}_2 \vee x_3 \vee \bar{x}_4), (\bar{x}_1 \vee \bar{x}_2 \vee x_3)$. Note that the variables in I do not appear in any of the clauses in $\Phi \setminus I$.

Our approach to solve MAXNAE2SAT$_O(n, 2)$ problem is summarized in Algorithm 1.

Algorithm 1. MAXNAE2SAT$_O(n, 2)$

Require: A MAXNAE2SAT$_O(n, 2)$ formula Φ with m clauses over n variables.

Ensure: An assignment nae-satisfying the maximum number of clauses.

Let O, S, and N denote the sets of the variables which repetition factor is bounded by one, two with same form, and two with different forms, respectively.

Define an array A of size n to store the truth values of variables.

while $(O \neq \emptyset)$ **do**

    Find an arbitrary clause $(x_i \vee x_j)$ such that $x_i \in O$.

    Obtain the independent clause set I starting with the clause $(x_i \vee x_j)$.

    Assign the truth values of the variables in I to make all the clauses nae-satisfied.

    Update Φ, O, S, and N by removing the variables and clauses of I.

    Set the truth values of the variables of I in array A.

while $(S \neq \emptyset)$ **do**

    Find the clauses $(x_i \vee x_j)$ and $(x_i \vee x_k)$ such that $x_i \in S$.

    Obtain the independent clause set I starting with the clauses $(x_i \vee x_j)$ and $(x_i \vee x_k)$.

    Assign the truth values of the variables in I to make maximum number of clauses nae-satisfied.

    Update Φ, S, and N by removing the variables and clauses of I.

    Set the truth values of the variables of I in array A.

while $(N \neq \emptyset)$ **do**

    Find the clauses $(x_i \vee x_j)$ and $(\bar{x}_i \vee x_k)$ such that $x_i \in N$.

    Obtain the independent clause set I starting with the clauses $(x_i \vee x_j)$ and $(\bar{x}_i \vee x_k)$.

    Assign the truth values of the variables in I to make maximum number of clauses nae-satisfied.

    Update Φ and N by removing the variables and clauses of I.

    Set the truth values of the variables of I in array A.

Return A.

Lemma 4. *The algorithm produces an optimal solution in the independent clause set starting with $(x_i \vee x_j)$, where $x_i \in O$.*

Proof. Pick an arbitrary literal $x_i \in O$ that appears in exactly one clause $C_i = (x_i \vee x_j)$. Set x_i as **true** and x_j as **false** in C_i. Find the independent clause set I by searching a clause where x_j or its complement appears. If such a clause does not exist, then $x_j \in O$; otherwise, there exists a clause $C_j = (x_j \vee x_k)$. The truth value of x_j is already set, so to make the clause C_j nae-satisfied, set the truth value of x_k accordingly. Continue the above process and set the truth values of the new variables such that the clauses considered in I during this process are nae-satisfied. Observe that the above process will stop while there is a new variable encounter that appears in only one clause (i.e., the new variable is in O). The algorithm can nae-satisfy all such clauses considered above in I by setting its truth values of variables. Update Φ along with the sets O, S, and N by removing all the nae-satisfied clauses and the variables in I. The algorithm considers all such independent clause set I and makes them nae-satisfied. As the clauses considered in I are independent clause set and the algorithm nae-satisfies I, with this assignment, the algorithm gives as good an assignment as the optimal one.

Observe that the updated 2CNF formula Φ' consists of the variables which occur exactly in two clauses. □

Lemma 5. *The algorithm produces an optimal solution in independent clause set starting with $(x_i \vee x_j)$, where $x_i \in S$.*

Proof. Pick an arbitrary literal $x_i \in S$ that appears in exactly two clauses, either in its pure form or complementary form. Let both the clauses be $C_i = (x_i \vee x_j)$ and $C_i' = (x_i \vee x_k)$. If $x_j = x_k$, then delete C_i' from Φ and set the value of x_i as **true** and x_j as **false**. Observe that both the clauses consist of the same variables, so deleting one clause does not affect the optimal solution. If $x_k = \bar{x}_j$, then regardless of how the algorithm assigns truth values to x_i, x_j, and x_k, one of the two clauses will fail to be nae-satisfied. Thus, the algorithm sets x_i to **true** and x_j to **false**. If both x_j and x_k are different variables, then the algorithm sets x_i as **true** and both x_j, and x_k as **false** to nae-satisfy C_i and C_i'. Now, the algorithm considers the clauses where x_j and x_k are present. As each variable appears exactly in two clauses in Φ', x_j and x_k must appear in either two clauses, say C_j and C_k, or in one clause, say C_{jk}.

Case (i) If x_j and x_k appear in clause C_{jk}, then the nae-satisfiability of the clause is already decided by the algorithm. If clause C_{jk} is nae-satisfied, then the independent clause set considered in this phase is nae-satisfied. Otherwise, only the last clause is not nae-satisfied. So, the algorithm gives as good an assignment as the optimal one with this assignment.

Case (ii) If x_j and x_k appear in two different clauses $C_j = (x_j \vee x_l)$ and $C_k = (x_k \vee x_m)$, then the algorithm makes both the clauses nae-satisfied. Observe that the truth values of x_j and x_k are already assigned. So, the algorithm nae-satisfies C_j and C_k by assigning the appropriate truth values for the literals x_l and x_m. Note that this process stops when both the new variables appear in one clause, say $C_{lm} = (x_l \vee x_m)$ (see Case (i)). Otherwise, the process continues by considering another two clauses where x_l and x_m appear (see Case (ii)). In the above process, the algorithm assigns the truth values of the variables, which nae-satisfies either all the clauses or all but the last clause considered in the independent clause set. So, the algorithm gives as good an assignment as the optimal one with this assignment. At the end of the process, update Φ', S, and N by removing all the clauses and the variables in this independent clause set. □

Lemma 6. *The algorithm produces an optimal solution in independent clause set starting with $(x_i \vee x_j)$, where $x_i \in N$.*

Proof. In this case, Φ' may consist of only clauses where each variable appears twice; one is pure, and the other is its complement. The algorithm repeats the similar procedure as in the proof of Lemma 5 and assigns the truth values of the variables for the remaining clauses in Φ' and gives as good an assignment as the optimal one in this case. □

Lemma 7. *The algorithm produces an optimal solution for the $MAXNAE2SAT_O(n, 2)$ problem.*

Proof. The algorithm achieves an optimal solution within each of the independent clause sets starting with O, S, and N, as established in Lemmas 4, 5, and 6 respectively. Since the variables appearing in each independent clause set are distinct and do not occur in any other set, the sum of the optimal solutions over all independent clause sets yields the overall optimal solution for the $MAXNAE2SAT_O(n, 2)$ instance Φ. □

Lemma 8. *The algorithm runs in linear time.*

Proof. Create an array A of size n to store the truth values of the n variables of Φ. The value of $A[i]$ is set to 0, 1, or 2 to indicate the truth value of the ith variable as **false**, **true**, or **unset**, respectively. Initially, all the locations of the array are initialized to 2, i.e., all the variables are unset. In one pass over Φ, by setting appropriate counters, we can identify the sets O, S, and N. The sets O, S, N, and the clauses in Φ are represented by linked lists. Furthermore, with each variable, there is a pointer to and from the positions of each of its occurrences in the clauses to which it belongs. Observe that each time a variable in O is considered, the algorithm finds a set of independent clause set I. It nae-satisfies I by assigning the appropriate truth values to the variables of I. It updates the truth values of the variables of I in array A. All the variables and clauses in I are removed from the input instance. Considering the above data structure, finding each clause takes constant time. The algorithm takes time proportional to the length of the clauses removed for the above process. After removing all the independent clause sets encountered, while checking the set O, the algorithm does the same, considering the sets S and N. Thus, the entire algorithm takes $O(|\Phi|)$ time. Since each variable appears at most twice in Φ and Φ is a 2CNF formula, $|\Phi| \leq n$. Thus, the algorithm takes $O(n)$ time. □

7 An FPT Algorithm

In this section, we design an **FPT** algorithm for the MAXNAE2SAT$_D(n,3)$ problem based on some reduction rules and branching rules. A reduction rule is a polynomial algorithm that transforms the original problem into a solution-preserving simpler problem. A branching rule, on the other hand, recursively splits the problem into multiple subinstances (branches) based on the possible assignments of selected variables or clauses. Each branch represents a restricted version of the original problem, and the overall algorithm explores these branches to find an optimal solution. The efficiency of the branching strategy is often analyzed in terms of the branching factor and the size of the search tree.

Let Φ be an instance of the MAXNAE2SAT$_D(n,3)$ problem with m clauses over n variables. As each variable appears in at most three clauses and each clause consists of two literals, it follows that $m \leq \frac{3}{2} \cdot n$. We define the parameterized version of the MAXNAE2SAT$_D(n,3)$ problem as follows:

Instance: A 2CNF formula Φ as an instance of the MAXNAE2SAT$_D(n,3)$ problem and a parameter $k \geq 1$.
Question: Does there exist an assignment for the 2CNF formula Φ satisfying at least k clauses?

We first describe some of the reduction rules for the MAXNAE2SAT$_D(n,3)$ problem as follows:

Reduction rule 1. If $\Phi \equiv (x_1 \vee x_2) \wedge (x_1 \vee x_2) \wedge \Phi'$, then $\Phi \leftarrow (x_1 \vee x_2) \wedge \Phi'$.

Correctness: Since both clauses are identical, assigning different truth values to x_1 and x_2 will nae-satisfy both. Thus, retaining only one copy preserves the nae-satisfiability structure. At a later stage, after all reduction and branching rules have been applied and Φ' is solved optimally, if the clause $(x_1 \vee x_2)$ is nae-satisfied in the optimal assignment, then $k \leftarrow (k-2)$. Otherwise, k will not change.

Reduction rule 2. If $\Phi \equiv (x_1 \vee x_1) \wedge \Phi'$, then $\Phi \leftarrow \Phi'$ and $k \leftarrow k$.

Correctness: The clause $(x_1 \vee x_1)$ will never be nae-satisfied.

Reduction rule 3. If $\Phi \equiv (x_1 \vee x_2) \wedge \Phi'$ and the repetition factor of both the variables x_1 and x_2 is bounded by one, then $\Phi \leftarrow \Phi'$ and $k \leftarrow (k-1)$.

Correctness: Note that the repetition factor of the variables x_1 and x_2 is one. Thus, they are present only in the clause $(x_1 \vee x_2)$. The clause $(x_1 \vee x_2)$ can be nae-satisfied by setting the truth values of the variables x_1 as **true** and x_2 as **false** or vice versa.

Reduction rule 4. If $\Phi \equiv (x_1 \vee x_2) \wedge (\bar{x}_1 \vee x_2) \wedge (x_1 \vee \bar{x}_2) \wedge \Phi'$, then $\Phi \leftarrow \Phi'$ and $k \leftarrow (k-2)$.

Correctness: The variables x_1 and x_2 appear exactly three times across the three clauses above. Thus, the repetition factor of each variable in Φ is bounded by three, implying that x_1 and x_2 do not occur in any other clause of Φ'. For any truth value assignment to x_1 and x_2, at most two of these three clauses can be nae-satisfied. Therefore, assign the truth values **true** to both x_1 and x_2. Under this assignment, the first clause is not nae-satisfied, but the second and third clauses are. There exist other nae-satisfying assignments as well, each satisfying exactly two out of the three clauses.

Reduction rule 5. If $\Phi \equiv (x_1 \vee x_2) \wedge (x_1 \vee \bar{x}_2) \wedge \Phi'$ or $\Phi \equiv (x_1 \vee x_2) \wedge (\bar{x}_1 \vee x_2) \wedge \Phi'$ and the repetition factor of both the variables x_1 and x_2 is two, then $\Phi \leftarrow \Phi'$ and $k \leftarrow (k-1)$.

Correctness: For any possible truth values of x_1 and x_2, both the clauses will never be nae-satisfied simultaneously. Thus, assign the truth values of either both the variables as **true** or both the variables as **false**. We can also assign the truth value of x_1 as **true** and x_2 as **false**. All of the above assignments nae-satisfy one of the two clauses

Reduction rule 6. If $\Phi \equiv (x_1 \vee x_2) \wedge (\bar{x}_1 \vee x_3) \wedge \Phi'$, then $\Phi \leftarrow (x_2 \vee x_3) \wedge \Phi'$ and $k \leftarrow (k-1)$.

Correctness: In the original formula $\Phi \equiv (x_1 \vee x_2) \wedge (\bar{x}_1 \vee x_3) \wedge \Phi'$, for any assignment to x_1, there exists a corresponding assignment to x_2 and x_3 such that both clauses are nae-satisfied if and only if $(x_2 \vee x_3)$ is nae-satisfied. Thus, replacing the two clauses with $(x_2 \vee x_3)$ reduces the number of clauses by one while maintaining the maximum number of nae-satisfiable clauses. Hence, the update $k \leftarrow (k-1)$ is valid.

Apply the above reduction rules starting from Rule 1 through 6 exhaustively on Φ. Once none of the reduction rules are applicable to Φ, it follows that Φ contains no pair of clauses with identical literals. Furthermore, no clause in Φ includes two literals corresponding to the same variable.

We now present two branching rules for the MAXNAE2SAT$_D(n, 3)$ problem, as described below:

Let x_i be a variable with a repetition factor of three in the given formula. If x_i appears in the formula with different forms, then we design the following branching rule.

Branching rule 1. If the formula $\Phi = G \wedge (x_i \vee y_1) \wedge (x_i \vee y_2) \wedge (\bar{x}_i \vee y_3)$, then there is a branching:

1. $(G \wedge (y_1 \vee y_3) \wedge (y_2 \vee y_3), k-1)$.

2. $(G', k-2)$, where G' is obtained from G by assigning all literals y_1, y_2, and y_3 to either **true** or assigning them all to **false**.

Correctness: Let $R = (y_1 \vee y_3) \wedge (y_2 \vee y_3)$. Observe that if an optimal assignment nae-satisfies s ($s = 1, 2$) clauses from R, then we can nae-satisfy $(s+1)$ clauses from $(\Phi \setminus G)$ but cannot nae-satisfy $(s+2)$ clauses. However, if an optimal assignment does not nae-satisfy any clause from R, we can still nae-satisfy two clauses from $(\Phi \setminus G)$ by setting x_i as **false**, if all the literals y_1, y_2, and y_3 are **true** and vice versa. Note that the case $\Phi = G \wedge (\bar{x}_i \vee y_1) \wedge (\bar{x}_i \vee y_2) \wedge (x_i \vee y_3)$ is similar to the above case.

If x_i appears thrice in the formula with the same form, then we design the following branching rule.

Branching rule 2. If $\Phi = G \wedge (x_i \vee y_1) \wedge (x_i \vee y_2) \wedge (x_i \vee y_3)$, then there is a branching:

1. $(G \wedge (y_1 \vee \bar{y}_3) \wedge (y_2 \vee \bar{y}_3), k-1)$.

2. $(G', k-2)$, where G' is obtained from G by assigning the truth value of the literals y_1, y_2 as **true** and y_3 as **false** or y_1, y_2 as **false**, and y_3 as **true**.

Observe that the correctness of Branching rule 2 follows using a combinatorial argument similar to the one used to the correctness of Branching rule 1.

Algorithm 2. MAXNAE2SAT$_D(n, 3)$

Require: A MAXNAE2SAT$_D(n, 3)$ formula Φ and a parameter k (number of clauses asked to nae-satisfy).

Ensure: true, if k clauses are nae-satisfied simultaneously; **false**, otherwise.

1: **while** ($k \neq 0$ or $\Phi \neq \emptyset$) **do**

2:     Apply Reduction rules starting from 1 to 6 on Φ.

3:     **if** (all the variables in Φ have repetition factors bounded by two) **then**

4:         Apply the algorithm discussed in Section 6 to solve it in linear time.

5:     Choose a variable x_i such that the repetition factor of x_i is three.

6:     Apply Branching rule 1 or 2 according to the form of x_i.

7: Return **true**, if k clauses are nae-satisfied; **false**, otherwise.

Lemma 9. *Let x_i be a variable with repetition factor three, and reduction rules 1 to 6 are not applicable to Φ. By applying Branching rule 1 or Branching rule 2, we have a $(2,4)$-branching, and the resulting formulas are instances of the MAXNAE2SAT$_D(n, 3)$ problem.*

Proof. Applying Branching rule 1 or 2, we can eliminate one clause and get $\Phi' = G \wedge (y_1 \vee y_3) \wedge (y_2 \vee y_3)$ or $\Phi' = G \wedge (y_1 \vee \bar{y}_3) \wedge (y_2 \vee \bar{y}_3)$. Note that the repetition factor of variable y_3 may become four. Branching on the variable y_3 gives MAXNAE2SAT$_D(n,3)$ formulas in both branches. The branches could be either $(1,3)$ or $(2,2)$, i.e., y_3 appears once in one branch, thrice in another, or twice in each branch. We denote the branching factor of the branching vector (t_i, t_j) by $\tau(t_i, t_j)$. Note that $\tau(2,2) < \tau(1,3)$ [5]. So, the worst case number of leaves in the branching is from the $(1,3)$-branching. As one clause is already nae-satisfied, the resulting branching is a $(2,4)$-branching in the worst case [5]. □

Lemma 10. *Algorithm 2 solves the MAXNAE2SAT$_D(n,3)$ problem in $O(n^2 \cdot 1.2721^k)$ time.*

Proof. Observe that all the reduction rules can be applied in polynomial time. By the algorithm discussed in Sect. 6, any MAXNAE2SAT$_O(n,2)$ formula can be solved in linear time. Hence, the running time of Steps 1 to 4 is $O(n^2)$. As per Lemma 9, Step 6 yields a $(2,4)$-branching. Thus, the running time of the branching is $\tau(2,4)^k < 1.2721^k$ [5]. Therefore, the running time of the algorithm is $O(n^2 \cdot 1.2721^k)$. □

8 Conclusion

This paper investigated variants of the MAXNAESAT problem, following a line of inquiry similar to the investigations of MAXSAT variants as detailed in [18]. We established a new inapproximability bound for the MAXNAE2SAT problem by presenting an alternative **APX-completeness** proof. Additionally, we showed that the MAXNAE2SAT problem is solvable in linear time when the repetition factor of each variable is bounded by two. For instances where the repetition factor is bounded by three, we designed an **FPT** algorithm. The parameter for this **FPT** algorithm is the number of nae-satisfied clauses in an optimal solution.

References

1. Alimonti, P., Kann, V.: Some APX-completeness results for cubic graphs. Theoret. Comput. Sci. **237**(1–2), 123–134 (2000)
2. Berman, P., Karpinski, M.: On some tighter inapproximability results. In: International Colloquium on Automata, Languages, and Programming, pp. 200–209 (1999)
3. Brakensiek, J., Huang, N., Potechin, A., Zwick, U.: On the mysteries of MAX NAE-SAT. SIAM J. Discret. Math. **39**(1), 267–313 (2025)
4. Edwards, K., McDermid, E.: A general reduction theorem with applications to pathwidth and the complexity of MAX 2-CSP. Algorithmica **72**(4), 940–968 (2015)
5. Fomin, F.V., Kratsch, D.: Exact Exponential Algorithms. Texts in Theoretical Computer Science. An EATCS Series. Springer (2010)
6. Garey, M.R., Johnson, D.S.: Computers and Intractability: A Guide to the Theory of NP-Completeness. W. H. Freeman Company, San Francisco (1979)

7. Gaspers, S., Sorkin, G.B.: Separate, measure and conquer: faster polynomial-space algorithms for MAX 2-CSP and counting dominating sets. ACM Trans. Algorithms (TALG) **13**(4), 1–36, (2017)
8. Goemans, M.X., Williamson, D.P.: Improved approximation algorithms for maximum cut and satisfiability problems using semidefinite programming. J. ACM (JACM) **42**(6), 1115–1145 (1995)
9. Håstad, J.: Some optimal inapproximability results. J. ACM (JACM) **48**(4), 798–859 (2001)
10. Kann, V., Lagergren, J., Panconesi, A.: Approximability of Maximum Splitting of k-sets and some other APX-complete Problems. Citeseer (1995)
11. Khot, S., Kindler, G., Mossel, E., O'Donnell, R.: Optimal inapproximability results for max-cut and other 2-variable CSPs? SIAM Jo. Comput. **37**(1), 319–357 (2007)
12. Lee, A., Xu, B.: Classifying approximation algorithms: understanding the APX complexity class. arXiv preprint arXiv:2111.01551 (2021)
13. Li, W., Xu, C., Wang, J., Yang, Y.: An improved branching algorithm for (n, 3)-maxsat based on refined observations. In: International Conference on Combinatorial Optimization and Applications, pp. 94–108 (2017)
14. Misra, N., Ordyniak, S., Raman, V., Szeider, S.: Upper and Lower Bounds for Weak Backdoor Set Detection. In: Järvisalo, M., Van Gelder, A. (eds.) SAT 2013. LNCS, vol. 7962, pp. 394–402. Springer, Heidelberg (2013). https://doi.org/10.1007/978-3-642-39071-5_29
15. Christos, H.: Papadimitriou. Addison-Wesley, Computational complexity (1994)
16. Papadimitriou, C.H., Yannakakis, M.: Optimization, approximation, and complexity classes. J. Comput. Syst. Sci. **43**(3), 425–440 (1991)
17. Porschen, S., Randerath, B., Speckenmeyer, E.: Linear time algorithms for some not-all-equal satisfiability problems. In: SAT, pp. 172–187 (2003)
18. Raman, V., Ravikumar, B., Rao, S.S.: A simplified np-complete MAXSAT problem. Inf. Process. Lett. **65**(1), 1–6 (1998)
19. Subramani, K., Gu, X.: Absorbing random walks and the NAE2SAT problem. Int. J. Comput. Math. **88**(3), 452–467 (2011)
20. Xian, A., Zhu, K., Zhu, D., Lianrong, P., Liu, H.: Approximating Max NAE-k-SAT by anonymous local search. Theoret. Comput. Sci. **657**, 54–63 (2017)
21. Zhang, J., Ye, Y., Han, Q.: Improved approximations for max set splitting and max NAE SAT. Discret. Appl. Math. **142**(1–3), 133–149 (2004)
22. Zwick, U.: Approximation algorithms for constraint satisfaction problems involving at most three variables per constraint. SODA **98**, 201–210 (1998)

Exact Algorithms for the Maximum k-Balanced Weighted Biclique Problem

Jingyi Liu, Jianxin Wang, Qilong Feng, and Feng Shi[✉]

School of Computer Science and Engineering, Central South University, Changsha 410083, China
fengshi@csu.edu.cn

Abstract. The Maximum k-Balanced Weighted Biclique problem looks for a biclique in the given vertex-weighted bipartite graph such that the weight of the biclique is maximized, and the gap between the weights of the two independent vertex sets of the biclique is at most the given value k. Within the paper, we propose an exact algorithm for the problem with a new perspective: feedback vertex set. Specifically, our approach begins by conducting branch operations on the vertices with large degree and the ones in the feedback vertex set of the considered bipartite graph G, then calls a polynomial-time algorithm proposed for a related problem to each resulting graph that is acyclic. Our algorithm is shown to have time complexity $O(\min\{1.325^n, 2^{2\delta(G)}\}n^4 W_{max}^4)$, where n and W_{max} are the number and maximum weight of the vertices in G, respectively, and $\delta(G)$ is the minimum cardinality (i.e., number of vertices) of a feedback vertex set for G. Furthermore, our algorithm can be adapted to solve the Maximum Balanced Biclique problem (i.e., $k = 0$ and G is unweighted) with time complexity $O(\min\{1.325^n, 2^{2\delta(G)}\}n^2 \log^6 n)$, which is better than the best-known time complexity $O(1.3803^n n^2)$, especially if G satisfies $\delta(G) \leq n/5$.

Keywords: k-balanced weighted biclique · k-balanced weighted independent set · balanced biclique · balanced independent set · feedback vertex set

1 Introduction

The Maximum Clique problem [6,20,30] is well-studied in graph theory and is equivalent to the Maximum Independent Set problem [27,31,32]. Consider an unweighted bipartite graph $G = (L, R; E)$ with two disjoint vertex sets L, R and an edge set $E \subset L \times R$. The *Maximum Vertex Biclique problem* (abbr. MVBCP) on G looks for a *biclique* (i.e., a subset $L' \subset L$ and one $R' \subset R$ such that the induced subgraph $G[L' \cup R']$ is a complete bipartite graph) with the maximum number of vertices, and the *Maximum Edge Biclique problem* (abbr. MEBCP) on G looks for a *biclique* with the maximum number of edges. It is well-known that the MVBCP is polynomial-time solvable by the technique of integer linear programming [10] or the algorithm for the Maximum Matching in bipartite graph [12]. However, the MEBCP was shown to be NP-hard [19], and there are lots of works on it, e.g. [7,10,15,17,18,22,24].

If considering an extra constraint on the MVBCP that $|L'| = |R'|$, then we have the *Maximum Balanced Biclique problem* (abbr. MBBCP), which looks for a maximum biclique (L', R') with $|L'| = |R'|$ (note that the sizes of such bicliques can be measured by the number of vertices or edges). It has been extensively studied due to applications including nanoelectronic system design [1,25], biclustering of gene expression data [9,33], co-clustering of social network [13], and programmable logic array folding in the VLSI theory [21]. The MBBCP is NP-hard [2,11]. Except for some heuristic algorithms including [1,14,25,29,34–36,38], there are several exact algorithms proposed for it. Tahoori [25] proposed a trivial recursive exact algorithm. Later, McCreesh and Prosser [16] gave a branch-and-bound algorithm, which was improved by Zhou et al. [39]. Then Chen et al. [8] presented an improved branch-and-bound algorithm with time complexity $O(1.3803^n n^2)$. To our best knowledge, this is the first exact algorithm for the MBBCP with a non-trivial theoretical guarantee.

Recently Zhao et al. [37] introduced the *Maximum k-Balanced Weighted Biclique Problem* (abbr. M-k-BWBCP), which is a generalization of the MBBCP. Given a weighted bipartite graph $G = (L, R; E; W)$ with weight function W on $L \cup R$, the M-k-BWBCP asks to find a biclique (L', R') that maximizes $W(L') + W(R')$ with $|W(L') - W(R')| \le k$. When G is unweighted and $k = 0$, it is equivalent to the MBBCP. Zhao et al. [37] proposed a branch-and-bound algorithm with time complexity $O(d_{max}^2 \beta)$, where d_{max} is the maximum degree of the vertices and β is the total number of maximal bicliques in G.

Contribution. As any biclique of a bipartite graph G corresponds to an independent set in its complement $\overline{G}$, the M-k-BWBCP is equivalent to the *Maximum k-Balanced Weighted Independent Set Problem* (abbr. M-k-BWISP), whose definition is almost the same as that of the M-k-BWBCP, except that the M-k-BWISP looks for an independent set, not a biclique. Within the investigation, we focus on the M-k-BWISP. Our original idea is to first obtain an approximate *feedback vertex set* (abbr. FVS) of the considered bipartite graph G, then enumerate all its subsets S, and for each, call an algorithm on the resulting forest $G - \text{FVS} - N(S)$ to get an independent set forming a k-balanced independent set with S. However, as the ratio of the number of vertices in the approximated FVS of G to the number of vertices in G may approach 1, we refine this idea as follows. Firstly, we branch on the vertices with degree at least 4 (specifically, assume it is in or not in an optimal solution of M-k-BWISP). Then for each resulting graph G' with the maximum degree at most 3, we find a minimum cardinality FVS in polynomial time. Since G' is bipartite and subcubic, the cardinality of the obtained FVS can be bounded by $3|V(G')|/8$ and the performance of enumerating all subsets of the FVS can be upper bounded. Finally, for each subset S of the obtained FVS, the algorithm on the forest $G' - \text{FVS} - N(S)$ is called to get an independent set that, together with S, forms a k-balanced independent set. By the revised idea, we give an exact algorithm for the M-k-BWISP and M-k-BWBCP, with time complexity $O(\min\{1.325^n, 2^{2\delta(G)}\}n^4 W_{max}^4)$, where n and W_{max} are the number and maximum weight of the vertices in the considered graph G, respectively, and $\delta(G)$ is the minimum cardinality of an FVS for G. In addition, we adapt the proposed algorithm to solve the MBBCP with time complexity $O(\min\{1.325^n, 2^{2\delta(G)}\}n^2 \log^6 n)$, which is better than the one given in [8] with time complexity $O(1.3803^n n^2)$.

The paper is organized as follows. Section 2 gives the related notions and formulations of the considered problems. Section 3 gives an algorithm for the Weighted Independent Set Enumeration on Forest Problem, which is critical to the discussion for the M-k-BWISP and M-k-BWBCP. Then Sect. 4 gives the algorithm for the M-k-BWISP and M-k-BWBCP, and extends it to the MBBCP.

2 Preliminary

Consider a weighted bipartite graph $G = (L, R; E; W)$ with two disjoint vertex sets L, R, an edge set $E \subseteq L \times R$, and a weight function $W : L \cup R \to \mathbb{Z}^+$. Given a subset S of $L \cup R$, let $L(S) = L \cap S$ and $R(S) = R \cap S$, $W(S) = \sum_{v \in S} W(v)$, and $G[S]$ be the subgraph of G induced by S. The *cardinality*, *weight*, and *vectorial weight* of S are $|S|$, $W(S)$, and $(W(L(S)), W(R(S)))$, respectively. Additionally, the *pairing weight* of S is defined as $W(L(S)) \cdot n \cdot W_{max} + W(R(S))$, where $n = |L \cup R|$, and W_{max} is the maximum weight that the vertices of G have. As $W(S) \le nW_{max}$, for any two subsets of $L \cup R$, their vectorial weights are the same iff their pairing weights are the same. If $L(S) \ne \emptyset$ and $R(S) \ne \emptyset$, and there is an edge between any $u \in L(S)$ and any $v \in R(S)$ in $G[S]$, then S is a *biclique* of G. If $W(L(S)) = W(R(S))$ holds as well then S is a *balanced weighted biclique* (abbr. BWBC); if $|W(L(S)) - W(R(S))| \le k$ then S is a *k-balanced weighted biclique* (abbr. k-BWBC). The following is the formulation of the *Maximum k-Balanced Weighted Biclique problem*.

Maximum k-Balanced Weighted Biclique Problem (abbr. M-k-BWBCP)
INPUT: a weighted bipartite graph $G = (L, R; E; W)$;
OUTPUT: a k-BWBC in G with the maximum weight.

Denote by $\overline{G} = (L, R; \overline{E}; W)$ the complement of G, which has the same vertex set $L \cup R$ and weight function W with G, and has an edge (u, v) between $u \in L$ and $v \in R$ iff $(u, v) \notin E$. Given a subset S of $L \cup R$, if there is no edge between any vertex of $L(S)$ and any one of $R(S)$ in $G[S]$, then S is an *independent set* (abbr. IS) of G (note that $L(S)$ and $R(S)$ may be empty). If $W(L(S)) = W(R(S))$ holds as well then S is a *balanced weighted independent set* (abbr. BWIS); if $|W(L(S)) - W(R(S))| \le k$ then S is a *k-balanced weighted independent set* (abbr. k-BWIS). Observe that any biclique of the considered bipartite graph G is an independent set of $\overline{G}$, and any k-BWBC in G is a k-BWIS of $\overline{G}$. Thus we have the following problem, which is equivalent to the M-k-BWBCP.

Maximum k-Balanced Weighted Independent Set Problem (abbr. M-k-BWISP)

Algorithm ALG-IS-COLLECT(T, v)
Input: A weighted rooted tree T, and a vertex v of T;
Output: Return $\mathcal{C}(v)$ and $\overline{\mathcal{C}}(v)$.
1. **if** v is a leaf **then** return $\mathcal{C}(v) = \{\{v\}\}$ and $\overline{\mathcal{C}}(v) = \{\emptyset\}$;
2. **for** each child c_v of v
3.   call the ALG-IS-COLLECT$(T[D[c_v]], c_v)$ to obtain $\mathcal{C}(c_v)$ and $\overline{\mathcal{C}}(c_v)$;
4. let $\mathcal{C}(v) = \overline{\mathcal{C}}(v) = \{\emptyset\}$;
// construct $\overline{\mathcal{C}}(v)$
5. **for** each child c_v of v
6.   let $\mathcal{C}' = \overline{\mathcal{C}}(v)$;
7.   **for** each element S of $\mathcal{C}'$
8.   **for** each element S' of $\mathcal{C}(c_v) \cup \overline{\mathcal{C}}(c_v)$
9.   **if** $\overline{\mathcal{C}}(v)$ has no element with the same pairing weight as $S \cup S'$ **then**
10.   add $S \cup S'$ into $\overline{\mathcal{C}}(v)$;
// construct $\mathcal{C}(v)$
11. **for** each child c_v of v
12.   let $\mathcal{C}' = \mathcal{C}(v)$;
13.   **for** each element S of $\mathcal{C}'$
14.   **for** each element S' of $\overline{\mathcal{C}}(c_v)$
15.   **if** $\mathcal{C}(v)$ has no element with the same pairing weight as $S \cup S' \cup \{v\}$ **then**
16.   add $S \cup S' \cup \{v\}$ into $\mathcal{C}(v)$;
17. return $\mathcal{C}(v)$ and $\overline{\mathcal{C}}(v)$.

Fig. 1. The algorithm to construct the collections $\mathcal{C}(v)$ and $\overline{\mathcal{C}}(v)$ for a vertex v of T.

INPUT: a weighted bipartite graph $G = (L, R; E; W)$;
OUTPUT: a k-BWIS in G with the maximum weight.

3 Algorithm for the WISEFP

In the section, we consider a simple case that the input bipartite graph G of the M-k-BWISP is a forest and give an algorithm to enumerate all ISs of G with different vectorial weights. Thus we propose a new problem whose formulation is given below. A collection $\mathcal{C}$ of ISs of G is called an *IS family* of G if for each IS S of G, $\mathcal{C}$ contains exactly one IS that has the same vectorial weight as S.

Weighted Independent Set Enumeration on Forest Problem (abbr. WISEFP)
INPUT: a weighted bipartite graph $F = (L, R; E; W)$ that is a forest;
OUTPUT: an IS family of F.

To simplify the discussion, we arbitrarily choose a vertex for each tree $T \in F$ as its root and have the following notations. Given a vertex $v \in V(T)$, let $D(v)$ be the set containing all the descendants of v, and $D[v] = D(v) \cup \{v\}$. Let $T[D[v]]$ be the tree induced by $D[v]$.

For the WISEFP, we first construct an IS family for each tree T in the input forest F. Then we consider feasible unions of elements from these IS families. More specifically, we initialize a collection $\mathcal{C}$ to store an IS family of F, starting

Algorithm ALG-WISEFP(F)
Input: A weighted bipartite graph F that is a forest;
Output: An IS family of F.

1. let $\mathcal{C} = \{\emptyset\}$;
2. **for** each tree T in F
3.   arbitrarily choose a vertex r as the root of T;
4.   call the ALG-IS-COLLECT(T, r) to obtain $\mathcal{C}(r)$ and $\overline{\mathcal{C}}(r)$;
5.   find an IS family $\mathcal{C}_T$ of T from $\mathcal{C}(r) \cup \overline{\mathcal{C}}(r)$;
6.   let $\mathcal{C}' = \mathcal{C}$;   // $\mathcal{C}'$ is used to bound the number of for loop iterations
7.   **for** each element S in $\mathcal{C}'$
8.   **for** each element S' in $\mathcal{C}_T$
9.   **if** $\mathcal{C}$ has no element that has the same pairing weight as $S \cup S'$ **then**
10.   add $S \cup S'$ into $\mathcal{C}$;
11. return $\mathcal{C}$.

Fig. 2. The algorithm for the WISEFP.

with an empty set. For each tree T in F, we call the algorithm ALG-IS-COLLECT (given in Fig. 1) to generate two IS collections, $\mathcal{C}(r)$ and $\overline{\mathcal{C}}(r)$, for the root r of T. These collections satisfy the following properties: (1) Each IS in $\mathcal{C}(r)$ contains r, and for every IS S of T that contains r, $\mathcal{C}(r)$ contains exactly one IS with the same vectorial weight as S; (2) Each IS in $\overline{\mathcal{C}}(r)$ does not contain r, and for every IS S of T that does not contain r, $\overline{\mathcal{C}}(r)$ contains exactly one IS with the same vectorial weight as S. Then we can find an IS family $\mathcal{C}_T$ of T from $\mathcal{C}(r) \cup \overline{\mathcal{C}}(r)$. Next, for each IS $S \in \mathcal{C}$ and IS $S' \in \mathcal{C}_T$, we check if their union can be added to $\mathcal{C}$. After considering all the trees in F, the resulting $\mathcal{C}$ is an IS family of F. The algorithm is summarized in Fig. 2.

Given a rooted tree T, the algorithm ALG-IS-COLLECT runs from the leaves to the root, to construct collections $\mathcal{C}(v)$ and $\overline{\mathcal{C}}(v)$ for each vertex $v \in V(T)$. If v is a leaf, then it is safe to let $\mathcal{C}(v) = \{\{v\}\}$, and $\overline{\mathcal{C}}(v) = \{\emptyset\}$. For the other general cases, we have the following lemma for the construction of $\mathcal{C}(\cdot)$ and $\overline{\mathcal{C}}(\cdot)$.

Lemma 1. *Given an internal vertex v of T and the collections $\mathcal{C}(v')$ and $\overline{\mathcal{C}}(v')$ for each child v' of v, $\mathcal{C}(v)$ and $\overline{\mathcal{C}}(v)$ can be constructed correctly.*

Lemma 2. *Given a tree T with root r, the algorithm* ALG-IS-COLLECT *can obtain $\mathcal{C}(r)$ and $\overline{\mathcal{C}}(r)$ with time complexity $O(n^4 W_{max}^4)$, where $n = |V(T)|$, and W_{max} is the maximum weight that the vertices of T have.*

Based on Lemma 2, we have the following theorem.

Theorem 1. *Given a weighted bipartite graph F that is a forest, the algorithm* ALG-WISEFP *can solve the WISEFP on F with time complexity $O(n^4 W_{max}^4)$, where $n = |V(F)|$, and W_{max} is the maximum weight that the vertices of F have.*

Thus if the input graph is a forest, the M-k-BWISP can be efficiently solved by finding an IS family of the forest and selecting a k-BWIS from it with the maximum pairing weight.

Corollary 1. *Given a weighted bipartite graph F that is a forest, the algorithm* ALG-WISEFP *can solve the M-k-BWISP on F with time complexity $O(n^4 W_{max}^4)$, where $n = |V(F)|$, and W_{max} is the maximum weight that the vertices of F have.*

4 Algorithms for the M-k-BWISP

In the section, we consider the general case that input bipartite graph G of the M-k-BWISP is cyclic. Based on the previous work, we aim to delete some vertices from G to make it acyclic. Thus we introduce the Minimum Feedback Vertex Set Problem and the related algorithms in the following.

Given a simple graph $G = (V, E)$, a vertex-subset $V' \subset V$ is a *feedback vertex set* (abbr. FVS) of G if $G-V'$ is acyclic, where $G-V'$ denotes the graph obtained by removing the vertices of V' and the incident edges from G.

Minimum Feedback Vertex Set Problem (abbr. MFVSP)
INPUT: a simple graph G;
OUTPUT: an FVS with the minimum cardinality.

The MFVSP can be defined on both weighted and unweighted graphs, as the goal is to find a feedback vertex set of minimum cardinality. Becker and Geiger [4, 5] gave the first 2-approximation algorithm with time complexity $O(m+n\log n)$, where n and m are the numbers of vertices and edges in the graph, respectively. Bafna et al. [3] later proposed a different 2-approximation algorithm running in $O(\min\{m \log n, n^2\})$. For graphs with maximum degree at most 3, Ueno et al. [28] showed that MFVSP is solvable in polynomial time, and Takaoka [26] gave an $O(n^2 \log^6 n)$-time algorithm.

Now assume that we have obtained an FVS S of G. Then for each subset S' of S that is assumed to be in an optimal solution to the M-k-BWISP on G, we remove the vertices of $N(S') \cup S$ from G to make it acyclic such that the algorithm ALG-WISEFP is applicable. However, we cannot evaluate the algorithm simply based on the above idea as $|S|$ cannot be upper bounded. Fortunately, the following lemma gives some insights.

Lemma 3. *(Corollary 2, [23]) Let G be a connected triangle-free graph with n vertices and m edges. If $m < \lfloor 13n/8 \rfloor$, then $a(G) \geq 5n/8$, where $a(G)$ is the number of vertices in the maximum induced subgraph of G that is a forest.*

This lemma implies that for any connected triangle-free graph G with $m < \lfloor 13n/8 \rfloor$ edges, the minimum cardinality of the FVS of G can be upper bounded by $n - 5n/8 = 3n/8$. Since a bipartite graph G contains no odd cycles, it is triangle-free. Assume that the bipartite graph G has p connected components and maximum degree at most 3. For each component G_i with n_i vertices, it has fewer than $\lfloor 13n_i/8 \rfloor$ edges. By Lemma 3, the minimum FVS size of G_i is at most $3n_i/8$. Summing over all components, the minimum FVS size of G is at most $3n/8$. Thus we can preprocess the vertices in G with degree at least four if any. Specifically, for each such vertex in G, we do branch operations based on

Algorithm ALG-k-BWISP
Input: A weighted bipartite graphs $G = (L, R; E; W)$, and a vertex set B;
Output: A k-BWIS of G with the maximum weight.
1. let $S^* = \emptyset$;
2. **while** G is cyclic and has a vertex v with degree at least 4
3.   do the following two branch operations:
4.   (branch-1). return ALG-k-BWISP$(G - v, B)$;
5.   (branch-2). return ALG-k-BWISP$(G - N[v], B \cup \{v\})$;
6. call ALG-FVS-3 to get an FVS S of G with minimum cardinality;
7. **for** each subset A of S // $A = \emptyset$ if $S = \emptyset$ and it is trivially an IS of G
// assume that the vertices of A are in an optimal solution
8.   **if** A is not an independent set of G **then** continue;
9.   call ALG-WISEFP$(G - (S \cup N(A)))$ and let $\mathcal{C}$ be the returned collection;
10.   **for** each element S' of $\mathcal{C}$
11.   **if** $|W(L(B \cup A \cup S')) - W(R(B \cup A \cup S'))| \leq k$ and $W(B \cup A \cup S') > W(S^*)$
          then
12.   let $S^* = B \cup A \cup S'$;
13. return S^*.

Fig. 3. The algorithm for the M-k-BWISP based on large-degree vertices and FVS.

whether it is in an optimal solution of the M-k-BWISP. Once the resulting graph is acyclic or has the maximum degree at most three, we call the exact algorithm ALG-FVS-3 in [26]. Note that the algorithm ALG-FVS-3 returns an empty set if the resulting graph is acyclic. Then we can reduce the M-k-BWISP on the resulting graph to the WISEFP. Now we are ready to present our algorithm named ALG-k-BWISP for the M-k-BWISP (given in Fig. 3).

Theorem 2. *Given a weighted bipartite graph $G = (L, R; E; W)$, the algorithm ALG-k-BWISP can solve the M-k-BWISP on G with time complexity $O(1.325^n n^4 W_{max}^4)$, where $n = |L \cup R|$, and W_{max} is the maximum weight that the vertices of G have.*

Remark that given an instance of the M-k-BWISP, we can call the $O(m + n \log n)$-time 2-approximation algorithm given in [4,5] or any algorithm with practical performance to get an FVS. If its cardinality is small enough, then we can call the algorithm ALG-WISEFP on the resulting acyclic graph.

Corollary 2. *Given a weighted bipartite graph $G = (L, R; E; W)$, the M-k-BWISP on G can be solved in time complexity $O(\min\{2^{2\delta}, 1.325^n\} n^4 W_{max}^4)$, where $n = |L \cup R|$, and δ is the minimum cardinality of the FVS for G.*

Remark that our algorithm for the M-k-BWISP problem also applies to the MBBCP as a special case when $k = 0$ and the input graph is unweighted. In particular, the related operations given for the WISEFP can be simplified, and we can propose an improved algorithm for the MBBCP. Due to space limit, we directly give the result.

Theorem 3. *Given a bipartite graph $G = (L, R; E)$, the MBISP on G can be solved in time complexity $O(\min\{2^{2\delta}, 1.325^n\} n^2 \log^6 n)$, where $n = |L \cup R|$, and δ is the minimum cardinality of an FVS for G.*

Acknowledgments. This work is supported in part by the National Natural Science Foundation of China under Grants 62472449 and 62332020, and the Hunan Provincial Natural Science Foundation of China under Grant 2025JJ50395.

References

1. Al-Yamani, A.A., Ramsundar, S., Pradhan, D.K.: A defect tolerance scheme for nanotechnology circuits. IEEE Trans. Circ. Syst. I Regul. Pap. **54**(11), 2402–2409 (2007)
2. Alon, N., Duke, R.A., Lefmann, H., Rödl, V., Yuster, R.: The algorithmic aspects of the regularity lemma. J. Algorithms **16**(1), 80–109 (1994)
3. Bafna, V., Berman, P., Fujito, T.: A 2-approximation algorithm for the undirected feedback vertex set problem. SIAM J. Discret. Math. **12**(3), 289–297 (1999)
4. Becker, A., Geiger, D.: Approximation algorithms for the loop cutset problem. In: Proceedings of the 10th Conference on Uncertainty in Artificial Intelligence (UAI), pp. 60–68. Elsevier (1994)
5. Becker, A., Geiger, D.: Optimization of pearl's method of conditioning and greedy-like approximation algorithms for the vertex feedback set problem. Artif. Intell. **83**(1), 167–188 (1996)
6. Bomze, I.M., Budinich, M., Pardalos, P.M., Pelillo, M.: The maximum clique problem. Handbook of Combinatorial Optimization: Supplement Volume A, pp. 1–74 (1999)
7. Chen, H., Liu, T.: Maximum edge bicliques in tree convex bipartite graphs. In: Proceedings of the 11th International Workshop on Frontiers in Algorithmics (FAW), pp. 47–55. Springer (2017)
8. Chen, L., Liu, C., Zhou, R., Xu, J., Li, J.: Efficient exact algorithms for maximum balanced biclique search in bipartite graphs. In: Proceedings of the 47th International Conference on Management of Data (SIGMOD), pp. 248–260 (2021)
9. Cheng, Y., Church, G.M.: Biclustering of expression data. In: Proceedings of the 8th International Conference on Intelligent Systems for Molecular Biology (ISMB), vol. 8, pp. 93–103 (2000)
10. Dawande, M., Keskinocak, P., Swaminathan, J.M., Tayur, S.: On bipartite and multipartite clique problems. J. Algorithms **41**(2), 388–403 (2001)
11. Garey, M.R., Johnson, D.S.: Computers and Intractability: A Guide to the Theory of np-Completeness, Freeman. Fundamental (1997)
12. Kőnig, D.: Gráfok és mátrixok. Matematikai és Fizikai Lapok **38**, 116–119 (1931)
13. Li, J., Sim, K., Liu, G., Wong, L.: Maximal quasi-bicliques with balanced noise tolerance: concepts and co-clustering applications. In: Proceedings of the 8th SIAM International Conference on Data Mining (SDM), pp. 72–83. SIAM (2008)
14. Li, M., Hao, J.-K., Wu, Q.: General swap-based multiple neighborhood adaptive search for the maximum balanced biclique problem. Comput. Oper. Res. **119**, 104922 (2020)
15. Lyu, B., Qin, L., Lin, X., Zhang, Y., Qian, Z., Zhou, J.: Maximum biclique search at billion scale. Proc. VLDB Endowment **13**(9), 1359–1372 (2020)

16. McCreesh, C., Prosser, P.: An exact branch and bound algorithm with symmetry breaking for the maximum balanced induced biclique problem. In: Proceedings of the 11th International Conference Integration of AI and OR Techniques in Constraint Programming (CPAIOR), pp. 226–234. Springer (2014)
17. Nussbaum, D., Pu, S., Sack, J.-R., Uno, T., Zarrabi-Zadeh, H.: Finding maximum edge bicliques in convex bipartite graphs. Algorithmica **64**, 311–325 (2012)
18. Pandey, A., Sharma, G., Jain, N.: Maximum weighted edge biclique problem on bipartite graphs. In: Proceedings of the 6th Conference on Algorithms and Discrete Applied Mathematics (CALDAM), pp. 116–128. Springer (2020)
19. Peeters, R.: The maximum edge biclique problem is NP-complete. Discret. Appl. Math. **131**(3), 651–654 (2003)
20. Peng, Y., Xu, Y., Zhao, H., Zhou, Z., Han, H.: Most similar maximal clique query on large graphs. Front. Comp. Sci. **14**, 1–16 (2020)
21. Ravi, S.S., Lloyd, E.L.: The complexity of near-optimal programmable logic array folding. SIAM J. Comput. **17**(4), 696–710 (1988)
22. Shaham, E., Yu, H., Li, X.: On finding the maximum edge biclique in a bipartite graph: a subspace clustering approach. In: Proceedings of the 16th SIAM International Conference on Data Mining (SDM), pp. 315–323. SIAM (2016)
23. Shi, L., Xu, H.: Large induced forests in graphs. J. Graph Theory **85**(4), 759–779 (2017)
24. Sözdinler, M., Özturan, C.: Finding maximum edge biclique in bipartite networks by integer programming. In: Proceedings of the 21st IEEE International Conference on Computational Science and Engineering (CSE), pp. 132–137. IEEE (2018)
25. Tahoori, M.B.: Application-independent defect tolerance of reconfigurable nanoarchitectures. ACM J. Emerg. Technol. Comput. Syst. **2**(3), 197–218 (2006)
26. Takaoka, A., Tayu, S., Ueno, S.: On minimum feedback vertex sets in bipartite graphs and degree-constraint graphs. IEICE Trans. Inf. Syst. **E96.D**(11), 2327–2332 (2013)
27. Tarjan, R.E., Trojanowski, A.E.: Finding a maximum independent set. SIAM J. Comput. **6**(3), 537–546 (1977)
28. Ueno, S., Kajitani, Y., Gotoh, S.: On the nonseparating independent set problem and feedback set problem for graphs with no vertex degree exceeding three. Discret. Math. **72**(1–3), 355–360 (1988)
29. Wang, Y., Cai, S., Yin, M.: New heuristic approaches for maximum balanced biclique problem. Inf. Sci. **432**, 362–375 (2018)
30. Wu, Q., Hao, J.-K.: A review on algorithms for maximum clique problems. Eur. J. Oper. Res. **242**(3), 693–709 (2015)
31. Xia, X., Peng, X., Liao, W.: On the analysis of ant colony optimization for the maximum independent set problem. Front. Comp. Sci. **15**(4), 1–3 (2021)
32. Xiao, M., Nagamochi, H.: Exact algorithms for maximum independent set. Inf. Comput. **255**, 126–146 (2017)
33. Yang, J., Wang, H., Wang, W., Yu, P.S.: An improved biclustering method for analyzing gene expression profiles. Int. J. Artif. Intell. Tools **14**(05), 771–789 (2005)
34. Yuan, B., Li, B.: A low time complexity defect-tolerance algorithm for nanoelectronic crossbar. In: Proceedings of the 1st International Conference on Information Science and Technology (ICIST), pp. 143–148. IEEE (2011)
35. Yuan, B., Li, B.: A fast extraction algorithm for defect-free subcrossbar in nanoelectronic crossbar. ACM J. Emerg. Technol. Comput. Syst. **10**(3), 1–19 (2014)
36. Yuan, B., Li, B., Chen, H., Yao, X.: A new evolutionary algorithm with structure mutation for the maximum balanced biclique problem. IEEE Trans. Cybern. **45**(5), 1054–1067 (2015)

37. Zhao, Y., Chen, Z., Chen, C., Wang, X., Lin, X., Zhang, W.: Finding the maximum k-balanced biclique on weighted bipartite graphs. IEEE Trans. Knowl. Data Eng. **35**(8), 7994–8007 (2023)
38. Zhou, Y., Hao, J.-K.: Tabu search with graph reduction for finding maximum balanced bicliques in bipartite graphs. Eng. Appl. Artif. Intell. **77**, 86–97 (2019)
39. Zhou, Y., Rossi, A., Hao, J.-K.: Towards effective exact methods for the maximum balanced biclique problem in bipartite graphs. Eur. J. Oper. Res. **269**(3), 834–843 (2018)

Approximation Algorithms for Individual Preference Facility Location

Shuilian Liu[1,2], Yicheng Xu[1,2(✉)], and Yong Zhang[1,2]

[1] Shenzhen Institutes of Advanced Technology, Chinese Academy of Sciences, Beijing, China
[2] University of Chinese Academy of Sciences, Beijing, China
yc.xu@siat.ac.cn

Abstract. We study the facility location problem in the context of individual fairness to propose the *Individual Preference Facility Location* (IPFL) problem. In the vanilla facility location problem, the goal is to select a subset of facilities to serve all clients while minimizing total opening and connection costs. IPFL aims to optimize the facility location objective while meeting individual preferences by requiring that each client is served by a facility within its fair radius. The fair radius is defined as the distance between a client and its τ-th nearest neighbor, where τ is a carefully designed parameter. IPFL balances facility load by opening more facilities in dense areas. However, a few clients may disproportionately affect the final costs or violate the individual preference constraints. To address this, we extend IPFL to its outlier variant, IPFLO, where up to m clients can remain unserved. As our contribution, we provide 2-approximation algorithms for both IPFL and IPFLO using a dual fitting technique.

Keywords: Facility location · Individual fairness · Outliers · Dual fitting

1 Introduction

The Uncapacitated Facility Location (UFL) problem is a well-studied clustering problem in operations research, where the input consists of a set of facilities with opening costs and a set of clients. The goal is to select a subset of facilities from the candidate facilities to serve all clients, minimizing both the total opening and connection costs. Under the assumption of $\mathbf{P} \neq \mathbf{NP}$, the work of [11] establishes that achieving an approximation factor better than 1.463 for the UFL is impossible. On the positive side, a series of approximation algorithms have been well-studied. Shmoys, Tardos, and Aardal [21] propose the first constant factor algorithm using a linear programming (LP)-rounding technique, achieving a ratio of 3.16. Many other approximation algorithms have also been proposed using techniques, such as local search [16], primal dual [14], dual fitting [13], and greedy augmentation [5,11]. Among them, Jain et al. [13] propose a dual fitting algorithm and use the factor-revealing LPs technique to achieve a 1.61-approximation. The best result to date is the 1.488-approximation algorithm proposed by Li [17], which combines the techniques of LP-rounding and dual fitting. Dual fitting is a versatile technique that can be adapted to solve other generalizations of the UFL. The UFL with outliers is introduced by Charikar et al. [6], who use the primal-dual technique to obtain a 3-approximation algorithm. Jain et al. [13] employ the dual fitting technique, improving the approximation ratio to 2.

Recently, fair clustering has become a crucial and widely discussed topic in machine learning and combinatorial optimization. Fairness is typically defined either for protected minority groups (e.g., ethnicity, gender) or for individuals. Various fair clustering problems have been proposed, for example, balanced clustering [8], colorful clustering [2], socially fair clustering [1,10], and individually fair clustering [15]. Chierichetti et al. [8] first introduce fairness into clustering, ensuring that each cluster must have an approximately equal proportion of representations. They propose the fairlet decomposition method to achieve an approximation balance in the output for the k-center, k-median clustering. Jung et al. [15] first propose individually fair clustering for centroid-based clustering, including k-center, k-median, and k-means. They define the fair radius for each point as the distance to its $\lceil \frac{n}{k} \rceil$-th nearest neighbor, where n denotes the total number of data points and k is the specified number of clusters. Individually fair clustering requires that each point must be assigned to a center within its fair radius.

While fair clustering has received significant attention, incorporating fairness into the UFL problem remains underexplored, with most studies on fair clustering focusing on the k-center, k-median, and k-means problems. In this paper, we introduce a novel model, the Individual Preference Facility Location (IPFL) problem, inspired by Jung et al. [15], which incorporates individual fairness into facility location. The goal is to select a subset of facilities from the candidate facilities to serve all clients, ensuring that each client is assigned to a facility within its fair radius. The fair radius is defined as the distance from a client to its τ-th nearest neighbor, where τ can represent a population threshold. We aim to minimize both the opening and connection costs while satisfying individual preference constraints.

Our Contributions. In this paper, we introduce the Individual Preference Facility Location (IPFL) problem, which incorporates individual fairness into facility location. In IPFL, individual preferences define clients' service expectations, ensuring that each client is served by a facility within its fair radius. To solve IPFL, we propose a dual-fitting-based algorithm that guarantees a 2-approximation. We also extend IPFL to its outlier variant, Individual Preference Facility Location with Outliers (IPFLO), allowing up to m clients to be excluded from service. We develop a 2-approximation algorithm for IPFLO using a similar approach.

1.1 Related Work

Jung et al. [15] first introduce the concept of individually fair clustering, which ensures that each point has a center within its $\lceil \frac{n}{k} \rceil$-th nearest neighbor. However, it has been observed that individually fair clustering does not always exist. They propose a 2-approximation algorithm in metric space, ensuring that every point is assigned to a center within at most twice their fair radius. Subsequently, a series of bi-criteria (β, γ)-approximation algorithms have been developed. These algorithms guarantee that each point is assigned to a center within a distance of β times its fair radius, while the clustering cost is at most γ times the optimal cost. Mahabadi et al. [18] consider individually fair clustering in the l_p-norm cost function. They use local search technique to achieve $(84, 7)$-approximation for k-median $(p = 1)$, $(O(1), 7)$-approximation for k-means $(p = 2)$, and $(O(\log n), 7)$-approximation for k-center $(p = \infty)$, respectively. Chakrabarty et al. [20] formulate individual fairness constraints in a linear programming model and improve to $(8, 8)$-approximation for k-median, $(4, 8)$-approximation for k-means, and $(2+\varepsilon, 8)$-approximation for k-center, respectively via LP-rounding. Vakilian et al. [22] improve the approximation ratio by reducing the problem of individually fair clustering to a median matroid problem, achieving $(7.081 + \varepsilon, 3)$-approximation for k-median, $(16+\varepsilon, 3)$-approximation for k-means, and $(3+\varepsilon, 3)$-approximation k-center. Chhaya et al. [7] introduce the first coreset for individually fair clustering based on a sensitivity score sampling method. They significantly improve the running time from $O(n^4)$ in [18] to $O(nkd + k^8 d^4 + (k \log n)^4)$. In parallel, Bateni et al. [3,4] develop a local-search algorithm, achieving an approximation factor of $(3, 8)$ for k-median. They also use a fast local search algorithm to provide an $(O(1), 6)$-approximation for k-means. More recently, Ebbens et al. [9] propose the state-of-the-art result for k-center: a $(2, 2)$-approximation with running time $O(n^2 \log n)$.

In real-world scenarios, a few clients may disproportionately influence the outcome. To address this, Han et al. [12] first introduce the individually fair k-center with outliers problem, where up to m points can remain unserved. They provide a 4-approximation using a local search technique. Very recently, Maity et al. [19] propose a 12-approximation algorithm for individually fair k-means with outliers and a 24-approximation solution for individually fair k-median with outliers.

2 Problem Definitions

We consider facility set $\mathcal{F}$ and client set $\mathcal{C}$ in a metric space $(\mathcal{F} \cup \mathcal{C}, c)$, where each facility $i \in \mathcal{F}$ has an opening cost f_i. There is a connection cost c_{ij} for each facility $i \in \mathcal{F}$ and client $j \in \mathcal{C}$. The facility location problem is as,

$$\min_{X \subseteq \mathcal{F}} \sum_{j \in \mathcal{C}} c(j, X) + \sum_{i \in X} f_i,$$

where $c(j, X) = \min_{i \in X} c_{ij}$.

We capture individual preferences in facility location by defining a fair radius function $r_{j,\tau}$ as the distance to its τ-th nearest neighbor, where τ denotes the population threshold. Formally, for each client $j \in \mathcal{C}$, we define a client ball $B(j, r) = \{j' \in \mathcal{C} : c_{jj'} \leq r\}$ and the fair radius $r_{j,\tau} = \min\{r : |B(j,r)| \geq \tau\}$. In previous works [3,7,9,15,18,20], τ is typically defined as $\lceil \frac{n}{k} \rceil$, where n represents the total number of clients and k denotes the specified number of facilities to be opened. IPFL aims to select a facility set $X \subseteq \mathcal{F}$ such that each client $j \in \mathcal{C}$ satisfies the individual preference constraints while minimizing the facility location objective.

Definition 1. *(Individual Preference Facility Location, IPFL). Given facility set $\mathcal{F}$ and client set $\mathcal{C}$, opening cost function $f : \mathcal{F} \to \mathbb{R}$, connection cost function $c : \mathcal{F} \times \mathcal{C} \to \mathbb{R}$, population threshold τ, IPFL is defined as,*

$$\min_{X \subseteq \mathcal{F}} \sum_{j \in \mathcal{C}} c(j, X) + \sum_{i \in X} f_i$$
$$\text{s.t.} \quad c(j, X) \leq r_{j,\tau}, \forall j \in \mathcal{C}.$$

Building upon the definition of IPFL, we now introduce its outlier variant, IPFLO, where up to m clients may remain unserved. IPFLO aims to find a facility set $X \subseteq \mathcal{F}$ and an outlier set $Z \subseteq \mathcal{C}$ such that all clients in $\mathcal{C} \setminus Z$ satisfy the individual preference constraints while minimizing the facility location objective.

Definition 2. *(Individual Preference Facility Location with Outliers, IPFLO). Given facility set $\mathcal{F}$ and client set $\mathcal{C}$, opening cost function $f : \mathcal{F} \to \mathbb{R}$, connection cost function $c : \mathcal{F} \times \mathcal{C} \to \mathbb{R}$, population threshold τ, a positive integer m, IPFLO is defined as,*

$$\min_{X \subseteq \mathcal{F}, Z \subseteq \mathcal{C} : |Z| \leq m} \sum_{j \in \mathcal{C} \setminus Z} c(j, X) + \sum_{i \in X} f_i$$
$$\text{s.t.} \quad c(j, X) \leq r_{j,\tau}, \forall j \in \mathcal{C} \setminus Z.$$

3 Dual Fitting Algorithms

In this section, we present a dual-fitting algorithm for IPFL, called Dual-Fitting IPFL, and extend it to solve IPFLO by proposing the Dual-Fitting IPFLO algorithm.

3.1 Dual-Fitting IPFL Algorithm

In this subsection, we propose a greedy-based algorithm for IPFL and analyze it using a dual-fitting technique. A natural greedy idea is to iteratively select a facility and a subset of clients, aiming to minimize the opening and connection costs while ensuring that each client connects to the facility within its fair radius. Let X be the set of facilities opened so far, and U denote the set of clients not yet assigned to facilities in X. At each iteration, we select a facility $i \in \mathcal{F} \setminus X$

and a subset of clients $Y \subseteq U$ that minimizes $(f_i + \sum_{j \in Y} c_{ij})/|Y|$, subject to the individual preference constraint $c_{ij} \leq r_{j,\tau}$ for every $j \in Y$. If such a client set Y cannot be found, the instance is infeasible. In other words, if a client does not have any candidate facility within its fair radius, the IPFL becomes infeasible. Otherwise, we add i to X, and remove Y from U. Repeat until $U = \varnothing$.

However, instead of permanently assigning a client to a facility upon selection, we can save connection costs by allowing to reassign the client to a newly opened facility in later iterations. Specifically, when opening a new facility i, we reassign any client $j \in \mathcal{C} \setminus U$ from its current facility in X to i whenever the reassignment reduces its connection cost, i.e., $c_{ij} < c(j, X)$. Let $(b)_+ = \max\{b, 0\}$. The total saving cost of reassigning clients to facility i is given by $\sum_{j \in \mathcal{C}-U}(c(j, X) - c_{ij})_+$. Thus, at each iteration, we select a facility $i \in \mathcal{F} \setminus X$ and a subset of clients $Y \subseteq U$ that minimizes $(f_i - \sum_{j \in \mathcal{C}-U}(c(j, X) - c_{ij})_+ + \sum_{j \in Y} c_{ij})/|Y|$, while ensuring that every client is assigned to a facility within its fair radius. If such a client set Y cannot be found, the instance is infeasible. Otherwise, we add facility i to the set X, and remove clients in Y from U. Repeat until $U = \varnothing$. Since $c(j, X) \leq r_{j,\tau}$, the reassignment ensures that the client j still satisfies the individual preference constraint.

To analyze the performance of the above greedy algorithm, we consider a linear programming for IPFL. The decision variables y_i and x_{ij} in the integer programming model represent whether facility i is opened and whether client j is assigned to facility i, respectively. We relax the integrality of y_i and x_{ij} to obtain the following linear programming (*IPFL-LP*) for IPFL. Constraint (1) ensures that each client is connected to at least one facility, while constraint (2) guarantees that if client j is assigned to facility i, then facility i must be opened. Additionally, constraint (3) requires that each client $j \in \mathcal{C}$ can only be assigned to facilities within its fair radius $r_{j,\tau}$.

$$\min \quad \sum_{i \in \mathcal{F}} \sum_{j \in \mathcal{C}} c_{ij} x_{ij} + \sum_{i \in \mathcal{F}} f_i y_i \qquad (IPFL-LP)$$

$$\text{s.t.} \quad \sum_{i \in \mathcal{F}} x_{ij} \geq 1 \qquad \forall j \in \mathcal{C} \qquad (1)$$

$$y_i - x_{ij} \geq 0 \qquad \forall i \in \mathcal{F}, j \in \mathcal{C} \qquad (2)$$

$$x_{ij} = 0 \qquad \forall i \in \mathcal{F}, j \in \mathcal{C} : c_{ij} > r_{j,\tau} \qquad (3)$$

$$x_{ij} \geq 0 \qquad \forall i \in \mathcal{F}, j \in \mathcal{C} \qquad (4)$$

$$y_i \geq 0 \qquad \forall i \in \mathcal{F} \qquad (5)$$

The dual of the above *IPFL-LP* is:

$$\max \sum_{j \in \mathcal{C}} v_j \qquad (IPFL-DP)$$

s.t.
$$v_j - w_{ij} + z_{ij} \leq c_{ij} \quad \forall i \in \mathcal{F}, j \in \mathcal{C} : c_{ij} > r_{j,\tau} \qquad (6)$$
$$v_j - w_{ij} \leq c_{ij} \quad \forall i \in \mathcal{F}, j \in \mathcal{C} : c_{ij} \leq r_{j,\tau} \qquad (7)$$
$$\sum_{j \in \mathcal{C}} w_{ij} \leq f_i \quad \forall i \in \mathcal{F} \qquad (8)$$
$$v_j \geq 0 \quad \forall j \in \mathcal{C} \qquad (9)$$
$$w_{ij} \geq 0 \quad \forall i \in \mathcal{F}, j \in \mathcal{C} \qquad (10)$$
$$z_{ij} \text{ free} \quad \forall i \in \mathcal{F}, j \in \mathcal{C} : c_{ij} > r_{j,\tau} \qquad (11)$$

The greedy idea described above can be restated as follows. Each client j has a bid v_j representing the amount they are willing to pay for connection cost and the share of facility cost. Let X represent the set of open facilities and U denote the set of unassigned clients. We increase the bid v_j from zero for all clients in U, following the approach used in Jain et al. [13]. When client j touches a facility i, i.e., $v_j = c_{ij}$, we consider two cases: $c_{ij} \leq r_{j,\tau}$ and $c_{ij} > r_{j,\tau}$. If $c_{ij} \leq r_{j,\tau}$ and facility i is in X, we connect client j to facility i, and remove j from U. If $c_{ij} \leq r_{j,\tau}$ and facility i is in $\mathcal{F} \setminus X$, we increase w_{ij} at the same rate as v_j, where the w_{ij} represents the portion of the facility cost paid by client j. If $c_{ij} > r_{j,\tau}$, keep $w_{ij} = 0$ to prevent the client from paying the cost of facility i. To preserve the feasibility of constraint (6), we dynamically adjust the penalty term z_{ij} in inverse proportion to v_j when the service distance between facility i and client j exceeds the fair radius. Since z_{ij} is initialized to zero, this reduction effectively amplifies the penalty for client-facility pairs violating the distance constraint. If the total payment of the clients in U equals the cost of facility i, i.e., $\sum_{j \in U} w_{ij} = f_i$, we add facility i to X, connect the clients with $w_{ij} > 0$ to it, and remove them from U. Note that $w_{ij} > 0$, only when $v_j > c_{ij}$ and $c_{ij} \leq r_{j,\tau}$. Once a client j first connects to a facility, its bid v_i stops increasing, and correspondingly, the variables w and z stop increasing.

Similarly, allowing the clients to switch to new facilities can reduce costs. Specifically, if a client j switches to a new facility i, its bid becomes $c(j, X)$. It pays an additional cost of $(c(j, X) - c_{ij})_+$ towards i. In this case, the total payment for facility i is $\sum_{j \in U} w_{ij} + \sum_{j \in \mathcal{C} \setminus U} (c(j, X) - c_{ij})_+$. If the total of payments equals the facility i opening cost f_i, i.e., $\sum_{j \in U} w_{ij} + \sum_{j \in \mathcal{C} \setminus U} (c(j, X) - c_{ij})_+ = f_i$, we add facility i to X, connect the clients with $w_{ij} > 0$ to i, and remove the clients from U. Moreover, the clients in $\mathcal{C} \setminus U$ with $c(j, X) - c_{ij} > 0$ switch to facility i. Repeat until $U = \varnothing$. The formal description of the algorithm is given in the Dual-Fitting IPFL algorithm.

We obtain the facilities solution set X using the Dual-Fitting IPFL algorithm. Each client is assigned to its nearest facility in X. Let $\sum_{j \in \mathcal{C}} c(j, X) + \sum_{i \in X} f_i$ denote the cost of the algorithm solution. We naturally obtain the following lemma.

Algorithm 1. Dual-Fitting IPFL Algorithm

Input: An instance $(\mathcal{F}, \mathcal{C}, f, c, \tau)$
Output: A facility set $X \subseteq \mathcal{F}$
1: $v, w, z \leftarrow 0$
2: $X \leftarrow \emptyset$
3: $U \leftarrow \mathcal{C}$
4: **while** $U \neq \emptyset$ **do**
5:     Uniformly increase v_j for all $j \in U$ until one of the following cases occurs:
6:     **Case 1:**
7:     **if** $\exists j \in U, i \in \mathcal{F} \setminus X$ such that $v_j = c_{ij}$ and $c_{ij} \leq r_{j,\tau}$ **then**
8:         Uniformly increase w_{ij} at the same rate as the increase of v_j
9:     **end if**
10:     **Case 2:**
11:     **if** $\exists i \in \mathcal{F} \setminus X$ such that $\sum_{j \in U} w_{ij} + \sum_{j \in \mathcal{C} \setminus U}(c(j,X) - c_{ij})_+ = f_i$ **then**
12:         $X \leftarrow X \cup \{i\}$
13:         **for** all $j \in U$ such that $w_{ij} > 0$ **do**
14:             $U \leftarrow U \setminus \{j\}$ and cease increasing v_j, w_{ij}, z_{ij}
15:         **end for**
16:     **end if**
17:     **Case 3:**
18:     **if** $\exists j \in U, i \in X$ such that $v_j = c_{ij}$ and $c_{ij} \leq r_{j,\tau}$ **then**
19:         $U \leftarrow U \setminus \{j\}$ and cease increasing v_j, w_{ij}, z_{ij}
20:     **end if**
21:     **Case 4:**
22:     **if** $\exists j \in U, i \in \mathcal{F}$ such that $v_j = c_{ij}$ and $c_{ij} > r_{j,\tau}$ **then**
23:         Uniformly decrease z_{ij} at the same rate as the increase of v_j
24:     **end if**
25: **end while**
26: **return** Facility set X

Lemma 1. *If the IPFL instance is feasible, $\sum_{j \in \mathcal{C}} c(j, X) + \sum_{i \in X} f_i = \sum_{j \in \mathcal{C}} v_j$.*

Proof. For each while loop, we either connect a client to the facility in X or open a new facility. If a client j connects to a facility in X, the left side rises by $c(j, X)$, while the right side rises by v_j, holding the equality. If a new facility i is opened, let W denote the set of clients with $w_{ij} > 0$. In this case, the left-hand side increases by $\sum_{j \in W} c(j, X) + f_i$, while the right-hand side increases by $\sum_{j \in W} v_j$. Since the equality holds at each iteration, it is preserved throughout the loop and in the final result. □

If the dual variables (v, w, z) corresponding to the Dual-Fitting IPFL algorithm are dual feasible, the dual objective $\sum_{j \in \mathcal{C}} v_j$ provides a lower bound for the optimal solution. For each $i \in \mathcal{F}$ and $j \in \mathcal{C}$, when $c_{ij} \leq r_{j,\tau}$, the dual constraints (7) and (8) ensure that $\sum_{j \in \mathcal{C}}(v_j - c_{ij})_+ \leq f_i$. In contrast, in our algorithm, at each loop, we maintain $\sum_{j \in U} w_{ij} + \sum_{j \in \mathcal{C} \setminus U}(c(j, X) - c_{ij})_+ = f_i$. The clients in $\mathcal{C} \setminus U$ who connect to the new facility i pay $(c(j, X) - c_{ij})_+$, whereas to satisfy dual constraints (7) and (8), they must pay $(v_j - c_{ij})_+$ for facility i. As a result, our algorithm produces an infeasible solution.

We construct a dual feasible solution using the dual variables (v, w, z). Once a client j first connects to a facility, its bid v_j ceases to increase. We refer to v_j as the stopping time for client j.

Lemma 2. *If the IPFL instance is feasible, consider the moment when client j first connects to a facility i at time v_j. Clients who have already connected to other facilities prior to time v_j will contribute at least $v_j - c_{ij} - 2c_{ik}$ to the opening cost of facility i.*

Proof. If client k has connected to facility h before time v_j, the payment to facility i is $c_{hk} - c_{ik}$. By the triangle inequality, we know that $c_{hj} \leq c_{ij} + c_{ik} + c_{hk}$. Since client j first connects to facility i and facility h opens before facility i, it follows that $v_j \leq c_{hj}$. Therefore, we obtain the inequality $c_{hk} - c_{ik} \leq v_j - c_{ij} - 2c_{ik}$.

Lemma 3. *Assume that all clients are ordered in non-decreasing order of $v_1 \leq v_2 \leq \cdots \leq v_n$. If the IPFL instance is feasible, for any client $j \in \mathcal{C}$, suppose that it first connects to a facility i at time v_j, we have*

$$\sum_{k=1}^{j-1}(v_j - c_{ij} - 2c_{ik}) + \sum_{k=j}^{n}(v_j - c_{ik}) \leq f_i.$$

Proof. The total opening cost for facility i comes from two groups: clients already connected to other facilities prior to time v_j and clients still unconnected at time v_j. For any client k with $v_k < v_j$, their payment to facility i is at least $v_j - c_{ij} - 2c_{ik}$. In contrast, for any client k with $v_k \geq v_j$ who has not yet been served at time v_j, their payment is $v_j - c_{ik}$. Summing the total payments for facility i gives the above inequality. □

Lemma 4. *Assume that all clients are ordered in non-decreasing order of $v_1 \leq v_2 \leq \cdots \leq v_n$. If the IPFL instance is feasible, we have*

$$\sum_{j \in \mathcal{C}}(v_j - 2c_{ij}) \leq f_i.$$

Proof. By summing the inequalities in Lemma 3 over all $j \in \mathcal{C}$, we obtain

$$\sum_{j=1}^{n}\left(\sum_{k=1}^{j-1}(v_j - c_{ij} - 2c_{ik}) + \sum_{k=j}^{n}(v_j - c_{ik})\right) \leq n \cdot f_i,$$

By rearranging the terms, we find that

$$n\sum_{j=1}^{n} v_j - \sum_{j=1}^{n}(j-1)c_{ij} - n\sum_{k=1}^{n} c_{ik} - \sum_{j=n}^{n}\sum_{k=1}^{j-1} c_{ik} \leq n \cdot f_i.$$

By exchanging the order of summation in the term $\sum_{j=n}^{n}\sum_{k=1}^{j-1} c_{ik}$, we can rewrite it as $\sum_{k=1}^{n}(n-k)c_{ik}$. Finally, by substituting the consistent notation into the derived equations, we obtain:

$$\sum_{j=1}^{n}(v_j - 2c_{ij}) \leq f_i.$$

□

Initialize $(\tilde{v}, \tilde{w}, \tilde{z})$ to 0, and for each $j \in \mathcal{C}$, set $\tilde{v}_j = v_j/2$. For each $i \in \mathcal{F}$ and $j \in \mathcal{C}$, if $c_{ij} \leq r_{j,\tau}$, set $\tilde{w}_{ij} = (\tilde{v}_j - c_{ij})_+$; otherwise, set $\tilde{z}_{ij} = -\tilde{v}_j$.

Lemma 5. *The solution $(\tilde{v}, \tilde{w}, \tilde{z})$ is a dual feasible solution when the IPFL instance is feasible.*

Proof. From Lemma 4, we have $\sum_{j \in \mathcal{C}}(v_j - 2c_{ij}) \leq f_i$. Rewriting the inequality, it holds that $\sum_{j \in \mathcal{C}}(2\tilde{v}_j - 2c_{ij}) \leq f_i$. Dividing both sides by 2, this leads to $\sum_{j \in \mathcal{C}}(\tilde{v}_j - c_{ij}) \leq f_i/2 \leq f_i$. Thus, the solution $(\tilde{v}, \tilde{w}, \tilde{z})$ is dual feasible. □

Theorem 1. *The Dual-Fitting IPFL algorithm is a 2-approximation algorithm when the IPFL instance is feasible.*

Proof. At each while loop, we either connect a client to a facility in X or add a new facility to X. In the first case, a client j connects to a facility $i \in X$ where $v_j = c_{ij} \leq r_{j,\tau}$, ensuring the individual preference constraint. In the second case, there are two types of clients connecting to the new facility i'. The first type of client j pays $w_{i'j}$ for facility i', where $w_{ij} > 0$ only if $v_j > c_{i'j}$ and $c_{i'j} \leq r_{j,\tau}$, ensuring the individual preference constraint is satisfied. The second type of client j pays $(c(j,X) - c_{i'j})_+$ for facility i' when switching connections. They reconnect to a closer facility and still satisfy the individual preference constraint. Let OPT denote the optimal solution. From the Lemmas 1 and 5, we obtain $\sum_{j \in \mathcal{C}} c(j,X) + \sum_{i \in X} f_i = \sum_{j \in \mathcal{C}} v_j = \sum_{j \in \mathcal{C}} 2\tilde{v}_j \leq 2OPT$. Thus, the Dual-Fitting IPFL algorithm is a 2-approximation algorithm for IPFL.

3.2 Extends Dual-Fitting IPFL Algorithm to IPFLO

In this section, we apply a similar approach to deal with IPFLO. We introduce o_j to indicate whether client j is selected as an outlier. Constraint (12) ensures that each client is either connected to a facility or designated as an outlier, while constraint (14) limits the number of outliers to at most m. The linear programming of IPFLO is as follows:

$$\min \sum_{i\in\mathcal{F}}\sum_{j\in\mathcal{C}} c_{ij}x_{ij} + \sum_{i\in\mathcal{F}} f_i y_i \qquad (IPFLO-LP)$$

$$\text{s.t.} \quad \sum_{i\in\mathcal{F}} x_{ij} + o_j \geq 1 \qquad \forall j \in \mathcal{C} \qquad (12)$$

$$y_i - x_{ij} \geq 0 \qquad \forall i \in \mathcal{F}, j \in \mathcal{C} \qquad (13)$$

$$\sum_{j\in\mathcal{C}} o_j \leq m \qquad (14)$$

$$x_{ij} = 0 \qquad \forall i \in \mathcal{F}, j \in \mathcal{C} : c_{ij} > r_{j,\tau} \qquad (15)$$

$$x_{ij} \geq 0 \qquad \forall i \in \mathcal{F}, j \in \mathcal{C} \qquad (16)$$

$$y_i \geq 0 \qquad \forall i \in \mathcal{F} \qquad (17)$$

The dual of the above *IPFLO-LP* is:

$$\max \sum_{j\in\mathcal{C}} v_j - m \cdot q \qquad (IPFLO-DP)$$

$$\text{s.t.} \quad v_j - w_{ij} + z_{ij} \leq c_{ij} \quad \forall i \in \mathcal{F}, j \in \mathcal{C} : c_{ij} > r_{j,\tau} \qquad (18)$$

$$v_j - w_{ij} \leq c_{ij} \qquad \forall i \in \mathcal{F}, j \in \mathcal{C} : c_{ij} \leq r_{j,\tau} \qquad (19)$$

$$\sum_{j\in\mathcal{C}} w_{ij} \leq f_i \qquad \forall i \in \mathcal{F} \qquad (20)$$

$$v_j \leq q \qquad \forall j \in \mathcal{C} \qquad (21)$$

$$v_j \geq 0 \qquad \forall j \in \mathcal{C} \qquad (22)$$

$$w_{ij} \geq 0 \qquad \forall i \in \mathcal{F}, j \in \mathcal{C} \qquad (23)$$

$$z_{ij} \text{ free} \qquad \forall i \in \mathcal{F}, j \in \mathcal{C} : c_{ij} > r_{j,\tau} \qquad (24)$$

Note that the integrality gap of *IPFLO-LP* is unbounded, which means it is impossible to prove a bounded approximation guarantee using *IPFLO-LP*.

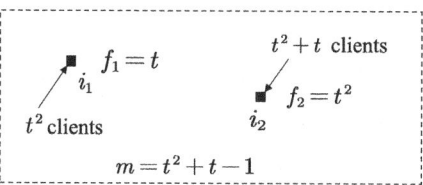

Fig. 1. The unbounded gap instance

Figure 1 depicts two isolated clusters in a metric space: the first cluster contains facility i_1 with an opening cost of t, co-located with t^2 clients, while the second cluster includes facility i_2 with a cost of t^2, co-located t^2+t clients. The two facilities are separated by an arbitrarily large distance. When the number of outliers is set to $m=t^2+t-1$, the integral optimal solution must open facility i_2, incurring a total cost of t^2. However, the LP relaxation allows for a fractional solution: opening $1-\frac{1}{t}$ of i_1 and $\frac{1}{t}$ of i_2. The LP solution achieves a cost of $2t-1$, while leaving t^2+t-1 clients unserved. Consequently, the integrality gap is $\frac{t^2}{2t-1}$, which becomes unbounded as $t \to \infty$.

To deal with this issue, we modify the instance by removing the most expensive facility from the optimal solution. For any given instance $I=(\mathcal{F},\mathcal{C},f,c,\tau)$, we first guess the most expensive facility i_e in the optimal solution, where its opening cost denotes as f_{i_e}. We then modify the instance by setting the opening cost of all facilities with costs greater than f_{i_e} to ∞, setting the cost of facility i_e to zero, and leaving the costs of all other facilities unchanged. By modifying the facility costs in this way, we obtain the modified instance $I'=(\mathcal{F},\mathcal{C},f',c,\tau)$. Let OPT and OPT' denote the optimal solution for instances I and I', respectively, such that $OPT = OPT' + f_{i_e}$.

We propose the Dual Fit IPFLO algorithm for the modified instance I'. Initially, we define the set $U=\mathcal{C}$ and assume that all clients are outliers. The set X will contain the facilities to be opened. The facility i_e is initially added to X. The dual variables (v,w,z) are initialized to 0, and we proceed by increasing the bid v_j for all clients $j \in U$ until one of the four cases from the Dual-Fitting IPFL algorithm occurs, at which point we handle it in a similar manner. If a client j is connected to a facility, it is removed from U and is no longer an outlier. However, since the individual preference constraints depend on the distribution of facilities. If more than m clients have no candidate facility within their fair radius, the instance is infeasible. Otherwise, we increase v_j for each $j \in U$ until $|U| \le m$. When we select the last facility i_l and connect clients to it to ensure that $|U| \le m$, we need to randomly add clients with $w_{i_l j} > 0$ to the set U, maintaining that $|U|=m$. Finally, we obtain the facility set X, the outlier set U, and dual variables (v,w,z).

We open the facilities in X and connect the clients in $\mathcal{C} \setminus U$ to the nearest facility in X. We separately account for the cost of the facility i_l. Since some clients who pay for the final facility i_l may be selected as outliers, the clients in $\mathcal{C}\setminus U$ associated with facility i_l might not fully pay its costs. Let $\sum_{j \in \mathcal{C}\setminus U} c(j,X) + \sum_{i \in X\setminus i_l} f_i$ denote the solution cost (excluding facility i_l) for instance I' obtained by the Dual Fit IPFLO algorithm. We set $q = \max_{j \in \mathcal{C}} v_j$ to ensure the feasibility of the constraint (21). Then we have the following inequality:

$$\sum_{j \in \mathcal{C}\setminus U} c(j,X) + \sum_{i \in X \setminus i_l} f_i \le \sum_{j \in \mathcal{C}\setminus U} v_j = \sum_{j \in \mathcal{C}} v_j - m \cdot q.$$

However, a similar problem arises with the IPFL problem. We obtain the dual variables (v,w,z), which is infeasible. It does not satisfy the dual constraints (19) and (20). To resolve this, we perform a similar process to find a dual feasible solution that serves as a lower bound. Initially, we set $(\tilde{v},\tilde{w},\tilde{z})=0$. For each client $j \in \mathcal{C}$, set $\tilde{v}_j = v_j/2$. For each $i \in \mathcal{F}$ and $j \in \mathcal{C}$, if $c_{ij} \le r_{j,\tau}$, we set

$\tilde{w}_{ij} = (\tilde{v}_j - c_{ij})_+$; otherwise, we set $z_{ij} = -\tilde{v}_j$. The only difference is that we set $q' = \max_{j \in \mathcal{C}} \tilde{v}_j$. This leads to:

$$2q' = 2\max_{j \in \mathcal{C}} \tilde{v}_j = \max_{j \in \mathcal{C}} v_j = q.$$

Thus, it holds that

$$\sum_{j \in \mathcal{C} \setminus U} c(j, X) + \sum_{i \in X \setminus i_l} f_i \leq \sum_{j \in \mathcal{C}} v_j - m \cdot q = \sum_{j \in \mathcal{C}} 2\tilde{v}_j - 2m \cdot q' \leq 2OPT'.$$

Theorem 2. *The Dual-Fitting IPFLO algorithm is a 2-approximation when the IPFLO instance is feasible.*

Proof. Since our algorithm follows the same procedure as the Dual-Fitting IPFL algorithm for opening facilities and connecting clients, both ensure that individual fairness constraints are satisfied. Moreover, the final solution satisfies $\sum_{j \in \mathcal{C} \setminus U} c(j, X) + \sum_{i \in X \setminus i_l} f_i + f_{i_e} + f_{i_l} \leq 2OPT' + 2f_{i_e} \leq 2OPT$. □

4 Conclusion

In this paper, we introduce individual fairness to the facility location problem and propose the Individual Preference Facility Location (IPFL). We present a 2-approximation for IPFL using a dual fitting technique, which employs a greedy strategy to bound the primal objective by the dual objective. By scaling the dual solution to feasibility, we establish a lower bound on the optimal solution. We extend IPFL to its outlier version, IPFLO, using a similar approach to achieve a 2-approximation. The feasibility of both IPFL and IPFLO depends on facility distribution and parameter τ. IPFL becomes infeasible if any client lacks a facility within their fair radius, whereas IPFLO relaxes this by permitting up to m clients to violate the individual preference constraints. However, IPFLO remains infeasible when more than m clients have no facilities in their fair radii. On the positive side, if the set of facilities and the set of clients are the same, a 2-approximation solution is always guaranteed for any instance of both IPFL and IPFLO. Finally, if we relax the fair radius by a factor, we can always find a feasible solution in any instance, which opens an interesting direction for developing bi-criteria approximation algorithms for both IPFL and IPFLO.

Acknowledgments. This work is supported by National Natural Science Foundation of China 12371321 and Guangdong Basic and Applied Basic Research Foundation 2024A1515030197.

References

1. Abbasi, M., Bhaskara, A., Venkatasubramanian, S.: Fair clustering via equitable group representations. In: Proceedings of the 2021 ACM Conference on Fairness, Accountability, and Transparency (FAccT), pp. 504–514 (2021)
2. Bandyapadhyay, S., Inamdar, T., Pai, S., Varadarajan, K.: A constant approximation for colorful k-center. In: 27th Annual European Symposium on Algorithms (ESA). Schloss-Dagstuhl-Leibniz Zentrum für Informatik (2019)
3. Bateni, M., Cohen-Addad, V., Epasto, A., Lattanzi, S.: Scalable and improved algorithms for individually fair clustering. In: Workshop on Trustworthy and Socially Responsible Machine Learning, NeurIPS (2022)
4. Bateni, M., Cohen-Addad, V., Epasto, A., Lattanzi, S.: A scalable algorithm for individually fair k-means clustering. In: International Conference on Artificial Intelligence and Statistics (AISTATS), pp. 3151–3159 (2024)
5. Charikar, M., Guha, S.: Improved combinatorial algorithms for the facility location and k-median problems. In: 40th Annual Symposium on Foundations of Computer Science (FOCS), pp. 378–388 (1999)
6. Charikar, M., Khuller, S., Mount, D.M., Narasimhan, G.: Algorithms for facility location problems with outliers. In: ACM-SIAM Symposium on Discrete Algorithms (SODA), vol. 1, pp. 642–651 (2001)
7. Chhaya, R., Dasgupta, A., Choudhari, J., Shit, S.: On coresets for fair regression and individually fair clustering. In: International Conference on Artificial Intelligence and Statistics (AISTATS), pp. 9603–9625 (2022)
8. Chierichetti, F., Kumar, R., Lattanzi, S., Vassilvitskii, S.: Fair clustering through fairlets. Adv. Neural Inf. Process. Syst. (NeurIPS) **30**, 1–9 (2017)
9. Ebbens, M., Funk, N., Höckendorff, J., Sohler, C., Weil, V.: A subquadratic time approximation algorithm for individually fair k-center. arXiv preprint arXiv:2412.04943 (2024)
10. Ghadiri, M., Samadi, S., Vempala, S.: Socially fair k-means clustering. In: Proceedings of the 2021 ACM Conference on Fairness, Accountability, and Transparency (FAccT), pp. 438–448 (2021)
11. Guha, S., Khuller, S.: Greedy strikes back: improved facility location algorithms. J. Algorithms **31**(1), 228–248 (1999)
12. Han, L., Xu, D., Xu, Y., Yang, P.: Approximation algorithms for the individually fair k-center with outliers. J. Global Optim. **87**(2), 603–618 (2023)
13. Jain, K., Mahdian, M., Markakis, E., Saberi, A., Vazirani, V.V.: Greedy facility location algorithms analyzed using dual fitting with factor-revealing LP. J. ACM **50**(6), 795–824 (2003)
14. Jain, K., Vazirani, V.V.: Approximation algorithms for metric facility location and k-median problems using the primal-dual schema and Lagrangian relaxation. J. ACM (JACM) **48**(2), 274–296 (2001)
15. Jung, C., Kannan, S., Lutz, N.: Service in your neighborhood: fairness in center location. In: 1st Symposium on Foundations of Responsible Computing (FORC), pp. 5:1–5:15 (2020)
16. Korupolu, M.R., Plaxton, C.G., Rajaraman, R.: Analysis of a local search heuristic for facility location problems. J. Algorithms **37**(1), 146–188 (2000)
17. Li, S.: A 1.488 approximation algorithm for the uncapacitated facility location problem. Inf. Computat. **222**, 45–58 (2013)
18. Mahabadi, S., Vakilian, A.: Individual fairness for k-clustering. In: International Conference on Machine Learning (ICML), pp. 6586–6596 (2020)

19. Maity, B., Das, S., Dasgupta, A.: Linear programming based approximation to individually fair k-clustering with outliers. arXiv preprint arXiv:2412.10923 (2024)
20. Negahbani, M., Chakrabarty, D.: Better algorithms for individually fair k-clustering. Adv. Neural. Inf. Process. Syst. **34**, 13340–13351 (2021)
21. Shmoys, D.B., Tardos, É., Aardal, K.: Approximation algorithms for facility location problems. In: Proceedings of the ACM Symposium on Theory of Computing (STOC), pp. 265–274 (1997)
22. Vakilian, A., Yalciner, M.: Improved approximation algorithms for individually fair clustering. In: International Conference on Artificial Intelligence and Statistics (AISTATS), pp. 8758–8779 (2022)

The Online Power Cover Problem on a Line

Zhonghao Liu[1], Man Xiao[1,2], Xiaofei Liu[3(✉)], and Weidong Li[1]

[1] School of Mathematics and Statistics, Yunnan University, Kunming, China
[2] School of Mathematics and Information, China West Normal University, Nanchong, China
[3] School of Information Science and Engineering, Yunnan University, Kunming, China
lxfjl2016@163.com

Abstract. In this paper, we study the online power cover problem on a line. Suppose L is a line on the plane, and S is a set of sensors on the line L, and each sensor can deploy a power and generate a covered area that can cover users. This problem is to find a minimum power assignment to cover a sequence of users on the line L arriving one by one. In this paper, we first prove that the lower bound is 2 for this problem even when $|S| = 2$. Then, we present an online algorithm with a competitive ratio that is no more than $|S|$ based on the greedy technique. Note that, this algorithm is the best possible online algorithm for this problem with $|S| = 2$. Finally, we consider a special case of this problem, in which $S = \{s_0, s_1, s_2\}$ and $d(s_0, s_1) = d(s_1, s_2)$, and present an online algorithm with a competitive ratio that is no more than $\sqrt{2} + 1$.

Keywords: Power cover problem · Online algorithm · Competitive ratio

1 Introduction

The power cover problem has been widely used in real-world applications such as wireless sensor networks, traffic control, and environment monitoring [3,15]. In this problem, we are given a user set U and a sensor set S. Each sensor $s \in S$ can adjust its power, and the covered area of sensor s with power $p(s)$ is a 'disk' $D(s, r(s))$ with center s and radius $r(s)$, where $r(s)$ is determined by $p(s)$ through the relation

$$p(s) = r(s)^\alpha,$$

where $\alpha \geq 1$ is a constant determined by the environment condition, which is called the attenuation factor of power. The power cover problem is to find a minimum power assignment to cover all users.

Bilò et al. [2] proved that the power cover problem is NP-hard for the case with $\alpha \geq 2$. Alt et al. [1] improved this result and proved that this problem is NP-hard for the case with $\alpha > 1$. For the case with $\alpha = 1$, Lev-Tov and

Peleg [5] presented a polynomial-time approximation scheme (PTAS) via the plane subdivision technique. Through detailed geometric analysis, Bilò et al. [2] proposed a PTAS for the case with $\alpha \geq 1$. There are many variants of the power cover problem, such as, the power cover problem with penalty [9], the partial power cover problem [6], the power multi-cover problem [4] and so on, where related results can be found in [8,10,11]. In some real-world scenarios, some special location relationships exist between the sensor and the user. When all sensors and users are on a line, the power cover problem can be solved using a polynomial-time algorithm [2]. Liang et al. [7] considered the partial power multi-cover problem, and presented two polynomial-time algorithms. Zhang et al. [14] proved that the K-prize-collecting coverage problem is NP-hard, and presented a PTAS.

In the above problems, the information of all users is known in advance, however, in the real world, users arrive one by one, $i.e.$, when a user arrives, we must immediately and irrevocably decide to how to cover it without any information about the further users, which is denoted as online. This type of problem is called an online power cover problem. The online matching problem is closely related to the online power cover problem, and Xiao et al. [12,13] studied the online bottleneck matching problem on a line or star. In this paper, we consider the online power cover problem on a line. We firstly prove that the lower bound is 2 for this problem even when only two sensors are in S. Secondly, we present an online algorithm based on the greedy technique, and prove that the competitive ratio is no more than $|S|$. Note that, this algorithm is the best possible online algorithm for this problem with $|S| = 2$. Finally, we consider a special case of this problem, in which the sensor set $S = \{s_0, s_1, s_2\}$ and $d(s_0, s_1) = d(s_1, s_2)$, where $d(a, b)$ is the Euclidean distance between points a and b, and present an online algorithm with a competitive ratio that is no more than $\sqrt{2} + 1$.

The remaining sections of this paper are organized as follows. In Sect. 2, we formally introduce the online power cover problem on a line, and prove the lower bound for this problem. In Sect. 3, we present the online algorithm based on the greedy technique. In Sect. 4, we consider the special case of this problem, in which the sensor set $S = \{s_0, s_1, s_2\}$ and $d(s_0, s_1) = d(s_1, s_2)$. Finally, we provide a brief conclusion.

2 Preliminaries

Given a line L on the plane, a set of sensors S arranged on the L, and a sequence $\mathcal{U} = u_1, u_2, \ldots, u_n$ of users on the L that arrive one by one, let L be a number axis, the position of each sensor s (or user u) can be identified by the coordinate x_s of s (or x_u of u) by the L. Each sensor $s \in S$ can deploy a power $p(s)$ and generate a covered area $D(s, r(s))$, where $D(s, r(s)) = [x_s - r(s), x_s + r(s)]$ is a 'disk' centered at s whose radius $r(s)$, and $r(s)$ is determined by $p(s)$ through the equation $p(s) = r(s)^\alpha$, where $\alpha \geq 1$ is an attenuation factor, which is a given constant. User u is covered by sensor s if the coordinate x_u of u satisfies that $x_u \in D(s, r(s))$. In our model, user in $\mathcal{U}$ ($= u_1, u_2, \ldots, u_n$) arrives one by one, and n is unknown until the last user arrives. When a user arrives, we need to cover it either using an existing disk or increasing the radius of some disk to cover it without any information about the further users. The online power cover problem on a line (OPCL) is to find a minimum power assignment $p : S \to \mathcal{R}_{\geq 0}$ such that all users are covered by the disks supported by p.

Theorem 1. *For the OPCL even when $\alpha = 1$ and $|S| = 2$, there is no online algorithm $\mathbb{A}$ with a competitive ratio that is strictly less than 2.*

Proof. Let $S = \{s_0, s_1\}$ be the sensor set, and without loss of generality, assume that the coordinate of s_0 is 0, and the coordinate of s_1 is 1. Let the coordinate of the first arriving user u_1 be $\frac{1}{2}$.

Case 1. Algorithm $\mathbb{A}$ assigns a power to s_0 to cover u_1, i.e. the power assignment is $p(s_0) = \frac{1}{2}$ and $p(s_1) = 0$. Then the last user u_2 arrives, where the coordinate of u_2 is $\frac{3}{2}$. Algorithm $\mathbb{A}$ has two ways to cover u_2. 1. Algorithm $\mathbb{A}$ assigns a power to s_0 to cover u_2, the power assignment is $p(s_0) = \frac{3}{2}$ and $p(s_1) = 0$; 2. Algorithm $\mathbb{A}$ assigns a power to s_1 to cover u_2, the power assignment is $p(s_0) = \frac{1}{2}$ and $p(s_1) = \frac{1}{2}$. Thus, the objective value $OUT^{\mathbb{A}}$ of the solution generated by $\mathbb{A}$ is at least 1, and

$$\frac{OUT^{\mathbb{A}}}{OPT} \geq \frac{1}{\frac{1}{2}} = 2,$$

where OPT is the objective value of the optimal power assignment p^* the corresponding offline problem, which $p^*(s_0) = 0$, $p^*(s_1) = \frac{1}{2}$ and $OPT = \frac{1}{2}$.

Case 2. Algorithm $\mathbb{A}$ assigns a power to s_1 to cover u_1, i.e. the power assignment is $p(s_0) = 0$ and $p(s_1) = \frac{1}{2}$. Then the last user u_2 arrives, where the coordinate of u_2 is $-\frac{1}{2}$. Algorithm $\mathbb{A}$ has two ways to cover u_2. 1. Algorithm $\mathbb{A}$ assigns a power to s_0 to cover u_2, the power assignment is $p(s_0) = \frac{1}{2}$ and $p(s_1) = \frac{1}{2}$; 2. Algorithm $\mathbb{A}$ assigns a power to s_1 to cover u_2, the power assignment is $p(s_0) = 0$ and $p(s_1) = \frac{3}{2}$. Thus, the objective value $OUT^{\mathbb{A}}$ of the solution generated by $\mathbb{A}$ is at least 1, and

$$\frac{OUT^{\mathbb{A}}}{OPT} \geq \frac{1}{\frac{1}{2}} = 2,$$

where OPT is the objective value of the optimal power assignment p^* the corresponding offline problem, which $p^*(s_0) = \frac{1}{2}$, $p^*(s_1) = 0$ and $OPT = \frac{1}{2}$.

Therefore, the theorem holds. □

3 The Online Algorithm for the OPCL

Based on the greedy technique, we present an online algorithm with a competitive ratio of no more than $|S|$.

Algorithm 1: Greedy algorithm

Input: A line L which is a number axis, a set of sensors S on the L, a sequence $\mathcal{U} = u_1, u_2, \ldots, u_n$ of users on the L that arrive one by one, and an attenuation factor α.
Output: A power assignment $\{p(s)\}_{s \in S}$ and the total power OUT.
1 Initialize, for any $s \in S$, set $p(s) = 0$, $U_s = \emptyset$, and let x_s be the coordinate of s.
2 Let u be the arriving user, and let x_u be its coordinate.
3 Set $s' := \arg\min_{s:s \in S} |x_s - x_u|$.
4 Set $U_{s'} := U_{s'} \cup \{u\}$ and $p(s') := \max\{p(s'), (|x_{s'} - x_u|)^\alpha\}$.
5 If there is another user arriving, go to Step 2; otherwise, set $OUT = \sum_{s \in S} p(s)$, and output $\{p(s)\}_{s \in S}$ and OUT.

Theorem 2. *Algorithm 1 is a $|S|$-competitive for the OPCL.*

Proof. Let
$$s_{\max} = \arg\max_{s:s \in S} p(s)$$
be the sensor assigning the maximum power by Algorithm 1. Let
$$u_{\max} = \arg\max_{u:u \in U_{s_{\max}}} |x_{s_{\max}} - x_u|$$
be the user in $U_{s_{\max}}$ that is the farthest from $s_{\max}$, where $U_{s_{\max}}$ is the vertex set covered by sensor $s_{\max}$ by Algorithm 1. Thus, we have
$$\begin{cases} p(s_{\max}) = (|x_{s_{\max}} - x_u|)^\alpha; \\ p(s_{\max}) \geq p(s),\ \forall s \in S. \end{cases} \quad (1)$$

Let p^* be an optimal power assignment for the corresponding offline OPCL, and its objective value is $OPT = \sum_{s:s \in S} p^*(s)$. In optimal power assignment p^*, user $u_{\max}$ must be covered by a disk supported by p^*, i.e., there exists a sensor $s' \in S$ satisfying that $p^*(s') \geq (|x_{s'} - x_{u_{\max}}|)^\alpha$. Based on Algorithm 1, we have $s_{\max} = \arg\min_{s:s \in S} |x_s - x_{u_{\max}}|$, and
$$p(s_{\max}) = (|x_{s_{\max}} - x_{u_{\max}}|)^\alpha \leq (|x_{s'} - x_{u_{\max}}|)^\alpha \leq p^*(s') \leq OPT, \quad (2)$$
where the first inequality follows from $\alpha \geq 1$, and the last inequality follows from $p^*(s) \geq 0$ for any $s \in S$.

Therefore, we have
$$OUT = \sum_{s:s \in S} p(s) \leq \sum_{i=1}^{|S|} p(s_{\max}) = |S| \cdot p(s_{\max}) \leq |S| \cdot OPT,$$
where the first inequality follows from inequality (1), and the second inequality follows from inequality (2). □

Based on Theorem 2, when $|S| = 2$, we have the competitive ratio of Algorithm 1 is 2. According to Theorem 1, the following theorem is obvious.

Theorem 3. *When $|S| = 2$, Algorithm 1 is the best possible online algorithm.*

4 The OPCL with $S = \{s_0, s_1, s_2\}$ and $d(s_0, s_1) = d(s_1, s_2)$

In this section, we consider a special case of the OPCL, in which $S = \{s_0, s_1, s_2\}$ and $d(s_0, s_1) = d(s_1, s_2)$, where $d(a, b)$ is the Euclidean distance between points a and b. Without loss of generality, we assume that the coordinate of s_0 is 0, the coordinate of s_1 is 1, and the coordinate of s_2 is 2. For this special case, we present an online algorithm with a competitive ratio that is no more than $\sqrt{2} + 1$, and present the details in Algorithm 2.

Algorithm 2:

Input: A line L which is a number axis, a set of sensors $S = \{s_0, s_1, s_2\}$ on the L, a sequence $\mathcal{U} = u_1, u_2, \ldots, u_n$ of users on the L that arrive one by one, and an attenuation factor α.

Output: A power assignment $\{p(s)\}_{s \in S}$ and the total power OUT.

1 Initialize, for any $i = 0, 1, 2$, set $p(s_i) = 0$, $U_i = \emptyset$, and let i be the coordinate of s_i.
2 Let u be the arriving user, and let x_u be its coordinate.
3 **if** $x_u < 1 - \frac{2^{\frac{1}{2\alpha}}}{2^{\frac{1}{2\alpha}} + 1}$ **then**
4 $\quad$ set $U_0 := U_0 \cup \{u\}$ and $p(s_0) := \max\{p(s_0), (|x_u|)^\alpha\}$.
5 **if** $1 - \frac{2^{\frac{1}{2\alpha}}}{2^{\frac{1}{2\alpha}} + 1} \leq x_u \leq 1 + \frac{2^{\frac{1}{2\alpha}}}{2^{\frac{1}{2\alpha}} + 1}$ **then**
6 $\quad$ $U_1 := U_1 \cup \{u\}$ and $p(s_1) := \max\{p(s_1), (|x_u - 1|)^\alpha\}$.
7 **else**
8 $\quad$ $U_2 := U_2 \cup \{u\}$ and $p(s_2) := \max\{p(s_2), (|x_u - 2|)^\alpha\}$.
9 If there is another user, go to Step 2; otherwise, set $OUT = p(s_0) + p(s_1) + p(s_2)$, and output $\{p(s)\}_{s \in S}$ and OUT.

Theorem 4. *The competitive ratio of Algorithm 2 is no more than $\sqrt{2} + 1$.*

Proof. For each $i = 0, 1, 2$, let $u_{i,\max} = \arg\max_{u : u \in U_i} |x_u - i|$ be the user in U_i that is the farthest from s_i, i.e., $p(s_i) = (|x_{u_{i,\max}} - i|)^\alpha$, and

$$OUT = p(s_0) + p(s_1) + p(s_2) = (|x_{u_{0,\max}}|)^\alpha + (|x_{u_{1,\max}} - 1|)^\alpha + (|x_{u_{2,\max}} - 2|)^\alpha.$$

Let p^* be an optimal power assignment of the corresponding offline problem, and based on p^*, we can partition U into three user sets U_0^*, U_1^* and U_2^*, where

$$\begin{cases} U_0^* = \{u \in U \,|\, |x_u|^\alpha \leq p^*(s_0)\}; \\ U_1^* = \{u \in U \,|\, |x_u - 1|^\alpha \leq p^*(s_1)\} \setminus U_0^*; \\ U_2^* = U \setminus (U_0^* \cup U_1^*). \end{cases} \quad (3)$$

let $u^*_{i,\max}$ be the user in U^*_i that is the farthest from s_i for each $i = 0, 1, 2$, then the objective value of p^* is

$$OPT = p^*(s_0) + p^*(s_1) + p^*(s_2) = (|x_{u^*_{0,\max}}|)^\alpha + (|x_{u^*_{1,\max}} - 1|)^\alpha + (|x_{u^*_{2,\max}} - 2|)^\alpha.$$

In Algorithm 2, $\frac{2^{\frac{1}{2\alpha}}}{2^{\frac{1}{2\alpha}}+1}$ is an important gap to assign the user. For convenience, we set

$$\gamma = \frac{2^{\frac{1}{2\alpha}}}{2^{\frac{1}{2\alpha}}+1} > \frac{1}{2},$$

where the inequality follows from $\alpha \geq 1$. We have

$$\frac{\gamma^\alpha}{(1-\gamma)^\alpha} = \frac{(\frac{2^{\frac{1}{2\alpha}}}{2^{\frac{1}{2\alpha}}+1})^\alpha}{(1 - \frac{2^{\frac{1}{2\alpha}}}{2^{\frac{1}{2\alpha}}+1})^\alpha} = (\frac{2^{\frac{1}{2\alpha}}}{1})^\alpha = \sqrt{2}. \quad (4)$$

Note that, this equation is used in large below quantities.

Then, we prove that $OUT \leq (\sqrt{2} + 1) \cdot OPT$, and show this by proving the following two cases.

Case 1. Either $0 = \arg\max_{i:s_i \in S} p^*(s_i)$ or $2 = \arg\max_{i:s_i \in S} p^*(s_i)$.

The proof of these two situations is similar by $d(s_0, s_1) = d(s_1, s_2)$, and without loss of generality, assume that $0 = \arg\max_{i:s_i \in S} p^*(s_i)$, i.e.,

$$|x_{u^*_{0,\max}}| \geq \max\{|x_{u^*_{1,\max}} - 1|, |x_{u^*_{2,\max}} - 2|\}.$$

Case 1a. If $|x_{u^*_{0,\max}}| < 1 - \gamma$, then we have $|x_{u^*_{2,\max}} - 2| < 1 - \gamma < \frac{1}{2}$. Based on Algorithm 2, we have $U_i = U^*_i$ for each $i = 0, 1, 2$ by $\frac{2^{\frac{1}{2\alpha}}}{2^{\frac{1}{2\alpha}}+1} = \gamma \geq \frac{1}{2}$, and $OUT = OPT$.

Case 1b. If $|x_{u^*_{0,\max}}| \geq 1 - \gamma$, when $x_{u^*_{0,\max}} \leq 0$, $u^*_{0,\max}$ is added to U_0 and $u_{0,\max} = u^*_{0,\max}$, i.e., $|x_{u_{0,\max}}| = |x_{u^*_{0,\max}}|$, where U_0 is the user set covered by s_0 generated by Algorithm 2. When $x_{u^*_{0,\max}} > 0$, we have $x_{u^*_{0,\max}} > 1 - \gamma = 1 - \frac{2^{\frac{1}{2\alpha}}}{2^{\frac{1}{2\alpha}}+1}$, and $u^*_{0,\max}$ is not added to U_0 by Algorithm 2, which implies that $U_0 \subseteq U^*_0$, i.e., $|x_{u_{0,\max}}| \leq |x_{u^*_{0,\max}}|$. Therefore, we have

$$|x_{u_{0,\max}}| \leq |x_{u^*_{0,\max}}|.$$

Then we consider that the coordinate $x_{u^*_{2,\max}}$ of $u^*_{2,\max}$. If $x_{u^*_{2,\max}} \leq 1 + \gamma$, $u^*_{2,\max}$ is not added to U_2 by Algorithm 2, which implies that $|x_{u_{2,\max}} - 2| \leq |x_{u^*_{2,\max}} - 2|$; Otherwise, i.e., $x_{u^*_{2,\max}} > 1 + \gamma$, we have that u^*_2 is added to U_2 by Algorithm 2, which implies that $|x_{u_{2,\max}} - 2| = |x_{u^*_{2,\max}} - 2|$. Therefore, we have

$$|x_{u_{2,\max}} - 2| \leq |x_{u^*_{2,\max}} - 2|.$$

Thus, we have

$$\frac{OUT}{OPT} = \frac{(|x_{u_0,\max}|)^\alpha + (|x_{u_1,\max} - 1|)^\alpha + (|x_{u_2,\max} - 2|)^\alpha}{(|x_{u_0^*,\max}|)^\alpha + (|x_{u_1^*,\max} - 1|)^\alpha + (|x_{u_2^*,\max} - 2|)^\alpha}$$

$$\leq \frac{(|x_{u_0^*,\max}|)^\alpha + (|x_{u_1,\max} - 1|)^\alpha + (|x_{u_{2^*},\max} - 2|)^\alpha}{(|x_{u_0^*,\max}|)^\alpha + (|x_{u_1^*,\max} - 1|)^\alpha + (|x_{u_2^*,\max} - 2|)^\alpha}$$

$$\leq 1 + \frac{(|x_{u_1,\max} - 1|)^\alpha}{(|x_{u_0^*,\max}|)^\alpha + (|x_{u_1^*,\max} - 1|)^\alpha + (|x_{u_2^*,\max} - 2|)^\alpha}$$

$$\leq 1 + \frac{(|x_{u_1,\max} - 1|)^\alpha}{(|x_{u_0^*,\max}|)^\alpha}$$

$$\leq 1 + \frac{\gamma^\alpha}{(1-\gamma)^\alpha}$$

$$= \sqrt{2} + 1,$$

where the last inequality follows from $1 - \gamma \leq x_{u_1,\max} \leq 1 + \gamma$ based on Algorithm 2 and $|x_{u_0^*,\max}| > 1 - \gamma$ based on the assumption; the last equality follows from equality (4).

Case 2. $1 = \arg\max_{i:s_i \in S} p^*(s_i)$, i.e.,

$$|x_{u_1^*,\max} - 1| \geq \max\{|x_{u_0^*,\max}|, |x_{u_2^*,\max} - 2|\}. \tag{5}$$

Then, we consider the relationship between $|x_{u_0^*,\max}|$ and $|x_{u_2^*,\max} - 2|$.

Similar to the proof of **Case 1b**, If $|x_{u_0^*,\max}| \geq 1 - \gamma$ and $x_{u_0^*,\max} > 0$, $u_{0,\max}^*$ is not added to U_0 by Algorithm 2, which implies that $|x_{u_0,\max}| \leq |x_{u_0^*,\max}|$; If $|x_{u_0^*,\max}| > 1 - \gamma$ and $x_{u_0^*,\max} < 0$, u_0^* is added to U_0 by Algorithm 2, which implies that $|x_{u_0,\max}| = |x_{u_0^*,\max}|$, i.e.,

$$\begin{cases} |x_{u_0,\max}| \leq |x_{u_0^*,\max}|, & \text{if } |x_{u_0^*,\max}| \geq 1 - \gamma; \\ |x_{u_2,\max} - 2| \leq |x_{u_2^*,\max} - 2|, & \text{if } |x_{u_2^*,\max} - 2| \geq 1 - \gamma. \end{cases} \tag{6}$$

where the second inequality is not hard to obtain from the first inequality by $d(s_0, s_1) = d(s_1, s_2)$. Without loss of generality, we assume that

$$|x_{u_0^*,\max}| \leq |x_{u_2^*,\max} - 2|.$$

Case 2a. If $|x_{u_1^*,\max} - 1| \leq \gamma$, then we consider two cases about $|x_{u_2^*,\max} - 2|$.

If $|x_{u_2^*,\max} - 2| < 1 - \gamma < \frac{1}{2}$, based on Algorithm 2, we have $U_i = U_i^*$ for each $i = 0, 1, 2$ by $\frac{2^{\frac{1}{2\alpha}}}{2^{\frac{1}{2\alpha}} + 1} = \gamma \geq \frac{1}{2}$, and $OUT = OPT$.

If $|x_{u_{2,\max}^*} - 2| \geq 1-\gamma$, when $|x_{u_{0,\max}^*}| < 1-\gamma$, we have that u_0^* is added to U_0 based on Algorithm 2, i.e., $|x_{u_{0,\max}}| = |x_{u_{0,\max}^*}|$. Thus, we have

$$\begin{aligned}\frac{OUT}{OPT} &= \frac{(|x_{u_{0,\max}}|)^\alpha + (|x_{u_{1,\max}} - 1|)^\alpha + (|x_{u_{2,\max}} - 2|)^\alpha}{(|x_{u_{0,\max}^*}|)^\alpha + (|x_{u_{1,\max}^*} - 1|)^\alpha + (|x_{u_{2,\max}^*} - 2|)^\alpha}\\ &\leq \frac{(|x_{u_{0,\max}^*}|)^\alpha + (|x_{u_{1,\max}} - 1|)^\alpha + (|x_{u_{2^*,\max}} - 2|)^\alpha}{(|x_{u_{0,\max}^*}|)^\alpha + (|x_{u_{1,\max}^*} - 1|)^\alpha + (|x_{u_{2,\max}^*} - 2|)^\alpha}\\ &\leq 1 + \frac{(|x_{u_{1,\max}} - 1|)^\alpha}{(|x_{u_{0,\max}^*}|)^\alpha + (|x_{u_{1,\max}^*} - 1|)^\alpha + (|x_{u_{2,\max}^*} - 2|)^\alpha}\\ &\leq 1 + \frac{(|x_{u_{1,\max}} - 1|)^\alpha}{(|x_{u_{2,\max}^*} - 2|)^\alpha}\\ &\leq 1 + \frac{\gamma^\alpha}{(1-\gamma)^\alpha}\\ &= \sqrt{2} + 1,\end{aligned}$$

where the first inequality follows from $|x_{u_{2,\max}^*} - 2| \geq |x_{u_{2,\max}} - 2|$ by inequality (6), the last inequality follows from $|x_{u_{1,\max}} - 1| \leq \gamma$ and $|x_{u_{2,\max}^*} - 2| \geq 1-\gamma$ based on the assumption; the equality follows from equality (4).

Case 2b. If $|x_{u_{1,\max}^*} - 1| > \gamma$, then we also consider two cases about $|x_{u_{2,\max}^*} - 2|$.

If $|x_{u_{2,\max}^*} - 2| \leq 1-\gamma$, we have that $|x_{u_{0,\max}^*}| \leq 1-\gamma$ by the assumption $|x_{u_{0,\max}^*}| \leq |x_{u_{2,\max}^*} - 2|$, and

$$\gamma - 1 < x_u < 3 - \gamma, \ \forall u \in U; \tag{7}$$

Otherwise, there exists a user u with the coordinate $|x_u| < \gamma - 1$ (or $|x_u| > 3-\gamma$), if $u \in U_0^*$ (or $u \in U_2^*$), we have that $1-\gamma < |x_u| \leq |x_{u_{0,\max}^*}| \leq |x_{u_{2,\max}^*} - 2|$ (or $1-\gamma < |x_u - 2| \leq |x_{u_{2,\max}^*} - 2|$), which contradicts $|x_{u_{2,\max}^*} - 2| \leq 1-\gamma$.

Based on inequality (7) and Algorithm 2, we have

$$\begin{cases} |x_{u_{0,\max}}| < 1-\gamma; \\ |x_{u_{2,\max}} - 2| < 1-\gamma. \end{cases} \tag{8}$$

Thus, we have

$$\begin{aligned}\frac{OUT}{OPT} &= \frac{(|x_{u_{0,\max}}|)^\alpha + (|x_{u_{1,\max}} - 1|)^\alpha + (|x_{u_{2,\max}} - 2|)^\alpha}{(|x_{u_{0,\max}^*}|)^\alpha + (|x_{u_{1,\max}^*} - 1|)^\alpha + (|x_{u_{2,\max}^*} - 2|)^\alpha}\\ &\leq \frac{(1-\gamma)^\alpha + \gamma^\alpha + (1-\gamma)^\alpha}{(|x_{u_{1,\max}^*} - 1|)^\alpha}\\ &\leq \frac{(1-\gamma)^\alpha + \gamma^\alpha + (1-\gamma)^\alpha}{\gamma^\alpha}\\ &= 1 + \frac{2 \cdot (1-\gamma)^\alpha}{\gamma^\alpha}\\ &= \sqrt{2} + 1,\end{aligned}$$

where the first inequality follows from inequality (8); the second inequality follows from the assumption that $|x_{u_{1,\max}^*} - 1| > \gamma$; and the last equality follows from equality (4).

If $|x_{u_{2,\max}^*} - 2| > 1 - \gamma$, we have $|x_{u_{2,\max}} - 2| \leq |x_{u_{2,\max}^*} - 2|$ based on inequality (6). Then we also consider two cases about $x_{u_{0,\max}^*}$. If $|x_{u_{0,\max}^*}| > 1 - \gamma$, we have $|x_{u_{0,\max}}| \leq |x_{u_{0,\max}^*}|$ based on inequality (6). If $|x_{u_{0,\max}^*}| \leq 1 - \gamma$, we have that $x_u \geq \gamma - 1$ by inequality (7), and based on the Algorithm 2, vertex $u \in U$ with $x_u < 1 - \gamma$ is added to U_0, i.e., $|x_{u_{0,\max}}| \leq 1 - \gamma < \gamma < |x_{u_{1,\max}^*} - 1|$, where $u_{0,\max} = \arg\max_{u:u \in U_0} |x_u|$ be the user in U_0 that is the farthest from s_0, the second inequality follows from $\gamma > \frac{1}{2}$, the last inequality follows from the assumption that $|x_{u_{1,\max}^*} - 1| > \gamma$. Therefore, we have

$$|x_{u_{0,\max}}| \leq \max\{|x_{u_{0,\max}^*}|, |x_{u_{1,\max}^*} - 1|\}.$$

Thus, we have

$$\frac{OUT}{OPT} = \frac{(|x_{u_{0,\max}}|)^\alpha + (|x_{u_{1,\max}} - 1|)^\alpha + (|x_{u_{2,\max}} - 2|)^\alpha}{(|x_{u_{0,\max}^*}|)^\alpha + (|x_{u_{1,\max}^*} - 1|)^\alpha + (|x_{u_{2,\max}^*} - 2|)^\alpha}$$

$$\leq \frac{(\max\{|x_{u_{0,\max}^*}|, |x_{u_{1,\max}^*} - 1|\})^\alpha + (|x_{u_{1,\max}} - 1|)^\alpha + (|x_{u_{2,\max}} - 2|)^\alpha}{(|x_{u_{0,\max}^*}|)^\alpha + (|x_{u_{1,\max}^*} - 1|)^\alpha + (|x_{u_{2,\max}^*} - 2|)^\alpha}$$

$$\leq 1 + \frac{(|x_{u_{1,\max}} - 1|)^\alpha}{(|x_{u_{0,\max}^*}|)^\alpha + (|x_{u_{1,\max}^*} - 1|)^\alpha + (|x_{u_{2,\max}^*} - 2|)^\alpha}$$

$$\leq 1 + \frac{\gamma^\alpha}{\gamma^\alpha}$$

$$= 2,$$

where the last inequality follows from the assumption that $|x_{u_{1,\max}^*} - 1| > \gamma$ and $|x_{u_{1,\max}} - 1| \leq \gamma$ by Algorithm 2.

Therefore, the theorem holds. □

5 Conclusions

In this paper, we study the online power cover problem on a line. We first prove that the lower bound is 2 for the OPCL even when $|S| = 2$. Then, we present an online algorithm with a competitive ratio that is no more than $|S|$ based on the greedy technique. Note that, this algorithm is the best possible online algorithm for the OPCL with $|S| = 2$. Finally, we consider a special case of the OPCL, in which $S = \{s_0, s_1, s_2\}$ and $d(s_0, s_1) = d(s_1, s_2)$, and present an online algorithm with a competitive ratio that is no more than $\sqrt{2} + 1$.

Acknowledgement. The work is supported in part by the National Natural Science Foundation of China [Grant No. 12461059] and Yunnan Fundamental Research Projects [Grants No. 202501AS070076, No. 202501AS070170, No. 202401AT070442, No. 202301AU070197].

Declaration. The authors declare that they have no known competing financial interests.

References

1. Alt, H., et al.: Minimum-cost coverage of point sets by disks. In: Proceedings of the Twenty-second Annual Symposium on Computational Geometry, pp. 449–458 (2006)
2. Bilò, V., Caragiannis, I., Kaklamanis, C., Kanellopoulos, P.: Geometric clustering to minimize the sum of cluster sizes. In: Brodal, G.S., Leonardi, S. (eds.) ESA 2005. LNCS, vol. 3669, pp. 460–471. Springer, Heidelberg (2005). https://doi.org/10.1007/11561071_42
3. Cai, Z., Chen, Q.: Latency-and-coverage aware data aggregation scheduling for multihop battery-free wireless networks. IEEE Trans. Wireless Commun. **20**(3), 1770–1784 (2021)
4. Huang, Z., Feng, Q., Wang, J., Xu, J.: PTAS for minimum cost multi-covering with disks. In: Proceedings of the 2021 ACM-SIAM Symposium on Discrete Algorithms (SODA), pp. 840–859. SIAM (2021)
5. Lev-Tov, N., Peleg, D.: Polynomial time approximation schemes for base station coverage with minimum total radii. Comput. Netw. **47**(4), 489–501 (2005)
6. Li, M., Ran, Y., Zhang, Z.: A primal-dual algorithm for the minimum power partial cover problem. J. Comb. Optim. **44**(3), 1913–1923 (2022)
7. Liang, W., Li, M., Zhang, Z., Huang, X.: Minimum power partial multi-cover on a line. Theoret. Comput. Sci. **864**, 118–128 (2021)
8. Liu, X., Dai, H., Li, S., Li, W.: k-prize-collecting minimum power cover problem with submodular penalties on a plane (in Chinese). Sci .Sinica Inf. **52**(6), 947–959 (2022)
9. Liu, X., Li, W., Dai, H.: Approximation algorithms for the minimum power cover problem with submodular/linear penalties. Theoret. Comput. Sci. **923**, 256–270 (2022)
10. Ran, Y., Huang, X., Zhang, Z., Du, D.-Z.: Approximation algorithm for minimum power partial multi-coverage in wireless sensor networks. J. Global Optim. **80**(3), 661–677 (2021). https://doi.org/10.1007/s10898-021-01033-y
11. Optimal coverage in wireless Sensor Networks. SOIA, vol. 162. Springer, Cham (2020). https://doi.org/10.1007/978-3-030-52824-9_9
12. Xiao, M., Li, W.: Online bottleneck matching on a star. In: International Conference on Algorithmic Aspects in Information and Management, pp. 123–133. Springer (2024)
13. Xiao, M., Zhao, S., Li, W., Yang, J.: Online bottleneck matching on a line. J. Comb. Optim. **45**(4), 108 (2023)
14. Zhang, H., Zheng, X., Liu, Z., Liu, X.: The k-prize-collecting coverage problem by aligned disks. Int. Trans. Oper. Res. (2025). https://doi.org/10.1111/itor.13608
15. Zhang, Z., Willson, J., Lu, Z., Wu, W., Zhu, X.: Approximating maximum lifetime k-coverage through minimizing weighted k-cover in homogeneous wireless sensor networks. IEEE/ACM Trans. Netw. **24**(6), 3620–3633 (2016)

The Subinterval Cover Problem

Kelin Luo[1](✉) [iD], Chenran Yang[2](✉) [iD], Zonghan Yang[3](✉) [iD], and Yuhao Zhang[3](✉) [iD]

[1] University at Buffalo, Buffalo, USA
kelinluo@buffalo.edu
[2] Moscow Institute of Physics and Technology, Dolgoprudny, Russia
ian.ch@phystech.edu
[3] Shanghai Jiao Tong University, Shanghai, China
{fstqwq,zhang_yuhao}@sjtu.edu.cn

Abstract. In this paper, we introduce the subinterval cover problem. Given a set of interval candidates, each associated with a specific cost factor, we need to select a subinterval from each candidate to cover the entire range. The cost of selecting a subinterval in a candidate interval is determined by multiplying the selected length by the cost factor (the weight). Our goal is to minimize the total cost. This problem has broad applications, such as drone delivery and robot motion planning.

Our first finding is that a natural LP formulation of the problem has an unbounded integrality gap, which highlights the challenge of studying this problem. Second, we develop polynomial-time dynamic programming algorithms for two special cases: 1) two distinct cost factors; 2) a constant number of distinct cost factors, and the interval candidates form a laminar set family: each pair of intervals is either disjoint, or one interval contains the other. As a corollary of our dynamic programming algorithm, we design a polynomial-time $\sqrt{w_{\max}/w_{\min}}$-approximation algorithm for the general cases, where $w_{\max}$ and $w_{\min}$ are the maximum and minimum weights of the intervals, respectively.

Keywords: Subinterval Cover · Approximation Algorithm · Dynamic Programming

1 Introduction

The Set Cover problem is a fundamental combinatorial optimization problem. Given a set of elements and a collection of subsets, we aim to select several subsets to cover all the elements. The objective can be minimizing the number of subsets or the total weights. This problem is recognized as NP-hard by [24], and various approximation algorithms have been proposed to find approximation solutions efficiently, such as the standard greedy algorithm, which admits an approximation ratio of $O(\log n)$ [15,27]. Fruitful studies also consider the Set Cover problem in geometric settings, as they already capture a broad range of real-world applications. We are given a universe of points in R^d, and a set of geometric shapes like disks. The objective is to cover the universe with a minimized number of shapes. The problem is NP-hard even for very simple shapes such as unit disks and squares [7,20,23]. However, in the one-dimensional special case, the shapes and the universe reduce to intervals, and the problem can be solved by a simple polynomial-time greedy algorithm, even in the weighted version (each candidate interval has an associated weight).

Our paper focuses on the one-dimensional scenario and explores the following perspective. In the set-cover style definition, when we discuss a candidate interval, the only two options are whether to select it or not. However, in many cases, we do not need to pay the complete cost to select the entire interval. The question is, can we provide an option to select a part of the interval with a lower price? This motivation stems from the following real-world scenario related to drone delivery.

> **Example (Drone Delivery).** We aim to deliver a package from location A to B (on a line) by drone. Each drone can serve a given area (an interval on the line). We need to select some drones and arrange for them to deliver the package and minimize the total usage of drones. If we use the Geometric Interval Set Cover problem to model this scenario, we can only consider the selection of the drone but omit the arrangement. However, in some cases, using a drone in a small subinterval of its serving area suffices, reducing the overall cost. To optimize the arrangement, we need to allow a partial selection of intervals in the theoretical model.

Towards this goal, we formalize the following *Subinterval Cover Problem (SIC)*, which allows algorithms to select an interval partially. Given n candidate intervals $\{1, 2, \ldots, n\}$ (denoted by $[n]$), each is defined within the range $[0,1)$ as $T_i = [L_i, R_i)$. Each interval can either cover a specific subinterval $[l_i, r_i)$ such that $[l_i, r_i) \subseteq T_i$, or remain unused. The cost of this operation is defined to be $W_i(l_i, r_i)$. The goal is to cover the range $[0, 1)$ while minimizing the total cost. (To ensure feasibility, we further require that $\bigcup_{i=1}^n T_i = [0, 1)$.) As a fairly natural starting point, we consider a linear cost function of each candidate interval. That is, we assume that each interval T_i has a nonnegative weight w_i, and the cost of an interval is calculated based on the length of the used subinterval $[l_i, r_i)$, i.e., $W_i(l_i, r_i) = w_i \cdot (r_i - l_i)$. The classic idea for solving interval set cover fails to solve the SIC problem, since they have not considered partial selection. Then, the main question we should ask is, can we also solve SIC in polynomial time?

1.1 Our Results

We first demonstrate a natural Integral Linear Programming (ILP) and its LP relaxation. The LP rounding method is a standard attempt to solve the problem. However, we show an unbounded integrality gap under our natural ILP formulation, even in the case of only three distinct weights in Sect. 3, which highlights the challenge of studying this problem.

Note that if every candidate interval has identical weights, the problem becomes trivial, and the optimal cost is always w_i. Therefore, to understand the challenge of the problem, we start with a minimal non-trivial special case of two weights, where only two distinct weights exist. Is it already tough to solve? We answer this question on the positive end with a polynomial-time dynamic programming algorithm.

Theorem 1. *There is an $O(n^7)$ algorithm for the SIC problem with 2 distinct weights.*

However, we find that our approach is hard to generalize even if the number of distinct weights only grows to three, unless we further introduce some assumptions. Our second positive result is a polynomial-time DP algorithm when there is a constant number of distinct cost factors and the interval candidates form a laminar family. For the complete proof and detailed algorithmic analysis, we refer the reader to the full version of this paper.

Theorem 2. *There is an $O(kn^{2k+3})$ algorithm for the SIC problem with k distinct weights, under the constraint that the intervals form a laminar set family.*

Without the laminar assumption, we still do not know whether the problem is NP-hard or polynomially solvable, even assuming the number of distinct weights is only three. A remarkable negative signal is that the integrality gap becomes infinity when the number of distinct weights becomes three. We leave it as an open question.

Furthermore, we extend our algorithms to get some approximation results in the general setting. The first is as follows.

Theorem 3. *There is a $\min\left(\sqrt{w_{\max}/w_{\min}}, w_{\max}/w_{2^{nd}\min}\right)$-approximation algorithm for the general SIC problem with time complexity $O(n^7)$.*

In addition, our approach to the subinterval cover problem can be extended to tackle the collaborative drone delivery problem. This problem, first introduced by Erlebach et al. [18], conceptualizes each candidate interval as a drone with a limited (and connected) travel range and a specified speed. The objective is to deliver a package from the starting point to the ending point with the minimum total time, with each drone being utilized once within its moving range. The problem aligns with our model if we restrict the problem to a path, except that each drone is associated with an initial position in its serving range (aligned with the candidate interval in our model). The difference is that the cost of scheduling a drone also depends on the moving distance starting from the initial position but not the selected subinterval. The only known solution for this scenario is an $O(n)$-approximation algorithm, which remains the best-known result in general graph settings. In the full version of this paper, we show that our solutions can be applied effectively in scenarios where each drone (or robot) has a predefined initial placement on the path, providing a 2-approximation reduction for these cases. By tolerating an extra approximation ratio of 2, any algorithmic results for the SIC problem can be applied to the drone delivery problem on paths.

1.2 More Related Work

The Subinterval Cover Problem (SIC) has a close relationship to the classical Set Cover Problem, particularly in scenarios involving predetermined ranges [14] or routes [1,16] in variants such as the Geometric Set Cover Problem. Various cases of the geometric set cover problem have been extensively studied in the literature, including both unweighted settings, where the goal is to minimize the number of selected subsets [8,15,17,19,24,27], and weighted settings, where the objective is to minimize the total weight [2,3,13,21,26]. For the unweighted geometric set cover problem in the 1D case, both the geometric set cover and hitting set problems (focusing on the dual perspective) can be solved in polynomial time using a simple greedy algorithm. However, in higher dimensions, these problems are known to be NP-complete [24], even for simple shapes, such as when the subset collection is induced by unit disks or squares. Aside from the known near-optimal solutions for the general set cover problem, such as the greedy algorithm [15,27], which iteratively selects the set that covers the largest number of uncovered elements until the entire universe is covered and achieves an approximation ratio of $O(\log n)$, better approximations can often be achieved in low-dimensional geometric settings using linear programming (LP)-based approaches [8,19]. The weighted version of the geometric set cover problem, where each object is associated with a weight, has also received considerable attention. In this variant, the goal is to select a subset of objects with the minimum total weight that covers the entire range. However, only a few instances are known to offer an approximation guarantee better than $O(\log n)$. For example, selecting unit disks in the plane admits a $O(1)$-approximation algorithm using dynamic programming [2] and the LP approach [26]. Nevertheless, the 1D weighted geometric set cover problem, also known as a special case of the weighted interval cover problem, can still be solved in polynomial time [3].

Our SIC problem extends the 1D weighted geometric set cover by incorporating the concept of subinterval selection and accounting for the associated subinterval costs, which has practical applications, such as in the collaborative delivery problem. The collaborative delivery problem, which uses multiple agents, such as drones, to deliver a package with a predesigned route, has gained significant attention in recent years. Initial research on collaborative delivery under energy constraints was pioneered by Chalopin et al. [11]. They considered a scenario where a set of energy-constrained drones such as battery budget, positioned on a straight line, must collectively transport an item from a source point to a target point. A polynomial-time algorithm was provided for determining whether drones can successfully deliver an item along a line when they have a uniform budget. However, the problem becomes (weakly) NP-complete when the drones have different budgets [11]. They also provided a quasi-pseudo-polynomial time algorithm [11], along with resource-augmented algorithms [10]. Subsequent studies have built upon this foundation, exploring various aspects of collaborative delivery under energy constraints [6,12].

In addition to energy-constrained collaboration, research has also explored heterogeneous collaboration, where drones have diverse energy consumption rates [4] and speeds [5]. Delivering one package with minimum total time and minimum energy consumption can both be solved in polynomial time [4,5,9]. However, delivering two packages has been proven to be NP-hard [9]. In this scenario, each drone's unlimited budget implies an unlimited movement range, with the only difference being their efficiency ratios. Additionally, 'natural' flying constraints have been investigated, where drones have designated serving ranges, necessitating collaboration due to geographical, regulatory, or other practical constraints [18,25]. Erlebach et al. [18] demonstrated that with drones having designated serving ranges, and varying speeds, delivering a package from a source to a destination with minimum completion time is a (weakly) NP-hard problem, even when the drones move along a line [18]. We observe that a 2-approximation algorithm for this problem can be provided by applying the solution for the SIC problem.

2 Preliminary

In this preliminary section, we introduce some basic structural properties for the optimal solution of the problem.

Definition 1 (Solution). *A solution is defined as a tuple (I, l, r): $I \subseteq [n]$ is the set of used intervals, and $[l_i, r_i) \subseteq T_i$ is the subinterval chosen for interval $i \in I$.*

Definition 2 (Disjoint). *A solution is considered disjoint if each point $x \in [0, 1)$ is covered by exactly one selected subinterval.*

Lemma 1. *A disjointed optimal solution exists for the SIC problem.*

Proof. Assume there is an optimal solution $\mathsf{OPT} = (I, l, r)$ where the selected subintervals of intervals $i, j \in I$ intersect. In this case, we can adjust the subintervals while maintaining the feasibility of the solution. Without loss of generality, let us assume $l_i \leq l_j$. We then consider two cases:

- If $l_i \leq l_j \leq r_i \leq r_j$: adjust l_j to start at r_i by setting $l_j \leftarrow r_i$;
- If $l_i \leq l_j < r_j \leq r_i$: remove interval j from the solution set.

The cost is non-increasing after either subinterval adjustment because the interval j always involves a shorter length. Since there is only a finite number of interval pairs, we can eliminate all intersections, ensuring that each point is covered by exactly one selected subinterval. This results in a disjoint solution.

Definition 3 (Discrete). *A solution is considered discrete if, for every $i \in I$, the selected subinterval $[l_i, r_i)$ satisfies the condition that both l_i and r_i belong to $\{L_1, \ldots, L_n, R_1, \ldots, R_n\}$.*

Lemma 2. *There exists a discrete and disjoint optimal solution for the SIC problem.*

Proof. By Lemma 1, the existence of a disjoint optimal solution is established. Suppose OPT = (I, l, r) is such a solution for the SIC problem. If all its non-empty subintervals of OPT satisfy the condition that every l_i and r_i are in the set $A = \{L_1, \ldots, L_n, R_1, \ldots, R_n\}$, then it is also a discrete solution. However, if some non-empty subintervals, for example, have r_i not aligned with any points in A, and given that all subintervals cover the entire range, there must be another non-empty selected subinterval of interval j adjacent to the selected subinterval $[l_i, r_i)$, such that $l_j = r_i$. Note that since both subintervals are non-empty and their ending points are not in set A, it holds that $L_i < r_i = l_j < R_i$ and $L_j < r_i = l_j < r_j \leq R_j$. Adjustments can be made to the subintervals while maintaining the solution's feasibility. Consider two cases:

- If $w_i \leq w_j$: Adjust r_i (and correspondingly l_j) to $\min\{r_j, R_i\}$. If $\min\{r_j, R_i\} = r_j$, then remove interval j from the solution set. If $\min\{r_j, R_i\} = R_i$, then the point r_i (and correspondingly l_j) falls within set A. In this case, the number of non-aligned endpoints of the subintervals is reduced by one, and the cost remains non-increasing.
- If $w_i > w_j$: Adjust r_i (and correspondingly l_j) to $\max\{l_i, L_j\}$. If $\max\{l_i, L_j\} = l_i$, then remove interval i from the solution set. If $\max\{l_i, L_j\} = L_j$, then the point r_i (and correspondingly l_j) falls within set A. In this case, the number of non-aligned endpoints of the subintervals is reduced by one, and the cost remains non-increasing.

Corollary 1. *Without loss of generality, we can assume a restriction in the problem that the selection of every l_i, r_i is discrete and disjoint.*

Given that there is only a finite number of non-aligned l and r values, and since each adjustment reduces this number by at least one, we can achieve an aligned *discrete* optimal solution OPT' through a finite number of steps.

Lemma 3. *In any SIC problem instance, the optimal solution is equivalent before and after merging any two intersecting least weight intervals.*

Proof. Consider two least weight intervals T_i and T_j with weights $w_i = w_j = w_{\min}$ and $T_i \cap T_j \neq \varnothing$. Suppose these intervals are merged into $T_k = T_i \cup T_j$ in a new instance. Let OPT, OPT' denote the optimal solution before and after merging, respectively.

- OPT' $\Rightarrow$ OPT: If k is used in an OPT', by the fact that $T_k = T_i \cup T_j$ we can construct corresponding disjoint $[l_i, r_i), [l_j, r_j)$ such that their union is $[l_k, r_k)$.
- OPT $\Rightarrow$ OPT': If i, j are used in two non-adjacent intervals, from the adjustment arguments in the proof of Lemma 2, we could always expand them so that $[l_i, r_i) \cup [l_j, r_j)$ is an interval while keeping the optimality.

Corollary 2. *Without loss of generality, we can assume that the least weight intervals in any instance are disjoint and $\forall i \in I, w_i = w_{\min}$, in OPT, the selected subinterval $[l_i, r_i)$ is equal to the full interval $[L_i, R_i)$.*

3 ILP and LP Formalization

To address the subinterval cover problem, we can leverage Integer Linear Programming (ILP) and its Linear Programming (LP) relaxation for modeling and solving this combinatorial optimization problem. The objective is to minimize the total cost of selected subintervals while ensuring complete coverage of the universe interval.

We begin by defining the decision variables for our ILP model. Let $x_{i,l,r}$ be a binary variable indicating that $x_{i,l,r} = 1$ if the interval i is used to cover the interval $[l, r)$, and $x_{i,l,r} = 0$ otherwise. These variables enable us to succinctly formulate the problem requirements: to efficiently cover the entire interval using selected subintervals from the intervals, while minimizing the total cost of the subintervals used. As w_i denotes the unit cost of the interval i, and $(r - l)$ represents the length of the subinterval selected within the interval i, the cost is calculated as $(r-l)w_i x_{i,l,r}$. Although the original model allows for the free selection of the subinterval $[l, r)$ from the intervals, with potentially infinite choices, Lemma 2 asserts that the number of feasible choices for a subinterval $[l, r)$ is $O(n^2)$. Specifically, define

$$A_i = \{[l, r) \mid L_i \leq l \leq r \leq R_i, l, r \in \{L_1, \ldots, L_n, R_1, \ldots, R_n\}\},$$

as the sets of feasible choices for $[l, r)$ with respect to interval i. We can then formulate an ILP as follows.

$$\text{minimize} \quad \sum_{i \in [n]} \sum_{[l,r) \in A_i} (r - l) w_i x_{i,l,r} \tag{1}$$

$$\text{subject to} \quad \sum_{[l,r) \in A_i} x_{i,l,r} \leq 1 \qquad \forall i \in [n] \tag{2}$$

$$\sum_{i \in [n], [l,r) \in A_i} x_{i,l,r} \geq 1 \tag{3}$$

$$x_{i,l,r} \in \{0, 1\} \qquad \forall i \in [n], [l, r) \in A_i \tag{4}$$

Constraints (2) prevent using interval i in two non-continuous subintervals. Constraints (3) ensure that every point within the interval is covered by some subintervals. We can relax the integrality Constraints (4) from $x_{i,l,r} \in \{0, 1\}$ to $x_{i,l,r} \in [0, 1]$ to have the relaxed LP formulation.

LP rounding is usually an effective way to design an approximation algorithm. However, we demonstrate in Sect. 3.1 with a simple example that there is an unbounded integrality gap for this natural LP formulation.

Before delving into the details, let us first provide an intuitive understanding of the difference between LP and ILP formulations: LP allows for the allocation of a total quantity of no more than 1 across multiple, say k ($k \geq 1$), non-adjacent subintervals within an interval T_i, while in ILP, k is 1. Figure 1 gives an example of the case where $k = 2$.

3.1 Existence of the Arbitrarily Large LP Integrality Gap

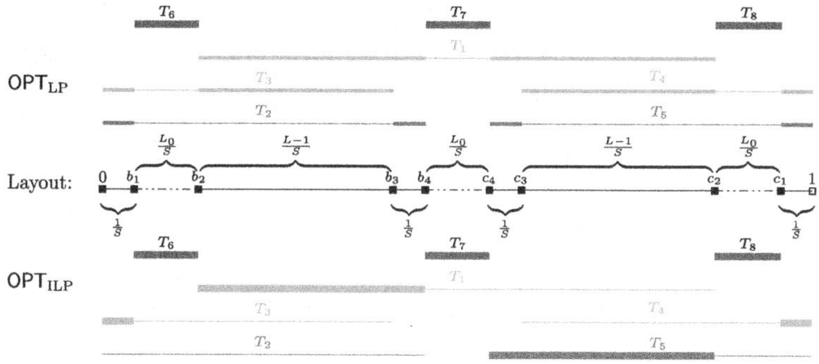

Fig. 1. OPT_{LP} and OPT_{ILP} for the same instance of 8 intervals with 3 distinct weights (marked as 0-Blue, $w_{\min}$-Orange, $w_{\max}$-Red). The thickest, second-thickest, and thinnest line segments represent the allocation of 1, $\frac{1}{2}$ (i.e., $T_{1...5}$ in OPT_{LP}) and 0, respectively.

For any maximum weight $w_{\max}$ and minimum non-zero weight $w_{\min} < w_{\max}$, we can construct an instance in Fig. 1, with 5 non-zero (weight) intervals $T_{1...5}$ and 3 zero (weight) intervals $T_{6...8}$, to demonstrate that the integrality gap is $\Omega(w_{\max}/w_{\min})$. Consequently, the gap can be arbitrarily large for appropriate values of $w_{\max}$ and $w_{\min}$.

Let $a = w_{\min}$, $c = w_{\max}$. Define $L = \left(\frac{c}{a}\right)^2$ and $L_0 = 1 + \frac{c(2L+2)}{a}$. The endpoints $\{0, b_1, b_2, b_3, b_4, c_4, c_3, c_2, c_1, 1\}$ are arranged sequentially from left to right. Before normalization, the total length is $S = 2(1 + L_0 + L) + L_0$, and the normalization factor is $\frac{1}{S}$. The relationships between points are defined as follows: $b_1 - 0 = 1 - c_1 = \frac{1}{S}$, $b_2 - b_1 = c_1 - c_2 = \frac{L_0}{S}$, $b_3 - b_2 = c_2 - c_3 = \frac{L-1}{S}$, $b_4 - b_3 = c_3 - c_4 = \frac{1}{S}$.

The non-zero intervals are placed as $T_1 = [b_2, c_2)$, $T_2 = [0, b_4)$, $T_3 = [0, b_3)$, $T_4 = [c_3, 1)$, $T_5 = [c_4, 1)$ and their weights are assigned as $w_{1,3,4} = a$, $w_{2,5} = c$. The corresponding ranges of the zero intervals $T_6 = [b_1, b_2)$, $T_7 = [b_4, c_4)$, $T_8 = [c_2, c_1)$, are called zero ranges and are represented by dashed segments. Each of $T_{1...5}$ is separated by a zero range, and the structure is symmetric around the center.

The length L_0 guarantees that, in any optimal solution, the subintervals $[l_1, r_1), \ldots, [l_5, r_5)$ do not cross any zero range: Since $a \cdot L_0 > c(2L + 2)$, if a subinterval with the minimum weight crosses a zero range, the cost will be higher than in the case where zero ranges are covered by $T_{6...8}$ and all the rest (with total length $2L + 2$) are covered by c-weight intervals. That is, in any optimal solution, zero ranges are covered by zero intervals.

To obtain OPT_{LP}, the strategy involves minimizing the use of T_2 and T_5 while maximizing the use of T_1, T_3, and T_4. The upper part of Fig. 1 shows one possible case.

$$\mathsf{OPT}_{\mathrm{LP}} = \frac{1}{S} \cdot 2 \cdot \frac{1}{2}\left(1 \cdot (a+c) + (La + (L-1)a + 1 \cdot c)\right) + 0 \cdot 3 \cdot \frac{L_0}{S} = \frac{2La + 2c}{S}.$$

For the ILP optimal solution, each of $T_{1...5}$ can only be used on at most one side of the zero range that separates it. Note $L > 1 > \frac{1}{S}$, so the range $[0, b_1)$ will be covered by T_3 but not T_2, and the range $[c_1, 1)$ will be covered by T_4 but not T_5, as shown in the lower part of Fig. 1,

$$\mathsf{OPT}_{\mathrm{ILP}} = \frac{1}{S}(La + Lc + 2 \cdot a + 0 \cdot 3L_0) = \frac{L(a+c) + 2a}{S}.$$

In this instance, the integrality gap

$$\frac{\mathsf{OPT}_{\mathrm{ILP}}}{\mathsf{OPT}_{\mathrm{LP}}} = \frac{L(a+c) + 2a}{2La + 2c} > \frac{Lc}{2La + 2c} = \frac{c}{2a + \frac{2c}{L}} = \frac{c}{a\left(2 + 2\left(\frac{a}{c}\right)\right)} \geq \frac{1}{4} \cdot \frac{c}{a}$$

$$= \Omega(w_{\max}/w_{\min}).$$

4 Algorithm Under Binary Weights

In this section, we introduce a dynamic programming algorithm designed to solve the subinterval cover problem, which involves two distinct weights.

The core of any dynamic programming strategy relies on defining of well-structured subproblems. We define subproblems to address the optimal solution to serve a specific range by a specific subset of intervals. Roughly speaking, the specific subset of intervals can be the first i-th intervals, and the specific range can be $[0, 0.5)$. Now consider the case we are solving i. We can first decide how to use i by enumerating the subinterval of i, say $[l_i, r_i) \subseteq [L_i, R_i)$. For example, it can be $[0.1, 0.2)$. The problem is, how do we cover $[0, 0.1)$ and $[0.2, 0.5)$ next? In general, we need to enumerate the subset of intervals, which serves $[0, 0.1)$, and which serves $[0, 2, 0.5)$, which takes exponential time. To circumvent this problem, we need a clever DP order and a fine-grained subproblem definition, which reduces the size of the subproblem back to polynomials while keeping the correctness. First, we introduce the DP order.

Lemma 4 (DP order). *There is a total order $\preceq$ on $[n]$ satisfying the following left-to-right rule and inside-to-outside rule.*

- *Left to right: $u \preceq v$ if $L_u < L_v$ and $R_u < R_v$;*
- *Inside to outside: $u \preceq v$ if $L_v \leq L_u$ and $R_u \leq R_v$.*

Proof. Without loss of generality, we assume that there are no two identical intervals for convenience, as the reflexivity is already given by inside-to-outside-type for identical intervals.

Construct a graph $G = (V, E)$ where $V = [n]$ and

$$E = \{(u, v) \mid u \neq v \wedge ((L_u \geq L_v \wedge R_u \leq R_v) \vee (L_u < L_v \wedge R_u < R_v))\}.$$

We are to prove that this graph is an acyclic tournament graph, which admits a feasible total order. A tournament graph is a graph with exactly one directed edge between each pair of distinct vertices, and the absence of 3-cycles in a tournament graph G ensures that G is acyclic [22].

For any pair of distinct vertices u, v, the condition $(L_u \geq L_v \wedge R_u \leq R_v)$ and $(L_u < L_v \wedge R_u < R_v))$ are mutually exclusive, hence there is exactly one edge between u, v, and by definition, G is a tournament graph. We denote each edge as either inside-to-outside-type (e_1) for the condition $(L_u \geq L_v \wedge R_u \leq R_v)$ or left-to-right-type (e_2) for $(L_u < L_v \wedge R_u < R_v)$. Due to the transitivity of e_1 and e_2, respectively, a 3-cycle cannot consist of edges of the same type. For the two remaining cases:

- $u \xrightarrow{e_1} v \xrightarrow{e_1} w \xrightarrow{e_2} u$, from the two e_1 edges, we infer $R_w \geq R_u \geq R_u$, which contradicts the e_2 edge condition that $R_w < R_u$.
- $u \xrightarrow{e_1} v \xrightarrow{e_2} w \xrightarrow{e_2} u$, from the two e_2 edges, we deduce $L_v < L_w < L_u$, which contradicts the e_1 edge condition that $L_u \geq L_v$.

Therefore, there are no 3-cycles in G, and there exists a valid total order of $[n]$.

Then, we are ready to define the subproblem in our DP algorithm. In general, the subproblem describes the optimal solution for using a subset of candidate intervals to cover a specific range. The trivial way is to define it for every subset, which is exponential. However, we restrict the subset to a family corresponding to the defined DP order, which can be captured by two parameters: 1) i: it should be in the first i-th candidate interval following the DP order. 2) $[C_l, C_r)$, its left endpoints should fall in the interval. Combining with the discrete property in Corollary 1, this restricted family of subsets makes the number of subproblems polynomial.

Definition 4. *Let $f_i(S, C)$ denote the minimum cost to cover the range $S = [S_l, S_r)$ using the first i intervals in the ordered sequence of intervals, where only intervals with left endpoints falling within $C = [C_l, C_r)$, that is, $C_l \leq L_i < C_r$. By Lemma 2, without loss of generality, we can assume that all C_l, C_r, S_l, S_r are the endpoints of the intervals.*

The next question is why we only need to focus on the special family, not every subset. The reason is the existence of a good monotonicity behind the optimal solution, corresponding to the left endpoints. Recall that L_i is the left endpoint of i, and l_i is the left endpoint of the selected subinterval for i. The monotonicity says for every $j \preceq i$, we will use j to cover some range on the left side of l_i, if and only if $L_j < l_i$. We conclude the lemma as follows. We remark that it is the most important property we observe in the two-weight special case. However, this property is hard to generalize even to the three-weight case.

Lemma 5 (Monotonicity of subinterval left endpoints). *For every subproblem $f_i(S, C)$, there exists an optimal solution $\mathsf{OPT} = (I, l, r)$ such that for every $i, j \in I$ with $j \preceq i$, we have $L_j < l_i \iff l_j < l_i$.*

Algorithm 1. Dynamic programming under binary weights restriction.

1: Sort and renumber all intervals according to Lemma 4 such that $1 \preceq 2 \preceq \cdots \preceq n$.
2: Basic case:
$$f_0(S,C) = \begin{cases} 0, & \text{if } S = \varnothing \\ \infty, & \text{otherwise} \end{cases}$$
3: For $i \geq 1$, recursively define $f_i(S,C)$ as follows:
$$f_i(S,C) = \min \begin{cases} f_{i-1}(S,C), \\ f_{i-1}\left([S_l,u),[C_l,u)\right) + w_i(v-u) + f_{i-1}\left([v,S_r),[u,C_r)\right), \\ \qquad \forall u \in [C_l,C_r), [u,v) \subseteq S \cap [L_i,R_i) \end{cases}$$

Proof. Since by definition $L_j \leq l_j$, the implication $l_j < l_i \Rightarrow L_j < l_i$ is immediate. Our task is to show that $L_j < l_i \Rightarrow l_j < l_i$.

For convenience, we remap the weights to 0 and 1, representing the lower and higher weights, respectively. By the assumption in Corollary 1, let $\mathsf{OPT} = (I,l,r)$ be a discrete and disjoint optimal solution to the subproblem. Define $T_j^1 = (T_j \setminus T_i)$ and $T_j^2 = (T_j \cap T_i)$. Given the specified order, $j \preceq i$ implies $R_j \leq R_i$. When $l_j \in T_j^1$, the lemma is upheld. Therefore, we exclusively consider $l_j \in T_j^2$, where T_j^2 belongs to T_i.

Corollary 2 necessitates that $w_i = w_j = 1$. With $w_i = 0$, we deduce $[l_j, r_j) \subseteq T_i$ from $l_j \in T_j^2$ and $l_j < r_j \leq R_j \leq R_i$. Note the subinterval $[l_j, r_j)$ is nonempty, we obtain $[l_i, r_i) \neq [L_i, R_i)$, which contradicts Corollary 2; in the case of $w_j = 0$, $[l_j, r_j) = [L_j, R_j)$. If the implication is false, i.e., $L_j < l_i$ while $l_j \geq l_i$, we also get a contradiction because $L_j = l_j$. The remaining scenario considers $w_i = w_j = 1$. Under this circumstance, when $l_j \geq l_i$, $r_i \leq l_j$ follows from the disjointness. Combining this with the hypothesis $l_i > L_j$ leads to $[l_i, r_i) \subseteq T_j^2$. Since $T_j^2 \subseteq T_i$, both $[l_i, r_i)$ and $[l_j, r_j)$ are in T_j^2, swapping the subintervals $[l_i, r_i)$ and $[l_j, r_j)$ is valid, facilitating an optimal solution that conforms to the lemma. The lemma then holds for every i by induction.

Lemma 6. *Algorithm 1 correctly solves every subproblem $f_i(S,C)$.*

Proof. The proof follows induction and Lemma 5. Assuming we have an optimal solution for every f_{i-1} subproblem, the algorithm considers every option that satisfies the monotonicity and takes the optimal one based on the optimal solutions for the subproblems. Consider the optimal solution of $f_i(S,C)$, assuming i is not selected, straightforwardly, it equals to $f_{i-1}(S,C)$. Assuming i is selected at $[l_i, r_i)$, we need to further know who should take care $[S_l, l_i)$ and $[r_i, S_r)$. By the nice monotonicity in Lemma 5, we know that the subset that serves the range $[S_l, l_i)$ must have the property $L_j < l_i$, and the subset that serves the range $[r_i, S_r)$ must have the property $L_j \geq l_i$. Therefore, if we know $[l_i, r_i)$, the optimal solution of $f_i(S,C)$ can be recovered by $f_{i-1}([S_l, l_i), [C_l, l_i))$ and $f_{i-1}([l_i, r_i), [l_i, C_r))$. Finally, by enumerating $[l_i, r_i)$, the DP approach can correctly solve $f_i(S,C)$.

Proof (Proof of Theorem 1). By Lemma 6, Algorithm 1 gives an optimal solution for the binary weights subinterval cover problem by $f_n([0,1),[0,1))$. The time complexity $O(n^7)$ is determined by the number of states, which is $O(n^5)$, and the number of transitions for each state, which is $O(n^2)$.

As a reduction to Algorithm 1, we have a min $\left(\sqrt{w_{\max}/w_{\min}}, w_{\max}/w_{2^{\text{nd}}\min}\right)$-approximation algorithm mentioned in Theorem 3.

Proof (Proof of Theorem 3). Let OPT be the optimal solution for the original problem. We now modify the weights of the original problem by setting weights $w_i \leq \theta$ to $w_i' = \theta$, where $\theta = \sqrt{w_{\max}/w_{\min}}$, and weights $w_i > \theta$ to $w_i' = w_{\max}$. Let OPT$'$ be the optimal solution for the modified problem. Every weight is scaled up to $\sqrt{w_{\max}/w_{\min}}$ times, so we have $\frac{\text{OPT}'}{\text{OPT}} \leq \sqrt{w_{\max}/w_{\min}}$. The modification leads to a binary-weight condition, applying Algorithm 1, we get OPT$'$.

Next, we apply the solution OPT$'$ to the original problem by reverting the weights back to their original values, resulting in a solution ALG for the original problem. It is evident that ALG $\leq$ OPT$'$, as the weights in the original problem are no greater than those in the modified problem.

Combining the above, $\frac{\text{ALG}}{\text{OPT}} \leq \frac{\text{OPT}'}{\text{OPT}} \leq \sqrt{w_{\max}/w_{\min}}$.

If $w_{\min} = 0$, we can only set $\theta = 0$ to avoid the $+\infty$ approximation ratio(when OPT only consists of 0-weight intervals). By similar procedure, the approximation ratio is $w_{\max}/w_{2^{\text{nd}}\min}$.

5 Conclusion and Future Work

In conclusion, we introduce a clean combinatorial optimization problem: the subinterval cover problem. We demonstrate that this problem is inherently challenging, even for approximation. To shed some light on future research, we delve into two specific scenarios: cases with two distinct weights and those with constant weights with laminar families. For these cases, we devise efficient polynomial-time exact algorithms.

However, extending our 2-weight algorithm presents a challenge. The key monotonic structure identified in Lemma 5 does not hold true when confronted with three distinct weights, as exemplified in the full version of this paper. Note that we have tried weaker monotonicity to maintain a polynomial-size DP definition, but all attempts fail when the number of distinct weights increases to three. Notably, the integrality gap becomes unbounded when three distinct weights are introduced. Consequently, we pose an intriguing open question:

– Is the problem NP-Hard, even only considering three distinct weights?

Furthermore, we explore the potential for generalizing the cost function for a wider range of applications. That is, the cost of selecting a subinterval within candidate interval i can be expressed as a function $W_i(l_i, r_i)$. As a simple example, the cost function can be a constant (it becomes the geometric interval set cover problem), and the problem is also solvable in polynomial time [3]. Discovering more well-motivated cost functions continues to be interesting for future studies.

References

1. Ali, J., Dyo, V.: Coverage and mobile sensor placement for vehicles on predetermined routes: a greedy heuristic approach. In: Proceedings of the 14th International Joint Conference on e-Business and Telecommunications (ICETE 2017) - WINSYS, pp. 83–88. SciTePress (2017). ISBN 978-989-758-261-5
2. Ambühl, C., Erlebach, T., Mihalák, M., Nunkesser, M..: Constant-factor approximation for minimum-weight (connected) dominating sets in unit disk graphs. In: International Workshop on Approximation Algorithms for Combinatorial Optimization, pp. 3–14. Springer (2006)
3. Atallah, M.J., Chen, D.Z., Lee, D.T.: An optimal algorithm for shortest paths on weighted interval and circular-arc graphs, with applications. Algorithmica **14**(5), 429–441 (1995)
4. Bärtschi, A., et al.: Energy-efficient delivery by heterogeneous mobile agents. arXiv preprint arXiv:1610.02361 (2016)
5. Bärtschi, A., Graf, D., Mihalák, M.: Collective fast delivery by energy-efficient agents. arXiv preprint arXiv:1809.00077 (2018)
6. Bärtschi, A., et al.: Collaborative delivery with energy-constrained mobile robots. Theoret. Comput. Sci. **810**, 2–14 (2020)
7. Berman, F., Leighton, F.T.., Snyder, L.: Optimal tile salvage. Technical Report 81-396, Purdue University, Department of Computer Sciences (1981)
8. Bronnimann, H.: Almost optimal set covers in finite VC-dimension. Discrete Comput. Geom. **14**, 263–279 (1995)
9. Carvalho, I.A., Erlebach, T., Papadopoulos, K.: On the fast delivery problem with one or two packages. J. Comput. Syst. Sci. **115**, 246–263 (2021)
10. Chalopin, J., Das, S., Mihal'ák, M., Penna, P., Widmayer, P.: Data delivery by energy-constrained mobile agents. In: Algorithms for Sensor Systems: 9th International Symposium on Algorithms and Experiments for Sensor Systems, Wireless Networks and Distributed Robotics, ALGOSENSORS 2013, Sophia Antipolis, France, September 5-6, 2013, Revised Selected Papers 9, pp. 111–122. Springer (2014)
11. Chalopin, J., Jacob, R., Mihalák, M., Widmayer, P.: Data delivery by energy-constrained mobile agents on a line. In: Esparza, J., Fraigniaud, P., Husfeldt, T., Koutsoupias, E. (eds.) ICALP 2014. LNCS, vol. 8573, pp. 423–434. Springer, Heidelberg (2014). https://doi.org/10.1007/978-3-662-43951-7_36
12. Chalopin, J., Das, S., Disser, Y., Labourel, A., Mihalák, M.: Collaborative delivery on a fixed path with homogeneous energy-constrained agents. Theoret. Comput. Sci. **868**, 87–96 (2021)
13. Chan, T.M., He, Q.: Faster approximation algorithms for geometric set cover. arXiv preprint arXiv:2003.13420 (2020)
14. Church, R., Velle, C.R.: The maximal covering location problem. Papers Regional Sci. **32**(1), 101–118 (1974)
15. Chvatal, V.: A greedy heuristic for the set-covering problem. Math. Oper. Res. **4**(3), 233–235 (1979)
16. Cruz Caminha, P.H., de Souza Couto, R., Maciel Kosmalski Costa, L.H., Fladenmuller, A., Dias de Amorim, M.: On the coverage of bus-based mobile sensing. Sensors **18**(6), 1976 (2018)
17. Das, G.K., Fraser, R., Lóopez-Ortiz, A., Nickerson, B.G.: On the discrete unit disk cover problem. In: International Workshop on Algorithms and Computation, pp. 146–157. Springer (2011)

18. Erlebach, T., Luo, K., Spieksma, F.C.: Package delivery using drones with restricted movement areas (2022). URL http://arxiv.org/abs/2209.12314. arXiv:2209.12314 [cs]
19. Even, G., Rawitz, D., Shahar, S.M.: Hitting sets when the VC-dimension is small. Inf. Process. Lett. **95**(2), 358–362 (2005)
20. Fowler, R.J., PaLerson, M.S., Tanimoto, S.L.: Optimal packing and covering the plane are NP-complete. Inf. Process. Lett. **12**(3), 133–137 (1981)
21. Har-Peled, S., Lee, M.: Weighted geometric set cover problems revisited. J. Comput. Geo. **3**(1), 65–85 (2012)
22. Harary, F., Moser, L.: The theory of round robin tournaments. Am. Math. Month. **73**(3):231–246 (1966). ISSN 0002-9890. https://doi.org/10.2307/2315334. URL https://www.jstor.org/stable/2315334. Publisher: Mathematical Association of America
23. Johnson, D.S.: The NP-completeness column: an ongoing guide. J. Algorithms **3**(2), 182–195 (1982)
24. Karp, R.M.: Reducibility among combinatorial problems. Springer (2010)
25. Krizanc, D., Narayanan, L., Opatrny, J., Pankratov, D.: The en route truck-drone delivery problem. arXiv preprint arXiv:2402.00829 (2024)
26. Pandit, S., Pemmaraju, S.V., Varadarajan, K.: Approximation algorithms for domatic partitions of unit disk graphs. In: International Workshop on Approximation Algorithms for Combinatorial Optimization, pp. 312–325. Springer (2009)
27. lavík, P.: A tight analysis of the greedy algorithm for set cover. In: Proceedings of the Twenty-Eighth Annual ACM Symposium on Theory of Computing, pp. 435–441 (1996)

Oblivious Robots Under Round Robin: Gathering on Rings

Alfredo Navarra and Francesco Piselli(✉)

Department of Mathematics and Computer Science, University of Perugia, Perugia, Italy
`alfredo.navarra@unipg.it, francesco.piselli@unifi.it`

Abstract. Robots with very limited capabilities are placed on the vertices of a graph and are required to move toward a single, common vertex, where they remain stationary once they arrive. This task is referred to as the GATHERING problem.

Most of the research on this topic has focused on feasibility challenges in the *asynchronous* setting, where robots operate independently of each other. A common assumption in these studies is that robots are equipped with *multiplicity detection*, the ability to recognize whether a vertex is occupied by more than one robot. Additionally, initial configurations are often restricted to ensure that no vertex hosts more than one robot. A key difficulty arises from the possible symmetries in the robots' placement relative to the graph's topology.

This paper investigates the GATHERING problem on Rings under a *sequential* scheduler, where only one robot at a time is active. While this sequential activation helps to break symmetries, we remove two common assumptions: robots do not have multiplicity detection, and in initial configurations, vertices can be occupied by multiplicities.

We prove that such a generalized GATHERING problem cannot be solved under any sequential schedulers. However, we provide a complete characterization of the problem when a sequential *Round Robin* scheduler is used, where robots are activated one at a time in a fixed cyclic order that repeats indefinitely. Furthermore, we fully characterize the DISTINCT GATHERING problem, the most used variant of GATHERING, in which the initial configurations do not admit multiplicities.

Keywords: Gathering · Ring · Sequential · Round Robin

1 Introduction

In the field of theoretical computer science, swarm robotics is one of the most investigated research areas. Robots are usually mobile units with full autonomy that, by operating individually, are able to establish some sort of collective

---

The work has been supported in part by the Italian National Group for Scientific Computation INdAM - GNCS Project, CUP E53C24001950001.

behavior in order to solve a common problem. Robots are considered in the abstract with their capabilities induced by an underlying model. Those capabilities are usually reduced to the minimum, in order to have a more flexible and fault-resistant model. In this context, some representative models are, for example, the Amoebot [11] and the SILBOT [6,20,21,23]. One of the most investigated models for a theoretical perspective in swarm robotics is certainly the $\mathcal{OBLOT}$ [15]. In this model, robots operate by executing Look-Compute-Move cycles. In each cycle, a robot obtains a snapshot of the system (Look), executes its algorithm to determine the destination of its next movement (Compute), and moves toward the computed destination (Move).

Within such a context, one of the most popular problems is the so-called GATHERING where robots, placed on the vertices of an anonymous graph, are required to reach a common vertex (not known in advance) from where they do not move anymore.

Apart for some impossibility results or basic conditions that guarantee the resolution of the GATHERING problem provided in [4,13], most of the literature focuses on specific topologies that are very symmetric, where the vertices can be partitioned into a few classes of equivalence. Since robots have few topological properties to exploit, the design of a resolution algorithm becomes challenging. Those topologies are: Trees [7,8], Regular Bipartite graphs [17], Finite Grids [7], Infinite Grids [12], Tori [18], Oriented Hypercubes [1], Complete graphs [3,4], Complete Bipartite graphs [3,4], Butterflies [2], and Rings [9,10,13,19].

In most of those studies, the robots operate under an *asynchronous* scheduler, where robots are activated independently of each other. Other works concern *synchronous* schedulers, where the robots share a common notion of time and a subset of them is activated at the same time. A very common assumption is to have robots endowed with the *multiplicity detection*. With this property, robots are able to recognize whether a vertex contains a *multiplicity*, i.e., if two or more robots are located at the same vertex.

Focusing on rings, these are vertex-transitive graphs where the robots' movements depend entirely on their relative positioning. So far, on rings, the GATHERING problem has been studied without considering multiplicities in initial configurations. Recently in [16], this version of the problem has been referred to as DISTINCT GATHERING, and it has been studied for robots moving on the Euclidean plane under a *Round Robin* scheduler. This is a specific type of *sequential* scheduler, where robots are activated one at a time, in a fixed periodic order. The more generic *sequential* scheduler, which requires only to activate one robot at a time, has been used in [14] to solve the *Universal Pattern Formation* (UPF) problem. In UPF, robots can start from configurations containing multiplicities, and the requirement is to move so as to form any given pattern.

1.1 Our Results

In this paper, we consider robots on rings operating under the $\mathcal{OBLOT}$ model with no additional assumptions. We provide an impossibility result for the GATHERING problem under a general *sequential* scheduler, and we present a full characterization for both GATHERING and DISTINCT GATHERING under the *round robin* scheduler, proposing an asymptotically time optimal algorithm.

An extended version of the paper, including all the proofs of the impossibility results, the correctness of the proposed algorithm, as well as a sample execution of the algorithm, can be found in [22].

2 Robot Model

We consider the standard $\mathcal{OBLOT}$ model of distributed systems of autonomous mobile robots. In $\mathcal{OBLOT}$, the system is composed of a set $\mathcal{R} = \{r_1, r_2, \ldots, r_k\}$ of computational *robots* that live and operate on a n-vertices anonymous *ring* without orientation. We refer to a maximal subset of consecutive empty (i.e., not occupied by robots) vertices of the ring as a *hole*, whereas, we refer to a maximal subset of consecutive occupied vertices as an *island*. Each vertex of the ring is initially empty, occupied by one robot, or occupied by more than one robot (i.e., a *multiplicity*).

Robots can be characterized according to many different settings. In particular, they have the following basic properties:

- **Anonymous:** they have no unique identifiers;
- **Autonomous:** they operate without a centralized control;
- **Dimensionless:** they are viewed as points, i.e., they have no volume nor occupancy restraints;
- **Disoriented:** they have no common sense of orientation nor handedness;
- **Oblivious:** they have no memory of past events;
- **Homogeneous:** they all execute the same deterministic algorithm with no type of randomization admitted;
- **Silent:** they have no means of direct communication.

Each robot in the system has sensory capabilities, allowing it to determine which vertices of the ring are occupied, relative to its location. Each robot refers to a Local Reference System (LRS) that might differ from robot to robot. Each robot has a specific behavior described according to the sequence of the following four states: Wait, Look, Compute, and Move. Such a sequence defines the computational activation cycle (or simply a cycle) of a robot. More in detail:

1. Wait: the robot is in an idle state and cannot remain as such indefinitely;
2. Look: the robot obtains a snapshot of the system containing the positions of the occupied vertices with respect to its LRS, by activating its sensors. Each robot is seen as a point in the graph occupying a vertex. The robots cannot deduce the number k of robots, whereas they see the size n of the ring;
3. Compute: the robot executes a local computation according to a deterministic algorithm $\mathcal{A}$ (we also say that the robot executes $\mathcal{A}$). This algorithm is the same for all the robots and its result is the destination of the movement of the robot. Such a destination is either the vertex where the robot is already located, or a neighboring vertex at one hop distance (i.e., only one edge per move can be traversed);

4. Move: if the computed destination is a neighboring vertex v, the robot moves to v; otherwise, it executes a *nil* movement (i.e., it does not move).

In the literature, the computational cycle is simply referred to as Look-Compute-Move (LCM) cycle, because when a robot is in the Wait state, we say that it is *inactive*. Thus, the LCM cycle only refers to the *active* states of a robot. It is also important to notice that since the robots are oblivious, without memory of past events, every decision they make during the Compute phase is based on what they are able to determine during the current LCM cycle. In particular, during the Look phase, the robots take a snapshot of the system and they use it to elaborate the information, building what is called the *view* of the robot. Regarding the Move phase of the robots, the movements executed are always considered to be instantaneous. Thus, the robots are only able to perceive the other robots positioned on the vertices of the graph, never while moving. Regarding the position of a robot on a vertex, it may happen that two or more robots are located on the same vertex, i.e., they constitute a multiplicity.

Another important feature that can greatly affect the computational power of the robots is the *time scheduler*. We say that an *epoch*, is the minimum time window within which each robot has been activated at least once. In general, three main schedulers are used:

- *Semi-Synchronous* ($\mathcal{SSYNC}$): the activations of the robots are logically divided in global rounds. In each round, one or more robots are activated, obtaining the same snapshot. Then, based on the information acquired from the snapshot, they compute and execute their move, completing their cycle by the next round;
- *Fully-Synchronous* ($\mathcal{FSYNC}$): all the robots are activated in every round, executing their LCM cycle in a synchronized way;
- *Asynchronous* ($\mathcal{ASYNC}$): the robots are activated independently of each other and the duration of each phase of the LCM cycle is finite but unpredictable. In this scheduler, robots have no common notion of time. Thus, their decisions can be based on obsolete observations of the system.

In the $\mathcal{FSYNC}$ case, a round coincides with one epoch. In the $\mathcal{SSYNC}$ and $\mathcal{ASYNC}$ cases, it is assumed the existence of an *adversary* which determines the computational cycle's timing and which robot(s) will be activated. This timing is assumed to be *fair*, that is, each robot is able to execute its LCM cycle within finite time and infinitely often. Without this fairness assumption, the adversary could prevent some robot from ever being activated. The duration of an epoch is then finite but unpredictable. In this work, we consider another type of scheduler:

- *Sequential* ($\mathcal{SEQ}$): robots are activated one at a time, fairness is guaranteed.

In particular, we focus on the so-called:

- *Round Robin* ($\mathcal{RR}$): the robots are activated one at a time in a predetermined order which repeats forever. Each robot is then activated exactly once in each epoch. Of course, $\mathcal{RR} \subset \mathcal{SEQ}$, and an epoch equals k rounds, with k being the number of robots in the system.

3 Problem Formulation and Impossibility Results

The problem we aim to solve is the GATHERING on a n-ring, defined as follows:

Definition 1 (GATHERING). *Given k robots $r_1, r_2, \ldots, r_k$ arbitrarily placed on a n-ring, it is required to reach a configuration in a finite number of epochs where exactly one vertex is occupied and from thereon no robot moves.*

We distinguish the general case from the one usually adopted in the literature where initial configurations do not admit multiplicities, that is:

Definition 2 (DISTINCT GATHERING). *Given k robots $r_1, r_2, \ldots, r_k$ on a n-ring with $k \leq n$, where each vertex is occupied by at most one robot, it is required to solve the GATHERING.*

We now define all the cases where GATHERING or DISTINCT GATHERING are unsolvable. First of all, the next lemma provides a useful property for both the impossibility results and for the designing of the proposed algorithm.

Lemma 1. *Let C be a configuration on a n-ring with exactly 2 robots placed on different vertices. Under $\mathcal{RR}$, there is no algorithm that solves the GATHERING by making the robots move away from each other.*

Note that, as a consequence of this result, even when considering a configuration with more than 2 robots occupying exactly two vertices, since the robots are not able to detect multiplicities, any algorithm should make the robots move toward the other occupied vertex.

The next theorem provides an impossibility result for any scheduler in $\mathcal{SEQ}$.

Theorem 1. GATHERING *on rings using a generic $\mathcal{SEQ}$ scheduler is impossible for k robots, with $k \geq 3$.*

Following the above theorem, we choose to work using the $\mathcal{RR}$ scheduler, which operates activating all the robots sequentially in each epoch, one per round, always maintaining the same sequence.

We now define the configurations from which solving the GATHERING under $\mathcal{RR}$ is impossible for any algorithm.

Definition 3 (*Unsolvable Configuration*). *Given a configuration C and a $\mathcal{RR}$ scheduler, C is said to be Unsolvable if, for any algorithm $\mathcal{A}$, there exists a sequence of activations imposed by the adversarial scheduler, that makes the GATHERING impossible to be solved.*

We denote by $\mathcal{UC}$ the set of unsolvable configurations. In the next theorems, we show which configurations belong to $\mathcal{UC}$ with respect to the GATHERING and the DISTINCT GATHERING problems.

Theorem 2. *Let C be a configuration of $k \geq 3$ robots on a n-ring, with:*

i) Only 2 consecutive vertices occupied;

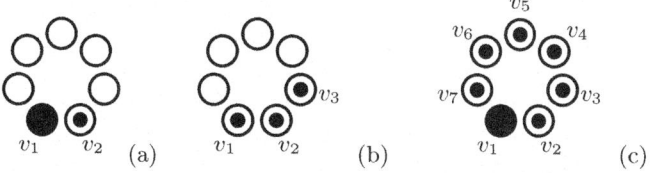

Fig. 1. Some unsolvable configurations for GATHERING. The robots are represented by black circles inside vertices. A full black vertex represents a multiplicity. Edges are not drawn for clarity. Labels associated with vertices are used only for analysis purposes: robots are not aware of them as the rings are, in fact, anonymous.

ii) Only 3 consecutive vertices occupied;
iii) All n vertices occupied.

For each of these types of configurations (cf., Fig. 1), there exists a $\mathcal{RR}$ scheduler that makes the GATHERING unsolvable.

Other configurations belonging to $\mathcal{UC}$ are presented in the following results.

Theorem 3. *Let C be a configuration on a 5-ring where $k \geq 5$ robots occupy 3 vertices. If a vertex occupied by a multiplicity is neighboring to another occupied vertex, then the* GATHERING *problem is unsolvable from C under a $\mathcal{RR}$ scheduler.*

Corollary 1. *Let C be a configuration of $k \geq 5$ robots occupying 4 vertices of a 5-ring. The* GATHERING *problem is unsolvable from C under a $\mathcal{RR}$ scheduler.*

All the configurations addressed by Theorem 2, Theorem 3, and Corollary 1, compose the set $\mathcal{UC}$ of unsolvable configurations for the GATHERING problem.

Concerning the DISTINCT GATHERING problem, configurations admitting only three consecutive vertices occupied as in case *ii)* of Theorem 2 are still unsolvable. The only other unsolvable configurations we detected are given by 4 robots on a 4-ring and by 5 robots on a 5-ring.

The set $\mathcal{UC}$ for the DISTINCT GATHERING problem is then composed of any n-ring with exactly 3 consecutive vertices occupied by 3 robots and by the configurations composed of a 4-ring or a 5-ring fully occupied.

According to the obtained impossibility results, in the next section we are going to fully characterize both the GATHERING and the DISTINCT GATHERING problems. We provide a unique resolution algorithm that works for any configuration $C \notin \mathcal{UC}$ for both problems.

4 Gathering on Rings

The algorithm presented in this paper is designed according to the methodology proposed in [5]. Let us now briefly summarize how an algorithm $\mathcal{A}$, conceived to solve a generic problem $\mathcal{P}$, can be designed using that methodology.

Recall that we are considering a model where the robots have very weak capabilities: they can only wake up, take a snapshot of the graph, and based on that observation they can take a deterministic decision. For those reasons, it is better to consider the problem $\mathcal{P}$ as composed of a series of sub-problems such that, each sub-problem, is simple enough to be solved by a "task" executed by one or more robots. Therefore, let us assume that the problem $\mathcal{P}$ is decomposed into simple tasks $T_1, T_2, \ldots, T_q$, where one of them is the *terminal* one, i.e., the one where the robots recognize that the problem $\mathcal{P}$ is solved and they do not execute any other move.

As we previously described, the robots operate following the LCM cycle, hence, they must be able to recognize which task they have to execute according to the configuration that they sense during the Look phase. This recognition can be executed by providing $\mathcal{A}$ with a predicate P_i for each task T_i. Once a robot wakes up and perceives that a certain predicate P_i is true, following $\mathcal{A}$ it knows that the task T_i must be executed in order to solve a sub-problem. Going into more detail, with predicates well-formed, algorithm $\mathcal{A}$ can be used in the Compute phase as follows: if once awakened, a robot r executing algorithm $\mathcal{A}$, detects that a certain predicate P_i is true, then r executes a move m_i associated with the task T_i. For this approach to be valid, each well-formed predicate must guarantee the following properties:

- $Prop_1$: each predicate P_i must be computable on the configuration C sensed by the robot in the Look phase;
- $Prop_2$: $P_i \wedge P_j = \mathtt{false}$, for each $i \neq j$; thanks to this property the robots are able to precisely recognize which task to execute, without ambiguity;
- $Prop_3$: for each possible configuration C sensed, there must exist a predicate P_i evaluated true by an activated robot.

To be recognized by the robots, each task T_i requires some *precondition* to be verified. Hence, for the definition of the predicates P_i, we need to define some *basic variables* that capture metric/numerical/ordinal/topological aspects of the configuration C sensed by the robots, that can be evaluated by each activated robot, based solely on the observations made during the Look phase.

Let us assume that $\mathtt{pre}_i$ is the composition of the variables characterizing the preconditions of T_i, for each $1 \leq i \leq q$. The predicate P_i can then be defined as follows:

$$P_i = \mathtt{pre}_i \wedge \neg(\mathtt{pre}_{i+1} \vee \mathtt{pre}_{i+2} \vee \cdots \vee \mathtt{pre}_q) \tag{1}$$

With this definition, we are sure that any predicate satisfies property $Prop_2$. This also implies a specific ordering of the tasks that must be followed. In fact the first predicate to be checked is the last one, i.e., $P_F = \mathtt{pre}_q$; if false, $P_{q-1} = \mathtt{pre}_{q-1} \wedge \neg \mathtt{pre}_q$ is checked. Those checks continue going "backward" (if predicates are false) until reaching $P_2 = \mathtt{pre}_2 \wedge \neg(\mathtt{pre}_3 \vee \mathtt{pre}_4 \vee \cdots \vee \mathtt{pre}_q)$. If also P_2 is false then, according to the designed task ordering, T_1 must be performed.

Let us now consider an execution of algorithm $\mathcal{A}$, where a task T_i is executed with respect to the current configuration C. The configuration C', generated by $\mathcal{A}$ after the execution of task T_i, must be assigned to a task T_j. Then, we can say

Table 1. The basic boolean variables used to define all the tasks' preconditions.

Var	Definition
b4	There is exactly one hole, which is constituted of at most 4 vertices
b5	There is exactly one hole, which is constituted of at least 5 vertices
f	There are no holes
h	There are exactly two holes, one constituted of just 1 vertex and the other constituted of more than 1 vertex
o1	All robots occupy exactly 1 vertex (GATHERING accomplished)
o2	All robots occupy exactly 2 neighboring vertices
o3	All robots occupy exactly 3 consecutive vertices
p	All robots occupy exactly 2 vertices separated by 1 empty vertex

that algorithm $\mathcal{A}$ can generate a transition from T_i to T_j. The set of all possible transitions of $\mathcal{A}$ determines a directed graph called *transition graph*. Note that the terminal task among $T_1, T_2, \ldots, T_q$, where the problem $\mathcal{P}$ is solved, must be a sink vertex in the transition graph.

4.1 High Level Description of Algorithm GATHE$\mathcal{RR}$ING

The proposed Algorithm GATHE$\mathcal{RR}$ING, is designed according to the methodology recalled previously. Following such an approach, the GATHERING problem is subdivided into a set of sub-problems. Solving all the sub-problems, leads the algorithm to solve the GATHERING.

The first sub-problem that needs to be solved by the robots consists in having a configuration where the robots occupy exactly two distinct islands with only one empty vertex separating them on one side, and more than one empty vertex separating them on the other side. Creating this configuration allows the robots to uniquely identify the same vertex where to solve the GATHERING. This vertex is the empty (single) one separating the two islands. In the following, such a vertex will be called *gathering vertex* and denoted by $\breve{g}v$.

A different sub-problem consists in moving all the robots toward the two vertices neighboring $\breve{g}v$, until only those two vertices are occupied. From that point, the next sub-problem requires to have only two neighboring vertices occupied by robots. To reach such a configuration, all the activated robots move toward $\breve{g}v$ until the only two vertices occupied by robots are $\breve{g}v$ itself and one of its neighbors. Finally, the last sub-problem makes the robots move toward $\breve{g}v$ until that is the only vertex occupied by robots, i.e., GATHERING is solved.

Table 2. Schematization of Algorithm GATHE$\mathcal{RR}$ING for solving both the GATHERING and the DISTINCT GATHERING problems.

Sub-Problem	Task	Precondition	Move	Transitions
$I_{\tilde{g}v}$	T_1	$true$	m_1	T_1, T_2, T_4, T_5
	T_2	b4 ∨ f	m_2	T_1, T_2, T_4
	T_3	b5	m_3	T_2, T_3, T_4
$N_{\tilde{g}v}$	T_4	h	m_4	T_4, T_5
$O_{\tilde{g}v}$	T_5	p	m_5	T_6, T_7
	T_6	o3	m_6	T_6, T_7
$Finalize$	T_7	o2	m_7	T_7, T_8
$Term$	T_8	o1	nil	T_8

4.2 Description of Sub-Problems and Tasks

In this section, we describe all the details of the five sub-problems we defined and the designed tasks to solve each of them. The basic variables we used to define the corresponding preconditions are shown in Table 1, whereas Table 2 shows the corresponding preconditions and transitions. For each task, the movements described are the ones presented in Table 3.

The first sub-problem $I_{\tilde{g}v}$ consists in creating a specific configuration where the robots are divided in two islands separated on one side by a hole of size 1, and on the other side by a hole of size at least 2. This configuration allows the robots to uniquely identify the gathering vertex $\tilde{g}v$, hence the name $I_{\tilde{g}v}$. To solve this sub-problem we need the robots to operate following the designed tasks. For $I_{\tilde{g}v}$, three tasks, T_1, T_2 and T_3, are possibly needed.

Task T_1. This task activates when the following predicate holds:

$$P_1 = (\text{pre}_1 \equiv true) \wedge \neg(\text{pre}_2 \vee \text{pre}_3 \vee \cdots \vee \text{pre}_8)$$

According to the defined preconditions, see Table 2, it follows that the configurations addressed by this task are those admitting at least two holes, and if they have exactly two holes, those holes are both of unitary size or both of size greater than one. In this case, the scheduled robot r wakes up and executes move m_1.

Only a robot neighboring to one of the biggest holes can move. There are various cases that can occur. In particular, if the islands are all of size 2 and there is no unique biggest hole, then r moves toward its closest empty vertex.

In a different scenario managed by task T_1, if $n-2$ vertices are occupied and not consecutive, r can move if it is neighboring to only one empty vertex; the executed movement is directed toward the occupied neighbor.

In any other case, any robot neighboring the biggest hole will move to the opposite direction with respect to such a hole with the goal of increasing its size.

Note that, if there are multiple holes with the same biggest size, they are all considered as the "biggest hole".

Table 3. Description of the moves from the point of view of a robot r.

Move	Description

m_1	*if* r is neighboring one of the biggest holes, *then*
	    if all the islands are of size 2, *then*
	        if there is no unique biggest hole, *then*
	            r moves toward its unique empty neighbor
	        else r moves toward its occupied neighbor
	    else if $n-2$ vertices are occupied but not consecutive, *then*
	        if r is neighboring only one empty vertex, *then*
	            r moves toward its occupied neighbor
	        else r moves away from the biggest hole

m_2	*if* both neighbors of r are occupied, *then*
	    if $n=6$ and the unique hole has size 1, *then*
	        if r is not on the farthest vertex from the hole, *then*
	            r moves away from the hole
	    else r moves toward the hole if any, or toward any direction

m_3	*if* r admits an empty neighbor x, *then* r moves toward x
m_4	*if* r is neighboring the biggest hole and one robot, *then* r moves toward its occupied neighbor
m_5	r moves toward the other occupied vertex
m_6	*if* r is neighboring an empty vertex, *then* r moves toward its occupied neighbor
m_7	r moves toward its occupied neighbor

Task T_2. This task is activated when there is exactly one hole in the graph, which size is at most 4, or when there is no hole at all. In the latter case, according to Theorem 2, of course only DISTINCT GATHERING can be solved. The following predicate holds:

$$P_2 = (\text{pre}_2 \equiv \text{b4} \lor \text{f}) \land \neg(\text{pre}_3 \lor \text{pre}_4 \lor \cdots \lor \text{pre}_8)$$

If the activated robot r is not neighboring to an empty vertex, then it is able to move executing m_2. In this case, r moves toward the closest empty vertex or, in the case of no holes, it moves toward any direction. The goal of this movement is to create a new hole of size 1.

Task T_3. Task T_3 is activated from a configuration with a unique hole, which is at least of size 5. The goal here is to create a configuration where there are two different islands of robots, where the smallest hole separating them is of size 1. In particular, task T_3 activates when the following predicate holds:

$$P_3 = (\text{pre}_3 \equiv \text{b5}) \land \neg(\text{pre}_4 \lor \text{pre}_5 \lor \cdots \lor \text{pre}_8)$$

An activated robot r moves executing move m_3 if it is neighboring to an empty vertex x. The movement is toward x, attempting to create a new hole of size 1.

The subsequent sub-problem requires to create a configuration where there are exactly two vertices occupied by the robots, separated by a hole of size 1 on one side, and a hole of size at least 2 on the other side. Focusing on the two occupied vertices neighboring $\check{g}v$, we call this sub-problem $N_{\check{g}v}$. This sub-problem requires the execution of task T_4.

Task T_4. When the robots form exactly two islands with the smallest hole separating them of size 1 and the other one of size at least 2, task T_4 activates. In particular, the predicate that holds is the following:

$$P_4 = (\text{pre}_4 \equiv \text{h}) \wedge \neg(\text{pre}_5 \vee \text{pre}_6 \vee \cdots \vee \text{pre}_8)$$

The goal of this task is to position all the robots on exactly two vertices, with one empty vertex in-between them. To achieve such a configuration, an activated robot r executes move m_4: if r is neighboring a robot and an empty vertex which is part of the biggest hole in the graph, then r moves toward its occupied neighbor. With such a movement, the islands gradually reduce their size until only two vertices are occupied.

Once $N_{\check{g}v}$ has been solved, the subsequent sub-problem consists in starting to occupy the gathering vertex $\check{g}v$, until only that vertex and one of its neighbors are occupied, hence the name $O_{\check{g}v}$. This sub-problem possibly requires the execution of tasks T_5 and T_6. Note that, until the start of those two tasks, vertex $\check{g}v$ is empty which is fundamental for the correct resolution of GATHERING.

Task T_5. This task is activated when all the robots are neighboring $\check{g}v$, with $\check{g}v$ still empty. Hence, the following predicate holds:

$$P_5 = (\text{pre}_5 \equiv \text{p}) \wedge \neg(\text{pre}_6 \vee \text{pre}_7 \vee \cdots \vee \text{pre}_8)$$

In this case, the activated robot r moves in the direction of the other occupied vertex executing move m_5. In doing so, r becomes the first robot to occupy $\check{g}v$.

Task T_6. After a robot moved according to move m_5 and $\check{g}v$ becomes occupied, if 3 consecutive vertices are occupied with $\check{g}v$ being the middle one, then task T_6 is executed and the following predicate holds:

$$P_6 = (\text{pre}_6 \equiv \text{o3}) \wedge \neg(\text{pre}_7 \vee \text{pre}_8)$$

During this task, only the external robots neighboring $\check{g}v$ are able to move executing move m_6: once activated, those robots move toward $\check{g}v$, moving forward with the GATHERING onto that vertex.

At the completion of $O_{\check{g}v}$, only two neighboring vertices are occupied, with one being what was before uniquely identified by the robots as the gathering

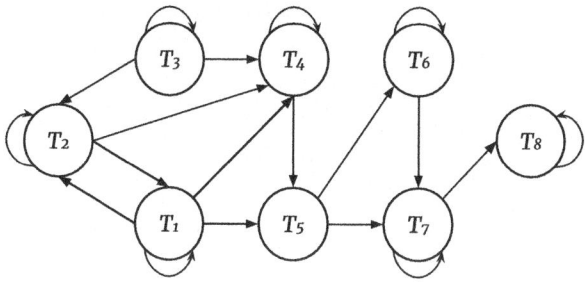

Fig. 2. Transition graph derived from Table 2.

vertex $\breve{g}v$. At this point, the two occupied vertices are indistinguishable but, since all the previous movements executed during tasks T_5 and T_6 moved the robots onto $\breve{g}v$, only the robots positioned on $\breve{g}v$'s neighbor still have to be activated by the adversary within the same epoch. This is guaranteed according to the $\mathcal{RR}$ schedule. Therefore, the subsequent sub-problem is the *finalization* one, denoted by *Finalize*, which requires the execution of task T_7.

Task T_7. Once only two neighboring vertices are occupied, task T_7 is activated and the following predicate holds:

$$P_7 = (\text{pre}_7 \equiv \text{o2}) \wedge \neg(\text{pre}_8)$$

The goal here is to move all the robots on the same vertex to solve GATHERING. Hence, the only move that can be executed by the robots is to move toward their occupied neighbor, i.e., once activated they execute move m_7.

Note that, although the configurations managed in this task are unsolvable in general (see Theorem 2), the movements dictated by Algorithm GATHE$\mathcal{RR}$ING allow to reach such configurations with a schedule that first activates only the robots on the same vertex. This avoids loops by solving the GATHERING before the robots on the other vertex are activated.

Once the *Finalize* sub-problem has been solved, there is nothing else to do, hence we called the last sub-problem *Term*, since it is the termination stage of the algorithm. The task associated with this sub-problem is task T_8.

Task T_8. When GATHERING has been solved, all the robots occupy the same vertex and task T_8 activates, with the following predicate holding:

$$P_8 = (\text{pre}_8 \equiv \text{o1})$$

During this task, the activated robots recognize that the problem has been solved. Thus, they only execute the *nil* movement that does not change the configuration.

5 Correctness

The following statements show that GATHE$\mathcal{RR}$ING solves GATHERING and DISTINCT GATHERING on rings.

Lemma 2. *Algorithm* GatheRRing *fulfills all the following properties:*

H_1: *The transition graph shown in Fig. 2 (and obtained from Table 2) is correct, i.e., for each task T_i, the tasks reachable from T_i are exactly those represented in the graph;*

H_2: *Apart for the self-loop in T_8, all the other loops in the transition graph, including self-loops, must be executed a finite number of times;*

H_3: *With respect to* Gathering *and* Distinct Gathering, *no unsolvable configuration is generated by* GatheRRing.

Theorem 4 (Correctness). GatheRRing *solves* Gathering *and* Distinct Gathering *for each configuration C not belonging to their corresponding sets $\mathcal{UC}$ in at most $n-3$ epochs.*

Sketch Proof. It is possible to show that all the properties H_1, H_2, and H_3 dictated by Lemma 2 hold for each task. Assume that C is provided as input to GatheRRing. According to $Prop_2$ and $Prop_3$, there exists a single task (say T_i) to be assigned to robots with respect to C. According to H_3, any configuration generated from T_i (say C') can be provided as input to GatheRRing. Moreover, by H_1 and H_2, we can consider C' belonging to some class (say T_j) different from T_i. According to this analysis, we can say that C' will evolve during the time by changing its membership from class to class according to the transition graph shown in Fig. 2. Although the execution of GatheRRing is infinite, properties $H_1 - H_3$ assure that any task apart for T_8 is completed within a finite number of LCM cycles. Task T_8, instead, which is reached within finite time (actually after at most $n-3$ epochs) allows only the *nil* movement, hence the reached configuration does not change anymore. □

6 Conclusion

We have studied the Gathering problem within rings. In particular, first we have shown that under a generic *sequential* ($\mathcal{SEQ}$) scheduler, Gathering is unsolvable. Then, we focused on solving the problem under the *Round Robin* ($\mathcal{RR}$) sequential scheduler, offering a complete characterization and proposing a resolution algorithm. The same algorithm also solves the Distinct Gathering problem when starting from any solvable configuration.

As a future work, it is worth investigating the Gathering problem, under $\mathcal{RR}$ or other schedulers within $\mathcal{SEQ}$, also on different graph topologies.

Acknowledgement. The authors would like to thank Serafino Cicerone, Alessia Di Fonso and Gabriele Di Stefano for fruitful discussions.

References

1. Bose, K., Kundu, M.K., Adhikary, R., Sau, B.: Optimal gathering by asynchronous oblivious robots in hypercubes. In: Gilbert, S., Hughes, D., Krishnamachari, B. (eds.) ALGOSENSORS 2018. LNCS, vol. 11410, pp. 102–117. Springer, Cham (2019). https://doi.org/10.1007/978-3-030-14094-6_7
2. Cicerone, S., Di Fonso, A., Di Stefano, G., Navarra, A.: Gathering of robots in butterfly networks. In: Proceedings of 26th International Symposium on Stabilization, Safety, and Security of Distributed Systems (SSS). LNCS, vol. 14931, pp. 106–120. Springer (2024). https://doi.org/10.1007/978-3-031-74498-3_7
3. Cicerone, S., Di Stefano, G., Navarra, A.: On gathering of semi-synchronous robots in graphs. In: Ghaffari, M., Nesterenko, M., Tixeuil, S., Tucci, S., Yamauchi, Y. (eds.) SSS 2019. LNCS, vol. 11914, pp. 84–98. Springer, Cham (2019). https://doi.org/10.1007/978-3-030-34992-9_7
4. Cicerone, S., Di Stefano, G., Navarra, A.: Gathering robots in graphs: the central role of synchronicity. Theor. Comput. Sci. **849**, 99–120 (2021). https://doi.org/10.1016/j.tcs.2020.10.011
5. Cicerone, S., Di Stefano, G., Navarra, A.: A structured methodology for designing distributed algorithms for mobile entities. Inf. Sci. **574**, 111–132 (2021). https://doi.org/10.1016/j.ins.2021.05.043
6. D'Angelo, G., D'Emidio, M., Das, S., Navarra, A., Prencipe, G.: Asynchronous silent programmable matter achieves leader election and compaction. IEEE Access **8**, 207619–207634 (2020)
7. D'Angelo, G., Di Stefano, G., Klasing, R., Navarra, A.: Gathering of robots on anonymous grids and trees without multiplicity detection. Theor. Comput. Sci. **610**, 158–168 (2016)
8. D'Angelo, G., Di Stefano, G., Navarra, A.: Gathering asynchronous and oblivious robots on basic graph topologies under the look-compute-move model. In: Search Theory: A Game Theoretic Perspective, pp. 197–222. Springer (2013)
9. D'Angelo, G., Navarra, A., Nisse, N.: A unified approach for gathering and exclusive searching on rings under weak assumptions. Distrib. Comput. **30**(1), 17–48 (2017)
10. D'Emidio, M., Di Stefano, G., Frigioni, D., Navarra, A.: Characterizing the computational power of mobile robots on graphs and implications for the euclidean plane. Inf. Comput. **263**, 57–74 (2018)
11. Derakhshandeh, Z., Dolev, S., Gmyr, R., Richa, A.W., Scheideler, C., Strothmann, T.: Brief announcement: amoebot - a new model for programmable matter. In: Proceedings of 26th ACM Symposium on Parallelism in Algorithms and Architectures (SPAA), pp. 220–222. ACM (2014)
12. Di Stefano, G., Navarra, A.: Gathering of oblivious robots on infinite grids with minimum traveled distance. Inf. Comput. **254**, 377–391 (2017)
13. Di Stefano, G., Navarra, A.: Optimal gathering of oblivious robots in anonymous graphs and its application on trees and rings. Distrib. Comput. **30**(2), 75–86 (2017)
14. Flocchini, P., Navarra, A., Pattanayak, D., Piselli, F., Santoro, N.: Oblivious robots under sequential schedulers: universal pattern formation. In: Proceedings of 32nd International Colloquium on Structural Information and Communication Complexity (SIROCCO). LNCS. Springer (2025)
15. Flocchini, P., Prencipe, G., Santoro, N. (eds.): Distributed Computing by Oblivious Mobile Robots. Synthesis Lectures on Distributed Computing Theory. Morgan & Claypool Publishers (2012)

16. Frei, F., Wada, K.: Brief announcement: distinct gathering under round robin. In: 38th International Symposium on Distributed Computing (DISC 2024). LIPIcs, vol. 319, pp. 48:1–48:8. Schloss Dagstuhl – Leibniz-Zentrum für Informatik (2024). https://doi.org/10.4230/LIPIcs.DISC.2024.48
17. Guilbault, S., Pelc, A.: Gathering asynchronous oblivious agents with local vision in regular bipartite graphs. Theor. Comput. Sci. **509**, 86–96 (2013)
18. Kamei, S., Lamani, A., Ooshita, F., Tixeuil, S., Wada, K.: Asynchronous gathering in a torus. In: 25th International Conference on Principles of Distributed Systems (OPODIS). LIPIcs, vol. 217, pp. 9:1–9:17. Schloss Dagstuhl - Leibniz-Zentrum für Informatik (2021)
19. Klasing, R., Kosowski, A., Navarra, A.: Taking advantage of symmetries: gathering of many asynchronous oblivious robots on a ring. Theor. Comput. Sci. **411**, 3235–3246 (2010)
20. Navarra, A., Piselli, F.: Asynchronous silent programmable matter: line formation. In: Proceedings of 25th International Symposium on Stabilization, Safety, and Security of Distributed Systems (SSS). LNCS, vol. 14310, pp. 598–612 (2023)
21. Navarra, A., Piselli, F.: Coating in SILBOT with one axis agreement. In: Proceedings of 26th International Symposium on Stabilization, Safety, and Security of Distributed Systems (SSS). LNCS, vol. 14931, pp. 177–192. Springer (2024). https://doi.org/10.1007/978-3-031-74498-3_13
22. Navarra, A., Piselli, F.: Oblivious robots under round robin: gathering on rings. arXiv preprint arXiv:2502.03939 (2025)
23. Navarra, A., Piselli, F.: Silent programmable matter: coating. In: 27th International Conference on Principles of Distributed Systems, (OPODIS). LIPIcs, vol. 286, pp. 25:1–25:17. Schloss Dagstuhl - Leibniz-Zentrum für Informatik (2023). https://doi.org/10.4230/LIPICS.OPODIS.2023.25

Finding a Set of Long Common Substrings with Repeats from m Input Strings

Tiantian Li[1], Lusheng Wang[2(✉)], and Daming Zhu[1]

[1] School of Computer Science and Technology, Shandong University, Qingdao, China
dmzhu@sdu.edu.cn
[2] Department of Computer Science, City University of Hong Kong, Hong Kong, China
cswangl@cityu.edu.hk

Abstract. In this paper, we propose two string problems, and study algorithms and complexity of various versions for those problems. Let $S = \{s_1, s_2, \ldots, s_m\}$ be a set of m strings. A *common substring of S* is a substring appearing in every string in S. Given a set of m strings $S = \{s_1, s_2, \ldots, s_m\}$ and a positive integer k, we want to find a set C of k common substrings of S such that the k common substrings in C appear in the same order and have no overlap among the m input strings in S, and the total length of the k common substring in C is maximized. This problem is referred to as the *longest total length of k common substrings from m input strings* ($LCSS(k, m)$ for short). The other problem we study here is called the *longest total length of a set of common substrings with length more than l from m input string* ($LSCSS(l, m)$ for short). Given a set of m strings $S = \{s_1, s_2, \ldots, s_m\}$ and a positive integer l, for $LSCSS(l, m)$, we want to find a set of common substrings of S, each is of length more than l, such that the total length of all the common substrings is maximized.

We show that both problems are NP-hard when k and m are variables.

We propose dynamic programming algorithms with time complexity $O(k\ n_1 n_2)$ and $O(n_1 n_2)$ to solve $LCSS(k, 2)$ and $LSCSS(l, 2)$, respectively, where n_1 and n_2 are the lengths of the two input strings. We then design an algorithm for $LSCSS(l, m)$ when every length $> l$ common substring appears once in each of the $m - 1$ input strings. The running time is $O(n_1^2 m)$, where n_1 is the length of the input string with no restriction on length $> l$ common substrings.

Finally, we propose a fixed parameter algorithm for $LSCSS(l, m)$, where each length $> l$ common substring appears $m - 1 + c$ times among the $m - 1$ input strings (other than s_1). In other words, each length $> l$ common substring may repeatedly appear at most c times among the $m - 1$ input strings $\{s_2, s_3, \ldots, s_m\}$. The running time of the proposed algorithm is $O((n_1 2^c)^2 m)$, where n_1 is the input string with no restriction on repeats.

The $LSCSS(l, m)$ is proposed to handle whole chromosome sequence alignment for different strains of the same species, where more than 98% of letters in core regions are identical.

Keywords: Dynamic programming · Algorithm · Common substrings · String

1 Introduction

Finding common sub-strings among given strings is an attractive algorithmic topic that has played a significant role in bio-sequence analysis, data compression and other bio-medical informatics [6,10]. A best-known problem in this topic is to find a longest common substring of two or more input strings, which is solvable in linear time via suffix tree of a concatenation of the input strings [14].

A sufficiently long common substring of multiple strings is meaningful to reflect the structural likeness of the strings and, useful in pursuing the structural likeness of the strings more professionally. In comparison of two or more bio-sequences, it is necessary to find multiple order consistent common substrings of the bio-sequences. This kind of common substrings are more likely conservative in letteritic processes of bio-sequences [3,9]. In order to find conservative areas in bio-sequences, various kinds of common or repeated substrings of input sequences have been proposed [1,8].

Many algorithms have been proposed in literature for finding common or repeated substrings from input strings. Linear time algorithms were proposed to find a given number of fixed-length common substrings in which each gap has a size upper bound [7]. A heuristic algorithm was proposed to find in a given string repeated substrings that consist of a given fraction of careless symbols [5]. In [4], an $O(n \log n)$ time algorithm was proposed to find among two given strings of n symbols a longest common substrings with one mismatch. An algorithm for the k-mismatch average common string problem that takes $O(n)$ space and $O(nlog(k)n)$ time was given in [11].

It is well known that the two related problems, the longest common subsequence and the sequence alignment problems of m input sequences can be solved in $O(n^m)$ time [6]. In pan-genome comparison, one need to align the whole chromosome of a set of strains of the same species [12,13]. Angiuoli and Salzberg have developed a tool, Mugsy, that can do multiple sequence alignment for the whole chromosomes of a set of strains of the same species [2]. They use suffix tree to seed an alignment with maximal unique matches.

In this paper, we propose two string problems, and study algorithms and complexity of various versions for those problems. Let $S = \{s_1, s_2, \ldots, s_m\}$ be a set of m strings. A *common substring of S* is a substring appearing in every string in S. Given a set of m strings $S = \{s_1, s_2, \ldots, s_m\}$ and a positive integer k, we want to find a set C of k common substrings of S such that the k common substrings in C appear in the same order and have no overlap among the m input strings in S, and the total length of the k common substring in C is maximized. This problem is referred to as the *longest total length of k common substrings from m input strings* ($LCSS(k,m)$ for short). The other problem we study here is called the *longest total length of a set of common substrings with length more than l from m input string* ($LSCSS(l,m)$ for short). Given a set of m strings

$S = \{s_1, s_2, \ldots, s_m\}$ and a positive integer l, we want to find a set of common substrings of S, each is of length more than l, such that the total length of all the common substrings is maximized.

We show that both problems are NP-hard when k and m are variables.

We propose dynamic programming algorithms with time complexity $O(k n_1 n_2)$ and $O(n_1 n_2)$ to solve $LCSS(k, 2)$ and $LSCSS(l, 2)$, respectively, where n_1 and n_2 are the lengths of the two input strings. We then design an algorithm for $LSCSS(l, m)$ when every length $> l$ common substring appears once in each of the $m - 1$ input strings. The running time is $O(n_1^2 m)$, where n_1 is the length of the input string with no restriction on length $> l$ common substrings.

Finally, we propose a fixed parameter algorithm for $LSCSS(l, m)$, where each length $> l$ common substring appears $m - 1 + c$ times among the $m - 1$ input strings (other than s_1). In other words, each length $> l$ common substring may repeatedly appear at most c times among the $m - 1$ strings. The running time of the proposed algorithm is $O((n_1 2^c)^2 m)$, where n_1 is the input string with no restriction on repeats.

The $LSCSS(l, m)$ is proposed to handle whole chromosome sequence alignment for different strains of the same species, where more than 98% of letters in core regions are identical. $LSCSS(l, m)$ can report a set of common substrings as the sketch of the alignment. Then consecutive common substrings with highly conservative small gaps are merged to form core regions. Very often, the small gap has two different letters at the ends and the other letters in the middle are identical.

2 A Dynamic Programming Algorithm for $LCSS(k, 2)$

In this section, we consider $LCSS(k, 2)$ with 2 input strings. Let $s_1 = s_1[1]s_1[2] \ldots s_1[n_1]$ and $s_2 = s_2[1]s_2[2] \ldots s_2[n_2]$ be the two input strings.

Let us consider the first i-th letters in s_1 and the first j-th letters in s_2. Assume that there are t common substrings for the first i-th letters in s_1 and the first j-th letter in s_2. We need to consider two cases.

Case 1: We define $T_t(i, j, 1)$ to be the maximum length of the t common substrings of s_1 and s_2 such that the t-th substring ends at positions i in s_1 and j in s_2, respectively. This implies that $s_1[i]$ and $s_2[j]$ must be identical.

Case 2: We define $T_t(i, j, 0)$ to be the maximum length of the t common substrings of the first i letters in s_1 and the first j letters in s_2 such that the t-th substring ends before position i in S_1 or before position j in S_2.

It is easy to see that $T_t(i, j, 1) = 0$ and $T_t(i, j, 0) = 0$ for $i = 0$ or $j = 0$ or $t = 0$. By definition, $T_t(i, j, 1) = 0$ if $s_1[i] \neq s_2[j]$. When $s_1[i] = s_2[j]$, i, j and $t > 0$, $T_t(i, j, 1) = T_{t-1}(i - 1, j - 1, 0) + 1$ if $s_1[i] = s_2[j]$ is the first letter of the t-th common substring. Otherwise, $s_1[i] = s_2[j]$ is an extension of the current t-th common substring ending at $s_1[i - 1]$ and $s_2[j - 1]$. Thus, we have $T_t(i, j, 1) = T_t(i - 1, j - 1, 1) + 1$ in this case.

Therefore, $T_t(i, j, 1)$ can be computed as in (1).

$$T_t(i,j,1) = \begin{cases} 0 & \text{if } i = 0 \text{ or } j = 0 \text{ or } t = 0; \\ 0 & \text{if } A[i] \neq B[j]; \\ \max\{T_{t-1}(i-1, j-1, 0), T_t(i-1, j-1, 1)\} + 1 & \text{if } A[i] = B[j], i, j \text{ and } t > 0. \end{cases} \quad (1)$$

When i, j and $t > 0$, $T_t(i, j, 0)$ is one of $T_t(i-1, j, 0)$, $T_t(i, j-1, 0)$, $T_t(i-1, j, 1)$, and $T_t(i, j-1, 1)$. Thus, $T_t(i, j, 0)$ can be computed as in (2).

$$T_t(i,j,0) = \begin{cases} 0 & i = 0 \text{ or } j = 0 \text{ or } t = 0; \\ \max_{r=0}^{1} \max\{T_t(i-1, j, r), T_t(i, j-1, r)\} & \text{otherwise.} \end{cases} \quad (2)$$

The length of an optimal solution is the largest number in the arrays $T_t(i, j, 0)$ and $T_t(i, j, 1)$. Standard backtracking process can report an optimal solution. The complete algorithm is given as Algorithm 1.

Algorithm 1. The dynamic programming for $LCSS(k, 2)$

Require: Two strings $s_1[1, n_1]$ and $s_2[1, n_2]$.
Ensure: The longest k common substrings with s_1 and s_2.
1: $T_t(i, j, 1) = T_t(i, j, 0) \leftarrow 0$, for $t = 0$ or $i = 0$ or $j = 0$;
2: **for** i from 1 to n_1 **do**
3:     **for** j from 1 to n_2 **do**
4:         **for** t from 1 to k **do**
5:             Get $T_t(i, j, 1)$ by Formula (1) and $T_t(i, j, 0)$ by Formula (2);
6:         **end for**
7:     **end for**
8: **end for**
9: **return** the longest k common substrings with s_1 and s_2 by a standard backtracking process.

In Algorithm 1, we need to compute $O(kn_1n_2)$ cells in the arrays $T_t(i, j, 0)$ and $T_t(i, j, 1)$. Obtaining the value of each cell requires O(1) time based on formulas (1) and (2). Thus, the total running time of Algorithm 1 is $O(kn_1n_2)$.

Theorem 1. *Algorithm 1 solves the longest k common substrings problem with two input strings in $O(kn_1n_2)$ time and $O(kn_1n_2)$ space, where n_1 and n_2 are the lengths of s_1 and s_2.*

We can easily extend the algorithm to solve $LCSS(k, m)$ for m input strings (by using $m+1$ indices to define $T_t()$'s) in $O(k \prod_{i=1}^{m} n_i)$ time and space, where string s_i has length n_i for each $i \in [1, m]$.

3 A Dynamic Programming Algorithm for $LSCSS(l, 2)$

In this section, we consider the longest total length of a set of common substrings with length more than l from 2 input strings. Let $s_1 = s_1[1]s_1[2]\ldots s_1[n_1]$ and $s_2 = s_2[1]s_2[2]\ldots s_2[n_2]$ be the two input strings. Let us consider the first i-th letters in s_1 and the first j-th letters in s_2. We need to consider two cases.

Case 1: Let $T(i, j, 1)$ be the largest total length of a set of common substrings with length $> l$ ending at $s_1[i]$ and $s_2[j]$. $T(i, j, 1) > 0$ implies $s_1[i] = s_2[j]$.

Case 2: Let $T(i,j,0)$ be the largest total length of a set of common substrings with length $> l$ for the first i letters in s_1 and the first j letters in s_2 ending before $s_1[i]$ in s_1 or before $s_2[j]$ in s_2.

For each pair of (i,j), we define $f(i,j)$ to be the largest integer r such that $s_1[i-q] = s_2[j-q]$ for $q = 0, 1, 2, \ldots, r-1$. We can compute $f(i,j)$ as follows:

$$f(i,j) = \begin{cases} 0 & \text{if } s_1[i] \neq s_2[j]; \\ f(i-1, j-1) + 1 & \text{if } s_1[i] = s_2[j]. \end{cases} \quad (3)$$

Therefore, it takes $O(n_1 n_2)$ time to compute all $f(i,j)$'s.

It is easy to see that $T(i,j,1) = 0$ and $T(i,j,0) = 0$ for $i = 0$ or $j = 0$. When i and $j > 0$, $T(i,j,0)$ is one of $T(i-1,j,0)$, $T(i,j-1,0)$, $T(i-1,j,1)$, and $T(i,j-1,1)$. Thus, $T(i,j,0)$ can be computed as in (4).

$$T(i,j,0) = \begin{cases} 0 & \text{if } i = 0 \text{ or } j = 0; \\ \max_{q=0}^{1} \max\{T(i-1,j,q), T(i,j-1,q)\} & \text{otherwise.} \end{cases} \quad (4)$$

By definition, $T(i,j,1) = 0$ if $f(i,j) \leq l$. When $f(i,j) > l$, $T(i,j,1)$ is the maximum of $T(i-x, j-x, 0) + x$ for $l+1 \leq x \leq f(i,j)$, where x is the length of the last common substring ending at $s_1[i]$. It is necessary to check every x for $l+1 \leq x \leq f(i,j)$. The reason is that the length $f(i,j)$ common substring ending at $s_1[i]$ may be decomposed into more than one common substring of length $> l$ in an optimal solution. Figure 1 gives an example.

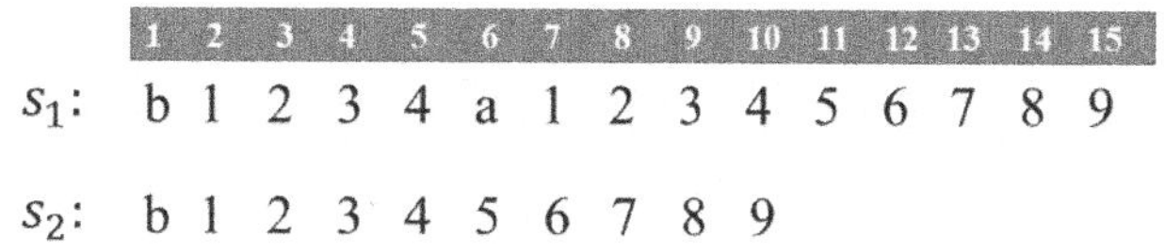

s_1: b 1 2 3 4 a 1 2 3 4 5 6 7 8 9

s_2: b 1 2 3 4 5 6 7 8 9

Fig. 1. The $LSCSS(4,2)$ for $s_1 = b\,1\,2\,3\,4\,a\,1\,2\,3\,4\,5\,6\,7\,8\,9$ and $s_2 = b\,1\,2\,3\,4\,5\,6\,7\,8\,9$, where $f(15, 10) = 9$, the substring 123456789 in s_1 needs to be decomposed into two substrings 1234 and 56789 in an optimal solution so that 1234 can be extended to the left to have $b1234$ as a length 5 common substring. The optimal value of $T(15, 10, 1)$ is obtained from $T(10, 5, 0) + 5 = 10$.

Therefore, $T(i,j,1)$ can be computed as in (5).

$$T(i,j,1) = \begin{cases} 0 & \text{if } i = 0 \text{ or } j = 0; \\ 0 & \text{if } f(i,j) \leq l; \\ \max_{x \in [l+1, f(i,j)]}\{T(i-x, j-x, 0) + x\} & \text{if } f(i,j) > l. \end{cases} \quad (5)$$

The length of an optimal solution is the largest number in the arrays $T(i,j,0)$ and $T(i,j,1)$. Standard backtracking process can report an optimal solution. Thus, we have an $O(n_1^2 n_2)$ running time algorithm to solve $LSCSS(l,2)$.

Now, we propose a fast method to find $T(i,j,1)$ for each $f(i,j)$ in $O(1)$ time. There are two cases for computing $T(i,j,1) = \max\{T(i-x, j-x, 0) + x | l+1 \leq x \leq f(i,j)\}$.

Case 1: $x = l + 1$, then $T(i, j, 1) = T(i - l - 1, j - l - 1, 0) + l + 1$.

Case 2: $l + 2 \leq x \leq f(i, j)$. Since position pair $(i - x, j - x)$ has distance x from i on s_1 and j on s_2, and it has distance $x - 1$ from $i - 1$ on s_1 and $j - 1$ on s_2, then

$$\max\{T(i - x, j - x, 0) + x | l + 2 \leq x \leq f(i,j)\}$$
$$= \max\{T(i - 1 - (x - 1), j - 1 - (x - 1), 0) + x - 1 | l + 1 \leq x - 1 \leq f(i - 1, j - 1)\} + 1$$
$$= T(i - 1, j - 1, 1) + 1.$$

Therefore, $T(i, j, 1) = \max\{T(i-1, j-1, 1)+1, T(i-l-1, j-l-1, 0)+l+1\}$. Then $T(i, j, 1)$ can be computed as (6).

$$T(i, j, 1) = \begin{cases} 0 & \text{if } i = 0 \text{ or } j = 0; \\ 0 & \text{if } f(i,j) \leq l; \\ \max\{T(i-1, j-1, 1)+1, T(i-l-1, j-l-1, 0)+l+1\} & \text{if } f(i,j) > l. \end{cases} \quad (6)$$

The length of an optimal solution is the largest number in the arrays $T(i, j, 0)$ and $T(i, j, 1)$. Standard backtracking process can report an optimal solution. The complete algorithm is given as Algorithm 2.

Algorithm 2. The dynamic programming for $LSCSS(l, 2)$

Require: Two strings $s_1[1, n_1]$ and $s_2[1, n_2]$.
Ensure: The largest total length of a set of length $> l$ common substrings for s_1 and s_2.
1: $T(i, j, 1) = T(i, j, 0) \leftarrow 0$, for $i = 0$ or $j = 0$;
2: **for** i from 1 to n_1 **do**
3:    **for** j from 1 to n_2 **do**
4:       Compute $f(i, j)$, $T(i, j, 0)$, and $T(i, j, 1)$ based on Formulas (3), (4), and (6), respectively;
5:    **end for**
6: **end for**
7: **return** the largest value among all $T(i, j, 0)$'s and $T(i, j, 1)$'s and the corresponding set of common strings (by standard backtracking).

In Algorithm 2, we need to compute $O(n_1 n_2)$ values of $T(i, j, q)$'s and each takes $O(1)$ times. Thus, we have an $O(n_1 n_2)$ running time algorithm to solve $LSCSS(l, 2)$.

Theorem 2. *Algorithm 2 solves the largest total length of a set of common substrings problem with two input strings in $O(n_1 n_2)$ time and $O(n_1 n_2)$ space, where n_1 and n_2 are the lengths of s_1 and s_2.*

4 $LSCSS(l, m)$ When Length $> l$ Common Substrings Appear once

Let $S = \{s_1, s_2, \ldots, s_m\}$ be the set of input strings. In this section, we consider a special case of $LSCSS(l, m)$, where every input string in $S - \{s_1\}$ does not contain repeated length $> l$ common substrings of S. Note that, length $> l$ common substrings of S may appear more than once in s_1.

Let $f(i)$ be the longest length of common substrings of S ending at position $s_1[i]$. For any $f(i) > l$, $s_1[i - f(i) + 1]\ldots s_1[i]$ is the corresponding common substring of S in s_1. We define $ve(i) = (ve(i)[2], ve(i)[1], \ldots, ve(i)[m])$, where $ve(i)[j]$ is a pair of indices $(ve(i)[j][1], ve(i)[j][2])$ indicating the *unique* starting and ending positions of the common substring $s_1[i - f(i) + 1]\ldots s_1[i]$ in s_j. If $f(i) \leq l$, we set $f(i) = 0$ since under this case $ve(i)$ does not exists (is not unique).

Let $d(i)$ be the maximum total length of a set of common substrings of S, each has length $> l$, such that the last selected common substring ends at $s_1[i]$. It is easy to see that $d(i) = 0$ for $f(i) \leq l$.

Reducing the Length of s_1: Assume that $f(i)$ is known, we convert s_1 into a new string s_1' containing positions with $f(i) > l$. Hereafter, when computing $d(i)$, s_1 always represents the new string s_1'.

Now, let us consider the computation of $d(i)$.

Let $s^i = s_1[i - f(i) + 1]\ldots s_1[i]$ be the longest common substring of S ending at $s_1[i]$.

Lemma 1. *There exists an optimal solution corresponding to $d(i)$ such that s^i form at most two common substrings of S.*

Proof. Let us look at the example given in Fig. 2. Assume that $l = 4$. The longest common substring of S ending at $s_1[18]$ is 123456789, which is of length 9. The optimal solution, corresponding to $d(18)$ contains two common substrings of S, $b1234$ and 56789 with total length 10, where 123456789 is decomposed into two common substrings 1234 and 56789, 1234 appears once more on the left of 123456789 in s_1 and s_3, and such an occurrence of 1234 can be extended to the left to include one more letter b.

Now, we understand that the reason that s^i may be decomposed into more than one common substring of S in an optimal solution is that the left most common substring (after decomposition) starting at $i - f(i) + 1$ may be extended to the left to form a longer common string of S. (See $b1234$ in the example.) Note that, only the left most common substring (after decomposition) has the chance to be extended to the left. The rest of common substrings (after decomposition) remain the same. Therefore, we can conclude that if s^i can be decomposed into more than one common substrings of S in an optimal solution corresponding to $d(i)$, then there exists an optimal solution such that s^i is decomposed into exactly two common substrings of S.

Assumption: When s^i form at most two common substrings of S for $d(i)$, the length of the second last common substring of S is always $l + 1$.

Note that, if the length of the second last common substring of S is greater than $l + 1$ we can re-decompose the last few letters to the last common substring of S ending at $s_1[i]$.

Let $s = s_1[i-x]s_1[i-x+1]\ldots s_1[i]$ be the last common substring of S that we are going to add to the solution for computing $d(i)$, where $x+1$ is the length of s. Let s' be the second last common substring of S that does not overlap with s in

```
         1  2  3  4  5  6  7  8  9  10 11 12 13 14 15 16 17 18
$s_1$:   b  1  2  3  4  a  a  a  a  1  2  3  4  5  6  7  8  9

$s_2$:   b  1  2  3  4  5  6  7  8  9  a  a  a  a

$s_3$:   b  1  2  3  4  1  2  3  4  5  6  7  8  9  b
```

Fig. 2. Here $m = 3$ and $l = 4$. The input strings are $s_1 = b\,1\,2\,3\,4\,a\,a\,a\,a\,1\,2\,3\,4\,5\,6\,7\,8\,9$, $s_2 = b\,1\,2\,3\,4\,5\,6\,7\,8\,9\,a\,a\,a\,a$ and $s_3 = b\,1\,2\,3\,4\,5\,6\,7\,8\,9\,b\,a\,a\,a\,a$. The two common substrings $b1234$ and 56789 form the optimal solution for $d(18)$.

s_1. To add s as the last common substring of S into the solution corresponding to $d(i)$, we need to check a condition.

Condition 1: s' does not overlap with s and s' is always before s in all the $m - 1$ input strings $s_2, s_3, \ldots, s_m$.

Assume that $f(i) > l$. To compute $d(i)$, we need to consider two cases according to Lemma 1.

Case 1: s^i appears in an optimal solution corresponding to $d(i)$ as the last common string of S.

Let $1 \leq i \leq n_1$, we define

$$C(i) = \{t\,|\,1 \leq t \leq i - f(i) \text{ and } \text{ve}(t)[k][2] < \text{ve}(i)[k][3] \text{ for } k = 2, 3, \ldots, m\}. \quad (7)$$

It takes $O(m)$ time to test if t should be added to $C(i)$ for each t. Note that $\text{ve}(t)[k][2] < \text{ve}(i)[k][1]$ for $k = 2, 3, \ldots, m$ ensures that the second last common substring of S ending at position t in s_1 has no overlap with the last common substring s^i over all the m input strings.

The computation of $d(i)$ is as follows:

$$d(i) = \max_{t \in C(i)} [d(t) + f(i)]. \quad (8)$$

It is worth to emphasize that we need to go through all $t \in C(i)$ instead of by looking at position $t = i - f(i)$ only.

An example is given in Fig. 3 by setting $l = 3$, where $d(5) = 5$, $d(9) = 4$ and $d(15) = 11$. Here $C(15) = \{4, 5, 9\}$, and the maximum value of $d(15) = d(5) + 6$, where $d(9)$ (position $t = 9 = 15 - f(15)$) does not give the maximum value of $d(15)$.

Case 2: s^i is decomposed into two common substrings in an optimal solution corresponding to $d(i)$.

Let $1 \leq i \leq n_1$ be the ending position of the last common substring of S in s_1 and t the ending position of the second last common substring of S in s_1. Then t should be in the range $1 \leq t \leq i - f(i) + l$.

We define $p(t, i)$ to be the maximal overlap length between s^t and s^i in all the m input strings. In s_1, the overlap length between s^t and s^i is $t - (i - f(i))$ when

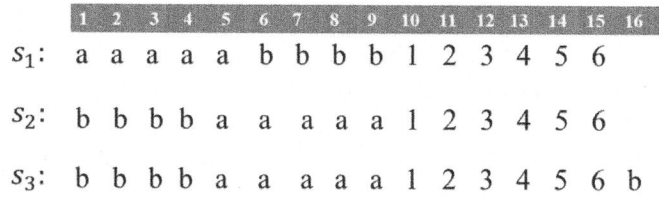

Fig. 3. An example with $l = 3$, where $s_1 = aaaaabbbb123456$, $s_2 = bbbbaaaaa123456$, and $s_3 = bbbbaaaaa123456b$.

$0 < t - (i - f(i)) < l + 1$. Thus, we set $x_1 = t - (i - f(i))$ if $0 < t - (i - f(i)) < l + 1$. Otherwise, $x_1 = 0$. In s_j (for $j = 2, 3, \ldots, m$), the overlap length between s^t and s^i is $ve(t)[j][2] - ve(i)[j][1] + 1$ when $0 < ve(t)[j][2] - ve(i)[j][1] + 1 < l + 1$. Thus, we set $x_j = ve(t)[j][2] - ve(i)[j][1] + 1$ if $0 < ve(t)[j][2] - ve(i)[j][1] + 1 < l + 1$. Otherwise, $x_j = 0$. Therefore,

$$p(t, i) = \max_{k=1}^{m} x_k. \tag{9}$$

Each $p(t, i)$ can be computed in $O(m)$ time.

Now, we define $list(i)$ to be the set of positions $1 \leq t \leq i - f(i) + l$ such that $0 < p(t, i) \leq l$ and $f(i) - p(t, i) \geq l + 1$. That is $list(i) = \{t \mid 0 < p(t, i) \leq l$ and $f(i) - p(t, i) \geq l + 1\}$. Note that $f(i) - p(t, i) > l$ ensures that the length of s^i is at least l. Moreover, the size of $list(i)$ is upper bounded by $O(n_1)$. To make sure that Condition 1 holds, we further define

$$L(i) = \{t \mid \in list(i) \text{ and } ve(t)[k][2] < ve(i)[k][2] + p(t, i)$$
$$\text{for } k = 2, 3, \ldots, m\}. \tag{10}$$

It takes $O(mn_1)$ time to construct both $list(i)$ and $L(i)$ for each i. The computation of $d(i)$ in Case 2 is as follows:

$$d(i) = \max_{t \in L(i)} [d(t) + f(i) - p(t, i)]. \tag{11}$$

Therefore, combine Case 1 and Case 2, $d(i)$ can be computed as in (12)

$$d(i) = \begin{cases} 0 & \text{if } i = 0; \\ 0 & \text{if } f(i) \leq l; \\ \max\{\max_{t \in L(i)}[d(t) + f(i) - p(t, i)], \\ \quad \max_{j \in C(i)}[d(j) + f(i)]\} & \text{if } f(i) > l. \end{cases} \tag{12}$$

The total length of an optimal solution is give as $\max_{i=1}^{n_1} d(i)$.

Lemma 2. *Given $f(i)$ and $ve(i)$, we can compute all $d(i)$'s in $O(n_1^2 m)$ time. Computation of $f(i)$ and $ve(i)$: Let $S = \{s_1, s_2, \ldots, s_m\}$ be the set of m input strings. We can form a new string C by adding a distinct special symbol to the end of each s_i and concatenating the m strings. Then we construct a suffix tree T for the newly created long string in $O(n)$ time and space, where $n = \sum_{i=1}^{m}(n_i + 1)$ is the total length of the m input strings.*

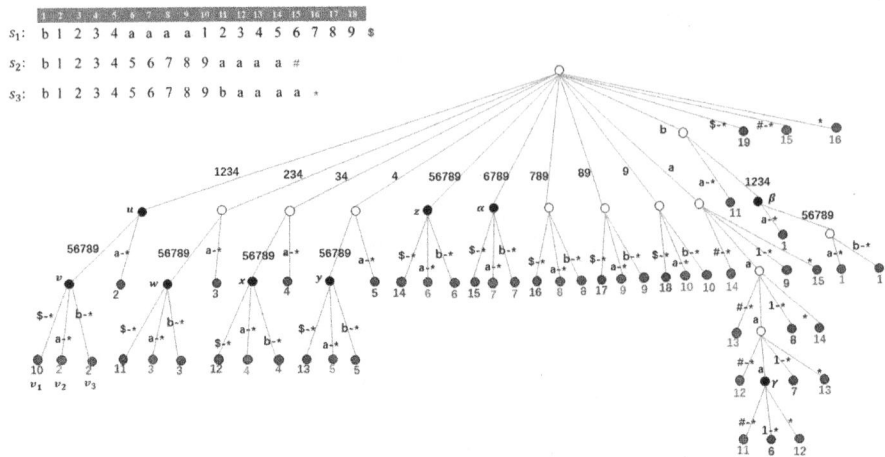

Fig. 4. The suffix tree for $s_1 = aaaaabbbb123456$, $s_2 = bbbbaaaaa123456$, and $s_3 = bbbbaaaaa123456b$.

An example is given Fig. 4. Each edge (u, v) in the suffix tree T is associated with a pair of integers (i, j) indicating the corresponding substring $C[i] \ldots C[j]$ on the edge. Thus, $j - i + 1$ is the length of the substring on edge (u, v).

Let u be a node in the suffix tree T. We use $l(u)$ to denote the total length of all edges from the root to u. All the values of $l(u)$'s can be computed in $O(n)$ time by doing a depth first search on T and the value of $l(v)$ can be obtained from its parent u as $l(v) = l(u)+$ length of edge (u, v). Each leaf u in T corresponds to a suffix $c(u)$ of C. The starting position $s(u)$ of $c(u)$ in C is $n - l(u) + 1$. We define $n_0 = -1$ for convenience. If $[\sum_{j=0}^{i}(n_j + 1)] + 1 \leq s(u) \leq [\sum_{j=0}^{i+1}(n_j + 1)] - 1$, then such a suffix $c(u)$ starts at a position in s_{i+1} for $i = 0, 1, 2, \ldots, m-1$. We update the the value of $s(u) = s(u) - \sum_{j=0}^{i}(n_j + 1)$ to indicate the true position in s_{i+1}.

Now every node u in T corresponds to a substring $c(u)$ of C and $c(u)$ can be obtained by going through the edges from the root to u. A leaf node u in T is labelled as s_i if $c(u)$ starts in s_i.

For each node u in T, let $T(u)$ be the subtree tree rooted at u and define $vec(u) = (vec(u)[2], vec(u)[3], \ldots, vec(u)[m])$ be a vector, where $vec(u)[i]$ is an integer (with initial value 0), and $set(u)$ be a set (with initial value $\emptyset$). If $vec(u)[i] > 0$, then there is a leaf node v in $T(u)$ such that v is labelled as s_i and $vec(u)[i] = s(v)$. We use $set(u)$ to store all the $s(v)$'s of leaf nodes in $T(u)$ that are labelled as s_1. We do a depth first search on T, and compute $set(u)$ and $vec(u)$ as follows:

Step 1 If u is a leaf with $l(u) > l$ and is labelled as s_i for $i \geq 2$, we set $vec(u)[i] = s(u)$. We add $s(u)$ to $set(u)$, if u is labelled as s_1.

Step 2 For internal node u with $l(u) > l$, let $u_1, u_2, \ldots, u_q$ be all the children of u in T. In the depth first search, after all the children $u_1, u_2, \ldots, u_q$ have been visited and their $set()$ and $vec()$ values have been computed, $set(u) = \cup_{i=1}^{q} set(u_i)$ and $vec(u)[i] = vec(u_j)[i]$ for the unique $1 \leq j \leq q$ such that $vec(u_j)[i] > 0$.

Step 3 If there is more than one $1 \leq j \leq q$ such that $vec(u_j)[i] > 0$, then the substring $c(u)$ appears more than once in s_i for some $i \geq 2$. We can stop the process and declare that some common substring of S appears more than once in some s_i for some $i \geq 2$. (Other ways to handle this case lead to different heuristics and will be discussed later.)

A node u is *valid* if $vec(u)[i] > 0$ for all $i = 2, 3, \ldots, m$ and $set(u) \neq \emptyset$. It is easy to see that $l(u) \leq n_1$ for any valid u and $c(u)$ is a common substring of S. We use a boolean variable $valid(u)$ to indicate if u is valid. The value of all $valid(u)$'s can be computed in $O(nm)$ time after $set(u)$'s and $vec(u)$'s for all nodes have been computed.

Now, let us focus on valid internal node in T. Let u be a valid node and u_1 is a child of u which is also valid. Then $T(u)$ contains a set of leaf nodes $unique(u)$ that are not in $T(u_1)$. Based on the assumption that each length $> l$ common substring of S appears once in s_i for $i = 2, 3, \ldots, m$, all the leaf nodes in $unique(u)$ are labelled as s_1. Thus, total number of valid nodes in T is at most n_1, each corresponds to a unique position in s_1. For each valid node u, we set $ve(u)[i][1] = vec(u)[i]$ and $ve(u)[i][2] = vec(u)[i] + l(u) - 1$.

Now, we are ready to compute $f(i)$. We initially set $f(i) = 0$ for $i = 1, 2, \ldots, n_1$. For each valid node u in T and every $x \in set(u)$, if $f(x + l(u) - 1) < l(u)$, we set $f(x + l(u) - 1) = l(u)$ and $ve(x + l(u) - 1) = ve(u)$. This process takes $O(n_1^2 m)$ time since the size of each $set(u)$ is at most n_1, there are at most n_1 valid nodes, and the size of $ve(i)$ is $O(m)$.

Note that, for some $1 \leq i \leq n_i$, the letter $s_1[i]$ in a common substring $s_1[x] \ldots s_1[i]$ of S may not be on an valid node in T. Instead, it may be on an edge of T. In this case, $f(i)$ is still 0 at present. Thus, for each $f(i) > 0$, we need to set $f(i - j) = f(i) - j$ for $j = 1, 2, \ldots, f(i) - (l+1)$ and $ve(i)$ can be updated accordingly.

Let us look at the example given Fig. 4. $set(v_1) = \{10\}, vec(v_1) = (0,0)$, $set(v_2) = \emptyset, vec(v_2) = (2,0)$ and $set(v_3) = \emptyset, vec(v_3) = (0,2)$, then $set(v) = \cup_{i=1}^{3} set(v_i) = \{10\}, vec(v) = (2,2)$ and $valid(v) = 1$, $ve(v) = ((2,10),(2,10))$. In the same way, all the other valid nodes can be identified (marked as black nodes) and the $set()$ and $ve()$ of each valid node can be computed. $f(5)$ and $ve(5)$ are related to valid nodes u and β where $set(u) = \{2,10\}, ve(u) = ((2,5),(2,5))$ and $set(\beta) = \{1\}, ve(\beta) = ((1,5),(1,5))$. For valid node u and $2 \in set(u)$, set $f(5) = 4$ and $ve(5) = ((2,5),(2,5))$, for valid node β and $1 \in set(\beta)$, since $f(5) < l(\beta) = 5$ then update $f(5) = 5$ and $ve(5) = ((1,5),(1,5))$. In the same way, we have $f(13) = 4$, $ve(13) = ((2,5),(2,5))$, $f(18) = 9$, $ve(18) = ((2,10),(2,10))$, $f(9) = 4$, $ve(9) = ((11,15),(12,16))$ and $f(4) = 4$, $ve(4) = ((1,4),(1,4))$. And for the other positions i, we set $f(i) = 0$ and consider $ve(i)$ does not exist since they are of no help in computing $d(i)$.

Lemma 3. *All the values of $f(i)$ and $ve(i)$ can be computed in $O(n_1^2 m)$ time.*

Theorem 3. *The algorithm solves $LSCSS(l,m)$ when length $> l$ common substrings of S appear once in $s_2, s_3, \ldots, s_m$ with running time $O(n_1^2 m)$.*

References

1. Alamro, H., Ayad, L.A., Charalampopoulos, P., Iliopoulos, C.S., Pissis, S.P.: Longest common prefixes with K-mismatches and applications. In: International Conference on Current Trends in Theory and Practice of Informatics, pp. 636–649. Springer (2018)
2. Angiuoli, S.V., Salzberg, S.L.: Mugsy: fast multiple alignment of closely related whole genomes. Bioinformatics **27**(3), 334–342 (12 2010)
3. Antoniou, P., Crochemore, M., Iliopoulos, C., Peterlongo, P.: Application of suffix trees for the acquisition of common motifs with gaps in a set of strings. In: International Conference on Language and Automata Theory and Applications (2007)
4. Flouri, T., Giaquinta, E., Kobert, K., Ukkonen, E.: Longest common substrings with k mismatches. Inf. Process. Lett. **115**(6–8), 643–647 (2015)
5. Grossi, R., Pietracaprina, A., Pisanti, N., Pucci, G., Upfal, E., Vandin, F.: MadMX: a novel strategy for maximal dense motif extraction. In: International Workshop on Algorithms in Bioinformatics, pp. 362–374. Springer (2009)
6. Gusfield, D.: Algorithms on stings, trees, and sequences: computer science and computational biology. ACM SIGACT News **28**(4), 41–60 (1997)
7. Iliopoulos, C.S., McHugh, J., Peterlongo, P., Pisanti, N., Rytter, W., Sagot, M.F.: A first approach to finding common motifs with gaps. Int. J. Found. Comput. Sci. **16**(06), 1145–1154 (2005)
8. Manzini, G.: Longest common prefix with mismatches. In: International Symposium on String Processing and Information Retrieval, pp. 299–310. Springer (2015)
9. Marsan, L., Sagot, M.F.: Extracting structured motifs using a suffix tree algorithms and application to promoter consensus identification. In: Proceedings of the Fourth Annual International Conference on Computational Molecular Biology, pp. 210–219 (2000)
10. Storer, J.A.: Data Compression: Methods and Theory. Computer Science Press, Inc (1987)
11. Thankachan, S.V., Apostolico, A., Aluru, S.: A provably efficient algorithm for the k-mismatch average common substring problem. J. Comput. Biol. **23**(6), 472–482 (2016)
12. Wang, D., Li, S., Guo, F., Ning, K., Wang, L.: Core-genome scaffold comparison reveals the prevalence that inversion events are associated with pairs of inverted repeats. BMC Genomics **18**(1), 1–13 (2017)
13. Wang, D., Li, J., Wang, L.: Comprehensive study of instable regions in pseudomonas aeruginosa and mycobacterium tuberculosis. Biomed. Eng. Online **17**(1), 1–14 (2018)
14. Zhang, Y.c., Che, M., Ma, J.: Analysis of the longest common substring algorithm. Comput. Simul. **12**, 025 (2007)

An LP-Rounding Based Algorithm for Soft Capacitated Facility Location Problem with Submodular Penalties

Hanyin Xiao, Jiaming Zhang, Zhikang Zhang[✉], and Weidong Li

School of Mathematics and Statistics, Yunnan University,
Kunming, People's Republic of China
zhikang2024@163.com

Abstract. The soft capacitated facility location problem (SCFLP) is a classic combinatorial optimization problem with widespread applications in operations research and computer science. Given a facility set $\mathcal{F}$ and a client set $\mathcal{D}$, each facility i has a capacity u_i and an open cost f_i, allowing to open multiple times, and each client j has a demand d_j. The SCFLP is to find a facility subset in $\mathcal{F}$ and connect each client to the facilities opened, such that the total cost including open cost and connection cost is minimized. SCFLP is NP-hard, motivating research into approximation algorithms. This paper considers a variant: the soft capacitated facility location problem with submodular penalties (SCFLPSP), where some clients may remain unserved at a penalty cost. We focus on the integer splittable case, where a client's demand can be served by multiple facilities in integer amounts. Using LP-rounding, we propose a $(\lambda R + 4)$-approximation algorithm, where $R = \frac{\max_{i \in \mathcal{F}} f_i}{\min_{i \in \mathcal{F}} f_i}, \lambda = \frac{R + \sqrt{R^2 + 8R}}{2R}$. Notably, when the open cost is uniform, the approximation ratio is 6.

Keywords: Soft capacitated facility location problem · Approximation algorithm · Submodular penalties · LP-rounding

1 Introduction

The facility location problem is a fundamental model in operations research and computer science, with numerous variants proposed over time [22]. This paper focuses on two key variants: the soft capacitated facility location problem (SCFLP) and the uncapacitated facility location problem with submodular penalties (UFLPSP).

In the SCFLP, each facility $i \in \mathcal{F}$ has a capacity u_i, representing the maximum service it can serve. Facilities can be opened multiple times, each incurring an additional opening cost and increasing capacity proportionally. For the uniform capacities case, Chudak and Shmoys [3] proposed a 3-approximation algorithm using LP-rounding. For the general case, Jain and Varizani [10] reduced

SCFLP to the uncapacitated facility location problem (UFLP) and used the primal-dual technique to obtain a 4-approximation algorithm. Arya et al. [2] improved this ratio to 3.732 via local search. Following the technique of Jain and Vazirani [10], Jain et al. [9] presented a 3-approximation algorithm. Mahdian et al. [20] further improved the bound to 2.89 via dual-fitting and later achieved the best-known ratio of 2 through reduction and a dual-fitting approach [21]. This bound is tight due to the integrality gap of 2 in the problem's LP relaxation. For more general settings, Alfandari [1] adapted the set-covering greedy heuristic, achieving a logarithmic approximation ratio of $(1+\epsilon)H(n)$, where H is the harmonic series and n is the number of clients. Han et al. introduced a $(3\tau+1)$-approximation algorithm for the τ-relaxed SCFLP, where the connection cost satisfied the τ-relaxed triangle inequality, a relaxation of the triangle inequality [6].

Recently, classical combinatorial optimization problems with submodular penalties have received increasing attention [8,15–19]. In the UFLPSP, a penalty function $h: 2^{\mathcal{D}} \to \mathbb{R}_+$ determines the cost of leaving certain clients unserved, which is submodular. Hayrapetyan et al. [8] showed that an LP-based α-approximation algorithm for UFLP leads to a $(1+\alpha)$-approximation algorithm for UFLPSP. Consequently, according to [12], a 2.488-approximation algorithm can be obtained. Chudak and Nagano [4] developed an efficient $(1+\varepsilon)(1+\alpha)$-approximation algorithm for UFLPSP via convex relaxation, leading to a faster $(2.488+\varepsilon)$-approximation. Du et al. [5] proposed a primal-dual 3-approximation algorithm. Li et al. [14] further improved the above approximation ratio to 2.375 by combining primal-dual techniques with greedy augmentation techniques. Currently, the best ratio for this problem is 2, obtained by Li et al. [13] using LP-rounding technique. They also provided a general framework for addressing the submodular penalty covering problem using this technique.

In this paper, we consider the soft capacitated facility location problem with submodular penalties (SCFLPSP), where each facility has a capacity and each client may remain unserved at a penalty cost. We focus on the integer splittable demand case. In the following, Sect. 2 introduces SCFLPSP and formulates the problem. Section 3 presents and analyzes an algorithm for the fractional splittable case. In addition to this, the integer splittable case can be extended by network flow algorithms. Section 4 concludes the study.

2 Preliminaries

2.1 Problem Statement

Given a set $\mathcal{F}$ of facilities and a set $\mathcal{D}$ of clients. Each client $j \in \mathcal{D}$ has a demand $d_j \in \mathbb{Z}_+$ and each facility has a capacity constraint, which serves at most $u_i \in \mathbb{Z}_+$ demand, where each facility can be opened more than once. The open cost for each facility $i \in \mathcal{F}$ is $f_i \in \mathbb{Z}_+$. The unit connection cost for connecting client $j \in \mathcal{D}$ to facility $i \in \mathcal{F}$ is c_{ij}, which is metric in $\mathcal{F} \cup \mathcal{D}$. For each client subset $S \subseteq \mathcal{D}$, there exists a penalty function $h(S)$, satisfying the following conditions:

(1) $h(\emptyset) = 0$; (2) $h(\cdot)$ is nondecreasing; (3) $h(\cdot)$ is submodular, i.e.,

$$h(S \cap T) + h(S \cup T) \leq h(S) + h(T), \quad \forall S, T \subseteq \mathcal{D}.$$

The soft capacitated facility location problem with submodular penalties (SCFLPSP) is to open a facility subset such that each client is either served by an open facility or penalized, with the objective to minimize the total cost, including open cost, connection cost and penalty cost. This paper focuses on the integer splittable case, where a client's demand can be served by multiple facilities in integer amounts. Throughout the paper, we use the following notations: $n = |\mathcal{D}|$ and $m = |\mathcal{F}|$. For simplicity, clients in $\mathcal{D}$ are indexed as $\{1, 2, \cdots, n\}$.

To formulate integer splittable SCFLPSP, we introduce integer variable x_{ij} to indicate the amount that the demand of client j served by facility i, integer variable y_i to indicate the number of times facility i is opened, and binary variable z_S to indicate whether client subset $S \subseteq \mathcal{D}$ is penalized. The integer linear program can be obtained as follows.

$$IP(1): \min \quad \sum_{i \in \mathcal{F}} f_i y_i + \sum_{i \in \mathcal{F}} \sum_{j \in \mathcal{D}} c_{ij} x_{ij} + \sum_{S \subseteq \mathcal{D}} h(S) z_S$$

$$\text{s.t.} \quad \sum_{i \in \mathcal{F}} \frac{x_{ij}}{d_j} + \sum_{S \subseteq \mathcal{D}: j \in S} z_S \geq 1, \quad \forall j \in \mathcal{D}, \tag{1}$$

$$\frac{x_{ij}}{d_j} \leq y_i, \quad \forall i \in \mathcal{F}, j \in \mathcal{D}, \tag{2}$$

$$\sum_{j \in \mathcal{D}} x_{ij} \leq u_i y_i, \quad \forall i \in \mathcal{F}, \tag{3}$$

$$x_{ij}, y_i \in \mathbb{Z}_+ \cup \{0\}, \quad \forall i \in \mathcal{F}, j \in \mathcal{D}, \tag{4}$$

$$z_S \in \{0, 1\}, \quad \forall S \subseteq \mathcal{D}, \tag{5}$$

where $h(S)$ is a nondecreasing submodular function and $h(\emptyset) = 0$. The constraint (1) means that the demand of each client is either satisfied or rejected for service. The constraint (2) means that if the fractional demand of a client is served by facility i, facility i must be open. The constraint (3) means that the total amount of demand served by facility i cannot exceed its capacity u_i.

The relaxed form $LP(1)$ of $IP(1)$ can be obtained by relaxing the integer constraints on the variables, with the main difference being that constraints (4) and (5) are replaced by the following constraints:

$$x_{ij}, y_i, z_S \geq 0, \quad \forall i \in \mathcal{F}, j \in \mathcal{D}, S \subseteq \mathcal{D}. \tag{6}$$

2.2 Single Node Soft Capacitated Facility Location Problem

In the following section, we use single node soft capacitated facility location problem (SNSCFLP) to determine how to open facilities. In the SNSCFLP, we are given only one client and a facility set F'. Each facility $i \in F'$ has an open cost f_i and a capacity u_i, where facilities can be opened more than once. The total demand from this client is D'. The distance from the client to facility i is denoted as c_i, only depending on the facilities. The objective of the SNSCFLP is to open some facilities to serve this client in order to minimize the total distance.

By introducing variables y'_i to indicate number of times facility i is opened, and x'_i to indicate the amount of demand served by facility i for this client. The SNSCFLP can be written as the following linear program relaxation.

$$LP(2): \min \quad \sum_{i \in F'} f_i y'_i + \sum_{i \in F'} c_i x'_i$$

$$\text{s.t.} \quad \sum_{i \in F'} x'_i \geq D', \tag{7}$$

$$x'_i \leq u_i y'_i, \quad \forall i \in F', \tag{8}$$

$$x'_i, y'_i \geq 0, \quad \forall i \in F'. \tag{9}$$

The optimal solution can be obtained from Algorithm 1. Compared to [11], the biggest difference lies in the absence of opening times constraints in SNSCFLP, that is, without the constraint of $y'_i \leq 1$ in the program $LP(2)$.

Algorithm 1: Greedy Algorithm

1 **initialization:** set $x_i^{(2)} = 0, y_i^{(2)} = 0, \forall i \in F'$.
2 select $i \in F'$ with the minimum value of $\frac{f_i}{u_i} + c_i$, and set $x_i^{(2)} = D'$, $y_i^{(2)} = \frac{D'}{u_i}$.

Lemma 1. *An optimal solution $(x^{(2)}, y^{(2)})$ for SNSCFLP can be obtained through Algorithm 1, and in the optimal solution, at most one facility $i \in F'$ with the opening times $y_i^{(2)} > 0$.*

Due to space constraints, we omit most of the proofs of the lemmas in this paper.

3 Algorithms and Analysis

In this section, we first design a polynomial-time algorithm for the fractional splittable case, where a client's demand can be served by multiple facilities in fractional amounts. Next, a feasible solution can be obtained using a network flow algorithm for integer splittable SCFLPSP. This solution performs at least as well as the fractional case, enabling an analysis of the approximation ratio.

For the fractional splittable SCFLPSP, we introduce variable x_{ij} to indicate the amount that the demand of client j served by facility i, variable y_i to indicate the number of times facility i is opened, and variable z_S to indicate whether client subset $S \subseteq \mathcal{D}$ is penalized. The mixed-integer program $MIP(1)$ can be obtained by replacing the part of the constraints (4) of $IP(1)$ on x_{ij} with the following constraints:

$$x_{ij} \geq 0, \quad \forall i \in \mathcal{F}, j \in \mathcal{D}, \tag{10}$$

where $h(S)$ is a nondecreasing submodular function and $h(\emptyset) = 0$. Observe that $LP(1)$ is also the linear problem relaxation of $MIP(1)$.

We next define the penalty set $\hat{P}$ based on the relationship between the solution of the convex relaxation of $LP(1)$ and the solution of LP(1).

From the literature [4], the definition of Lovász extension function $h'(\cdot)$ of any given submodular function $h(\cdot)$ is

$$LP(4): h'(z) = \max \sum_{j \in \mathcal{D}} \eta_j z_j$$

$$\text{s.t.} \sum_{j \in S} \eta_j \leq h(S), \quad \forall S \subseteq \mathcal{D},$$

$$\eta_j \geq 0, \quad \forall j \in \mathcal{D},$$

where $z = (z_j)_{j \in \mathcal{D}} \in \mathbb{R}^n_+$. Moreover, we have the following properties of $h'(\cdot)$. Let $I(S)$ be the indicator vector of subset $S \subseteq \mathcal{D}$, i.e., $I(S) \in \{0, 1\}^n$ and its jth component is 1 if $j \in S$, and 0 otherwise for each $j \in \mathcal{D}$.

Property 1. $h'(I(S)) = h(S)$ and $h'(\mathbf{0}) = h(\emptyset)$.

Property 2. For any real number $a \geq 0$ and any vector $\boldsymbol{w} \geq 0$, $h'(a\boldsymbol{w}) = ah'(\boldsymbol{w})$.

Property 3. $h'(\cdot)$ is a monotonically increasing convex function.

Let $(x^{(1)}, y^{(1)}, z^{(1)})$ denote an optimal solution to $LP(1)$, used only for analysis. Define a new variable $\bar{z} = (\bar{z}_1, \bar{z}_2, \cdots, \bar{z}_n)$, where $\bar{z}_j = 1 - \sum_{i \in \mathcal{F}} \frac{x_{ij}^{(1)}}{d_j}$ for any client $j \in \mathcal{D}$. By Lemma 2.2 [13], the following lemma is immediately derived.

Lemma 2. $\bar{z}$ *constructed by* $(x^{(1)}, y^{(1)}, z^{(1)})$ *satisfies*

$$h'(\bar{z}) \leq \sum_{S \subseteq \mathcal{D}} h(S) z_S^{(1)}.$$

There is a convex relaxation for program $LP(1)$ presented as follows. We define the convex relaxation of $LP(1)$ as $LP(5)$. The main difference is to replace $\sum_{S \subseteq \mathcal{D}} h(S) z_S$ and $\sum_{S \subseteq \mathcal{D}: j \in S} z_S$ in $LP(1)$ by $h'(z)$ and z_j, where $h'(z)$ is defined in $LP(4)$.

Lemma 3. $(x^{(1)}, y^{(1)}, \bar{z})$ *constructed by* $(x^{(1)}, y^{(1)}, z^{(1)})$ *is an optimal solution of program* $LP(5)$.

Algorithm 2 defines penalized client set $\hat{P}$. Define a parameter $\delta = \frac{1}{\lambda R + 4} \in [0, 1]$, where $R = \frac{\max_{i \in \mathcal{F}} f_i}{\min_{i \in \mathcal{F}} f_i}$, $\lambda = \frac{R + \sqrt{R^2 + 8R}}{2R}$.

Notice that, in Algorithm 2, we can get $z_j^{(5)} = 1 - \sum_{i \in \mathcal{F}} \frac{x_{ij}^{(5)}}{d_j}$ for any $j \in \mathcal{D}$, because $(x^{(5)}, y^{(5)}, z^{(5)})$ is an optimal solution of $LP(5)$,

The following lemma bounds the cost of penalized client set.

Algorithm 2: Establishing the penalized client set
---
1 solve the program $LP(5)$ to obtain an optimal solution $(x^{(5)}, y^{(5)}, z^{(5)})$.
2 set the penalized client set $\hat{P} = \{j \in \mathcal{D} : z_j^{(5)} \geq \delta\}$, $\delta = \frac{1}{\lambda R + 4}$, $\lambda = \frac{R + \sqrt{R^2 + 8R}}{2R}$.
3 **for** $S \subseteq \mathcal{D}$ **do**
4 $\quad$ set $\hat{z}_S = 1$ if $S = \hat{P}$, and $\hat{z}_S = 0$ otherwise.

Lemma 4. *The cost of the penalized client set defined by Algorithm 2 is no more than* $\frac{1}{\delta} h'(z^{(5)})$.

Proof. Consider replacing z_j with $z_j^{(5)}$ for any $j \in \mathcal{D}$ in program $LP(4)$. Without loss of generality, we assume that $0 \leq z_1^{(5)} \leq z_2^{(5)} \leq \cdots \leq z_n^{(5)} \leq 1$, where n is the number of clients. Furthermore there exists exactly l nonzero and positive numbers that are not equal such that $0 < z_{k_1}^{(5)} < z_{k_2}^{(5)} < \cdots < z_{k_l}^{(5)} < 1$ in $\{z_j^{(5)}\}_{j=1}^n$ which satisfy $\cup_{j=1}^n \{z_j^{(5)}\} = (\cup_{i=k_1}^{k_l} \{z_i^{(5)}\}) \cup \{0\}$, and for each $j \in \mathcal{D}$ such that $z_j^{(5)} > 0$, there is a unique index $\phi(j) \in \{1, \cdots, l\}$ such that $z_j^{(5)} = z_{k_{\phi(j)}}^{(5)}$. Let $S_i = \{j \in \mathcal{D} : z_j^{(5)} \geq z_{k_i}^{(5)}\}, i = 1, 2, \cdots, l$. Then these sets meet that $S_l \subset S_{l-1} \subset \cdots \subset S_1 \subset \mathcal{D}$. Let $z_{k_0}^{(5)} = 0$.

By the Lemma 3.1 in [7], we can establish an optimal solution $\eta^{(4)}$ of program $LP(4)$, i.e. $\eta^{(4)} = (\eta_1^{(4)}, \cdots, \eta_n^{(4)})$, where $\eta_j^{(4)} = h(\{j, j+1, \cdots, n\}) - h(\{j+1, j+2, \cdots, n\}), j = 1, \cdots, n$.

Combining the optimality of $\eta^{(4)}$ and the properties of $h'(\cdot)$, we have

$$h(S_i) = h'(I(S_i)) = \sum_{j \in S_i} \eta_j^{(4)}, \forall S_i \subset \mathcal{D}, i = 1, \cdots, l. \tag{11}$$

Thus, we can get

$$h(\hat{P}) = h'(I(\hat{P})) = \sum_{j \in \hat{P}} \eta_j^{(4)} = \sum_{j \in \{j \in \mathcal{D}: z_j^{(5)} \geq \delta\}} \eta_j^{(4)} \leq \sum_{j \in \{j \in \mathcal{D}: z_j^{(5)} \geq \delta\}} \frac{z_j^{(5)}}{\delta} \eta_j^{(4)}$$

$$\leq \frac{1}{\delta} \sum_{j \in \mathcal{D}} z_j^{(5)} \eta_j^{(4)} = \frac{1}{\delta} h'(z^{(5)}).$$

The first equation holds because of Property 1 of $h'(\cdot)$; the second equation holds because of the optimality of $\eta^{(4)}$. □

After defining the penalized client set, we only need to deal with the remaining client set $D = \mathcal{D} \setminus \hat{P}$ and facility set $F = \{i \in \mathcal{F} : y_i > 0\}$. Thus, we construct the linear program $LP(6)$.

$$LP(6): \min \quad \sum_{i\in F} f_i y_i + \sum_{i\in F}\sum_{j\in D} c_{ij} x_{ij}$$

$$\text{s.t.} \quad \sum_{i\in F} \frac{x_{ij}}{d_j} \geq 1-\delta, \quad \forall j \in D, \tag{12}$$

$$\frac{x_{ij}}{d_j} \leq y_i, \quad \forall i \in F, j \in D, \tag{13}$$

$$\sum_{j\in D} x_{ij} \leq u_i y_i, \quad \forall i \in F, \tag{14}$$

$$x_{ij}, y_i \geq 0, \quad \forall i \in F, j \in D. \tag{15}$$

The dual program of $LP(6)$ is

$$DP(2): \max \quad \sum_{j\in D}(1-\delta)\alpha_j$$

$$\text{s.t.} \quad \frac{\alpha_j}{d_j} \leq c_{ij} + \frac{\beta_{ij}}{d_j} + \gamma_i, \quad \forall i \in F, j \in D,$$

$$\sum_{j\in D} \beta_{ij} \leq f_i - u_i \gamma_i, \quad \forall i \in F,$$

$$\alpha_j, \beta_{ij}, \gamma_i \geq 0, \quad \forall i \in F, j \in D.$$

Let $(x^{(5)}, y^{(5)}, z^{(5)})$ and $(x^{(6)}, y^{(6)})$ be the optimal solutions of $LP(5)$ and $LP(6)$, respectively. From $LP(6)$, the following lemma holds.

Lemma 5. $(x^{(5)}, y^{(5)})$ *is the feasible solution of* $LP(6)$. *Moreover, we have*

$$cost_{LP(6)}(x^{(6)}, y^{(6)}) \leq cost_{LP(6)}(x^{(5)}, y^{(5)}).$$

Let $(\alpha^{(2)}, \beta^{(2)}, \gamma^{(2)})$ be an optimal solution of $DP(2)$. Now, we establish some clusters centered around client $j \in D$ and some facilities near by the center (see Algorithm 3). Let C denote the current cluster centers, U denote the candidate center set, N_k denote the neighbors of $k \in U$, $F_j = \{i \in F : x_{ij}^{(6)} > 0\}$.

Note that, in Algorithm 3, when facility i allocate to any cluster (k, N_k), $k \in C$, we increase the demand associated with this cluster by adding to it all of the fractional demand served by facility i in $(x^{(6)}, y^{(6)})$. Thus, the amount of demand in the cluster (k, N_k) is $D_k = \sum_{i\in N_k}\sum_{j\in D} x_{ij}^{(6)}$.

For convenience, in the following, we define N_j as follows: for $j \in C$, N_j is the set obtained at the end of the algorithm, and for $j \notin C$, N_j denotes the set when j is deleted from U in the Algorithm 3 Step 8. Additionally, let $j^*(i)$ denote the cluster center of facility $i \in F$. Finally, we can obtain some clusters (k, N_k), $k \in C$ with the following properties.

Property 4. The cluster N_k, $k \in C$ is a partition of facility set F, that is, $F = \cup_{k\in C} N_k$ and $N_j \cap N_k = \emptyset$ for any $j \neq k \in C$.

Algorithm 3: Clustering

1 **initialization:** $C = \emptyset, U = D, N_k = F_k$.
2 **while** $U \neq \emptyset$ **do**
3   $\quad$ select $k \in U$ with the minimum of $\frac{\alpha_k^{(2)}}{d_k}$;
4   $\quad$ let $C = C \cup \{k\}, U = U \backslash \{k\}$;
5   $\quad$ update $N_j = N_j \backslash (N_k \cup \{i \in N_j : c_{ij} > c_{ik}\})$ for any $j \in U$;
6   $\quad$ update $U = U \backslash \{j \in U : \sum_{i \in N_j} \frac{x_{ij}^{(6)}}{d_j} < \frac{1}{\lambda}(1-\delta)\}$, where $\lambda = \frac{R + \sqrt{R^2 + 8R}}{2R}$, $R = \frac{\max_{i \in \mathcal{F}} f_i}{\min_{i \in \mathcal{F}} f_i}$.
7 **if** $U = \emptyset$ **then**
8   $\quad$ set $\widetilde{F} = F - \cup_{k \in C} N_k$;
9   $\quad$ allocate $i \in \widetilde{F}$ to the nearest cluster center $k \in C$, i.e. $k = \arg\min_{j \in C} c_{ij}$.

Property 5. For any $k \in C$, we have $\sum_{i \in N_k} y_i^{(6)} \geq \sum_{i \in N_k} \frac{x_{ik}^{(6)}}{d_j} \geq \frac{1}{\lambda}(1-\delta)$.

Next, we will construct a solution $(\hat{x}, \hat{y})$ for $LP(6)$ by Algorithm 4 and Algorithm 5. For any $i \in \mathcal{F}$ and $j \in \mathcal{D}$, $\hat{x}_{ij}$ and $\hat{y}_i$ are initialized to 0.

Algorithm 4 determines the opening times of facilities among the clusters by constructing a SNSCFLP instance for any $(k, N_k), k \in C$.

Algorithm 4: Opening facilities

1 **for** $k \in C$ **do**
2   $\quad$ construct a SNSCFLP instance: the only client is the center k with the total demand $D_k = \sum_{i \in N_k} \sum_{j \in D} \frac{x_{ij}^{(6)}}{1-\delta}$, note that $\frac{x_{ij}^{(6)}}{1-\delta} \leq d_j$ because $(x^{(6)}, y^{(6)})$ is an optimal solution of $LP(6)$; the facility set is N_k, each facility $i \in F$ with the opening cost f_i and the capacity u_i; the cost of facility i to the center is $c_i = c_{ik}$;
3   $\quad$ solve the SNSCFLP instance to obtain an optimal solution (w^k, v^k) by Algorithm 1.
4 **for** $k \in C$ **do**
5   $\quad$ **for** $i \in N_k$ **do**
6   $\quad\quad$ set $\hat{y}_i = \lceil v_i^k \rceil$.

We round up the opening times of facilities in each cluster (k, N_k) of the SNSCFLP instance, potentially adding one extra opening for each facility $i \in N_k$ with $v_i^k > 0$. By Lemma 1, only one facility in cluster (k, N_k) is opened, with an open cost of f^k. The following lemma bounds the extra cost.

Lemma 6. *For any $k \in C$, the extra open cost of the facility opened in the SNSCFLP instance defined in the cluster (k, N_k) is no more than $\frac{\lambda R}{1-\delta} \sum_{i \in N_k} f_i y_i^{(6)}$.*

Algorithm 5: Connecting clients

1 **for** $k \in C$ **do**
2 $\quad$ solve the following system to obtain solution $\hat{x}_{ij}$:
$\quad\{\hat{x}_{ij} : \sum_{i \in N_k} \hat{x}_{ij} = \sum_{i \in N_k} \frac{x_{ij}^{(6)}}{1-\delta}, \forall j \in D; \sum_{j \in D} \hat{x}_{ij} = w_i^k, \forall i \in N_k\}.$

We open enough facilities to satisfy the demand of clients in Algorithm 4. Next, we construct a connection $\hat{x}_{ij}$ to connect client $j \in D$ to facility $i \in F$.

Using Algorithm 2 $\sim$ Algorithm 5, We derive a feasible solution $(\hat{x}, \hat{y}, \hat{z})$ for $MIP(1)$. In Algorithm 4, we determine how to open the facilities to meet the demand of clients. It suffices to compute a minimum cost integer connection $\overline{x}$ from clients in D to F. Clearly, $(\overline{x}, \hat{y}, \hat{z})$ is a feasible solution for $IP(1)$ with the total cost no more than the total cost of $(\hat{x}, \hat{y}, \hat{z})$. Suppose $opt1$ is the optimal value of $MIP(1)$. For convenience, in the later sections, denote $\bar{N}_j = F_j \setminus N_j$. The following two lemmas are essentially implied in [11].

Lemma 7. *For any $j \in C, i \in F$, if $x_{ij}^{(6)} > 0$, then $c_{ij^*(i)} \leq \frac{\alpha_j^{(2)}}{d_j}$.*

Lemma 8. *For any $j \notin C, i \in F$, if $x_{ij}^{(6)} > 0$, then we have*

(1) If $i \in N_j$, then $c_{ij^*(i)} \leq c_{ij} + c_{i^*(j)j} + \frac{\alpha_j^{(2)}}{d_j}$, where $i^*(j) = \arg\min_{i \in \bar{N}_j}\{c_{ij}\}$;

(2) If $i \in \bar{N}_j$, then $c_{ij^*(i)} \leq \frac{\alpha_j^{(2)}}{d_j}$.

We need to estimate the optimal value of each SNSCFLP instance in the cluster (k, N_k) since SNSCFLP serves as a subproblem for determining the opening times of facilities.

Lemma 9. *For each cluster (k, N_k), let O_k be the optimal value for the corresponding SNSCFLP instance. We have*

$$O_k \leq \frac{1}{1-\delta}\left(\sum_{i \in N_k} f_i y_i^{(6)} + \sum_{j \in D}\sum_{i \in N_k} c_{ik} x_{ij}^{(6)}\right).$$

The following lemma estimates the total cost of any cluster (k, N_k).

Lemma 10. *For each (k, N_k), we have*

$$\sum_{i \in N_k} f_i \hat{y}_i + \sum_{j \in D}\sum_{i \in N_k} c_{ij} \hat{x}_{ij} \leq \frac{\lambda R + 1}{1-\delta}\sum_{i \in N_k} f_i y_i^{(6)} + \frac{1}{1-\delta}\sum_{j \in D}\sum_{i \in N_k} c_{ij} x_{ij}^{(6)}$$
$$+ \frac{2}{1-\delta}\sum_{j \in D}\sum_{i \in N_k} c_{ik} x_{ij}^{(6)}.$$

Theorem 1. *For the fractional splittable SCFLPSP, there exists a $(\lambda R + 4)$-approximation algorithm.*

Proof. Using Lemma 10 and summing all $k \in C$, we have

$$\sum_{i \in F} f_i \hat{y}_i + \sum_{j \in D} \sum_{i \in F} c_{ij} \hat{x}_{ij} \le \frac{\lambda R + 1}{1 - \delta} \sum_{i \in F} f_i y_i^{(6)} + \frac{1}{1 - \delta} \sum_{j \in D} \sum_{i \in F} c_{ij} x_{ij}^{(6)}$$
$$+ \frac{2}{1 - \delta} \sum_{j \in D} \sum_{i \in F} c_{ij^*(i)} x_{ij}^{(6)}.$$

Considering the last term of the RHS in the above equation, we need to estimate the distance from i to $j^*(i)$, that is, $c_{ij^*(i)}$. We divide all unpenalized clients D into two sets: the center set C and the non-center set $D \backslash C$, thus we have

$$\sum_{j \in D} \sum_{i \in F} c_{ij^*(i)} x_{ij}^{(6)} = \sum_{j \in C} \sum_{i \in F} c_{ij^*(i)} x_{ij}^{(6)} + \sum_{j \in D \backslash C} \sum_{i \in F} c_{ij^*(i)} x_{ij}^{(6)}.$$

By the Lemma 7 and Lemma 8, we have

$$\sum_{j \in D} \sum_{i \in F} c_{ij^*(i)} x_{ij}^{(6)} \le \sum_{j \in D} \sum_{i \in F} \frac{\alpha_j^{(2)}}{d_j} x_{ij}^{(6)} + \sum_{j \in D \backslash C} \sum_{i \in N_j} c_{ij} x_{ij}^{(6)} + \sum_{j \in D \backslash C} \sum_{i \in N_j} c_{i^*(j)j} x_{ij}^{(6)}.$$

Moreover, if $j \in D \backslash C$, then we get $\sum_{i \in N_j} \frac{x_{ij}^{(6)}}{d_j} < \frac{1}{\lambda}(1 - \delta)$. Since $c_{i^*(j)j}$ is the minimum in $\bar{N}_j$ which is less than the average value in $\bar{N}_j$. Thus,

$$\sum_{i \in N_j} c_{i^*(j)j} x_{ij}^{(6)} \le \frac{(1-\delta)d_j}{\lambda} c_{i^*(j)j} = \frac{(1-\delta)d_j}{\lambda} \min_{i \in \bar{N}_j} \{c_{ij}\}$$
$$\le \frac{1-\delta}{\lambda} \frac{\sum_{i \in \bar{N}_j} c_{ij} x_{ij}^{(6)}}{\sum_{i \in \bar{N}_j} \frac{x_{ij}^{(6)}}{d_j}} \le \frac{1}{\lambda - 1} \sum_{i \in \bar{N}_j} c_{ij} x_{ij}^{(6)}.$$

Thus, the total cost of connection and open of the solution $(\hat{x}, \hat{y}, \hat{z})$ is

$$\sum_{i \in F} f_i \hat{y}_i + \sum_{j \in D} \sum_{i \in F} c_{ij} \hat{x}_{ij}$$
$$\le \frac{\lambda R + 1}{1 - \delta} \sum_{i \in F} f_i y_i^{(6)} + \frac{1}{1 - \delta}(1 + \frac{2}{\lambda - 1}) \sum_{j \in D} \sum_{i \in F} c_{ij} x_{ij}^{(6)} + \frac{2}{1 - \delta} \sum_{j \in D} \sum_{i \in F} \frac{\alpha_j^{(2)}}{d_j} x_{ij}^{(6)}.$$

Since $(x^{(6)}, y^{(6)})$ is an optimal solution of $LP(6)$, meaning $\sum_{i \in F} \frac{x_{ij}^{(6)}}{d_j} = (1 - \delta)$, the total cost of the solution $(\hat{x}, \hat{y}, \hat{z})$ is

$$cost_{MIP(1)}(\hat{x}, \hat{y}, \hat{z}) = \sum_{i \in F} f_i \hat{y}_i + \sum_{j \in D} \sum_{i \in F} c_{ij} \hat{x}_{ij} + h(\hat{P})$$

$$\leq \frac{2}{1-\delta} \sum_{j \in D} \alpha_j^{(2)} \sum_{i \in F} \frac{x_{ij}^{(6)}}{d_j} + \frac{\lambda R + 1}{1-\delta} \sum_{i \in F} f_i y_i^{(6)} + \frac{1}{1-\delta}(1 + \frac{2}{\lambda - 1}) \sum_{j \in D} \sum_{i \in F} c_{ij} x_{ij}^{(6)}$$

$$+ h(\hat{P})$$

$$= 2 \sum_{j \in D} \alpha_j^{(2)} + \frac{\lambda R + 1}{1-\delta} \sum_{i \in F} f_i y_i^{(6)} + \frac{1}{1-\delta}(1 + \frac{2}{\lambda - 1}) \sum_{j \in D} \sum_{i \in F} c_{ij} x_{ij}^{(6)} + h(\hat{P}).$$

Since $DP(2)$ is the dual program of $LP(6)$, by the weak duality theorem,

$$\sum_{j \in D} \alpha^{(2)} \leq \frac{1}{1-\delta}(\sum_{i \in F} f_i y_i^{(6)} + \sum_{i \in F} \sum_{i \in D} c_{ij} x_{ij}^{(6)}).$$

Furthermore,

$$cost_{MIP(1)}(\hat{x}, \hat{y}, \hat{z}) \leq \frac{2}{1-\delta}(\sum_{i \in F} f_i y_i^{(6)} + \sum_{i \in F} \sum_{i \in D} c_{ij} x_{ij}^{(6)}) + \frac{\lambda R + 1}{1-\delta} \sum_{i \in F} f_i y_i^{(6)}$$

$$+ \frac{1}{1-\delta}(1 + \frac{2}{\lambda - 1}) \sum_{j \in D} \sum_{i \in F} c_{ij} x_{ij}^{(6)} + h(\hat{P})$$

$$= \frac{\lambda R + 3}{1-\delta} \sum_{i \in F} f_i y_i^{(6)} + \frac{1}{1-\delta}(3 + \frac{2}{\lambda - 1}) \sum_{i \in F} \sum_{j \in D} c_{ij} x_{ij}^{(6)} + h(\hat{P}).$$

Note that $R = \frac{\max_{i \in F} f_i}{\min_{i \in F} f_i}$ and $\lambda = \frac{R + \sqrt{R^2 + 8R}}{2R}$. Thus, $\lambda R = \frac{2}{\lambda - 1}$, we have

$$cost_{MIP(1)}(\hat{x}, \hat{y}, \hat{z}) \leq \frac{\lambda R + 3}{1-\delta}(\sum_{i \in F} f_i y_i^{(6)} + \sum_{i \in F} \sum_{j \in D} c_{ij} x_{ij}^{(6)}) + h(\hat{P})$$

$$= \frac{\lambda R + 3}{1-\delta} cost_{LP(6)}(x^{(6)}, y^{(6)}) + h(\hat{P}).$$

By the Lemma 5, we have

$$cost_{LP(6)}(x^{(6)}, y^{(6)}) \leq cost_{LP(6)}(x^{(5)}, y^{(5)}) = \sum_{i \in F} f_i y_i^{(5)} + \sum_{i \in F} \sum_{j \in D} c_{ij} x_{ij}^{(5)}.$$

Combining Lemma 4 and Lemma 3, and $\delta = \frac{1}{\lambda R + 4}$, we can get

$$cost_{MIP(1)}(\hat{x}, \hat{y}, \hat{z}) \leq \frac{\lambda R + 3}{1-\delta}(\sum_{i \in F} f_i y_i^{(5)} + \sum_{i \in F} \sum_{j \in D} c_{ij} x_{ij}^{(5)}) + \frac{1}{\delta} h'(z^{(5)})$$

$$\leq \max\{\frac{\lambda R + 3}{1-\delta}, \frac{1}{\delta}\}(\sum_{i \in F} f_i y_i^{(5)} + \sum_{i \in F} \sum_{j \in D} c_{ij} x_{ij}^{(5)} + h'(z^{(5)}))$$

$$= (\lambda R + 4)(\sum_{i \in \mathcal{F}} f_i y_i^{(1)} + \sum_{i \in \mathcal{F}} \sum_{j \in \mathcal{D}} c_{ij} x_{ij}^{(1)} + h'(\bar{z})).$$

By the Lemma 2,

$$cost_{MIP(1)}(\hat{x},\hat{y},\hat{z}) \leq (\lambda R + 4)(\sum_{i \in \mathcal{F}} f_i y_i^{(1)} + \sum_{i \in \mathcal{F}} \sum_{j \in \mathcal{D}} c_{ij} x_{ij}^{(1)} + \sum_{S \subseteq \mathcal{D}} h(S) z_S^{(1)})$$
$$= (\lambda R + 4) cost_{LP(1)}(x^{(1)}, y^{(1)}, z^{(1)})$$
$$\leq (\lambda R + 4) opt1.$$

As a consequence of Theorem 1, we conclude the following result.

Theorem 2. *For the integer splittable SCFLPSP, a $(\lambda R + 4)$-approximation algorithm can be obtained.*

4 Conclusion

This paper investigates the soft capacitated facility location problem with submodular penalties (SCFLPSP), allowing some clients to remain unserved at a penalty cost. We focus on the integer splittable demand case, where each client's demand is met by multiple facilities in integer amounts. Based on LP-rounding, we propose a $(\lambda R + 4)$-approximation algorithm, where $R = \frac{\max_{i \in \mathcal{F}} f_i}{\min_{i \in \mathcal{F}} f_i}$, $\lambda = \frac{R + \sqrt{R^2 + 8R}}{2R}$. In particular, when the open cost is uniform, the approximation ratio is 6. To validate this result, we first develop an approximation algorithm for the fractional splittable case, then extend it to the integer splittable case. Our analysis reveals certain loose inequalities, suggesting room for improvement in the approximation ratio through more refined techniques.

Acknowledgments. The work was supported in part by the Project of the Natural Science Foundation of Yunnan Province of China Grants No. 202501AS070076, No. 202501AS070170 and Research and Innovation Program of Yunnan University [Grants No. TM-23236822, No. TM-23236819, No. KC-242410224].

References

1. Alfandari, L.: Improved approximation of the general soft-capacitated facility location problem. RAIRO Oper. Res. 41(1), 83–93 (2007)
2. Arya, V., Garg, N., Khandekar, R., et al.: Local search heuristic for k-median and facility location problems. In: Proceedings of the Thirty-third Annual ACM Symposium on Theory of Computing. STOC 2001, pp. 21–29 Association for Computing Machinery, New York, NY, USA (2001). https://doi.org/10.1145/380752.380755
3. Chudak, F.A.; Shmoys, D.B. Improved approximation algorithms for the capacitated facility location problem. In: Proceedings of the 10th Annual ACM-SIAM symposium on Discrete algorithms. SODA 1999, pp. 875–876 Baltimore, MD, USA (1999). https://doi.org/10.5555/314500.315061

4. Chudak, F. A., Nagano, K.: Efficient solutions to relaxations of combinatorial problems with submodular penalties via the Lovász extension and non-smooth convex optimization. In: Proceedings of the Eighteenth Annual ACM-SIAM Symposium on Discrete Algorithms. SODA 2007, pp. 79–88 Society for Industrial and Applied Mathematics, USA (2007). https://doi.org/10.5555/1283383.1283393
5. Du, D., Lu, R., Xu, D.: A primal-dual approximation algorithm for the facility location problem with submodular penalties. Algorithmica **63**, 191–200 (2012)
6. Fernandes, C.G., Meira, L., Miyazawa, F.K., Pedrosa, L.: A systematic approach to bound factor-revealing LPs and its application to the metric and squared metric facility location problems. Math. Program. **153**, 655–685 (2015)
7. Fujishige, S.: Submodular Functions and Optimization, vol. 58. Elsevier (2005)
8. Hayrapetyan, A., Swamy, C., Tardos, É.: Network design for information networks. In: Proceedings of the Sixteenth Annual ACM-SIAM Symposium on Discrete Algorithms. SODA 2005, vol. 5, pp. 933–942 Society for Industrial and Applied Mathematics, USA (2005). https://doi.org/10.5555/1070432.1070567
9. Jain, K., Mahdian, M., Saberi, A.: A new greedy approach for facility location problems. In: Proceedings of the Thiry-fourth Annual ACM Symposium on Theory of Computing. STOC 2002, pp. 731–740 Association for Computing Machinery, New York, NY, USA (2002). https://doi.org/10.1145/509907.510012
10. Jain, K., Vazirani, V.V.: Approximation algorithms for metric facility location and k-median problems using the primal-dual schema and Lagrangian relaxation. J. ACM **48**(2), 274–296 (2001)
11. Levi, R., Shmoys, D.B., Swamy, C.: LP-based approximation algorithms for capacitated facility location. Math. Program. **131**(1), 365–379 (2012)
12. Li, S.: A 1.488 approximation algorithm for the uncapacitated facility location problem. Inf. Comput. **222**, 45–58 (2013)
13. Li, Y., Du, D., Xiu, N., Xu, D.: Improved approximation algorithms for the facility location problems with linear/submodular penalties. Algorithmica **73**, 460–482 (2015)
14. Li, Y., Du, D., Xiu, N., et al.: A combinatorial 2.375-approximation algorithm for the facility location problem with submodular penalties. Theor. Comput. Sci. **476**, 109–117 (2013)
15. Liu, X., Li, W.: An approximation algorithm for the-prize-collecting multicut problem in trees with submodular penalties. Math. Struct. Comput. Sci. **34**(3), 193–210 (2024)
16. Liu, X., Li, W.: Combinatorial approximation algorithms for the submodular multicut problem in trees with submodular penalties. J. Comb. Optim. **44**(3), 1964–1976 (2022)
17. Liu, X., Li, W.: Approximation algorithms for the multiprocessor scheduling with submodular penalties. Optim. Lett. **15**(6), 2165–2180 (2021). https://doi.org/10.1007/s11590-021-01724-1
18. Liu, X., Dai, H., Li, S., Li, W.: The k-prize-collecting minimum power cover problem with submodular penalties on a plane. SCIENTIA SINICA Informationis **52**(6), 947–959 (2022)
19. Liu, X., Li, W., Dai, H.: Approximation algorithms for the minimum power cover problem with submodular/linear penalties. Theoret. Comput. Sci. **923**, 256–270 (2022)
20. Mahdian, M., Ye, Y., Zhang, J.: Improved approximation algorithms for metric facility location problems. In: Jansen, K., Leonardi, S., Vazirani, V. (eds.) APPROX 2002. LNCS, vol. 2462, pp. 229–242. Springer, Heidelberg (2002). https://doi.org/10.1007/3-540-45753-4_20

21. Mahdian, M., Ye, Y., Zhang, J.: A 2-Approximation algorithm for the soft-capacitated facility location problem. In: Arora, S., Jansen, K., Rolim, J.D.P., Sahai, A. (eds.) Approximation, Randomization, and Combinatorial Optimization. Algorithms and Techniques. RANDOM APPROX 2003 2003. LNCS, vol. 2764. Springer, Berlin, Heidelberg (2003). https://doi.org/10.1007/978-3-540-45198-3_12
22. Xiao, H., Zhang, J., Zhang, Z., Li, W.: A survey of approximation algorithms for the universal facility location problem. Mathematics **13**(7), 1023 (2025)

Less-Excludable Mechanism for DAOs in Public Good Auctions

Jing Chen and Wentao Zhou[✉]

Department of Computer Science and Technology, Tsinghua University,
Beijing 100084, China
jchencs@tsinghua.edu.cn, zhouwt24@mails.tsinghua.edu.cn

Abstract. With the rise of smart contracts, decentralized autonomous organizations (DAOs) have emerged in public good auctions, allowing "small" bidders to gather together and enlarge their influence in high-valued auctions. However, models and mechanisms in the existing research literature do not guarantee *non-excludability*, which is a main property of public goods. As such, some members of the winning DAO may be explicitly prevented from accessing the public good. This side effect leads to regrouping of small bidders within the DAO to have a larger say in the final outcome. In particular, we provide a polynomial-time algorithm to compute the best regrouping of bidders that maximizes the total bidding power of a DAO. We also prove that such a regrouping is less-excludable, better aligning the needs of the entire DAO and the nature of public goods.

Next, notice that members of a DAO in public good auctions often have a positive externality among themselves. Thus we introduce a *collective factor* into the members' utility functions. We further extend the mechanism's allocation for each member to allow for *partial access* to the public good. Under the new model, we propose a mechanism that is incentive compatible in generic games and achieves higher social welfare as well as less-excludable allocations.

Keywords: Mechanism design · DAO · Public good auction · Non-excludability

1 Introduction

The combination of blockchains and auctions has become increasingly prevalent in recent years, because blockchain features such as decentralization, data immutability and smart contracts naturally provide convenience for auctions. A representative example is the emergence of Decentralized Autonomous Organizations (DAOs) [5,6] and their application in public good auctions.

DAO is a new organizational framework in which management and operation rules are encoded in smart contracts on the blockchain. It can unite a group of people for a common goal in a decentralized format, such as fund-raising and auctions [13,22]. By 2022, the scale of DAOs had reached 1.6 million participants, collectively managing over 13,000 DAOs with a total value of $16 billion [14]. By 2024, the total treasury of DAOs reached $37.6 billion [9], a phenomenon that cannot be ignored in the real world.

In both on-chain and off-chain auctions, DAOs can be treated as bidders. Their internal structures allow them to gather more power from individuals and participate in the auction in a more efficient way. DAOs are especially closely associated with auctions of public goods, as public goods have two key characteristics, *non-rivalry* and *non-excludability* [15], which align well with the motivation of DAOs and the permissionless and egalitarian nature of blockchains. It's thus easy to form DAOs for such purposes. For example, in 2021, when Sotheby's auctioned off one of the original copies of the United States Constitution, more than 17,000 people formed the ConstitutionDAO and raised more than $47 million to join the auction by smart contracts deployed on Ethereum. Although the ConstitutionDAO narrowly failed to win the auction, this new form of participation has been recognized by traditional auction houses.

The participation of DAOs in an auction requires a two-level mechanism as suggested by [4]. The upper-level mechanism runs by the auctioneer and treats each DAO as an individual bidder participating in the auction. The lower-level mechanism runs by each DAO to aggregate the DAO's members and act on behalf of them. This includes collecting the members' bids to form a total bid that is submitted to the upper-level mechanism and, in case of winning the auction in the latter, distributing the access to the item and the corresponding total payment among members. The mechanism introduced by [4] is incentive compatible (IC) for each member of a DAO and is approximately optimal for the social welfare of the auction. However, it has to give up non-excludability in order to achieve IC and other good properties, which means that even when a DAO wins the auction, some small members of it still cannot get access to the item. This unfortunate side-effect puts the original intention of DAOs at question and leaves space for small members to form another (smaller) DAO among themselves so that they can collectively have a larger say in the larger DAO. Although game theory often assumes players' independent behavior, a world of DAOs intrinsically leverages the collective power of small members and one should take such collective behavior into consideration when studying DAOs.

1.1 Our Contributions

As non-excludability is a crucial characteristic of public goods, we try to highlight it in our study and balance it with other conflicting goals. With the model and the mechanism of [4] introduced in Sect. 2, we shall leverage the collective behavior of a DAO's members in two ways.

In Sect. 3, from an algorithmic point of view, we consider how the members of a DAO in the original mechanism may regroup themselves to build a better substructure within the DAO. Interestingly, such a collusive behavior doesn't undermine the power of the entire DAO but cooperatively helps the DAO to make a higher bid, besides improving its non-excludability. For the grouping problem, we construct an efficient algorithm and show that the optimal subgroups within a DAO can be computed in polynomial time. An open question is: why stopping at two levels? If some small members are still excluded from accessing the public good given the subgroups, they may continue this procedure to form a smaller DAO within the subgroups within the original DAO, etc., until they are influential enough in the small group and their power gets amplified by their groups in higher levels. Indeed, the grouping problem can have many levels. But since people usually do not think that far in their reasoning and actions, a two-layer grouping problem is our main focus in this study and we leave the hierarchical consideration to the future.

In Sect. 4, from a game-theoretic point of view, we generalize the standard quasi-linear utility model and add a collective factor to a DAO's members' utility function. This reflects the fact that members of a DAO in a public good auction may have a positive externality among themselves and care about whether others also get access to the item. Moreover, we extend the auction's allocation space from binary access to allowing for partial access, so that some small members who were completely prevented from accessing the item can now be given access to some extent. For example, a museum holding the auctioned public good (in case of an antique) may be entirely free to some members of the DAO and free for several months every year to other members. In reality, the format of the partial access can be flexible depending on the nature of the public good, such as lotteries, time-sharing or space-sharing. In the generalized model, we show a new mechanism that satisfies individual rationality, budget balance, and equal treatment, which are properties of the previous mechanism in [4], as well as incentive compatibility in *generic games*.[1] Compared with the previous mechanism, our mechanism improves both the non-excludability and the social welfare of the auction. This is a first step towards achieving full non-excludability in DAOs' public good auctions, an open problem that deserves further studies.

1.2 Related Work

Public Good Auctions. Many previous studies have shown by experiments that when public goods are involved, individuals are influenced by social factors and cannot be entirely modeled as self-interested agents [11,17]. The literature on the social efficiency of public good auctions is also extensive. For example, [12] proved that a winner-pays-only mechanism was not sufficiently effective, while a full payment mechanism offered certain advantages in the context of public goods fund-raising auctions. For a non-divisible public good which can be provided by different players at different costs, [16] defined a non-cooperative game and proposed an auction mechanism that is efficient, fair and incentive compatible, but does not have a uniquely determined payment for each player. The cooperative form was further studied by [10]. These studies took non-excludability as a given property and focused on other aspects of public good auctions. We instead highlight the impact of non-excludability and consider to what degree

---

[1] Recall that in game theory a generic game is where there is no critical tie.

it can be achieved, both algorithmically and via incentive-compatible mechanisms. Recently [18] modified the utility function in binary network public goods games to reflect various prosocial motivations such as altruism and collectivism that influence individual decision making. The collective utilities we consider for DAOs' members are inspired by their work, while our objectives are different.

DAOs in Auctions. Public good auctions with DAOs as participants were considered by [4] as a two-level mechanism. DAOs have their operation rules encoded in smart contracts and act accordingly in collecting member bids, computing the total bid and computing the final allocation and prices for the members. Thus only the members of DAOs are strategic, not the DAOs in between members and the auctioneer. In order to achieve incentive compatibility, their mechanism has to explicitly forbid some members of the winning DAO from accessing the good, thus violating non-excludability. In [19], the author considered a similar two-level model but focused on analyzing the equilibrium structure for DAOs participating in first- and second-price auctions, and only studied the scenario where a single DAO competed with an individual bidder. Moreover, [21] summarized many open problems about DAOs, including the design and analysis of incentive mechanisms for providing different types of public goods. Our work builds upon [4] and aims instead for less-excludable public good auctions.

Cooperative Game Theory. When resolving the conflict between incentive compatibility and non-excludability, we took inspiration from [3] for bankruptcy, even though the problems considered are different. Finally, the subgroups of DAOs' members resemble in spirit coalition forming in cooperative game theory [7,10,20], but we took a pure algorithmic approach to compute the optimal grouping. It will be worthy to further explore the strategic behavior when forming subgroups and establish deeper connections with concepts such as the core.

2 Preliminaries

Using the model in [4], an auctioneer is selling a single public good to DAOs and individual bidders. Since an individual bidder can be seen as a DAO with only one member, in the rest of the paper the auction is conducted with m DAOs as bidders. For a DAO G, we may slightly abuse notations and let G also denote the set of its members. Let $n_G = |G|$. When G is clear from the context, n is used for short and $G = [n]$. Each member $i \in [n]$ has a true value v_i for the good, which can also be seen as its individual *willingness to pay* (WTP). It makes a bid b_i to the DAO. Let v_{-i} and b_{-i} be the true value vector and the bid vector for *all members of all DAOs except* i.

The two-level mechanism $M = (M_u, M_\ell)$ of [4] can be described as follows. The upper-level mechanism M_u is a second-price mechanism run by the auctioneer, where each DAO acts as a bidder, the highest bidder wins the good and pays the second highest bid. Each DAO G makes a bid $WTP_{total,G}$ —the *willingness to pay* by the entire DAO, which is computed by the DAO from its members' individual bids and is not necessarily the sum of the members' true values or bids. The auctioneer collects all DAOs' bids and then informs each DAO G of its allocation $X_{total,G} \in \{0,1\}$ and price $P_{total,G}$. Again when G is clear from the context, the subscript G is removed from the notations.

The lower-level mechanism M_ℓ is run by each DAO G and contains three parts: an aggregation function $WTP_{total}(b_1, \ldots, b_n)$ that summarizes all members' bids into a total bid of the DAO and submits to the upper-level mechanism; an access function $x(b_1, \ldots, b_n, X_{total}, P_{total}) = (x_1, \ldots, x_n)$, where $\forall i \in [n]$, $x_i \in \{0,1\}$ represents whether member i receives access to the good, with $x_i = 0$ whenever $X_{total} = 0$; and a cost-sharing function $p(b_1, \ldots, b_n, X_{total}, P_{total}) = (p_1, \ldots, p_n)$, where $\forall i \in [n]$, p_i represents the amount that member i should pay. Notice that the public good can only be allocated to at most one *winning DAO* by the upper-level mechanism, but all members of the winning DAO are able to access it unless explicitly excluded from doing so. The utility function for each member i can be written as $u_i = v_i \cdot x_i - p_i$. The social welfare of the auction can be written as $SW = \sum_{i \in [n]} v_i \cdot x_i$, where the sum is taken over the members of the winning DAO. The optimal social welfare is defined as $OPT\_SW = \max_G \sum_{i \in G} v_i$, which assumes access by all members of the DAO that receives the good, inline with non-excludability.

The mechanism design problem focuses on the lower-level mechanism, more specifically the three functions WTP_{total}, x, and p. Here are some ideal properties for a mechanism to pursue: for every DAO G and for all instances of the auction,

1. *Incentive compatibility:* $u_i(v_i, b_{-i}) \geq u_i(b_i, b_{-i})$, $\forall i \in [n]$, $\forall b_i, b_{-i}$.
2. *Budget balance:* $\sum_{i \in [n]} p_i = P_{total}$.
3. *Individual rationality:* $u_i(v_i, b_{-i}) \geq 0$, $\forall i \in [n]$, $\forall b_{-i}$.
4. *Equal treatment:* if $b_i = b_j$, then $x_i = x_j$ and $p_i = p_j$, $\forall i, j \in [n]$, $i \neq j$.
5. *Non-excludability:* if $X_{total} = 1$ then $x_i = 1$ $\forall i \in [n]$.

The mechanism M_ℓ of [4] satisfies all properties above except non-excludability, and approximately maximizes the social welfare. More specifically, given a DAO G, without loss of generality the members are ordered according to their bids non-increasingly, so that $b_1 \geq b_2 \geq \cdots \geq b_n$. According to M_ℓ, G's bid is defined as $WTP_{total} = \max_{i \in [n]}\{i \cdot b_i\}$. If G wins in the upper mechanism (i.e., $X_{total} = 1$), then find $i^* = \max_{i \in [n]}\{i | i \cdot b_i \geq P_{total}\}$. The access function and the cost-sharing function are as follows: $\forall i \in [n]$,

$$x_i = \begin{cases} 1, & i \leq i^* \\ 0, & i > i^* \end{cases} \quad \text{and} \quad p_i = \begin{cases} \frac{P_{total}}{i^*}, & i \leq i^* \\ 0, & i > i^* \end{cases}. \tag{1}$$

If DAO G does not win the auction, then $x_i = 0$ and $p_i = 0$ $\forall i \in [n]$. As shown by [4], this mechanism is an H_ℓ-approximation to the optimal social welfare, namely $\frac{OPT\_SW}{SW} \leq H_\ell$ for all instances of the auction and truthful bids, where $\ell = \max_G |G|$ is the size of the largest DAO and $H_\ell = \sum_{i=1}^{\ell} \frac{1}{i}$ is the ℓ-th harmonic number.

The above highlights the advantages of the mechanism. However, its access function indicates that some members with low bids in the winning DAO may still be prevented from accessing the public good, giving up non-excludability. This makes it attempting for some members of a DAO to gather together to form subgroups, collectively pretending to be a single member of the DAO in order to gain greater influence in the outcome of the auction; see Fig. 1 for such an example. In particular, the other key characteristic of public goods, non-rivalry, indicates that the good can be consumed by multiple people without reducing the amount available to others.[2] Thus excluding some members from access is only a mean to achieve incentive compatibility and other properties above. Examining the problem from a different angle and highlighting the nature of public goods, we consider how non-excludability can be improved while other properties may be weakened.

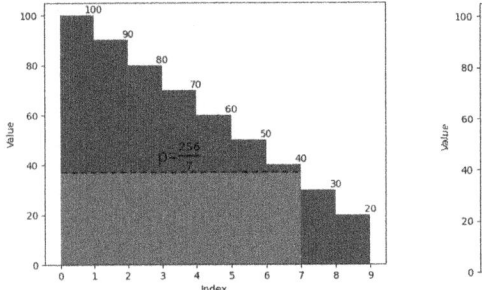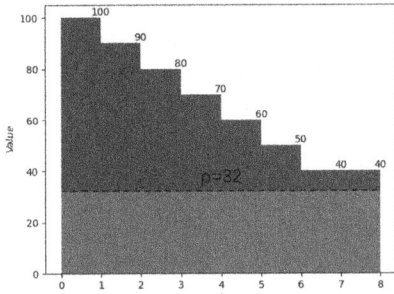

Fig. 1. In this example, a DAO has 9 members with valuations $100, 90, \ldots, 20$. Thus its $WTP_{total} = 300$. If it wins in the upper-level second-price auction with $P_{total} = 256$, then according to the mechanism, $i^* = 7$ and $p_i = \frac{256}{7}$, $\forall i \leq 7$. As the left figure shows, the vertical axis represents the amount of bid and the horizontal axis represents the member index. The shaded gray area represents $P_{total} = i^* \cdot \frac{P_{total}}{i^*}$. The last two members with valuations of 30 and 20 have no access to the good despite of being part of the DAO. However, if these two members form a subgroup with willingness to pay $WTP' = \max\{1 \times 30, 2 \times 20\} = 40$, then as the right figure shows, $i^* = 8$ and $WTP'_{total} = 320$, actually improving the entire DAO's willingness to pay. With the same price P_{total}, this subgroup can get access and the payment is 32. Back in the subgroup, members can now get access and the payment is 16 each, with non-excludability improved.

3 Subgroups in a DAO and Their Organizations

From an algorithmic perspective, in this section we consider how the members of a DAO may divide themselves into subgroups (e.g., some of them may form sub-DAOs) to gain more access to the public good. As we'll show, such a collective

---

[2] For example, an antique in a museum can be seen by one visitor without reducing how it shows up to other visitors.

behavior improves both non-excludability and the bidding power of the entire DAO. Since we are only interested in algorithms instead of strategic mechanisms here, we consider all members to bid at their true values, that is, $b_i = v_i$. Wlog, the members of a DAO are ordered according to their values non-increasingly, namely $v_1 \geq v_2 \geq \cdots \geq v_n$. Below we first introduce the model for a two-level DAO structure with subgroups and use an example to show that the DAO's total willingness to pay can be greatly improved by subgroups.

3.1 Two-Level DAOs

Inside a DAO G, its n members can be further partitioned into k subgroups that are disjoint from each other, referred to as a *grouping* of G and denoted by $\mathcal{G} = \{G^1, \ldots, G^k\}$, where $\bigcup_{j \in [k]} G^j = G$ and $G^j \cap G^{j'} = \emptyset, \forall j \neq j'$. Let $n^j = |G^j|$ for each $j \in [k]$. When focusing on a subgroup G^j, its members' valuations are also referred to as $v_1^j, \ldots, v_{n^j}^j$. Again wlog, the members of G^j are ranked by their valuations non-increasingly, namely $v_1^j \geq v_2^j \geq \cdots \geq v_{n^j}^j$. So a member $a \in [n]$ has the a-th highest valuation among all members of G and, if a belongs to the subgroup G^j and has the t-th highest valuation in G^j, then $v_a = v_t^j$.

Each subgroup's valuation (or rather, its willingness to pay) is determined by the valuations of its members, and the DAO's valuation/willingness-to-pay is determined by those of its subgroups. Equal treatment will be maintained across all subgroups and within each subgroup. More specifically, for each subgroup G^j, its valuation is $WTP^j = \max_{i \in [n^j]} \{i \cdot v_i^j\}$.

From the DAO's perspective, there are k "members", $G^1, \ldots, G^k$, whose valuations are $WTP^1, \ldots, WTP^k$. Again these subgroups are ordered according to their valuations non-increasingly, namely $WTP^1 \geq WTP^2 \geq \cdots \geq WTP^k$. The entire DAO's willingness to pay is now $WTP_{total} = \max_{j \in [k]} \{j \cdot WTP^j\}$, also written as $WTP_{total}(\mathcal{G})$ in order to emphasize the grouping $\mathcal{G}$.

If DAO G wins the item at price P_{total} with bid WTP_{total}, then for each of the two levels in the DAO, the access rule and the cost-sharing rule follow the original mechanism. That is, firstly the subgroups of $\mathcal{G}$ get accesses and cost-shares according to the original mechanism as if they were individual members of the DAO. Then for each subgroup G^j that has access to the item with a price p, the same rules apply to its members as if they form a separate DAO.

As the example in Fig. 2 shows, by forming subgroups, more members can contribute and the willingness to pay of the entire DAO may be greatly improved. Let the maximum willingness to pay of DAO G over all possible groupings be $opt\_WTP = \max_{k=1}^{n} \max_{\mathcal{G} = \{G^1, \ldots, G^k\}} WTP_{total}(\mathcal{G})$.

In Sects. 3.2 and 3.3, we show that the optimal grouping whose willingness to pay reaches opt_WTP enjoys a nice structure and provide a polynomial time algorithm for it. We prove the optimality of the algorithm and show that it improves non-excludability compared with the original DAO without subgroups.

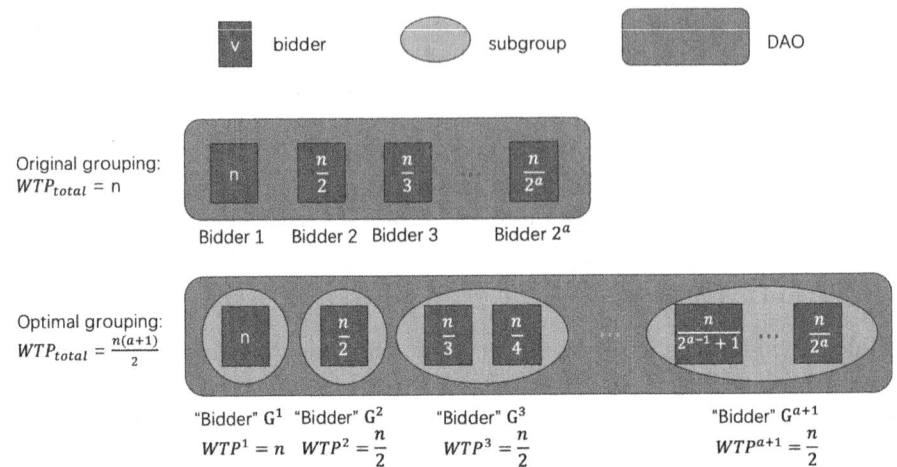

Fig. 2. In this example, the DAO has $n = 2^a$ members with valuations of $n, \frac{n}{2}, \frac{n}{3}, \ldots, \frac{n}{2^a-1}, \frac{n}{2^a}$. Without any subgroup, $WTP_{total} = n$. If the members are partitioned into the following $a+1$ subgroups, $\{1\}, \{2\}, \{3, 4\}, \ldots, \{2^{a-1}+1, \ldots, 2^a = n\}$, with corresponding valuations $\{n\}, \{\frac{n}{2}\}, \{\frac{n}{3}, \frac{n}{4}\}, \ldots, \{\frac{n}{2^{a-1}+1}, \ldots, \frac{n}{2^a}\}$, then the total willingness to pay of the DAO becomes $\frac{n(a+1)}{2}$, improved by a logarithmic factor.

3.2 The Optimal Grouping Algorithm

We first identify some properties for the optimal grouping.

Definition 1. *Given a grouping $\mathcal{G} = \{G^1, \ldots, G^k\}$, if for some G^i, $G^j \in \mathcal{G}$, $G^i \neq G^j$, $\exists$ members $a, c \in G^i$, $b \in G^j$, such that the indices satisfy $a < b < c$, then we say that the grouping $\mathcal{G}$ has a **crossing**. A grouping $\mathcal{G}$ is **continuous** if it does not have any crossing.*

Notice that a continuous grouping $\mathcal{G}$ would look like this: $\{1, 2, \ldots, N^1\}, \{N^1 + 1, N^1 + 2, \ldots, N^2\}, \ldots, \{N^{k-1} + 1, N^{k-1} + 2, \ldots, N^k = n\}$, where $N^j = n^1 + n^2 + \cdots + n^j$. That is, it consists of k continuous intervals for the members of the DAO. A contiguous grouping enjoys two natural ways to order its subgroups. On the one hand, the subgroups can be ordered according to their own valuations non-increasingly like before, which makes it easy to compute $WTP_{total}(\mathcal{G})$. On the other hand, they can be ordered by their members' valuations non-increasingly (i.e., by their members' indices non-decreasingly), which makes it easy to describe and analyze. By default the analysis below goes with the first ordering, unless explicitly mentioned. Due to space constraints, proofs are omitted and can be found in the full version of this paper [8].

Algorithm 1: Optimal Grouping Algorithm

Input: $v_1, \ldots, v_n$, where $v_1 \geq \cdots \geq v_n$
Output: opt_WTP, k, and the optimal grouping $\mathcal{G} = \{G^1, \ldots, G^k\}$

1. initialize $possible\_CV \leftarrow \emptyset$;
2. **for** $i = 1, \ldots, n$ **do**
3.   initialize $WTP \leftarrow 0$;
4.   **for** $j = i, \ldots, n$ **do**
5.   update $WTP \leftarrow \max\{WTP, (j - i + 1) \cdot v_j\}$;
6.   add WTP to $possible\_CV$; // The valuation of $\{i, i+1, \ldots, j\}$.
7. initialize $opt\_WTP \leftarrow 0$, $opt\_CV \leftarrow 0$, $k \leftarrow 0$;
8. **for** CV in $possible\_CV$ **do**
9.   initialize $CI \leftarrow 0$, $subgroup\_size \leftarrow 1$;
10.   **for** $i = 1, \ldots, n$ **do**
11.   **if** $subgroup\_size \cdot v_i \geq CV$ **then**
12.   update $CI \leftarrow CI + 1$, $subgroup\_size \leftarrow 1$;
13.   **else**
14.   update $subgroup\_size \leftarrow subgroup\_size + 1$;
15.   update $WTP_{total} \leftarrow CI \cdot CV$;
16.   **if** $WTP_{total} > opt\_WTP$ **then**
17.   update $opt\_WTP \leftarrow WTP_{total}$, $opt\_CV \leftarrow CV$, $k \leftarrow CI$;

// Given opt_CV, find the optimal grouping
18. initialize $N \leftarrow [0]$; // A list with one element 0
19. initialize $subgroup\_size \leftarrow 1$;
20. **for** $i = 1, \ldots, n$ **do**
21.   **if** $subgroup\_size \cdot v_i \geq opt\_CV$ **then**
22.   append i to the end of N and update $subgroup\_size \leftarrow 1$;
23.   **else**
24.   update $subgroup\_size \leftarrow subgroup\_size + 1$;
25. **for** $j = 1, \ldots, k$ **do**
26. $\hat{G}^j \leftarrow \{N[j-1] + 1, N[j-1] + 2, \ldots, N[j]\}$;
27. $\hat{G}^{k+1} \leftarrow \{N[k] + 1, N[k] + 2, \ldots, n\}$;
28. $\mathcal{G} \leftarrow \{\hat{G}^1, \ldots, \hat{G}^{k-1}, \hat{G}^k \cup \hat{G}^{k+1}\}$;

Theorem 1. *For any DAO G and any grouping $\mathcal{G} = \{G^1, \ldots, G^k\}$ of G, there exists a continuous grouping $\overline{\mathcal{G}} = \{\overline{G}^1, \ldots, \overline{G}^k\}$ such that $WTP_{total}(\overline{\mathcal{G}}) \geq WTP_{total}(\mathcal{G})$.*

Corollary 1. *For any DAO G, there exists a continuous grouping $\mathcal{G}$ that achieves its maximum willingness to pay, namely $WTP_{total}(\mathcal{G}) = opt\_WTP$.*

By Corollary 1, it suffices to consider continuous groupings in order to compute opt_WTP. Our Algorithm 1 above indeed does so. To intuitively see how it works, given a grouping $\mathcal{G}$, we further define the *critical index* of $\mathcal{G}$ as

$CI = \arg\max_{j \in [k]}\{j \cdot WTP^j\}$, breaking ties in favor of the largest j. The corresponding valuation WTP^{CI} is referred to as the *critical value* of $\mathcal{G}$ and denoted by CV. Thus $WTP_{total} = CI \cdot CV$.

Algorithm 1 enumerates *all possible subgroups* and their valuations (the subgroup's willingness to pay). The key is that, although the number of possible groupings is exponential, the number of distinct subgroups in all continuous groupings is only $O(n^2)$, and their valuations are all the possible critical values for the optimal continuous grouping. For each such valuation, assuming for a moment it is the critical value CV of the optimal grouping, we use a greedy algorithm to form as many subgroups as possible whose valuations are greater than or equal to CV. The number of such subgroups is then the critical index CI of the resulting grouping, and $WTP_{total} = CI \cdot CV$ is a possible value for *opt_WTP*. We find *opt_WTP* by taking the maximum among all WTP_{total}'s, and then the corresponding CV leads to the optimal grouping $\mathcal{G}$.

3.3 Analysis of the Optimal Grouping Algorithm

Theorem 2. *Given any DAO G, Algorithm 1 finds a continuous grouping $\mathcal{G}$ whose valuation is opt_WTP with time complexity $O(n^3)$.*

The optimal grouping guarantees that the DAO's WTP_{total} is maximized. Compared to the ungrouped case, it helps the DAO to bid higher and win in more scenarios over other DAOs. We further show that even if the original ungrouped WTP_{total} is enough to win the item, the optimal grouping guarantees that its allocation is less-excludable compared to the original one.

Definition 2 (less-excludable allocation). *For a winning DAO G and its two possible allocations $x = (x_1, \ldots, x_n)$ and $x' = (x'_1, \ldots, x'_n)$, if $x_i \geq x'_i \; \forall i \in [n]$, then x is **less-excludable** than x'.*

Under the current model, $x_i \in \{0, 1\} \; \forall i \in [n]$ and the comparison can be further made on groupings. The ungrouped case is a natural degenerated grouping $\mathcal{G}^*$ with a single subgroup $\{1, 2, \ldots, n\}$, and the individually grouped case $\{\{1\}, \{2\}, \ldots, \{n\}\}$ is another degenerated grouping which is equivalent to $\mathcal{G}^*$.

Definition 3 (less-excludable grouping). *For a DAO G and its two groupings $\mathcal{G}_1$, $\mathcal{G}_2$, under both of which G wins, denote the allocation sets as A_1, A_2, representing the sets of bidders who have access to the item under the groupings, respectively. If $A_1 \supseteq A_2$, then we say that $\mathcal{G}_1$ is **less-excludable** than $\mathcal{G}_2$.*

Theorem 3. *For a winning DAO G, the optimal grouping $\mathcal{G}$ by Algorithm 1 is less-excludable than the degenerated grouping $\mathcal{G}^*$.*[3]

---

[3] Notice that there is no worst-case gap between grouping and non-grouping in terms of their WTP, because non-grouping is a special case of grouping and in some scenarios *opt_WTP* is achieved in such a case.

It is easy to see that the two-level DAO with the optimal grouping satisfies budget balance and individual rationality. It also maintains equal treatment among subgroups and within the same subgroup, as well as allowing more access for small bidders. However, it doesn't satisfy incentive compatibility as a bidder with a high valuation may benefit by underbidding —by doing so it may end up in a larger subgroup and pay a lower price. This reflects the common free-rider problem when considering public goods, where individuals have the incentive to directly enjoy the access to the public good without making sufficient contributions to the collective [1,2]. In the following section, we consider whether improving non-excludability and achieving incentive compatibility can be better aligned in a different model.

4 Collective Utilities and a New Mechanism for DAOs

Recall that the motivation of DAOs is to leverage the collective strength of the members to achieve a goal based on a shared vision. Especially towards public goods, prosocial behaviors of individuals should be considered, as suggested by [18,23]. Therefore we introduce a positive externality into every member's utility who has access to the item, the social welfare $SW = \sum_{i \in [n]} v_i \cdot x_i$ of the DAO, scaled by a factor $\alpha \geq 0$. However, a player i may not want to pay more than its true valuation v_i no matter how much it cares about others. Thus a budget constraint is needed to reflect the fact that v_i is really player i's "willingness to pay". In addition, we enrich the access space by allowing partial accesses to the good, so that for each member i of a DAO, $x_i \in [0,1]$ instead of $\{0,1\}$. Accordingly, for each member $i \in [n]$,

$$u_i = \begin{cases} v_i \cdot x_i - p_i + \alpha \cdot SW, & x_i > 0 \text{ and } p_i \leq v_i \\ -\infty, & p_i > v_i \\ -p_i, & x_i = 0 \end{cases}.$$

Note that the social welfare SW also includes $v_i \cdot x_i$, so the utility function guarantees that for any α, player i cares more about its own value than it cares about others. Moreover, player i doesn't care about whether others get access or not if itself has no access (that is, $x_i = 0$). Other ways of bringing in a collective factor are of course possible and worth examining in future studies.

4.1 A New Mechanism

Below we introduce a new lower-level mechanism M_1, with an example in Fig. 3 illustrating how it works more intuitively. The upper-level mechanism M_u remains to be the second-price mechanism. Given a DAO G and its members' bids $b_1, \ldots, b_n$ ordered non-increasingly, the aggregation function of M_1 is

$$WTP_{total}(b_1, \ldots, b_n) = \max_{i \in [n]} \{\sum_{j=1}^{i} \min\{b_j, \ (1+\alpha) \cdot b_i\}\},$$

which is computable in $O(n^2)$ time.

If DAO G does not win the auction in M_u, then $x_i = 0$, $p_i = 0$ $\forall i \in [n]$. If DAO G wins the auction with price P_{total}, we have $WTP_{total} \geq P_{total}$. M_1 will compute two payments, P_1 and $P_2 = \frac{P_1}{1+\alpha}$. Let $k_1 = \max_{i \in [n]}\{i | b_i \geq P_1\}$ and $k_2 = \max_{i \in [n]}\{i | b_i \geq P_2\}$, with $k_1 \leq k_2$. They denote the numbers of members who gain full access and who have at least some partial access, respectively. Given a group of parameters (P_1, P_2, k_1, k_2), the access and the cost-sharing functions are as follows: $\forall i \in [n]$,

$$x_i = \begin{cases} 1, & i \leq k_1 \\ \frac{b_i}{P_1}, & k_1 < i \leq k_2 \\ 0, & i > k_2 \end{cases} \quad \text{and} \quad p_i = \begin{cases} P_1, & i \leq k_1 \\ b_i, & k_1 < i \leq k_2 \\ 0, & i > k_2 \end{cases}.$$

Note that a uniform pricing standard is guaranteed even for partial accesses: $\forall i \in [n]$, $x_i = \frac{p_i}{P_1}$. The restriction on small members' participation after k_2 helps avoid exacerbating free-riding behavior, which could negatively impact the overall effectiveness of the mechanism. The necessity of such a restriction has also been mentioned in works like [12].

A group of parameters (P_1, P_2, k_1, k_2) is *feasible* if it satisfies budget balance, $\sum_{i \in [n]} p_i = P_{total}$. Algorithm 2 below is used by M_1 to compute the optimal feasible parameters with the least excludability. The time complexity of the

---

Algorithm 2: Optimal Parameters

Input: $b_1, \ldots, b_n$, where $b_1 \geq \cdots \geq b_n$, and P_{total}
Output: P_1, P_2, k_1, k_2

1 initialize $k_2 \leftarrow n+1$, $size \leftarrow 0$;
2 **while** $P_{total} > size$ **do**
3   update $k_2 \leftarrow k_2 - 1$, $size \leftarrow 0$;
4   **for** $j = 1, \ldots, k_2$ **do**
5   update $size \leftarrow size + \min(b_j, (1+\alpha)b_{k_2})$;

6 initialize $P_{remain} \leftarrow P_{total} - k_2 \cdot b_{k_2}$;
7 **if** $P_{remain} \leq 0$ **then**
8   update $k_1 \leftarrow k_2$, $P_1 \leftarrow \frac{P_{total}}{k_2}$;
9 **else**
10   update $k_1 \leftarrow k_2 - 1$;
11   **while** $P_{remain} > 0$ **do**
12   **if** $P_{remain} \leq k_1 \cdot (b_{k_1} - b_{k_1+1})$ **then**
13   update $P_1 \leftarrow b_{k_1+1} + \frac{P_{remain}}{k_1}$;
14   **break**;
15   **else**
16   update $P_{remain} \leftarrow P_{remain} - k_1 \cdot (b_{k_1} - b_{k_1+1})$, $k_1 \leftarrow k_1 - 1$;

17 $P_2 \leftarrow \frac{P_1}{1+\alpha}$;

algorithm is $O(n^2)$. As shown in [8], it actually achieves the highest utility for every member among all feasible parameters.

4.2 Analysis of The Mechanism

Mechanism M_1 has another advantage: the value of α in the model can be any nonnegative number without affecting the mechanism's operation. In fact, α has a practical meaning and can be seen as a measure of the degree of collectivism. When $\alpha = 0$, M_1 degenerates to M_ℓ and has all its properties such as incentive compatibility. In this case, each member only cares about its individual access and the final allocation only has binary accesses. As α increases, the proportion of the collective factor in the utility function becomes larger, and the threshold for members to get access becomes lower. This results in more bids from members being aggregated, leading to a higher overall bid. For any strictly positive α, we have the following main theorem for mechanism M_1.

Theorem 4. *Under the collective-utility model and $\forall \alpha > 0$, mechanism M_1 satisfies individual rationality, budget balance, and equal treatment. It is also incentive compatible in generic games. Given a winning DAO G, M_1 achieves a higher willingness to pay for G, a higher social welfare, and a less-excludable allocation compared with M_ℓ.*

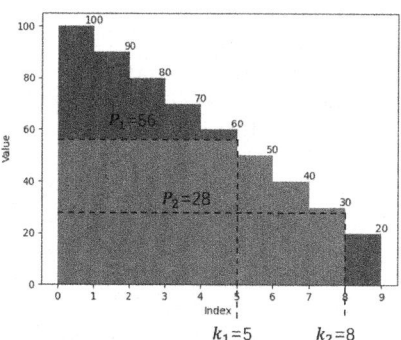

Fig. 3. An example for the mechanism M_1, with $\alpha = 1$ and members in the DAO being the same as Fig. 1. According to the aggregation function, $WTP_{total} = 460$. Assume the DAO wins the auction and $P_{total} = 400$. The gray area is used to determine the access and the payment of each member. It can also be viewed as a pool, with the horizontal axis representing the width and the vertical axis representing the height. The capacity of the pool (the size of the gray area) should equal the total payment P_{total} of the DAO and P_1 can be seen as the water line. The intuition of Algorithm 2 is to find the *widest* pool whose capacity reaches P_{total} with the required parameters.

5 Future Directions

Our study tackles non-excludability of public goods for DAOs from an algorithmic perspective and a mechanism design perspective. From both angles we show that the collective behavior of a DAO's members has a rich structure and could improve the social efficiency and the non-excludability of the DAO when treated properly. Many other open problems are also interesting, besides the ones already mentioned earlier. For instance, our mechanism can be extended to repeated games by introducing a time factor and relaxing the budget constraint to be time-dependent, transforming the setting into a multi-stage game with observations. Through myopic best-response analysis, we find that the system frequently enters cycles and exhibits stable states. Also, analyzing the mechanism via solution concepts other than incentive compatibility (e.g., regrets), and considering different auction settings such as multi-item Vickrey auctions for public goods are all intriguing avenues for future research.

Acknowledgments. We thank Jichen Li and Zihao Zhen for helpful discussions in the early stage of this work and several anonymous reviewers for valuable comments. This work is partially supported by Beijing Advanced Innovation Center for Future Blockchain and Privacy Computing.

References

1. Albanese, R., Van Fleet, D.D.: Rational behavior in groups: the free-riding tendency. Acad. Manag. Rev. **10**(2), 244–255 (1985)
2. Andreoni, J.: Why free ride?: strategies and learning in public goods experiments. J. Public Econ. **37**(3), 291–304 (1988)
3. Aumann, R.J., Maschler, M.: Game theoretic analysis of a bankruptcy problem from the Talmud. J. Econ. Theory **36**(2), 195–213 (1985)
4. Bahrani, M., Garimidi, P., Roughgarden, T.: When bidders are DAOs. In: Bonneau, J., Weinberg, S.M. (eds.) AFT 2023. LIPIcs, vol. 282, pp. 1–21. Schloss Dagstuhl – Leibniz-Zentrum für Informatik, Dagstuhl, Germany (2023). https://doi.org/10.4230/LIPIcs.AFT.2023.21
5. Buterin, V.: Bootstrapping a decentralized autonomous corporation: part I (2013). https://bitcoinmagazine.com/technical/bootstrapping-a-decentralized-autonomous-corporation-part-i-1379644274. Accessed 14 Apr 2025
6. Buterin, V.: DAOs, DACs, DAs and more: an incomplete terminology guide (2014). https://blog.ethereum.org/2014/05/06/daos-dacs-das-and-more-an-incomplete-terminology-guide. Accessed 14 Apr 2025
7. Chalkiadakis, G., Elkind, E., Wooldridge, M.: Computational Aspects of Cooperative Game Theory. Morgan & Claypool Publishers (2011)
8. Chen, J., Zhou, W.: Less-excludable mechanism for DAOs in public good auctions. arXiv preprint arXiv:2504.11854 (2025). https://doi.org/10.48550/arXiv.2504.11854
9. DeepDAO: Homepage. https://deepdao.io/organizations. Accessed 14 Apr 2025
10. Dehez, P.: Cooperative provision of indivisible public goods. Theor. Decis. **74**(1), 13–29 (2013)

11. Fehr, E., Fischbacher, U.: The nature of human altruism. Nature **425**(6960), 785–791 (2003)
12. Goeree, J.K., Maasland, E., Onderstal, S., Turner, J.L.: How (not) to raise money. J. Polit. Econ. **113**(4), 897–918 (2005)
13. Hassan, S., Filippi, P.D.: Decentralized autonomous organization. Internet Policy Rev. **10**(2) (2021)
14. Jenkinson, G.: Remote work triggers move to DAOs in the post-pandemic world: survey. Cointelegraph (2022)
15. Kaul, I., Grunberg, I., Stern, M.A.: Defining global public goods. In: Global Public Goods: International Cooperation in the 21st Century, pp. 2–19. Oxford University Press (1999)
16. Kleindorfer, P.R., Sertel, M.R.: Auctioning the provision of an indivisible public good. J. Econ. Theory **64**(1), 20–34 (1994)
17. Ledyard, J.O., et al.: Public Goods: A Survey of Experimental Research. California Inst. of Technology, Division of the Humanities and Social Sciences (1994)
18. Li, J., et al.: Altruism, collectivism and egalitarianism: on a variety of prosocial behaviors in binary networked public goods games. In: Agmon, N., An, B., Ricci, A., Yeoh, W. (eds.) AAMAS 2023, pp. 609–624. ACM, London, United Kingdom (2023). https://doi.org/10.5555/3545946.3598691
19. Rachmilevitch, S.: Auctions with multi-member bidders (2022), working paper
20. Ray, D., Vohra, R.: Coalition formation. In: Handbook of Game Theory with Economic Applications, pp. 239–326. Elsevier (2015)
21. Tan, J.Z., et al.: Open problems in DAOs. arXiv preprint arXiv:2310.19201 (2023). https://doi.org/10.48550/arXiv.2310.19201
22. Wang, S., Ding, W., Li, J., Yuan, Y., Ouyang, L., Wang, F.Y.: Decentralized autonomous organizations: concept, model, and applications. IEEE Trans. Comput. Soc. Syst. **6**(5), 870–878 (2019)
23. Wang, Z., Szolnoki, A., Perc, M.: Evolution of public cooperation on interdependent networks: the impact of biased utility functions. Europhys. Lett. **97**(4), 48001 (2012)

TBDS: Transaction-Based Data Sharing

Wu Xin[1], Hongyin Chen[1], Xiaoqi Dong[2], Jichen Li[1,3], Xiaotie Deng[1,3(✉)],
Zhonghai Wu[1], and Bin Xiao[4]

[1] Peking University, Beijing, China
{xinwu,chenhongyin,limo923,xiaotie,wuzh}@pku.edu.cn
[2] Renmin University of China, Beijing, China
xiaoqi.dong@ruc.edu.cn
[3] Zhongguancun Laboratory, Beijing, China
[4] The Hong Kong Polytechnic University, Kowloon, Hong Kong
csbxiao@connect.polyu.hk

Abstract. Data sharing is widely regarded as a critical approach to unlocking the value of data, encompassing data rights management, processing, and profit allocation. However, existing methods often suffer from fragmented designs, incompatible assumptions, and security challenges, perpetuating the issue of data silos. To address these limitations, we propose Transaction-Based Data Sharing (TBDS), a comprehensive and modular solution. TBDS adopts a formal-language-based data rights management module, maintaining compatibility across modules and enabling efficient permission verification and state updates. It leverages blockchain and Trusted Execution Environments (TEEs) to ensure secure operations while handling remote attestation internally, eliminating reliance on external parties. Furthermore, TBDS introduces a transaction-driven workflow, where smart contracts facilitate seamless interactions between modules. Finally, we design a chain-based profit allocation mechanism that balances incentives and privacy, ensuring fair and transparent distribution of rewards among participants. We also show that this mechanism improves participants' revenue. With these innovations, TBDS not only addresses the challenges of fragmented designs and security risks but also establishes a scalable and incentive-compatible framework, paving the way for more efficient and secure data-sharing ecosystems.

Keywords: Data Sharing · Transaction · Security · Blockchain · TEE

1 Introduction

Data-driven applications [5] like artificial intelligence [35], big data analytics [13], and large language models [22,42] are becoming cornerstones of modern industries and economies. Data is increasingly recognized as one of the most critical

---

W. Xin and H. Chen—Co-first authors.
This work was supported by NSFC/RGC Joint Research Scheme under Grant 62261160391 and Grant N_PolyU529/22, as well as by Chinaunicom Software.

production resources. However, the value of data lies not in its mere existence but in its accessibility and usability. Data sharing, typically presented as the process of data rights management, data processing, and profit allocation is widely regarded as a promising approach to unlocking this value [34]. By encouraging collaboration and participation from a diverse range of entities, from individual users to businesses and industries, data sharing is expected to enhance the value of data and deliver mutual benefits to all participants.

Practical data-sharing techniques often fail to engage diverse participants, reinforcing the problem of data silos [12,33]. This limitation arises from concerns over infringing on the rights of participants [3], which include data *owners* who own the data, and data *operators* who manage data on behalf of data owners. These rights include control, privacy, and fair value from data usage. For example, data owners face privacy risks when their data is shared without adequate safeguards [15,30,38], while both owners and operators might not get fair gains in data sharing [2,9,17,21]. Such imbalances deter participation and perpetuate data silos, undermining efforts to optimize data value.

Existing efforts to address these challenges often focus on isolated aspects of data rights management [4,11,20], data processing [7,15,26,36–38,40] and profit allocation [2,9,17,21,25]. However, these fragmented approaches fail to simply combine a full data-sharing protocol due to two critical limitations: incompatible security assumptions (e.g. remote attestation vs. Byzantine Fault Tolerance, continuous interaction vs. one-time access) and incompatible designs (API and data structure). These incompatibilities compromise both security integrity and implementation feasibility, exacerbating the reluctance of participants to engage in data sharing and perpetuating the issue of data silos.

To tackle the persistent challenge of data silos caused by fragmented designs, we propose Transaction-Based Data Sharing (TBDS), a modular architecture, comprising three distinct modules corresponding to the key phases of data sharing. These modules communicate and coordinate through message-based interactions, ensuring flexibility and scalability. At its core, TBDS is transaction-driven: Each transaction represents a specific execution, either carried out by a single module or collaboratively by multiple modules through message exchanges.

To accommodate various types of data processing and profit allocation modules, we design the data rights management module using a formal language. This formal language maintains a list of participants, including both data owners and operators, as well as a list of data and their relationships with participants, collectively representing the state of data rights. It provides entities for permission verification, enabling queries from other modules to complete their executions effectively. Additionally, the language supports internal transactions, allowing state updates within the module.

We propose the TBDS protocol, which integrates blockchain and Trusted Execution Environment (TEE) to enable secure and efficient data sharing. In this system, modules are implemented as smart contracts on the blockchain. Operators must stake funds in the profit allocation contract to participate, with their rewards or penalties determined by their behavior. To ensure data security, operators execute data processing algorithms within a TEE. TEE security relies on remote attestation, which traditionally requires an external party to verify attestations and handle failures, making integration with other systems challenging. However, our system addresses this internally. Each execution within the TEE submits its remote attestation to the blockchain, producing one of two outcomes that trigger a transaction to the profit allocation contract: if the attestation fails, signaling malicious behavior, the operator is penalized through slashing. This approach eliminates the need for additional security assumptions, highlighting the effect of our system's architecture.

We propose a profit allocation mechanism for sustainable privacy-preserving data sharing. Our mechanism employs a Stackelberg game model to optimize stakeholder compensation while maintaining data privacy through TEEs. Data owners retain full approval/rejection authority over sharing requests. Operators gain dual incentives: immediate data utilization value; recurring revenue from subsequent transactions. The core innovation is our recursive profit distribution model, where each transaction triggers automated, chain-based revenue redistribution traced by smart contracts. This architecture achieves privacy-incentive equilibrium while ensuring transparency and fairness in collaborative data economies. For the first time, to the best of our knowledge, we show that our mechanism improves the revenue of all participants through the data-sharing process.

In all, our contributions are: (1) TBDS, a modular design for data sharing centered on transactions, ensuring feasibility across all steps of the data-sharing process, (2) a data rights management module built on formal languages, enabling efficient compatibility with other modules, (3) the data sharing protocol based on blockchain and TEEs to ensure secure data sharing and, (4) a chain-based profit allocation mechanism with all participants' revenue improvement.

2 Related Work

Data sharing is widely regarded as the most critical approach to addressing data silos [12,33], involving three modules: data rights management, data processing, and profit allocation. Existing studies often focus on these modules in isolation.

In terms of data rights management, digital rights management systems were studied extensively even before data sharing became a prominent practice [28]. Since the introduction of GDPR [3], the emphasis has shifted towards safeguarding data privacy and protecting users' data rights [4]. Moreover, data trading platforms [11,20] have integrated mechanisms to facilitate the transfer of data ownership or usage rights.

For data processing, significant research has been conducted on privacy-preserving methods that enable the use of private data while protecting user rights. This includes advancements in fields like federated learning [29,38,40] and differential privacy [1,14,15]. Additionally, secure and privacy-focused data processing methods have been developed using blockchain [7,10,26], Multi-Party Computation (MPC) [8,19,32], and trusted execution environments (TEEs) [11,30,36,37].

For profit allocation or sharing, the primary methods include those based on egalitarian [39], those based on marginal gain (such as Individual Profit-Sharing [39], The Labour Union Game Profit-Sharing [18], The Shapley Game Profit-Sharing [6,16,21,27]), and those based on marginal loss (such as The Fair-Value Game [18]). Other methods of profit allocation are also mentioned in the context of data pricing and incentive mechanisms for data sharing, such as query-based pricing [17,23–25], machine learning model-based pricing [9,27]. The incentive mechanism in federated learning adopts an approximate method similar to the Shapley Game Profit-Sharing approach [41].

Despite these advancements, none of the existing studies address the entire data-sharing process holistically. This fragmented approach creates vulnerabilities that undermine the overall effectiveness of these solutions. For example, inadequate privacy protections expose data owners to data misuse, while the absence of fair value distribution mechanisms or incentive mechanisms discourages participation from both data owners and operators. Furthermore, the reliance on conflicting assumptions and incompatible designs (see examples in Sect. 3.4) renders the integration of separate solutions infeasible and insecure. This lack of a unified approach exacerbates participants' reluctance to engage in data-sharing systems, thereby perpetuating the persistence of data silos. Additionally, current data-sharing protocols fail to measure their incentive effects, leaving it unclear whether participants are sufficiently motivated to join and sustain these systems.

3 Transaction-Based Data Sharing Framework

We propose the Transaction-Based Data Sharing (TBDS) framework, a modular approach addressing limitations of existing fragmented solutions. TBDS organizes data rights management, secure data processing, and profit allocation as distinct modules coordinated through transaction-driven operations. This structure allows us to systematically address challenges in data sharing systems.

3.1 Participants

In TBDS, the participants include *data owners* and *data operators*. Each data owner holds ownership of a list of structured data which we call *data items* and she can designate some rights, which is a kind of permission, of any data item she owns to data operators. Data operators with operating rights possess the capabilities for data collection, storage, management, processing, and sales. The data operators include participants in all (most) existing data-sharing systems/works, such as data controller, processor, demander, provider, etc.

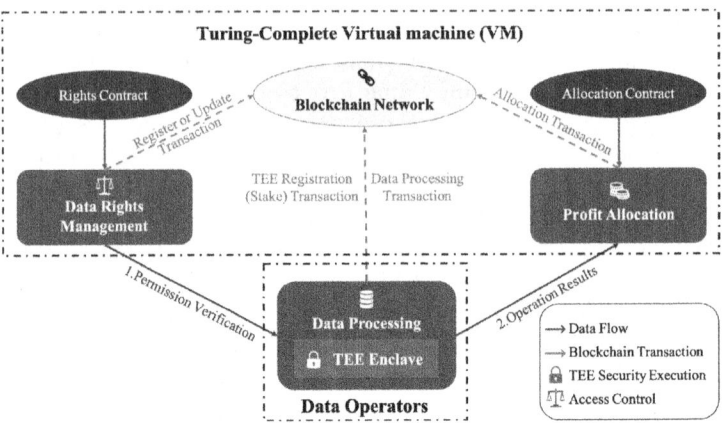

Fig. 1. Modular framework of TBDS.

3.2 Modular Framework

The TBDS framework (see in Fig. 1) consists of three modules: Data Rights Management, Data Processing, and Profit Allocation. We assume that there is a Turing-complete virtual machine (VM) running in the system, which receives and processes messages and sends output as messages back. In distributed system this is commonly implemented as smart contracts in a blockchain network [31]. The Data Rights Management module and the profit allocation module run in the VM, and data operators maintain the data processing module.

The Data Rights Management module functions as a database that maintains information about participants, data items, and their relationships. It includes an algorithm called *Permission Check*, which verifies whether certain participants have the required permissions to execute a protocol involving specific data items. Participants or other modules can interact with the Data Rights Management module to retrieve information, perform permission checks, or update the database state. Any state changes follow a defined protocol and are subject to a permission check to validate the issuer's authority to make those changes.

The Data Processing module is a protocol to process and utilize data items. The protocol would get the permission check from the Data Rights Management module and only execute when passing the permission check.

The Profit Allocation module is a monetary system that manages accounts for all participants. Each account holds a balance, which participants can withdraw in accordance with the module's rules. Other modules can trigger updates to the Profit Allocation module. For instance, when a data owner assigns rights to a data operator, the owner may receive rewards by transferring a portion of the operator's balance to their own account. Additionally, the Profit Allocation module implements a slashing mechanism, where participants stake a portion of their balance in the system. If a participant engages in malicious behavior within the data-sharing ecosystem, their staked balance can be reduced as a penalty.

3.3 Transaction-Driven Progress

The state of the data-sharing system is the collective states of all its modules. We define a transaction as the execution of a protocol within the system, whether it occurs within a single module or involves the collaboration of multiple modules. Each transaction can modify the state of one or more modules. The operations of the data-sharing system are represented as a sequential list of transactions, which collectively drive the system's progress. It should be noted that the term "transaction" used here is distinct from its conventional meaning in blockchain contexts, where it refers to an execution within the system.

3.4 Goal and Challenges

We aim to enable feasible and secure data sharing through our modular, transaction-driven framework. The key challenge lies in ensuring system-level compatibility of security assumptions across modules.

While our modular approach allows for integrating components from different solutions, significant challenges emerge when combining existing fragmented systems. For example, TEE-based data processing requires remote attestation, but existing rights management modules lack interfaces to validate these attestations. When attestation fails, profit allocation modules need protocols to implement penalties, which current solutions do not provide. Additionally, temporal access patterns complicate integration. Data processing designed for one-time access versus continuous interaction requires different state management approaches across modules. These fundamental incompatibilities necessitate well-defined module interactions and careful design considerations, which our transaction-driven approach addresses by providing a unified interaction framework that maintains security guarantees across module boundaries.

4 Data Rights Management Protocol

The data rights management protocol plays a crucial role in ensuring that data sharing within the TBDS framework is conducted in a secure, transparent, and fair manner. By defining a structured model for participants, data, and their relationships in this section, the protocol facilitates efficient permission management and supports seamless integration of various data sharing operations.

Formally, we define a *Data Rights Management DR* as a septuple $DR = (P, D, K, Ops, R, f, \delta)$:

1. **Participant Set (P):** P is the set of all participants in the TBDS including both data owners and operators.
2. **Data Set (D):** D is the set of all data items.
3. **Keyword Set (K):** K is the set of all keywords that describe the relationships between participants and data. Examples include "ownership", "operation", etc.

4. **Operation Set** (Ops): Ops is the set of all operations in the TBDS (e.g., grantOperation, stopOperation, changeOperation, useDataForAnalysis, useDataForSharing, etc.).
5. **Relationship Function** (R): The Relationship Function takes as input a vector of participants $\tilde{P} = (p_1, p_2, p_3, \cdots)$ where $p_1, p_2, p_3 \cdots \in P$, a set of data items $\tilde{D} \subseteq D$, and output a set of keywords $R(\tilde{P}, \tilde{d}) \subseteq K$, showing the relationship of $\tilde{P}$ and $\tilde{D}$. For instance, $R((p_1), \{d \in D\}) = \{\text{ownership}\}$ means that participant p_1 has the ownership over data item d and $R((p_1, p_2), \{d\}) = \{\text{operation}\}$ means that p_1 grant the operation rights of data item d to the p_2. The vector of (p_1, p_2) is not redundant compared to only saying that p_2 has the operation rights since the structure of rights granting might be useful for profit allocation (Sect. 6).
6. **Permission Checking Function** (f): The permission function determines whether a given operation is allowed based on the relationship keywords. It takes as input a vector of participants $\tilde{P}$, a set of data items $\tilde{D} \subseteq D$, an operation $op \in Ops$ and a relationship function R, and outputs $f(\tilde{P}, \tilde{D}, op, R) =$ yes if the operation op is permitted under the current relationship between $\tilde{P}$ and $\tilde{D}$. Otherwise, $f(\tilde{P}, \tilde{D}, op, R) =$ no.
7. **Internal Transaction** (δ): The internal transaction updates the overall system state when an operation is executed. First, we define the state of the system as $S = (P, D, R)$. The internal transaction takes the state of the system S and the operation op as input, updates and outputs a new state $\delta(S, op) = S' = (P', D', R')$ of the system, that reflects any changes to the participants, data, or their relationships caused by executing op.

5 TBDS Based on Blockchain and TEE

We implement the TBDS protocol by integrating blockchain smart contracts with Trusted Execution Environment (TEE) technology to create a secure and feasible data-sharing ecosystem (see Fig. 1). In our architecture, Data Rights Management and Profit Allocation modules run as smart contracts in the blockchain's virtual machine, while data operators maintain TEE-based Data Processing modules.

5.1 Blockchain-Based Data Rights Management and Profit Allocation

We implement the Virtual Machine in the TBDS framework using blockchain technology, which naturally serves as a monetary system for profit allocation and ensures Byzantine Fault Tolerance (BFT). Data rights management and profit allocation modules are implemented as smart contracts in the blockchain.

Data Rights Management Contract
The data rights management contract implements our data rights management protocol presented in Sect. 4. It maintains and updates the database state

$S = (P, D, R)$. Data owners register new data items by sending transactions to the smart contract, which records the item along with its initial relationship (typically ownership). To grant operating rights to a data operator, a data owner initiates a grantOperation transaction. The contract then verifies the request by calling the permission function $f((p_i, p_j), \{d\}, op, R)$. which returns yes only if the operator p_j is authorized to perform op on data d. Once an operation is authorized, the contract updates the state using the internal transaction $S' = \delta(S, ((p_i, p_j), \{d\}, op)$.

Example 1: Data Ownership Definition. Let $p_1 \in P$ be a participant and $d_1 \in D$ be a data item. We define data ownership by assigning the keyword ownership to the relationship between p_1 and d_1: $R((p_1, p_1), \{d_1\}) = \{\text{ownership}\}$.

This indicates that participant p_1 is the owner of data d_1.

Example 2: Granting Data Operating Rights. Suppose participant p_1 is the owner of data d_1. Initially, participant p_2 has no rights over d_1: $R((p_1, p_2), \{d_1\}) = \varnothing$.

Now, if p_1 decides to grant operating rights to p_2 (using the operation grantOperation $\in Ops$), the permission function should first verify that the owner is authorized to perform this grant: $f((p_1, p_2), \{d_1\}, \text{grantOperation}, R) = \text{yes}$. Then, the internal transaction δ is applied to update the state. Formally, we write: $S' = \delta(S, ((p_1, p_2), \{d_1\}, \text{grantOperation}))$, with the intended effect (the keyword operation represents the right to operate data): $R'((p_1, p_2), \{d_1\}) = R((p_1, p_2), \{d_1\}) \cup \{\text{operation}\} = \{\text{operation}\}$. This update reflects that p_2 now has the right to operate d_1.

Profit Allocation Contract. The profit allocation module is implemented as a smart contract that manages accounts, distributes rewards, and enforces penalties to create a sustainable incentive mechanism for data sharing:

1. **Staking Requirement:** Data operators stake funds to participate in the system. Their stake acts as collateral for their operations.
2. **Slashing Mechanism:** The contract implements rules to detect and penalize malicious behavior. When violations are detected, the system slashes the operator's stake and redistributes the penalty to honest participants.
3. **Reward Distribution:** If the attestation succeeds, the profit allocation contract computes rewards based on the operator's contribution. The rewards are determined using a fair, game-theoretic model (e.g., Shapley value) and distributed accordingly.

5.2 TEE-Based Data Processing

We integrate TEE as the data processing module to ensure the security of data processing. TEE has various implementations (such as Intel SGX, ARM TrustZone, or AMD SEV), and we adopt a generalized approach including two phases:

isolated execution of code and generation of verifiable remote attestation that can be validated within the TBDS system.

To execute an operation $op \in Ops$ (e.g., useDataForAnalysis) on data d in the TEE, the code execution includes checking the permission and processing of the data:

1. **Permission Verification:** The TEE calls the function $f((p_i, p_j), \{d\}, op, R)$ to verify that the operation is permitted. If the function returns yes, the process continues.
2. **Data Processing:** The TEE takes the data d (whether encrypted or not) as input and processes it within the isolated environment according to the specified operation op.

After the execution, the operator obtains both the result and remote attestation from the TEE.

Note that our primary concern is not necessarily the correctness of the output results, but rather ensuring that the data was processed according to the specified permissions and was not misused or processed in unauthorized ways. The remote attestation serves as cryptographic proof that the data processing adhered to the permitted operations and security constraints.

5.3 Cross-Module Security and Feasibility Integration

Our TBDS implementation unifies security assumptions across modules through two key mechanisms:

1. **Transaction-Driven Module Interaction:** Blockchain transactions facilitate standardized communication between modules, eliminating API incompatibilities and ensuring state transitions in one module can trigger appropriate responses in others.
2. **Integrated Attestation Verification:** TEE attestation reports are verified directly by the Data Rights Management contract, which then triggers conditional transactions to the Profit Allocation module—either slashing penalties for failed attestations or distributing profits for successful ones.

This approach ensures both feasibility and security by standardizing module interactions and handling security verifications and penalties within our framework rather than relying on external systems.

5.4 Transactions in Data Sharing

We use blockchain transactions to implement TBDS transactions, providing a secure mechanism for state transitions across modules. These typically include: data rights management database register/update, TEE registration, data processing transaction, and deposit/retrieval of money in profit allocation.

Data Rights Management Database Register/Update. Register/Update the Data Rights Management database when a data owner registers new data or modifies existing data rights.

1. **Operation Request:** The participant p_i submits a transaction to the Data Rights Management Contract with d and R_{init} (Initial or updated relationship keywords (e.g., {ownership} or {operation})).
2. **Permission Verification:** The contract verifies that p_i is authorized to register or update the data rights by checking the current state and permissions.
3. **Internal Transaction:** Upon successful verification, the contract invokes the internal transaction

$$S' = \delta\Big(S, \big((p_i, p_j), \{d\}, \texttt{register/update}\big)\Big),$$

which updates the relationship, for example,

$$R'\big((p_i, p_j), \{d\}\big) = R\big((p_i, p_j), \{d\}\big) \cup R_{\text{init}}.$$

The updated state S' reflects the new or modified relationship. The Result indicates whether the operation succeeded. An on-chain transaction record confirming the update. The integrated transaction is represented as:

$$T = \big((p_i, p_j), \{d\}, \texttt{register/update}, \text{Result}, S'\big).$$

TEE Registration. The TEE registration transaction prepares the system for secure remote attestation by requiring operators to commit collateral before accessing TEE services.

1. **Operation Request:** The data operator p initiates a TEE registration (stake) transaction by sending a request to the Profit Allocation Contract specifying Δ (The amount of funds to be staked).
2. **Verification:** The contract verifies that p has sufficient balance and that the stake meets protocol requirements.
3. **State Update:** The contract locks the amount Δ as collateral and updates the participant's account balance.

The updated account balance for p with Δ staked. An on-chain record of the stake transaction. The integrated transaction is represented as:

$$T = \big(p, \Delta, \texttt{stake}, \text{Result}, S'\big).$$

This transaction also registers attestation parameters needed for secure data processing. We omit these TEE-specific technical details for generality.

Data Processing Transaction. Process a data operation within the TEE environment.

1. **Operation Request:** The operator p_j submits a data processing transaction with d, op, and the current permissions.
2. **Permission Verification:** The system calls $f\big((p_i, p_j), \{d\}, op, R\big)$ to verify that op is permitted.
3. **TEE Invocation:** Upon a yes from the permission check, the operator triggers the TEE module. The TEE executes op on d in a secure environment, performing remote attestation beforehand.
4. **Attestation and Verification:** The operator submits the remote attestation result to the Data Rights Management contract for verification. If the attestation fails, the contract sends a slashing transaction to the Profit Allocation contract, which penalizes the operator by reducing their staked funds.
5. **Internal Transaction:** After successful processing, the contract updates the state via $S' = \delta\big(S, ((p_i, p_j), \{d\}, op)\big)$.

The operation result along with the updated state S'. An on-chain transaction record documenting the processing. The integrated transaction is

$$T = \big((p_i, p_j), \{d\}, op, \text{Result}, S'\big).$$

Deposit/Retrieval of Money in Profit Allocation. Manage monetary transactions in the Profit Allocation module by handling deposits, withdrawals, and reward distributions. The transaction type is either `deposit` or `retrieve`.

1. **Operation Request:** Participant p initiates a monetary transaction by submitting a request with Δ and specifying the transaction type.
2. **Verification:** The Profit Allocation Contract verifies the request—ensuring sufficient funds for a retrieval or processing the deposit.
3. **State Update:** The contract updates p's account balance:
   - For a `deposit`, the balance increases by Δ.
   - For a `retrieve`, the balance decreases by Δ when funds is sufficient.
4. **Reward Distribution:** Concurrently, the contract processes any reward distributions triggered by related data operations.

The updated account balance for participant p. An on-chain record of the deposit/retrieval transaction. The integrated transaction is represented as:

$$T = \big(p, \Delta, \texttt{deposit/retrieve}, \text{Result}, S'\big).$$

6 Profit Allocation Mechanism

We propose a chained allocation mechanism, modeled as a Stackelberg game, that optimizes data-sharing ecosystems. In this mechanism, data owners keep full control and earn recurring, smart-contract royalties from each downstream resale, and data operators, after the initial purchase, receive direct revenues plus shrinking residual shares from later trades. We also examine the joint equilibrium of privacy protection, fair profit allocation, and scalable transaction volume in the mechanism.

6.1 Chained Allocation Mechanism

We introduce a chain-based data-trading and value-distribution rule to incentivize data utilization and maximize social welfare. The system involves a single data owner (user) u and a sequence of data operators (companies) $c_1, c_2, \ldots$. The current operator c_k keeps a share $(1-q)$. The owner u automatically receives the remaining fraction q. The k-th resale (with $k \geq 1$) delivers an incremental payment q^k to the owner and nets the operator c_k a margin $q^{k-1}(1-q)$. The owner may at any time veto further resales; in addition, the system sets a hard cap N on the maximum resale depth to curb excessive propagation. This simple hierarchy guarantees audited provenance, predictable earnings for every operator, and a transparent, owner-centric flow of royalties.

6.2 Stackelberg Game Model

Assume that the payment for each trade is 1, and there is a growth rate of the value of the data γ at each time of data sharing. The interaction between owners and operators can be modeled as a Stackelberg game, where the Nash equilibrium satisfies:

- Data owners set the revocation probability d to trade off future royalty income against the marginal privacy loss β; the complete utility function that embeds β is presented in Appendix A. The expected cumulative profit function is $P_u = \sum_{k=1}^{N} \left[(1+\gamma)^{k-1} q^k\right] + \sum_{t=N+1}^{\infty} \left[(1+\gamma)^{t-1}(1-d)^{t-N} \cdot q^t\right]$.
- Data operators optimize dividend ratio q to maximize profits while ensuring owner participation. The profit for the operator, considering the risk of the user stopping authorization, is as follows:

$$P_c = \sum_{k=1}^{N} \left[(1+\gamma)^{k-1} q^{k-1}(1-q)\right] \\ + \sum_{t=N+1}^{\infty} \left[(1+\gamma)^{t-1}(1-d)^{t-N} \cdot q^{t-N-1}(1-q)\right]. \quad (1)$$

6.3 Equilibrium Analysis and Simulation

Theorems 1 and 2 characterize the equilibrium of the Stackelberg game in static scenarios ($\gamma = 0$) and dynamic scenarios ($\gamma > 0$).

Theorem 1. *In the static scenario, given any user's risk cost $\beta > 0$ and default trading time N, the unique equilibrium of the game is $q = 0$ and $d = 0$. However, if the system force $q > \beta^{\frac{1}{N}}$, then the equilibrium of the game is:*

1. *When $N = 0$: $q \to 1$ (q is close to 1 but not equal to 1), $d = 1$, leading to $P_c = 1 - \beta^{\frac{1}{4}}$.*
2. *When $N = 1$: $q \in [\sqrt{\beta}, 1)$, $d = \frac{q - \sqrt{\beta}}{q(1 - \sqrt{\beta})}$, leading to $P_c = 1 - \sqrt{\beta}$.*
3. *When $N \geq 2$: $q = \beta^{\frac{1}{N}}$, $d = 0$, leading to $P_c = 1 - \beta^{\frac{N}{N+1}}$.*

Theorem 2. *In a dynamic scenario (with growth rate $\gamma > 1$) given any user's risk cost $\beta \in (0, 1)$, the unique equilibrium is $q \to 1$ (q is close to 1 but not equal to 1) and $d \to 1$.*

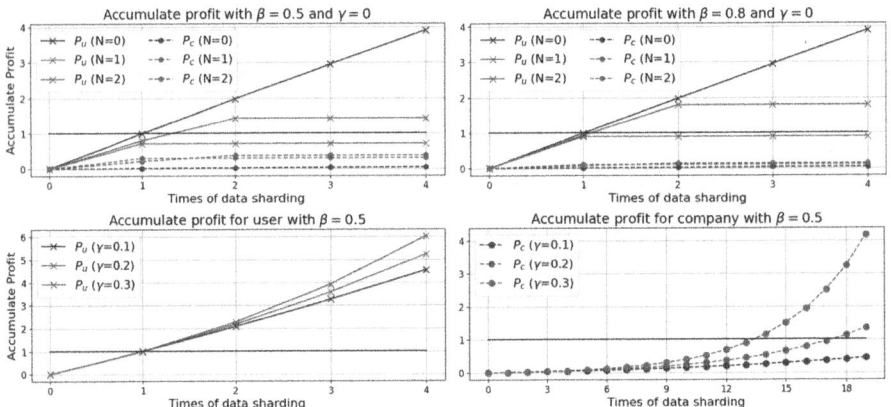

Fig. 2. Accumulated profits of users and company in equilibrium with the time of data shardings. The upper graphs depict the static scenario for cases where $\beta = 0.5$ and $\beta = 0.8$. The lower two graphs show the profits of users and companies with growth rates, specifically for the case where $\beta = 0.5$.

Figures in Fig. 2 show the simulation result of our payment allocation mechanism. From the graphs, it can be observed that in the static scenario, data sharing is only allowed to continue when $N = 0$, which results in the accumulated profit of the user continuously increasing. However, when $N = 1$ or $N = 2$, the user will immediately stop data sharing (causing the green line to cease growing at $N = 1$ and the red line at $N = 2$). Meanwhile, in the absence of a data value growth rate, the company's profit increases slowly and remains consistently below 1.

In contrast, in the dynamic scenario, as shown in the lower two graphs, although the company allocates most of its profit to the user, the company's accumulated profit exceeds one as the number of data sharing instances increases. This indicates that in the presence of data value growth, our mechanism encourages the company to continuously engage in data sharing, effectively addressing the issue of data silos.

7 Conclusion

We present Transaction-Based Data Sharing (TBDS), a modular and transaction-driven solution to address data silos. By leveraging formal languages, blockchain, and TEEs, TBDS ensures compatibility, security, and fairness in data sharing. Our approach integrates data rights management, processing, and profit allocation into a unified framework, fostering collaboration and maximizing data value. While TBDS unifies security assumptions across modules, achieving Universal Composability (UC) security remains an open problem for ensuring that security properties are preserved in arbitrary protocol interactions.

References

1. Abadi, M., et al.: Deep learning with differential privacy. In: Proceedings of the 2016 ACM SIGSAC Conference on Computer and Communications Security, pp. 308–318 (2016)
2. Agarwal, A., Dahleh, M., Sarkar, T.: A marketplace for data: an algorithmic solution. In: Proceedings of the 2019 ACM Conference on Economics and Computation, pp. 701–726 (2019)
3. Albrecht, J.P.: How the GDPR will change the world. Eur. Data Prot. L. Rev. **2**, 287 (2016)
4. Altman, M., Chong, S., Wood, A.: Formalizing privacy laws for license generation and data repository decision automation. Proc. Priv. Enhancing Technol. **2020**(2), 1–19 (2020)
5. Analytics, M.: The age of analytics: competing in a data-driven world. McKinsey Global Institute Research (2016)
6. Augustine, J., Chen, N., Elkind, E., Fanelli, A., Gravin, N., Shiryaev, D.: Dynamics of profit-sharing games. Internet Math. **11**(1), 1–22 (2015)
7. Azaria, A., Ekblaw, A., Vieira, T., Lippman, A.: Medrec: using blockchain for medical data access and permission management. In: 2016 2nd International Conference on Open and Big Data (OBD), pp. 25–30. IEEE (2016)
8. Bonawitz, K.: Towards federated learning at scale: Syste m design. arXiv preprint arXiv:1902.01046 (2019)
9. Chen, L., Koutris, P., Kumar, A.: Model-based pricing: do not pay for more than what you learn! In: Proceedings of the 1st Workshop on Data Management for End-to-end Machine Learning, pp. 1–4 (2017)
10. Chi, J., et al.: A secure and efficient data sharing scheme based on blockchain in industrial internet of things. J. Netw. Comput. Appl. **167**, 102710 (2020)
11. Dai, W., Dai, C., Choo, K., Cui, C., Zou, D., Jin, H.: SDTE: a secure blockchain-based data trading ecosystem. IEEE Trans. Inf. Forensics Secur. **15**, 725–737 (2019)

12. De Gregorio, G., Ranchordás, S.: Breaking down information silos with big data: a legal analysis of data sharing. In: Legal Challenges of Big Data, pp. 204–231. Edward Elgar Publishing (2020)
13. Dean, J., Ghemawat, S.: Mapreduce: simplified data processing on large clusters. Commun. ACM **51**(1), 107–113 (2008)
14. Dwork, C.: Differential privacy. In: International Colloquium on Automata, Languages, and Programming, pp. 1–12. Springer (2006)
15. Dwork, C.: Differential privacy: a survey of results. In: International Conference on Theory and Applications of Models of Computation, pp. 1–19. Springer (2008)
16. Ghorbani, A., Zou, J.: Data shapley: equitable valuation of data for machine learning. In: International Conference on Machine Learning, pp. 2242–2251. PMLR (2019)
17. Ghosh, A., Roth, A.: Selling privacy at auction. In: Proceedings of the 12th ACM Conference on Electronic Commerce, pp. 199–208 (2011)
18. Gollapudi, S., Kollias, K., Panigrahi, D., Pliatsika, V.: Profit sharing and efficiency in utility games. In: 25th Annual European Symposium on Algorithms (ESA 2017). Schloss-Dagstuhl-Leibniz Zentrum für Informatik (2017)
19. Hard, A., et al.: Federated learning for mobile keyboard prediction. arXiv preprint arXiv:1811.03604 (2018)
20. He, Y., Zhu, H., Wang, C., Xiao, K., Zhou, Y., Xin, Y.: An accountable data trading platform based on blockchain. In: IEEE INFOCOM 2019-IEEE Conference on Computer Communications Workshops (INFOCOM WKSHPS), pp. 1–6. IEEE (2019)
21. Jia, R., et al.: Towards efficient data valuation based on the shapley value. In: The 22nd International Conference on Artificial Intelligence and Statistics, pp. 1167–1176. PMLR (2019)
22. Kasneci, E., et al.: ChatGPT for good? on opportunities and challenges of large language models for education. Learn. Individ. Differ. **103**, 102274 (2023)
23. Koutris, P., Upadhyaya, P., Balazinska, M., Howe, B., Suciu, D.: Querymarket demonstration: pricing for online data markets. Proc. VLDB Endowment **5**(12), 1962–1965 (2012)
24. Koutris, P., Upadhyaya, P., Balazinska, M., Howe, B., Suciu, D.: Toward practical query pricing with querymarket. In: Proceedings of the 2013 ACM SIGMOD International Conference on Management of Data, pp. 613–624 (2013)
25. Koutris, P., Upadhyaya, P., Balazinska, M., Howe, B., Suciu, D.: Query-based data pricing. J. ACM (JACM) **62**(5), 1–44 (2015)
26. Liang, X., Zhao, J., Shetty, S., Liu, J., Li, D.: Integrating blockchain for data sharing and collaboration in mobile healthcare applications. In: 2017 IEEE 28th Annual International Symposium on Personal, Indoor, and Mobile Radio Communications (PIMRC), pp. 1–5. IEEE (2017)
27. Liu, J., Lou, J., Liu, J., Xiong, L., Pei, J., Sun, J.: Dealer: an end-to-end model marketplace with differential privacy. Proc. VLDB Endowment **14**(6) (2021)
28. Liu, Q., Safavi-Naini, R., Sheppard, N.P.: Digital rights management for content distribution. In: Conferences in Research and Practice in Information Technology Series, vol. 34, pp. 49–58. Citeseer (2003)
29. Liu, Y., Kang, Y., Xing, C., Chen, T., Yang, Q.: A secure federated transfer learning framework. IEEE Intell. Syst. **35**(4), 70–82 (2020)
30. Luo, Y., Fan, J., Deng, C., Li, Y., Zheng, Y., Ding, J.: Accountable data sharing scheme based on blockchain and SGX. In: 2019 International Conference on Cyber-Enabled Distributed Computing and Knowledge Discovery (CyberC), pp. 9–16. IEEE (2019)

31. Luu, L., Chu, D.H., Olickel, H., Saxena, P., Hobor, A.: Making smart contracts smarter. In: Proceedings of the 2016 ACM SIGSAC Conference on Computer and Communications Security, pp. 254–269 (2016)
32. McMahan, B., Moore, E., Ramage, D., Hampson, S., y Arcas, B.A.: Communication-efficient learning of deep networks from decentralized data. In: Artificial Intelligence and Statistics, pp. 1273–1282. PMLR (2017)
33. Patel, J.: Bridging data silos using big data integration. Int. J. Database Manage. Syst. **11**(3), 01–06 (2019)
34. Richter, H., Slowinski, P.R.: The data sharing economy: on the emergence of new intermediaries. IIC Int. Rev. Intellect. Property Competition Law **50**, 4–29 (2019)
35. Vaswani, A.: Attention is all you need. In: Advances in Neural Information Processing Systems (2017)
36. Wang, Y., et al.: SPDS: a secure and auditable private data sharing scheme for smart grid based on blockchain. IEEE Trans. Industr. Inf. **17**(11), 7688–7699 (2020)
37. Xie, H., Zheng, J., He, T., Wei, S., Hu, C.: TEBDS: a trusted execution environment-and-blockchain-supported IoT data sharing system. Futur. Gener. Comput. Syst. **140**, 321–330 (2023)
38. Yang, Q., Liu, Y., Chen, T., Tong, Y.: Federated machine learning: concept and applications. ACM Trans. Intell. Syst. Technol. (TIST) **10**(2), 1–19 (2019)
39. Yang, S., Wu, F., Tang, S., Gao, X., Yang, B., Chen, G.: On designing data quality-aware truth estimation and surplus sharing method for mobile crowdsensing. IEEE J. Sel. Areas Commun. **35**(4), 832–847 (2017)
40. Yang, T., et al.: Applied federated learning: improving google keyboard query suggestions. arXiv preprint arXiv:1812.02903 (2018)
41. Yu, H., et al.: A fairness-aware incentive scheme for federated learning. In: Proceedings of the AAAI/ACM Conference on AI, Ethics, and Society, pp. 393–399 (2020)
42. Zhao, W.X., et al.: A survey of large language models. arXiv preprint arXiv:2303.18223 (2023)

Pure Nash Equilibria of Weighted Picking Sequence Protocol is WEF1 for Two Agents

Rufan Bai[1], Huahua Miao[2], Xiaowei Wu[2], Cong Zhang[2(✉)], and Shengwei Zhou[2]

[1] Southeast University, Nanjing, China
rfbai@seu.edu.cn
[2] University of Macau, Macau S.A.R., China
{yc47994,xiaoweiwu,yc27429,yc17423}@um.edu.mo

Abstract. We consider the problem of allocating a set of indivisible goods to a set of strategic agents with arbitrary weights. While *truthful* mechanism is hard to guarantee any non-trivial fairness, Amanatidis et al. (WINE 2021 and MOR 2024) studied the fairness guarantees of the equilibria of the *Round-Robin* mechanism. They show that when all agents have the same weight, every pure Nash equilibrium of Round-Robin leads to an envy-free up to one item (EF1) allocation, with respect to agents' *true valuations*. In this paper, we investigate the weighted setting where agents are asymmetric and study the *weighted picking sequence protocol*, which is a natural extension of *Round-Robin* in the weighted setting. More specifically, we show that the *weighted picking sequence protocol* always has pure Nash equilibria and all the corresponding allocations are weighted EF1 with respect to the *true valuation functions* for two agents.

Keywords: Fair division · Weighted · Pure Nash equilibrium

1 Introduction

The problem of fair allocation considers how to divide a set of resources among a group of agents in a fair way. The study of the problem can be traced back to 1948, in which Steinhaus [25] first formalized fairness and introduced the notion of proportionality (PROP), i.e., each of n agents receives at least $1/n$ of the total value. Besides, another prominent fairness notion is envy-freeness (EF) [16,17], where no agent wants to exchange her bundle of items with some other agent to increase the utility. When resources are divisible, EF allocations always exist [1]. When resources are indivisible, both notions are impossible to guarantee, e.g., when allocating one item to two agents. To measure fairness in the discrete context, a number of weaker fairness notions have been proposed, among which envy-freeness up to one good (EF1) [11,21] is one of the most prevalent. An allocation is said to be EF1 if the envy between any two agents can be eliminated after removing one item from the bundle of the envied agent.

From the algorithmic perspective, plentiful polynomial-time algorithms have been proposed to compute EF1 allocations [21,22]. However, most of these algorithms are designed under the assumption that agents report their true valuations, thereby overlooking the strategic behavior that emerges when agents can manipulate to benefit. In particular, we assume that the agents are *strategic*, who have the potential to misreport their valuations to achieve higher utility in the final allocation. In the area of fair allocation, truthful mechanisms that incentivize strategic agents to reveal their true information seem to be necessary, since the fairness guarantee of a mechanism typically holds with respect to agents' *reported* preferences rather than their *true* preferences. Unfortunately, truthful mechanisms fail to guarantee any meaningful fairness notion unless monetary transfers are allowed, as demonstrated by Amanatidis et al. [3]. As an alternative, Amanatidis et al. [4] studied the fairness guarantee from an equilibrium perspective: whether there exists a (non-truthful) mechanism that allows each of pure Nash equilibria (PNE) to induce a fair allocation with respect to the agents' *true* valuations? They showed that the simple mechanism *Round-Robin* is such a mechanism with respect to EF1 fairness for the additive setting. In follow-up work, Amanatidis et al. [5] extended this result beyond the additive setting by demonstrating that the fairness guarantees of the *Round-Robin* mechanism at equilibrium hold even for agents with cancelable valuation functions.

In this paper, we extend this research by considering the *weighted* fair division, where agents may have unequal entitlements—a natural consideration in scenarios such as inheritance division or resource quotas. The weighted generalization of EF1, termed *weighted envy-freeness up to one good (WEF1)* [13], demands a fairness between any two agents proportional to their agent weights. The Weighted Picking Sequence (WPS) mechanism, a weighted variant of Round-Robin, computes WEF1 allocations [13,31] for the additive setting. However, its robustness to strategic behavior remains unexplored.

1.1 Our Results

In this paper, we study the fairness guarantee of the mechanism for the case of two strategic agents. We prove that for two agents with general weights, who may misreport their valuation functions, *every* PNE of the WPS mechanism guarantees WEF1 with respect to agents' true valuations. This result generalizes the equilibrium fairness of Round-Robin to weighted contexts. Our findings underscore WPS as a rare example of a simple, practical mechanism that harmonizes strategic incentives and fairness in weighted environments.

Main Result (Theorem 3). *For any fair allocation instance with two asymmetric agents, every PNE of the WPS mechanism is WEF1 with respect to the valuation functions v_1 and v_2.*

To the best of our knowledge, our work is the first to study the fairness properties of PNE in the weighted allocation setting. Our result shows that even when agents strategically misreport their bids in an attempt to maximize their utilities, the resulting equilibrium allocation always satisfies the WEF1 fairness guarantee. This result extends the classic Round-Robin fairness outcomes to strategic settings with asymmetric weights, demonstrating that, with the possibility of non-truthful bidding, each equilibrium outcome remains fair when judged by the agents' true valuations—a stark contrast to prior works where fairness and truthfulness are often at odds in the absence of monetary transfers.

Our analysis is built upon a series of novel ideas and interesting observations. We first show that when an agent bids truthfully, she is guaranteed a fair allocation relative to the other agent by hypothetically removing either the first or the second item allocated to the opponent (Lemmas 1 and 2). Next, we employ a proof by contradiction. Suppose that, contrary to our claim, there exists a PNE in which the resulting allocation is not WEF1. In this case, there must exist an agent who can unilaterally deviate from the equilibrium strategy to obtain a strictly higher utility than in the equilibrium outcome. This would provide the agent with a profitable unilateral deviation. However, by definition, no unilateral deviation should be able to improve an agent's outcome in any PNE. This contradiction establishes that every PNE must yield an allocation that is WEF1.

1.2 Other Related Works

The problem of fairly allocating indivisible items to non-strategic agents has been studied extensively. Here we only introduce some closely related works and refer to the recent surveys [2,7,26] for a more comprehensive overview.

Weighted Picking Sequence. Motivated by real-world applications where agents may not be equally obliged, Chakraborty et al. [13] proposed the *weighted* setting where agents are asymmetric. They introduced the fairness notion of *weighted envy-freeness up to one item* (WEF1) and proposed the *weighted picking sequence* (WPS) protocol for computing WEF1 allocations, which can be viewed as a weighted version of *Round-Robin*. In addition, Chakraborty et al. [14] studied general picking sequence algorithms and the corresponding monotonicity and fairness guarantees. Recently, the algorithms has been extended to compute WEF1 allocations for chores by Wu et al. [31] and Springer et al. [24]. Notably, Wu et al. [31] introduced a novel analysis framework from a continuous perspective for analyzing the WPS algorithms, providing new analytical tools for understanding fair allocation mechanisms in weighted settings. The continuous-perspective analysis technique has also been applied to other weighted fair allocation problems [8,18]. In this paper, we will also adopt this continuous perspective to show that every PNE of the WPS protocol for two agents is WEF1.

Truthful (and Fair) Allocation. For the strategic setting, Caragiannis et al. [12] and Markakis and Psomas [23] were the first to consider the question of whether it is possible to have mechanisms that are truthful and satisfy some fairness requirements. Amanatidis et al. [3] resolved this question for two agents, showing there is no truthful mechanism with fairness guarantees under any meaningful fairness notion. As a result, subsequent papers considered truthful mechanism design under restricted valuation function classes [9,10,19,20,27,28,30]. There is also research that bypasses the strong negative result by relaxing the truthfulness requirement [29,32].

1.3 Organization

The remainder of this paper is organized as follows: in Sect. 2 we formally introduce the model, the necessary notations, and the fairness notions. In Sect. 3 we present our result that establishes the fairness of the equilibrium for two strategic agents. We conclude the paper with some open questions in Sect. 4.

2 Preliminaries

We consider the problem of allocating a set of indivisible goods to a set of strategic agents. We use $M = \{e_1, \ldots, e_m\}$ to denote the set of items and $N = \{1, \ldots, n\}$ to denote the set of agents, respectively. Every agent $i \in N$ has an additive valuation function $v_i : 2^M \to \mathbb{R}^+ \cup \{0\}$, where we use $v_i(e)$ to denote $v_i(\{e\})$ for convenience of notation. We consider the weighted setting where each agent $i \in N$ has a weight $w_i > 0$ and $\sum_{i \in N} w_i = 1$. When $w_i = 1/n$ for all $i \in N$, we call the instance an unweighted instance. We use $\mathbf{w} = (w_1, \ldots, w_n)$ and $\mathbf{v} = (v_1, \ldots, v_n)$ to denote the weights and valuation functions of agents, respectively. For ease of notation we use $A + e$ and $A - e$ to denote $A \cup \{e\}$ and $A \setminus \{e\}$, respectively, for any $A \subseteq M$ and $e \in M$. We use $[k]$ to denote $\{1, \ldots, k\}$. An allocation $\mathbf{A} = (A_1, \ldots, A_n)$ is an n-partition of the items M such that $A_i \cap A_j = \emptyset$ for all $i \neq j$ and $\cup_{i \in N} A_i = M$, where agent i receives bundle A_i. An instance of our problem is denoted as $\mathcal{I} = (M, N, \mathbf{w}, \mathbf{v})$. We first introduce the fairness notion of *weighted envy-freeness up to one item* (WEF1) for the allocation of goods.

Definition 1 (WEF1). *An allocation $\mathbf{A}$ is called weighted envy-free up to one item (WEF1) if for any two agents $i, j \in N$, either $A_j = \emptyset$, or there exists an item $e \in A_j$ such that*

$$\frac{v_i(A_i)}{w_i} \geq \frac{v_i(A_j - e)}{w_j}.$$

Since our work mainly focuses on two agents, in the following we use notions specifically for the case of two agents, e.g., v_1, v_2 for the valuations, and (A_1, A_2) for the allocation.

2.1 Mechanisms and Equilibria

We study allocation mechanisms that ensure fairness without payments to agents. An allocation mechanism $\mathcal{M}$ is an algorithm that takes input from agents and allocates all goods to them. We consider the strategic setting where the agents might not be truthful, i.e., their reported input may be different from the true valuation functions. Specifically, we assume that each agent i reports a bid vector b_i. We use $b_i(e)$ to denote the bid value of item e for agent i. This extends to $b_i(S) = \sum_{h \in S} b_i(h)$ for any subset $S \subseteq M$.

A mechanism $\mathcal{M}$ takes a bid profile $\mathbf{b} = (b_1, b_2)$ as input and returns an allocation $\mathcal{M}(\mathbf{b})$. In the strategic setting, an agent will misreport her valuation, i.e., $b_i \neq v_i$, if such deviation is beneficial. Given an allocation $\mathbf{A} = (A_1, A_2)$, we slightly abuse the notation by writing $v_i(\mathbf{A})$ to denote $v_i(A_i)$.

Definition 2 (Best Response). *Let $\mathcal{M}$ be an allocation mechanism for two weighted agents, and consider a bid profile $\mathbf{b} = (b_1, b_2)$. We say that b_1 is a best response to b_2 if for every alternative bid b' we have*

$$v_i(\mathcal{M}(b', b_2)) \leq v_i(\mathcal{M}(b_1, b_2)).$$

Similarly, b_2 is a best response to b_1 if for all b' we have $v_i(\mathcal{M}(b_1, b')) \leq v_i(\mathcal{M}(b_1, b_2))$.

Definition 3 (Pure Nash Equilibrium for Two Agents). *We say that a bid profile (b_1, b_2) is a pure Nash equilibrium (PNE) if each agent's bid is a best response to the bid of the other agent.*

2.2 Weighted Picking Sequence Protocol

In this section, we present the *Weighted Picking Sequence Protocol* proposed by Chakraborty et al. [13]. The algorithm runs in rounds while in each round it picks the agent i with the minimum *weighted picking times* and assigns them their most desired remaining item (with respect to the bid b_i). When all agents have the same weight $1/n$, the mechanism coincides with the classic *Round-Robin* mechanism. See Mechanism 1 for detailed information. We refer to the process of allocating a good to an agent (lines 4–7) as a round of the mechanism.

Mechanism 1: Weighted-Picking-Sequence(b_1, b_2)

1 $P \leftarrow M$;   // Unallocated Items
2 For all $i \in [2]$, $A_i \leftarrow \emptyset$, $t_i \leftarrow 0$;
3 **while** $P \neq \emptyset$ **do**
4 $\quad i \leftarrow \operatorname*{argmin}_{i \in [2]} \{\frac{t_i}{w_i}\}$;   // break tie by the smaller weight
5 $\quad e \leftarrow \operatorname*{argmax}_{e \in P} \{b_i(e)\}$;   // break tie by the smallest index
6 $\quad A_i \leftarrow A_i + e$, $P \leftarrow P - e$;
7 $\quad t_i \leftarrow t_i + 1$;
8 **return** (A_1, A_2)

Let $\mathsf{WPS}(\mathbf{b})$ denote the allocation returned by Mechanism 1 under reported preference $\mathbf{b}$. Following Lemma 3.5 in [13], we directly have the following lemma.

Lemma 1. *For two different agents i, j, if b_i is the truthful bid of agent i (i.e., $b_i = v_i$), then in the allocation $\mathcal{A}$ returned by WPS(b_i, b_j), agent i is WEF towards agent j after removing the first item picked by agent j, i.e., we have $v_i(A_i)/w_i \geq v_i(A_j - e)/w_j$ where e is the first item picked by agent j during the WPS mechanism.*

2.3 Existence of PNE

Aziz et al. [6] proved that every picking sequence mechanism admits a PNE when agents have strict preferences.

Theorem 1 (Follows from [6]). *For any instance $(M, N, \mathbf{w}, \mathbf{v})$, where each good has a positive value and no agent has equal valuations for any two goods, the WPS mechanism has at least one PNE.*

In the following, we show that the result holds even when some items have zero value and the preferences are not strict. Clearly, we can add the same positive value to the value of every item on every agent, so that all goods have positive values. For every agent i, we can further add an infinitesimal positive value to every item based on their identity to break ties, so that the resulting new valuation function v_i' induces a strict preference. It can be shown that the constructed function satisfies the following properties.

(1) v_i' induces a strict preference ranking over M, which is consistent with the preference ranking induced by v_i;
(2) for any $S, T \subseteq M$, if $v_i(S) > v_i(T)$ then $v_i'(S) > v_i'(T)$.

In other words, the added noises are sufficiently small to break ties (Property (1)) but will not affect the strict preferences between two bundles in the original valuation function (Property (2)). See Lemma A.1 and Theorem A.3 of [4] for a formal proof of the same reduction.

Theorem 2. *For any instance $\mathcal{I} = (M, \{1, 2\}, \mathbf{w}, \mathbf{v})$, the WPS mechanism has at least one PNE.*

Proof. By the above argument, we can construct for every agent i a new valuation function v_i' under which all values are positive and the preference over the items is strict. Let $\mathcal{I}' = (M, \{1, 2\}, \mathbf{w}, \mathbf{v}')$ be the new instance. Then by Theorem 1, $\mathcal{I}'$ admits at least one PNE in the WPS mechanism. Let $\mathbf{b} = (b_1, b_2)$ be such a PNE, and (A_1, A_2) be the allocation returned by WPS$(\mathbf{b})$.

Suppose for contradiction that $\mathbf{b}$ is not a PNE for $\mathcal{I}$. Then some agent, say agent 1, can deviate to b_1' so that the allocation (A_1', A_2') returned by WPS(b_1', b_2) satisfies $v_1(A_1') > v_1(A_1)$. By Property (2) shown above, $v_1(A_1') > v_1(A_1)$ implies $v_1'(A_1') > v_1'(A_1)$, which contradicts the fact that $\mathbf{b}$ is a PNE of $\mathcal{I}'$. □

3 Fairness of PNE for Two Agents

In this section, we show that for any instance $\mathcal{I} = (M, \{1,2\}, \mathbf{w}, \mathbf{v})$, all PNE of the WPS mechanism are WEF1 with respect to the truthful valuation functions. Recall that in the WPS mechanism, agents take turns to pick items following a specific order that depends on the weights of agents. We call the ordering of agents the *picking sequence*, which has length m and starts with $(1, 2)$. Throughout this section, we assume without loss of generality that $w_1 < w_2$[1]. Under this assumption, the picking sequence begins with $(1, 2, 2)$. In other words, for any instance with $w_1 < w_2$, agent 1 always picks first, followed by two consecutive picks by agent 2. This fixed prefix enables a natural but important observation of the allocation, which we formalize as Lemma 2.

Lemma 2. *Consider two agents 1 and 2 with $w_1 < w_2$. If b_1 is the truthful bid of agent 1 (i.e., $b_1 = v_1$), then the allocation $\mathcal{A}$ returned by $\mathsf{WPS}(b_1, b_2)$ is WEF from agent 1's perspective after removing the second item picked by agent 2, i.e., we have $v_1(A_1)/w_1 \geq v_1(A_2 - e)/w_2$ where e is the second item picked by agent 2 during the WPS mechanism.*

While the lemma shows the similarity with Lemma 1 in the previous section, we remark that there are some subtle differences:

- Lemma 1 holds for both agents $1, 2$ while Lemma 2 only holds for agent 1;
- Lemma 1 requires the removal of the first item picked by agent 2 while Lemma 2 requires the removal of the second.

In other words, combining both lemmas, we have that agent 1 is WEF towards agent 2, after removing any of the first two items picked by agent 2. Such observation is crucial for showing that any PNE is WEF1 for agent 1. Before we give the formal proof of Lemma 2, we first present an example to illustrate the execution of the mechanism and its properties.

Example 1. Consider an instance with $n = 2$ agents, with weights $w_1 = 0.4$ and $w_2 = 0.6$, and $m = 7$ items. The picking sequence of the instance is given by $(1, 2, 2, 1, 2, 1, 2)$. Suppose that agent 1 reports her true valuation profile v_1 and agent 2 reports b_2 (which might not be truthful). Their bid profiles are shown in Table 1. Running the WPS protocol gives an allocation $A_1 = \{e_1, e_4, e_6\}$ and $A_2 = \{e_2, e_3, e_5, e_7\}$, which is WEF1 for agent 1 because after removing any of the first two items picked by agent 2, i.e., $e \in \{e_2, e_3\}$ from A_2, $v_1(A_1)/w_1 = 33.75$ is larger than $v_1(A_2 - e)/w_2 = 20$. However, removing the third item e_5 picked by agent 2 does not lead to a WEF allocation because $v_1(A_1)/w_1 = 33.75$ is less than $v_1(A_2 - e_5)/w_2 \approx 34.17$.

We remark that it is possible that $b_2 \neq v_2$. For example, if $v_2 = (4, 9, 8, 7, 6, 5, 10)$, then under $\mathsf{WPS}(v_1, b_2)$, the (true) value of agent 2 is $v_2(A_2) = 33$, while if she reports truthfully then her value is only $v_2(\{e_7, e_2, e_4, e_6\}) = 31$. Since such

---

[1] Note that the case of $w_1 = w_2$ reduces to the unweighted case, which has been considered in [4].

Table 1. Instance showing that if agent 1 reports truthfully, the returned allocation of the weighted picking sequence protocol is WEF1 for agent 1 after removing the first or second item picked by agent 2.

	e_1	e_2	e_3	e_4	e_5	e_6	e_7
v_1	10.5	10	10	2	1.5	1	0.5
b_2	10	9	8	7	6	5	4

deviation is beneficial, she will misreport under the PNE. Therefore, the WPS mechanism might not be truthful, even for two agents.

Proof of Lemma 2: Let $A_1 = \{e_1, \ldots, e_k\}$ and $A_2 = \{e'_1, \ldots, e'_{k'}\}$ be the bundles agents 1 and 2 receive in WPS(v_1, b_2), respectively. We index the items in the bundles A_1 and A_2 based on the order the items are chosen by the corresponding agents. Therefore, we have

$$v_1(e_1) \geq v_1(e_2) \geq \cdots \geq v_1(e_k) \quad \text{and} \quad b_2(e'_1) \geq b_2(e'_2) \geq \cdots \geq b_2(e'_{k'}).$$

In the following we show that $\frac{v_1(A_1)}{w_1} \geq \frac{v_1(A_2 - e'_2)}{w_2}$. We define a continuous non-increasing function $\rho : [0, k/w_1] \to \mathbb{R}^+ \cup \{0\}$ such that

$$\rho(\alpha) = v_1(e_z), \text{ when } \alpha \in \left(\frac{z-1}{w_1}, \frac{z}{w_1}\right] \text{ for some } z \in \{1, 2, \ldots, k\}.$$

Intuitively, as the allocation proceeds, $\rho(\alpha)$ tracks the changing valuations of the items received by agent 1. In particular, for each item e_z in the bundle allocated to agent 1, the function ρ takes the value $v_1(e_z)$ corresponding to the item's position in the allocation process. The value of ρ decreases as we progress through the allocation because items with lower valuations are allocated later, thus ensuring the non-increasing nature of ρ. Similarly, we define $\rho' : (0, k'/w_2] \to \mathbb{R}^+ \cup \{0\}$ to be a continuous function for bundle A_2 (under v_1) as follows:

$$\rho'(\alpha) = v_1(e'_z), \text{ when } \alpha \in \left(\frac{z-1}{w_2}, \frac{z}{w_2}\right] \text{ for some } z \in \{1, 2, \ldots, k'\}.$$

By definition of ρ and ρ', we have

$$\frac{v_1(A_1)}{w_1} = \int_0^{\frac{k}{w_1}} \rho(\alpha) d\alpha, \quad \text{and}$$
$$\frac{v_1(A_2 - e'_2)}{w_2} = \int_0^{\frac{1}{w_2}} \rho'(\alpha) d\alpha + \int_{\frac{2}{w_2}}^{\frac{k'}{w_2}} \rho'(\alpha) d\alpha.$$

Next, we use two technical lemmas established by Wu et al. [31].

Lemma 3 (Lemma 3.4 of [31]). *We have $k/w_1 \geq (k'-1)/w_2$.*

Intuitively, the lemma holds due to the following reason. Recall that when WPS generates the picking sequence, the mechanism selects the agent with the smallest weighted picking times $s_i := \frac{t_i}{w_i}$ at each round. At the k'-th (the last) appearance of agent 2 in the sequence, we must have $\frac{k'-1}{w_2} < \frac{k}{w_1}$ because otherwise agent 1 should be chosen.

Lemma 4 (Lemma 3.5 of [31]). *For all $\alpha \in \left(0, \frac{k'-1}{w_2}\right]$, we have $\rho(\alpha) \geq \rho'(\alpha + \frac{1}{w_2})$.*

Recall that $\rho(\alpha)$ records the value of the item agent 1 is picking when the weighted picking times s_1 of agent 1 grows to α. Roughly speaking, the lemma holds because the event that s_1 grows to α must happen before s_2 grows to $\alpha + \frac{1}{w_2}$, by the construction of the picking sequence.

Now we are ready to show that $\frac{v_1(A_1)}{w_1} \geq \frac{v_1(A_2 - e_2')}{w_2}$. Combining Lemmas 3 and 4, we have

$$\int_{\frac{1}{w_1}}^{\frac{k}{w_1}} \rho(\alpha) d\alpha \geq \int_{\frac{1}{w_1}}^{\frac{k'-1}{w_2}} \rho(\alpha) d\alpha$$
$$\geq \int_{\frac{1}{w_1}}^{\frac{k'-1}{w_2}} \rho'\left(\alpha + \frac{1}{w_2}\right) d\alpha = \int_{\frac{1}{w_1}+\frac{1}{w_2}}^{\frac{k'}{w_2}} \rho'(\alpha) d\alpha. \tag{1}$$

Recall that we have

$$\frac{v_1(A_1)}{w_1} = \int_0^{\frac{1}{w_1}} \rho(\alpha)d\alpha + \int_{\frac{1}{w_1}}^{\frac{k}{w_1}} \rho(\alpha)d\alpha \quad \text{and}$$
$$\frac{v_1(A_2-e_2')}{w_2} = \int_0^{\frac{1}{w_2}} \rho'(\alpha) d\alpha + \int_{\frac{2}{w_2}}^{\frac{1}{w_1}+\frac{1}{w_2}} \rho'(\alpha) d\alpha + \int_{\frac{1}{w_1}+\frac{1}{w_2}}^{\frac{k'}{w_2}} \rho'(\alpha) d\alpha.v$$

Note that since $w_1 < w_2$ we have $\frac{1}{w_1} + \frac{1}{w_2} > \frac{2}{w_2}$.
Given Inequality (1), to show $\frac{v_1(A_1)}{w_1} \geq \frac{v_1(A_2-e_2')}{w_2}$, it suffices to show

$$\int_0^{\frac{1}{w_1}} \rho(\alpha)d\alpha \geq \int_0^{\frac{1}{w_2}} \rho'(\alpha) d\alpha + \int_{\frac{2}{w_2}}^{\frac{1}{w_1}+\frac{1}{w_2}} \rho'(\alpha) d\alpha. \tag{2}$$

By Lemma 4, we have

$$\int_0^{\frac{1}{w_2}} \rho'(\alpha) d\alpha + \int_{\frac{2}{w_2}}^{\frac{1}{w_1}+\frac{1}{w_2}} \rho'(\alpha) d\alpha \leq \int_0^{\frac{1}{w_2}} \rho'(\alpha) d\alpha + \int_{\frac{1}{w_2}}^{\frac{1}{w_1}} \rho(\alpha) d\alpha.$$

On the other hand, since item e_1 is the most valuable item under v_1, for $\alpha \in (0, \frac{1}{w_1}]$ we have

$$\rho(\alpha) = v_1(e_1) \geq v_1(e_1') \geq \rho'(\alpha),$$

which implies that

$$\int_0^{\frac{1}{w_1}} \rho(\alpha)d\alpha = \int_0^{\frac{1}{w_2}} \rho(\alpha)\,d\alpha + \int_{\frac{1}{w_2}}^{\frac{1}{w_1}} \rho(\alpha)\,d\alpha \geq \int_0^{\frac{1}{w_2}} \rho'(\alpha)\,d\alpha + \int_{\frac{1}{w_2}}^{\frac{1}{w_1}} \rho(\alpha)\,d\alpha.$$

Therefore Inequality (2) holds, which (when combined with Inequality (1)) implies that $\frac{v_1(A_1)}{w_1} \geq \frac{v_1(A_2 - e'_2)}{w_2}$ and completes the proof. ∎

Theorem 3. *For any fair allocation instance $\mathcal{I} = (M, \{1,2\}, \mathbf{w}, \mathbf{v})$, every PNE of the WPS mechanism is WEF1 with respect to the valuation functions v_1, v_2.*

We prove Theorem 3 by showing that the returned allocation is WEF1 to agent 1 and 2, in Lemma 5 and Lemma 7, respectively.

Lemma 5. *For any fair allocation instance $\mathcal{I} = (M, \{1,2\}, \mathbf{w}, \mathbf{v})$, every PNE of the WPS mechanism is WEF1 to agent 1 (with respect to her true valuation function v_1).*

Proof. Suppose by contradiction that there exists a PNE $\mathbf{b} = (b_1, b_2)$ such that in the allocation (A_1, A_2) returned by WPS($\mathbf{b}$), agent 1 envies agent 2, even after removing the most valuable good from agent 2's bundle. Let $e^* \in \arg\max e \in A_2 \{v_1(e)\}$ be the most valuable item in A_2 in the perspective of agent 1 (under v_1). Since (A_1, A_2) is not WEF1 to agent 1, we have

$$\frac{v_1(A_1)}{w_1} < \frac{v_1(A_2 - e^*)}{w_2},$$

which implies

$$w_2 \cdot v_1(A_1) < w_1 \cdot v_1(A_2 - e^*).$$

Using that $w_2 = 1 - w_1$, and by reordering the inequality, we obtain

$$v_1(A_1) < w_1 \cdot v_1(M - e^*). \tag{3}$$

Now we consider another situation in which agent 1 reports her true valuation profile v_1. Let $A'_1 = \{e_1, \ldots, e_k\}$ and $A'_2 = \{e'_1, \ldots, e'_{k'}\}$ be the bundles agents 1 and 2 receive in WPS(v_1, b_2), respectively. Again, the items are indexed following the order they are included in the corresponding bundle, and therefore we have

$$v_1(e_1) \geq v_1(e_2) \geq \cdots \geq v_1(e_k) \quad \text{and} \quad b_2(e'_1) \geq b_2(e'_2) \geq \cdots \geq b_2(e'_{k'}).$$

By Lemmas 1 and 2, the allocation is WEF1 from the perspective of agent 1 and we have

$$\frac{v_1(A'_1)}{w_1} \geq \frac{v_1(A'_2 - e'_1)}{w_2} \quad \text{and} \quad \frac{v_1(A'_1)}{w_1} \geq \frac{v_1(A'_2 - e'_2)}{w_2}.$$

By a similar argument as before, we have

$$v_1(A'_1) \geq w_1 \cdot v_1(A'_1 \cup (A'_2 - e'_1)) = w_1 \cdot v_1(M - e'_1), \text{ and}$$
$$v_1(A'_1) \geq w_1 \cdot v_1(M - e'_2).$$

Therefore, we have
$$v_1(A_1') \geq w_1 \cdot (v_1(M) - \min\{v_1(e_1'), v_1(e_2')\}).$$

In the following, we use the above lower bound to show that $v_1(A_1') > v_1(A_1)$, which implies that reporting v_1 in WPS gives a strictly larger value for agent 1, and thus is a contradiction with b_1 being a best response to b_2. First, we show a simple but very important observation that is crucial to derive the contradiction.

Lemma 6. *At least one of items $\{e_1', e_2'\}$ is in bundle A_2.*

Proof. Recall that the picking sequence of WPS begins with $(1, 2, 2)$ (since $w_1 < w_2$). Let h_1, h_2 be the first two items picked by agent 2 in WPS(b_1, b_2). Therefore h_1, h_2 are two of the three most valuable items under b_2. Similarly, since e_1', e_2' are the first two items picked by agent 2 in WPS(v_1, b_2), they are two of the three most valuable items under b_2. Since there cannot be four items among the top three, we must have $|\{h_1, h_2\} \cap \{e_1', e_2'\}| \geq 1$, which implies at least one of items $\{e_1', e_2'\}$ is in $\{h_1, h_2\} \subseteq A_2$. □

Recall that e^* is the most valuable item in A_2 under v_1. Lemma 6 implies that we have either $v_1(e_1') \leq v_1(e^*)$ or $v_1(e_2') \leq v_1(e^*)$, both of which implies that
$$v_1(e^*) \geq \min\{v_1(e_1'), v_1(e_2')\}.$$

Therefore by Inequality (3) we have
$$v_1(A_1) < w_1 \cdot (v_1(M) - \min\{v_1(e_1'), v_1(e_2')\}) \leq v_1(A_1'),$$

which leads to a contradiction with (b_1, b_2) being a PNE. □

Finally, we show that the allocation under a PNE is also WEF1 to agent 2.

Lemma 7. *For any fair allocation instance $\mathcal{I} = (M, \{1, 2\}, \mathbf{w}, \mathbf{v})$, every PNE of the WPS mechanism is WEF1 to agent 2 (with respect to her true valuation function v_2).*

Proof. Suppose towards a contradiction that this is not the case. That is, there exists a PNE $\mathbf{b} = (b_1, b_2)$ such that in the allocation (A_1, A_2) returned by WPS$(\mathbf{b})$ agent 2 weighted envies agent 1, even after removing the most valuable good from agent 1's bundle.

Let e_1 be the first item picked by agent 1 in WPS$(\mathbf{b})$, which is the most valuable item under b_1. Let $e^* \in \arg\max_{e \in A_1}\{v_2(e)\}$ be the most valuable item in A_1 in the perspective of agent 2 (under v_2). Since (A_1, A_2) is not WEF1 to agent 2, we have that
$$\frac{v_2(A_2)}{w_2} < \frac{v_2(A_1 - e^*)}{w_1},$$

which implies that
$$v_2(A_2) < w_2 \cdot v_2(M - e^*) \leq w_2 \cdot v_2(M - e_1),$$

where the last inequality holds since e^* is the most valuable in A_1 under v_2.

Now consider the situation that agent 2 reports her true valuation function v_2.

Let (A'_1, A'_2) be the allocation returned by $\text{WPS}(b_1, v_2)$. Since the picking sequence of WPS depends only on (w_1, w_2), agent 1 is still the first agent that is chosen to pick an item. Therefore the first item picked by agent 1 in $\text{WPS}(b_1, v_2)$ is still e_1, the most valuable item under b_1. By Lemma 1, agent 2 is WEF1 towards agent 1 up to the first item picked by agent 1. Thus we have

$$\frac{v_2(A'_2)}{w_2} \geq \frac{v_2(A'_1 - e_1)}{w_1},$$

which implies

$$v_2(A'_2) \geq w_2 \cdot v_2(M - e_1) > v_2(A_2).$$

Therefore we have a contradiction that when agent 2 reports her true valuation v_2, her value $v_2(A'_2)$ is strictly larger than $v_2(A_2)$, her value when she reports b_2, violating the fact that **b** is a PNE of the WPS mechanism. □

4 Conclusion

In this work, we establish a fundamental connection between the equilibrium properties of the Weighted Picking Sequence mechanism and its fairness guarantees in the strategic setting. We prove that for any two-agent fair allocation instance, every PNE of the mechanism gives a WEF1 allocation with respect to the agents' true valuations. Our findings extend Amanatidis et al.'s equilibrium analysis framework [4] for the classic Round-Robin protocol to the weighted case, establishing the first equilibrium fairness guarantees for two weighted agents.

Our work leaves several interesting questions open. The most natural open question is whether the WEF1 guarantee can be guaranteed for any number of weighted agents. We conjecture that every PNE of WPS mechanism is WEF1 for any number of agents with respect to their true valuation functions. Second, exploring the weighted fairness of equilibria beyond additive valuations (e.g., submodular or cancelable functions) would substantially expand the mechanism's applicability. Third, it would be valuable to extend our results from WEF1 (which coincides with WEF(1,0)) to the generalized fairness notion of $\text{WEF}(x, 1-x)$ for all $x \in [0,1]$ proposed by [15]. Finally, we believe that it would also be interesting to explore the fairness of PNE for classic fair allocation mechanisms, e.g., Round-Robin or general picking sequences, for the allocation of chores or mixed items.

Acknowledgments. Xiaowei Wu is funded by the Science and Technology Development Fund (FDCT), Macau SAR (file no. 0147/2024/RIA2, 0014/2022/AFJ, 0085/2022/A, 001/2024/SKL and SKL-IOTSC-2024-2026). Rufan Bai is funded by the National Natural Science Foundation of China ((No. 62203123, No.62402103) and the Natural Science Foundation of Jiangsu Province(Grants No BK20241273).

References

1. Alon, N.: Splitting necklaces. Adv. Math. **63**(3), 247–253 (1987)
2. Amanatidis, G., et al.: Fair division of indivisible goods: recent progress and open questions. Artif. Intell. **322**, 103965 (2023)
3. Amanatidis, G., Birmpas, G., Christodoulou, G., Markakis, E.: Truthful allocation mechanisms without payments: characterization and implications on fairness. In: EC, pp. 545–562. ACM (2017)
4. Amanatidis, G., Birmpas, G., Fusco, F., Lazos, P., Leonardi, S., Reiffenhäuser, R.: Allocating indivisible goods to strategic agents: pure Nash equilibria and fairness. Math. Oper. Res. **49**(4), 2425–2445 (2024)
5. Amanatidis, G., Birmpas, G., Lazos, P., Leonardi, S., Reiffenh, R.: äuser. Round-robin beyond additive agents: existence and fairness of approximate equilibria. In: EC, pp. 67–87. ACM (2023)
6. Aziz, H., Goldberg, P., Walsh, T.: Equilibria in sequential allocation. In: Rothe, J. (ed.) ADT 2017. LNCS (LNAI), vol. 10576, pp. 270–283. Springer, Cham (2017). https://doi.org/10.1007/978-3-319-67504-6_19
7. Aziz, H., Li, B., Moulin, H., Wu, X.: Algorithmic fair allocation of indivisible items: a survey and new questions. SIGecom Exch. **20**(1), 24–40 (2022)
8. Aziz, H., Li, B., Moulin, H., Wu, X., Zhu, X.: Almost proportional allocations of indivisible chores: computation, approximation and efficiency. Artif. Intell. **331**, 104118 (2024)
9. Babaioff, M., Ezra, T., Feige, U.: Fair and truthful mechanisms for dichotomous valuations. Proc. AAAI Conf. Artif. Intell. **35**, 5119–5126 (2021)
10. Barman, S., Verma, P.: Truthful and fair mechanisms for matroid-rank valuations. In: AAAI, pp. 4801–4808. AAAI Press (2022)
11. Budish, E.: The combinatorial assignment problem: approximate competitive equilibrium from equal incomes. J. Polit. Econ. **119**(6), 1061–1103 (2011)
12. Caragiannis, I., Kaklamanis, C., Kanellopoulos, P., Kyropoulou, M.: On low-envy truthful allocations. In: Rossi, F., Tsoukias, A. (eds.) ADT 2009. LNCS (LNAI), vol. 5783, pp. 111–119. Springer, Heidelberg (2009). https://doi.org/10.1007/978-3-642-04428-1_10
13. Chakraborty, M., Igarashi, A., Suksompong, W., Zick, Y.: Weighted envy-freeness in indivisible item allocation. ACM Trans. Econ. Comput. **9**(3), 18:1-18:39 (2021)
14. Chakraborty, M., Schmidt-Kraepelin, U., Suksompong, W.: Picking sequences and monotonicity in weighted fair division. Artif. Intell. **301**, 103578 (2021)
15. Chakraborty, M., Segal-Halevi, E., Suksompong, W.: Weighted fairness notions for indivisible items revisited. ACM Trans. Econ. Comput. **12**(3), 9:1-9:45 (2024)
16. Foley, D.: Resource allocation and the public sector. Yale Economic Essays, pp. 45–98 (1967)
17. Gamow, G., Stern, M.: Puzzle-Math. Viking Press (1958)
18. Garg, J., Murhekar, A., Qin, J.: Weighted EF1 and PO allocations with few types of agents or chores. In: IJCAI, pp. 2799–2806. ijcai.org (2024)
19. Ghodsi, M., HajiAghayi, M., Seddighin, M., Seddighin, S., Yami, H.: Fair allocation of indivisible goods: beyond additive valuations. Artif. Intell. **303**, 103633 (2022)
20. Halpern, D., Procaccia, A.D., Psomas, A., Shah, N.: Fair division with binary valuations: one rule to rule them all. In: Chen, X., Gravin, N., Hoefer, M., Mehta, R. (eds.) WINE 2020. LNCS, vol. 12495, pp. 370–383. Springer, Cham (2020). https://doi.org/10.1007/978-3-030-64946-3_26

21. Lipton, R.J., Markakis, E., Mossel, E., Saberi, A.: On approximately fair allocations of indivisible goods. In: EC, pp. 125–131. ACM (2004)
22. Markakis, E: Approximation algorithms and hardness results for fair division with indivisible goods. Trends in Computational Social Choice, pp. 231–247 (2017)
23. Markakis, E., Psomas, C.-A.: On worst-case allocations in the presence of indivisible goods. In: Chen, N., Elkind, E., Koutsoupias, E. (eds.) WINE 2011. LNCS, vol. 7090, pp. 278–289. Springer, Heidelberg (2011). https://doi.org/10.1007/978-3-642-25510-6_24
24. Springer, M., Hajiaghayi, M., Yami, H.: Almost envy-free allocations of indivisible goods or chores with entitlements. In: AAAI, pp. 9901–9908. AAAI Press (2024)
25. Steinhaus, H.: The problem of fair division. Econometrica **16**, 101–104 (1948)
26. Suksompong, W.: Weighted fair division of indivisible items: a review. Inf. Process. Lett. **187**, 106519 (2025)
27. Suksompong, W., Teh, N.: On maximum weighted Nash welfare for binary valuations. Math. Soc. Sci. **117**, 101–108 (2022)
28. Suksompong, W., Teh, N.: Weighted fair division with matroid-rank valuations: monotonicity and strategyproofness. Math. Soc. Sci. **126**, 48–59 (2023)
29. Tao, B., Yang, M.: Fair and almost truthful mechanisms for additive valuations and beyond. arXiv preprint: arXiv:2306.15920 (2023)
30. Viswanathan, V., Zick, Y.: A general framework for fair allocation under matroid rank valuations. In: EC, pp. 1129–1152. ACM (2023)
31. Wu, X., Zhang, C., Zhou, S.: Weighted EF1 allocations for indivisible chores. In: EC, p. 1155. ACM (2023)
32. Xiao, M., Ling, J.: Algorithms for manipulating sequential allocation. In: AAAI, pp. 2302–2309. AAAI Press (2020)

A Comparative Study of Waitlist Mechanisms: Deferral Versus Pay-Per-Offer

Zhou Chen[1], Qi Qi[2(✉)], Hao Sun[2], and Muyang Zhao[2]

[1] School of Economics and Management, Southeast University, Nanjing, China
zchenaq@connect.ust.hk
[2] Gaoling School of Artificial Intelligence, Renmin University of China, Beijing, China
{qi.qi,haosun,myzhao13}@ruc.edu.cn

Abstract. This paper investigates waitlist mechanisms for public housing allocation, introducing the pay-per-offer mechanism as a novel alternative to the deferral-based approach. Through a Markov decision process, dynamically arriving items are allocated to waiting agents with heterogeneous values for different items and diverse outside options. Key contributions include analysis of optimal strategies for agents in these mechanisms, evaluation of how evaluation metrics vary based on the distribution of outside options and waitlist parameters. We provide valuable insights into the design and impact of various waitlist mechanisms.

Keywords: Public housing · waitlist mechanism · deferral · pay-per-offer

1 Introduction

The rising cost of housing in major cities has made homeownership increasingly unattainable for many residents. Public housing, provided by governments globally, aims to improve living standards for low-income households. Access to affordable housing is crucial for the well-being of many individuals and families worldwide. Public rental housing (PRH) in Hong Kong is a good example of such programs. However, the efficient allocation of limited public housing resources remains a significant challenge due to the mismatch between supply and demand. Waitlist mechanisms are commonly used to manage this allocation process, but their performance varies widely. Therefore, understanding and optimizing the design of waitlist mechanisms is very important.

We investigate the problem of matching public houses using waitlist mechanisms, with a focus on two specific approaches: *the waitlist with deferral* and the *waitlist with pay-per-offer*. The waitlist with deferral allows applicants to reject housing offers and wait for subsequent opportunities, while the pay-per-offer mechanism introduces a financial incentive by requiring applicants

---

The authors are listed in alphabetical order.

to pay a fee to remain on the waiting list. The latter approach aims to address inefficiencies by discouraging applicants from rejecting offers.

In Hong Kong's PRH program, eligible applicants receive at most three housing offers, that is the waitlist with 2 deferrals, and we want to see if allowing a different number of chances for applicants would be more efficient or switching into a pay-per-offer waitlist will help improve the system. These mechanisms may offer advantages over traditional waitlist systems, but their effectiveness and fairness depend on a variety of factors, including the outside option distribution and waiting list parameters. Therefore, it is important to study the performance of different waitlist mechanisms under various conditions to determine their suitability for different settings.

To analyze these mechanisms, we develop a discrete-time dynamic programming model that formulates the waitlist process as a Markov Decision Process. We examine the equilibrium state of the system and evaluate the performance of both mechanisms under various parameters. Our research aims to provide insights and suggestions for improving the waitlist mechanism for public housing allocation, which can have a significant impact on low-income residents in many cities worldwide.

Specifically, we study the waitlist mechanism for matching houses, introducing a novel pay-per-offer option alongside the existing deferral mechanism, considering dynamic arrivals of both agents and houses. By characterizing each agent's optimal strategy and the market clearance condition, we provide insights into the equilibrium outcome of the system. We also analyze how the equilibrium and various metrics change with the deferral number or price charged for each offer. Our results demonstrate that the distribution of agents' outside options plays a crucial role in determining the effectiveness of the waitlist mechanism. Additionally, we introduce two new metrics, "house in pool" and "queue length," alongside traditional metrics like social welfare and idle waiting time. These metrics provide a better understanding of system equilibrium, highlighting resource utilization and overall system population, thereby enriching our analysis.

1.1 Related Literature

The focus of this paper is primarily on the public housing [5]. However, this problem is not limited to public housing and has far-reaching implications for various field, like school choice, university admission, resident allocation [7,16], dormitory room allocation [12], refugee allocation [3,4], labor market [13], job or resource assignment [8]. Numerous studies have been conducted on the allocation of public resources over the years [1,11,14]. The allocation of public housing presents more dynamic features that require further investigation [17]. Recent research has focused on addressing this issue. In the context of kidney transplantation, several papers have discussed a dynamic matching model with one-sided private information [2,18,19]. Research on public housing allocation explores various waitlist mechanisms with uncertain private values, optimizing for reducing waiting time, maximizing agent welfare, and ensuring strategy-proofness and efficiency through real data analysis [6,10,15].

One most relevant study in the field is [5], which introduces a dynamic model for public housing allocation. In this model, both agents and houses arrive over time and each agent has an outside option type representing her level of need for a public house. The equilibrium state is defined by a stable condition where individual optimality and market clearance are met, effectively capturing the problem's dynamic nature. Therefore, we adopt the model pioneered by [5], who were among the first to introduce the concept of "waitlist with deferral." However, their investigation was limited to specific deferral numbers (zero deferral and infinite deferrals), while subsequent work by [9] examined a more generalized waitlist mechanism, considering varying deferral numbers. They investigated how the equilibrium system is influenced by different deferral numbers. These research focused on non-price-based approaches. In our study, we innovatively introduce a price mechanism into the waitlist system, replacing the traditional free deferral opportunities with a cost-based purchasing approach. Given the cost associated with each opportunity, our model diverges from [9], where the value of a house was considered a one-time utility. Instead, we refine the model by treating the value of a house as a one-period utility due to the expense incurred with each opportunity. While our model shares similarities with previous research, our study's main innovation lies in the broader exploration of mechanisms, particularly in introducing a price mechanism to the waitlist and introducing two new metrics: "houses in pool" and "queue length". These additions enable a comprehensive comparison between non-price-based and price-based waitlist mechanisms.

2 Model

We consider the strategic behavior of agents in a public housing allocation system, following the model proposed by [5]. We examine a discrete-time system operating over an infinite horizon, where both agents and new houses arrive at each time step j. We consider a supply-demand ratio denoted by μ, ranging between 0 and 1. We normalize the number of agents to 1 for simplicity.

In the waitlist system, upon entering, agents undergo a waiting period before receiving a house offer. Upon receiving an offer, agents can choose to either accept or reject it, potentially waiting for the next period. The value of the house to each agent i in period j is a random variable $v_{i,j}$, drawn independently and identically from a two-point distribution with high and low values $v_H = 1$ and $v_L(< 1)$, respectively, with probabilities p and $1 - p$. At the start of each period, every agent has a probability δ of remaining in the system, while $1-\delta$ probability entails exiting due to factors like higher income, rendering them ineligible for the public housing. Eventually, agents exit the system either by accepting a house offer or leaving empty-handed. When agents exit the system without accepting a house offer, they receive an outside option, denoted as α_i. The outside option represents the alternative they might pursue outside the public housing system, such as a private housing option or some other form of opportunity. The outside option value α_i is drawn from a distribution with cumulative distribution function $F(\cdot)$ and probability density function $f(\cdot)$, where we assume that α_i is upper bounded by $v_H = 1$. This outside option serves as a benchmark for agents' decisions when considering whether to accept or reject a house offer.

2.1 Two Types of Waitlist Mechanisms

- *Waitlist with k deferrals:* Upon reaching the top position and receiving a house offer for the first time, the agent assesses his/her personal value $v_{i,j}$ and decides whether to accept or reject it. If he/she rejects the offer and remains in the system, the agent evaluates $v_{i,j+1}$ in the next period and faces the same decision. The agent can reject up to k times while retaining his/her top position on the waitlist. If the agent rejects the house for the $(k+1)$th time, he/she loses his/her top position and must exit the system.
- *Paying a price c for one more offer:* Upon reaching the top position, an agent receives a random offer free of charge. The agent will assess her personal value $v_{i,j}$ for the house and decide to accept or reject the offer. If rejecting, the agent then chooses between purchasing another offer in the subsequent period at a price c or leaving the system unmatched. There is no limit on the total number of offers an agent can purchase.

2.2 Equilibrium

We assume that agents' behavior in the market is influenced by the collective actions of all agents rather than just individual actions. This approach allows us to analyze the market as a whole and study the concept of an equilibrium state. Specifically, we model each agent's decision-making process as a Markov decision process (MDP), following the approach used in [5]. Our focus is on investigating the system's properties under an equilibrium state.

In order to model the MDP, it is necessary to specify the states, the set of actions for each state, and the probabilities of transition between different states. Let $\tau \geq 0$ denote the idle waiting time, which is determined by the equilibrium state. When a new agent enters the market, she must wait for either $\lfloor \tau \rfloor$ periods (with probability $\lfloor \tau \rfloor + 1 - \tau$) or $\lfloor \tau \rfloor + 1$ periods (with probability $\tau - \lfloor \tau \rfloor$) before receiving her first offer. The state space is defined as $S = -\lfloor \tau \rfloor - 1, -\lfloor \tau \rfloor, \cdots, -1, 0, 1, \cdots, k$ for k deferral mechanism, or $S = -\lfloor \tau \rfloor - 1, -\lfloor \tau \rfloor, \cdots, -1, 0, \cdots$ for pay-per-offer mechanism. The states represent the number of periods an agent has to wait before the first chance of being offered a house (if negative) or the number of offers an agent has rejected in the current round. For states $s < 0$, the only action is to wait.

- In the waitlist mechanism with k deferrals, for states $s \geq 0$, the available actions are to accept or reject the offer. If $s < k$, the agent matches if she chooses to accept, and otherwise she moves to state $s+1$ with probability δ in the next period. If $s = k$, the agent matches if she accepts the offer. If agent rejects, she leaves the system.

In the waitlist with pay-per-offer, when $s \geq 0$, the available actions are to accept the offer, reject and buy one more offer, or reject and leave the system. The agent moves to state $s+1$ with probability δ only if she chooses the second action.

In the equilibrium, each agent has an optimal strategy to maximize her expected benefit, which corresponds to the MDP induced by the market with an idle waiting time of τ. When all agents act optimally, the result is the stationary state of the market, where the average number of agents matched in each period is equal to the supply of houses μ. Let $\pi(\alpha)$ represent the probability that an agent with an outside option α is matched to a house before leaving the system. The fraction of agents who get matched is denoted by

$$\bar{\pi} = \mathbb{E}_{\alpha \sim F}[\pi(\alpha)].$$

In an equilibrium, all agents use their optimal strategies, and $\bar{\pi} = \mu$.

2.3 Performance Evaluation Metrics

This paper aims to assess the performance of various mechanisms based on the following metrics.

- *Idle waiting time* τ. It indicates the number of periods a new agent must wait before reaching the top position and receiving a house offer for the first time. A crucial factor in the functioning of housing markets and a significant consideration for policymakers, it is often prominently displayed on government websites, and agents in the market actively respond to this information. The duration of idle waiting time is an important factor for agents, as they are incentivized to avoid extended waiting periods before finding a suitable match. Similarly, the government seeks to manage the idle waiting time to ensure efficient market operation and minimize the negative consequences of prolonged wait times.
- *Social welfare* SW. This metric evaluates the benefits brought to the agents in each period by the mechanism, which measures the level that the agents' living conditions are improved:

$$SW = \frac{1}{\mu} \mathbb{E}_\alpha[u(\alpha)],$$

where $u(\alpha) = \pi(\alpha)(v(\alpha) - \alpha)$ is the expected benefit for an agent with outside option α from joining the system, and $v(\alpha)$ is the matched value for agent with outside option α:

$$v(\alpha) = \frac{1}{\pi(\alpha)}[Pr(\text{matched a high value house}) + v_L \cdot Pr(\text{matched a low value house})].$$

We introduce two novel metrics: queue length and houses in pool:

- *Queue length L*. This metric describes the expected length of the queue in the equilibrium.
- *Houses in pool N*. It shows how many agents are provided with an offer in one period. As the rate of new arriving house remains constant, it reflects the quantity of houses available in the market, essentially representing the number of top positions being offered.

Queue length denotes the number of agents awaiting housing, reflecting the overall demand within the system. However, it does not fully capture equilibrium dynamics, as not all agents receive service. It's akin to a long queue at a store where only a few customers are served. This is where houses in pool becomes relevant. When the "houses in pool" metric is 1, it means that there is always one house left vacant and treated as an offer. However, if this number rises to 2, it indicates that top two agents will be offered and the number of vacant houses has doubled. A high number of houses in pool implies resource wastage, as these houses remain unoccupied, highlighting the potential waste of resources due to repeated housing opportunities.

3 Optimal Strategy and Metrics for Waitlist with Deferrals

This section analyzes the optimal strategy for agents in a waitlist mechanism with k deferrals. In this dynamic programming problem, where the value function for state s is denoted by $V : S \to \mathbb{R}$ and the action set at state s is denoted by A_s. The resulting state after taking action a at state s is represented by s_a. Regarding the agent with an outside option α:

$$V_\alpha(s) = \delta \cdot \mathbb{E}_v[\max_{a \in A_s}\{(\frac{v-\alpha}{1-\delta})\mathbf{1}_{\{matched\}} + V(s_a)(1 - \mathbf{1}_{\{matched\}})\}]. \quad (1)$$

Recall that δ is the probability that an agent is eligible and could stay in the system, which means that δ is also a discount factor, v is the value of the offered house for each period, and $\mathbf{1}_{\{matched\}}$ is an indicator function that takes the value of 1 if the agent is matched with a house and 0 otherwise. In the same time, each agent has the expected life span $\frac{1}{1-\delta}$ periods (assuming the life span is independent of the historical actions), so once matched, agent will receive $\frac{v-\alpha}{1-\delta}$ because agent will get net benefit $v - \alpha$ in each period. In the case where $s < 0$, the agent can only choose to wait. In the waitlist mechanism with deferrals, for states $s \geq 0$, agents have two options: accept or reject the offer. If their state is less than k, accepting the offer results in a match, while rejecting it moves them to the next period with a state of $s+1$. However, if their state is exactly k, accepting the offer results in a match, while rejecting leads to exiting the system. In [9], the value of a house is treated as a one-time utility, meaning that once an agent accepts a house, they receive the full value of the house immediately. However, in our model, we treat the value of a house as a one-period utility, reflecting the fact that agents incur costs with each opportunity they consider.

This adjustment is motivated by the practical consideration that agents face ongoing expenses (e.g., waiting costs, opportunity costs) while remaining in the system, and thus the value of a house should be evaluated in terms of its utility over a single period rather than as a lump-sum benefit.

In the following, we use the notation with subscript k to denote the evaluation criteria corresponding to the waitlist mechanism with k deferrals. For instance, the idle waiting time for the waitlist mechanism with k deferrals is τ_k. This notation allows us to easily differentiate between evaluation criteria for different numbers of deferrals. By solving the dynamic programming problem, we obtain results similar to those in [9], as summarized in the following:

Proposition 1. *For a waitlist with k deferrals, let $x^* = \frac{v_L(1-\delta+\delta p)-\delta p}{1-\delta}$. Then:*

(a) *If the outside option $\alpha \leq x^*$, the agent will accept any house immediately when offered.*
(b) *If $x^* < \alpha \leq v_L$, the agent will only accept a high-value house for the first k offers and accept any house for the last chance.*
(c) *If $\alpha > v_L$, the agent will only accept a high-value house.*

Proof. First we claim that if an agent's optimal strategy is to reject the house with value v_L at state 0, then she will never accept such a house at states 1, ..., $k-1$. Assume the agent rejects the house with value v_L at state $s-1$ and accepts the house with value v_L at state s. This implies that $\frac{v_L-\alpha}{1-\delta} < \delta[p\frac{1-\alpha}{1-\delta} + (1-p)\frac{v_L-\alpha}{1-\delta}]$, where the left-hand side represents the utility of accepting the current house with value v_L at state $s-1$, and the right-hand side represents the expected utility of waiting for the next period. Since this inequality is independent of the period index, it indicates that at state s, accepting a house with value v_L is inferior to rejecting it and waiting to accept any house in the next period. This contradicts the optimality of accepting at state s.

From the theory of dynamic programming, equation (1) has a unique solution. When $0 \leq s \leq k-1$, $V_\alpha(s) = \delta\mathbb{E}_v[\max\{\frac{v-\alpha}{1-\delta}, V_\alpha(s+1)\}]$; When $s = k$, $V_\alpha(s) = \delta\mathbb{E}_v[\max\{\frac{v-\alpha}{1-\delta}, 0\}]$.

(i) Assume that an agent will accept any house when being offered for the first time, then from the equation

$$V_\alpha(0) \geq V_\alpha(1) \Rightarrow \frac{v_L - \alpha}{1-\delta} \geq \delta(p\frac{1-\alpha}{1-\delta} + (1-p)\frac{v_L - \alpha}{1-\delta}),$$

we have $\alpha \leq \frac{v_L(1-\delta+\delta p)-\delta p}{1-\delta}$.

(ii) Assume that an agent will reject a house with value v_L at period 0,..., $k-1$ and accepting a house with value v_L at period k induces utility $\frac{v_L-\alpha}{1-\delta}$, and it is nonnegative if and only if $\alpha \geq v_L$.

Combining (i)(ii), we obtain the results provided in Proposition 1.

Based on this proposition, we can also establish the following theorem for a waitlist with k deferrals:

Theorem 1. *For waitlist with k deferrals:*

(a) If $(1-\delta(1-p))F(v_L) - (1-\delta)F(x^*) - \delta p \geq 0$, the idle waiting time τ_k is decreasing in k. Specifically, when $F(\cdot)$ is concave, this condition is satisfied.
(b) If $(1-\delta(1-p))F(v_L) - (1-\delta)F(x^*) - \delta p \leq 0$, the idle waiting time τ_k is increasing in k. Specifically, when $F(\cdot)$ is convex, this condition is satisfied.
(c) L_k is increasing in k.
(d) N_k is increasing in k.
(e) When $F(\cdot)$ is a concave function, SW_k is increasing in k.

Proof. In the steady state, we assume that, at the end of each period, there are totally $n_1 + ... n_k$ agents staying at the top position, where $n_i (1 \leq i \leq k)$ denotes the number of agents that has used i deferral chances, i.e., those have rejected i random offers. In addition, we assume that in each period, l_k agents reach the top position for the first time. So we have:

$$\begin{cases} \delta l_k (1 - F(x^*))(1-p) = n_1, \\ \delta n_i (1-p) = n_{i+1}, \quad 1 \leq i \leq k-1, \\ \delta(n_1 + ... + n_{k-1})p + \delta n_k \left(\frac{F(v_L) - F(x^*)}{1 - F(x^*)} + \frac{1 - F(v_L)}{1 - F(x^*)} p \right) + \delta l_k (F(x^*) + p(1 - F(x^*))) = \mu, \\ \delta(L_k + 1) - \mu - \delta n_k \frac{1 - F(v_L)}{1 - F(x^*)}(1-p) = L_k, \end{cases}$$

where the first two equations show the dynamics of agents in different states, the third equation implies that in each period μ agents match, and the last equation implies the stationary length of the queue.

Solving the above linear equations, we obtain n_i, l_k and L_k. By definition, $N_k = \sum_{i=1}^{k} n_i + l_k$. So we have:

$$L_k = \frac{\delta - \mu - (\delta(1-p))^{k+1}(1 - F(v_L))\delta^{\tau_k}}{1 - \delta}.$$

$$N_k = \frac{1 - \delta(1-p)(F(x^*) + (\delta(1-p))^k - F(x^*)(\delta(1-p))^k)}{1 - \delta(1-p)} \delta^{\tau_k}.$$

$$SW_k = \delta^{\tau_k} [\int_0^{x^*} \delta(v_L(1-p) + p - t) dF(t)$$
$$+ \int_{x^*}^{v_L} \left(\frac{\delta p(1 - (\delta(1-p))^{k+1})(1-t)}{1 - \delta(1-p)} + (\delta(1-p))^{k+1}(v_L - t) \right) dF(t)$$
$$+ \int_{v_L}^{1} \frac{\delta p(1 - (\delta(1-p))^{k+1})(1-t)}{1 - \delta(1-p)} dF(t)].$$

In the steady state, the fraction of agents who match equals μ, i.e.,

$$\bar{\pi}^k = \mathbb{E}_\alpha[\pi^k(\alpha)] = \mu. \tag{2}$$

$$\mu = \delta^{\tau_k} \left[\delta F(x^*) + \frac{\delta p + (1-\delta)(\delta(1-p))^{k+1}}{1 - \delta(1-p)} (F(v_L) - F(x^*)) + \delta p \frac{1 - (\delta(1-p))^{k+1}}{1 - \delta(1-p)} (1 - F(v_L)) \right]$$
$$= \delta^{\tau_k} \left[\frac{\delta(1-\delta)(1-p)(1 - (\delta(1-p))^k)}{1 - \delta(1-p)} F(x^*) + (\delta(1-p))^{k+1} F(v_L) + \delta p \frac{1 - (\delta(1-p))^{k+1}}{1 - \delta(1-p)} \right].$$

Let $s = (\delta(1-p))^k$, then we rewrite μ as

$$\mu = \delta^{\tau_k}\left[\frac{\delta(1-\delta)(1-p)(1-s)}{1-\delta(1-p)}F(x^*) + \delta(1-p)sF(v_L) + \delta p\frac{1-\delta(1-p)s}{1-\delta(1-p)}\right] \tag{3}$$

and

$$N_k = \frac{\mu[1-\delta(1-p)(F(x^*)+s-F(x^*)s)]}{\delta(1-\delta)(1-p)(1-s)F(x^*) + \delta(1-p)s(1-\delta(1-p))F(v_L) + \delta p(1-\delta(1-p)s)},$$

$$L_k = \frac{\delta-\mu}{1-\delta} - \frac{(1-p)(1-F(v_L))(1-\delta(1-p))\mu s}{(1-\delta)[(1-\delta)(1-p)(1-s)F(x^*) + (1-p)s(1-\delta(1-p))F(v_L) + p(1-\delta(1-p)s)]}.$$

Let $g_k(s) := \frac{\delta(1-\delta)(1-p)(1-s)}{1-\delta(1-p)}F(x^*) + \delta(1-p)sF(v_L) + \delta p\frac{1-\delta(1-p)s}{1-\delta(1-p)}$, then $\mu = \delta^{\tau_k}g_k$. Next, we see how g_k changes with s:

$$g_k'(s) = \frac{\delta(1-p)}{1-\delta(1-p)}\left[-(1-\delta)F(x^*) + (1-\delta(1-p))F(v_L) - \delta p\right].$$

If $(1-\delta(1-p))F(v_L) - (1-\delta)F(x^*) - \delta p \geq 0$, $g_k(s)$ is increasing in s; If $(1-\delta(1-p))F(v_L) - (1-\delta)F(x^*) - \delta p \leq 0$, $g_k(s)$ is decreasing in k. Together with the fact that s is decreasing in k. In the former case, g_k is decreasing in k, thus δ^{τ_k} is increasing in k and τ_k is decreasing in k. Similarly, τ_k is increasing in k in the latter case.

In the same time, we have

$$\frac{dN_k}{ds} = -\frac{\mu(1-p)(1-\delta(1-p))[F(v_L)(1-\delta(1-p)F(x^*)) - F(x^*)(1-\delta(1-p))]}{(\delta(1-\delta)(1-p)(1-s)F(x^*) + \delta(1-p)s(1-\delta(1-p))F(v_L) + \delta p(1-\delta(1-p)s))^2} \leq 0,$$

where the numerator is non-positive because $v_L \geq x^*$ and $1-\delta(1-p)F(x^*) \geq 1-\delta(1-p)$. So N_k is increasing in k.

$$\frac{dL_k}{ds} = -\frac{\delta(1-p)(1-F(v_L))(1-\delta(1-p))\mu(1-\delta)(1-p)F(x^*) + p}{(1-\delta)\delta[(1-\delta)(1-p)(1-s)F(x^*) + (1-p)s(1-\delta(1-p))F(v_L) + p(1-\delta(1-p)s)]^2} < 0,$$

so L_k is increasing in k.

When $F(\cdot)$ is a concave function, τ_k is decreasing in k, and

$$SW_k = \mu + \delta^{\tau_k}\cdot\{(1-v_L)(1-p)F(x^*) - (1-p)^{k+1}\delta^k(1-v_L)(F(v_L)-F(x^*)) - \int_0^{x^*}\alpha f(\alpha)d\alpha$$

$$- \frac{p+(1-p)^{k+1}\delta^k(1-\delta)}{1-(1-p)\delta}\int_{x^*}^{v_L}\alpha f(\alpha)d\alpha - \frac{p\left[1-((1-p)\delta)^{k+1}\right]}{[1-(1-p)\delta]\left[1-\delta^{\tau_k}((1-p)\delta)^{k+1}\right]}\int_{v_L}^1\alpha f(\alpha)d\alpha\}$$

$$= \underbrace{\mu + \delta^{\tau_k}\cdot\{(1-v_L)(1-p)F(x^*) - \int_0^{x^*}\alpha f(\alpha)d\alpha - \frac{p}{1-(1-p)\delta}\int_{x^*}^1\alpha f(\alpha)d\alpha}_{constant}$$

$$- \frac{(1+p)^{k+1}\delta^k(1-\delta)}{1-(1-p)\delta}\cdot\left[F(v_L) - F(x^*) + \frac{\int_{x^*}^1\alpha f(\alpha)d\alpha}{1-x^*} - \frac{\int_{v_L}^1\alpha f(\alpha)d\alpha}{1-v_L}\right]\}.$$

τ_k is decreasing in k, so δ^{τ_k} is increasing in k, and $-(1+p)^{k+1}\delta^k(1-\delta)$ is also increasing in k, so SW_k is increasing in k too.

While our analysis yields insights that are superficially similar to those in [9], our refined treatment of house value as a one-period utility (as opposed to a one-time utility) allows us to uncover more nuanced dynamics in the system. Specifically, the first two conclusions of our theorem suggest that increasing k (the number of allowed rejections) has varying effects on idle waiting time, depending on the agents' outside options. This aligns with the findings in [9], but our model captures the additional dimension of ongoing costs, which influences agents' decisions in a more realistic way.

The remaining three conclusions of our theorem formalize and extend the insights from [9]. Specifically, we prove that increasing k leads to more agents staying longer in the system, which increases both the queue length and the number of houses in the pool. Unlike [9], which primarily provides intuitive insights, our model rigorously accounts for the trade-offs agents face due to the costs associated with each opportunity. Through a formal theoretical analysis, we demonstrate that despite the longer waiting times and larger queues, increasing k results in higher social welfare. This is because the additional opportunities incentivize agents to wait for better matches, leading to more valuable allocations and greater overall welfare.

By providing a formal proof of these dynamics, our work strengthens the theoretical foundation of [9] and offers a more rigorous understanding of the system's behavior.

4 Optimal Strategy and Metrics for Waitlist with Pay-Per-Offer

In this section, we give analysis for the waitlist with pay-per-offer. We also modeled it as a dynamic programming problem. For agents with an outside option α:

$$V_\alpha(s) = \delta \cdot \mathbb{E}_v[\max_{a \in A_s}\{\frac{v-\alpha}{1-\delta}\mathbf{1}_{\{matched\}} + V(s_a)(1-\mathbf{1}_{\{matched\}}) - c\}], \quad (4)$$

When $s \geq 0$, agent can choose from three options: accept and leave the system, reject the offer and buy another offer, or reject the offer and leave the system.

Similar to the analysis in the previous section, we omit the proof. When the fixed cost for buying an extra offer chance, denoted as c, satisfies the condition $c \leq \frac{(1-v_L)p}{1-\delta}$, then the optimal strategy for the agent is given by:

Proposition 2. *For waitlist with pay-per-offer at price c:*

(a) *If outside option $\alpha \leq y^* = \frac{v_L(1-\delta+\delta p)-\delta p}{1-\delta} + c\delta$, the agent will accept any house immediately when offered;*

(b) *If $y^* < \alpha \leq z^* = 1 - \frac{c(1-\delta)}{p}$, the agent will only accept a high-value house, and buy a new offer if the current one is low value;*

(c) If $\alpha > z^*$, for the first offer, the agent will only accept a high-value house, and reject it and leave the system if it is low value.

Based on this proposition, we can prove the following theorem:

Theorem 2. *For waitlist with pay-per-offer at price c:*

(a) If $F(\cdot)$ is convex, then τ_c is decreasing in c.
(b) If $F(\cdot)$ is concave, then τ_c is increasing in c.
(c) N_c is decreasing in c.
(d) L_c is decreasing in c.

When c increases, more people with low outside option will accept any house offered at first time and more people with high outside option will not buy any offer, thus less people will buy the offer. For people around y^* will have a higher match probability because originally they will buy and when c increases they will accept offer immediately. For people around z^*, they originally will buy the offer and when c increases they will not buy anymore, therefore the match probability decreases. If the outside option is concave, the people around y^* have a larger density, so the match probability's increase effect is more influential, and the waiting time will increase to help reach the equilibrium condition where the supply of housing is equal to the demand. Conversely, the people around z^* have higher density and will decrease the idle waiting time for the system in the equilibrium.

Increasing c leads to less people staying longer in the system, which helps decrease the queue length and houses in pool, as higher prices for agents motivate them to leave early thus less crowded.

5 Conclusion

In conclusion, this paper examines the performance of two different mechanisms for matching public houses in the presence of waiting lists: a waitlist mechanism with deferral and with pay-per-offer. The evaluation metrics are analyzed in relation to the outside option distribution and waiting list parameters. Our research offers valuable insights into the design and impact of various waitlist mechanisms for public housing allocation.

Acknowledgements. This work was supported by National Natural Science Foundation of China (No.62172012, No.62472428, No.72401252), the Fundamental Research Funds for the Central Universities, and the Research Funds of Renmin University of China (No. 22XNKJ07, No. 23XNH028), and Major Innovation Planning Interdisciplinary Platform for the "Double-First Class" Initiative, Renmin University of China.

References

1. Abdulkadiroglu, A., Sönmez, T.: Matching markets: theory and practice. Adv. Econ. Econom. **1**, 3–47 (2013)
2. Agarwal, N., Ashlagi, I., Rees, M., Somaini, P., Waldinger, D.: An empirical framework for sequential assignments: The allocation of deceased donor kidneys. Technical report, MIT (2017)
3. Ahani, N., Andersson, T., Martinello, A., Teytelboym, A., Trapp, A.C.: Placement optimization in refugee resettlement. Oper. Res. **69**(5), 1468–1486 (2021)
4. Andersson, T., Ehlers, L.: Assigning refugees to landlords in Sweden: efficient, stable, and maximum matchings. Scand. J. Econ. **122**(3), 937–965 (2020)
5. Arnosti, N., Shi, P.: Design of lotteries and wait-lists for affordable housing allocation. Manage. Sci. **66**(6), 2291–2307 (2020)
6. Bloch, F., Cantala, D.: Dynamic assignment of objects to queuing agents. Am. Econ. J. Microecon. **9**(1), 88–122 (2017)
7. Bronfman, S., et al.: Assigning Israeli medical graduates to internships. Israel J. Health Policy Res. **4**, 1–7 (2015)
8. Budish, E.: The combinatorial assignment problem: approximate competitive equilibrium from equal incomes. J. Polit. Econ. **119**(6), 1061–1103 (2011)
9. Chen, Z., Qi, Q., Wang, C., Wang, W.: What is the optimal deferral number in waitlist mechanism. Available at SSRN 3203280 (2018)
10. Leshno, J.D.: Dynamic matching in overloaded systems. Working paper (2017)
11. Maschler, M., Zamir, S., Solan, E.: Game Theory. Cambridge University Press (2020)
12. Perach, N., Polak, J., Rothblum, U.G.: A stable matching model with an entrance criterion applied to the assignment of students to dormitories at the technion. Internat. J. Game Theory **36**, 519–535 (2008)
13. Roth, A.E.: The evolution of the labor market for medical interns and residents: a case study in game theory. J. Polit. Econ. **92**(6), 991–1016 (1984)
14. Roth, A.E., Sotomayor, M.: Two-sided matching. Handb. Game Theory Econom. App. **1**, 485–541 (1992)
15. Schummer, J.: Influencing waiting lists. Technical report, Working paper, Kellogg School of Management, 2015. 41 (2016)
16. Shi, P.: Assortment planning in school choice. Technical report, mimeo (2016)
17. Sönmez, T., Ünver, M.U.: Matching, allocation, and exchange of discrete resources. In: Handbook of social Economics, vol. 1, pp. 781–852. Elsevier (2011)
18. Su, X., Zenios, S.A.: Patient choice in kidney allocation: a sequential stochastic assignment model. Oper. Res. **53**(3), 443–455 (2005)
19. Su, X., Zenios, S.A.: Recipient choice can address the efficiency-equity trade-off in kidney transplantation: a mechanism design model. Manage. Sci. **52**(11), 1647–1660 (2006)

Optimal Repurchasing Contract Design for Efficient Utilization of Computing Resources

Zhengyan Deng[1(✉)], Yusen Zheng[2], Chenliang Sheng[3], and Shaowen Qin[4]

[1] Jiangnan University, Wuxi, China
6240910002@stu.jiangnan.edu.cn
[2] Peking University, Beijing, China
yusen@stu.pku.edu.cn
[3] Hefei University of Technology, Hefei, China
2023212716@mail.hfut.edu.cn
[4] Flinders University, Adelaide, Australia
shaowen.qin@flinders.edu.au

Abstract. The rapid advancement of AI and other emerging technologies has triggered exponential growth in computing resources demand. Faced with prohibitive infrastructure costs for large-scale computing clusters, users are increasingly resorting to leased computing resources from third-party providers. However, prevalent overestimation of operational requirements frequently leads to substantial underutilization of the computing resources. To mitigate such inefficiency, we propose a contract-based incentive framework for computing resources repurchasing. Comparing to auction mechanisms, our design enables providers to reclaim and reallocate surplus computing resources through market-driven incentives. Our framework operates in a multi-parameter environment where both clients' idle resource capacities and their unit valuations of retained resources are private information, posing a significant challenge to contract design. Two scenarios are considered based on whether all clients possess the same amount of idle resource capacity. By transforming the contract design problem into solving a mathematical program, we obtain the optimal contracts for each scenario, which can maximize the utility of computing resources providers while ensuring the requirements of incentive compatibility (IC) and individual rationality (IR). This innovative design not only provides an effective approach to reduce the inefficient utilization of computing resources, but also establishes a market-oriented paradigm for sustainable computing ecosystems.

Keywords: Computing resources repurchasing · Contract design · Individual rationality · Incentive compatibility

---

Full version of the paper can be found at https://arxiv.org/abs/2504.14823.
This research was supported by the National Science Foundation of China (NSFC) under grant numbers No. 12471339.

© The Author(s), under exclusive license to Springer Nature Singapore Pte Ltd. 2025
V. Chau et al. (Eds.): IJTCS-FAW 2025, LNCS 15828, pp. 264–278, 2025.
https://doi.org/10.1007/978-981-96-8312-3_20

1 Introduction

The rapid development of the digital economy and advancements in artificial intelligence (AI) have positioned computing resources as a key driver of modern productivity. According to the "2023âĂŞ2024 China Artificial Intelligence Computing Resources Development Assessment Report", published by IDC, the scale of China's intelligent computing resource reached 260 EFLOPS in 2022 and is projected to exceed 1,117 EFLOPS by 2027, with a remarkable compound annual growth rate (CAGR) of 33.9%[1]. This rapid growth has been accompanied by a sharp increase in the demand for high-performance GPUs. The imbalance between supply and demand has led to rising prices and frequent shortages, making it increasingly challenging to acquire high-performance GPUs. In response, an increasing number of clients are turning to the computing resource leasing market.

Traditional centralized cloud computing infrastructures are increasingly unable to meet these growing computational requirements. Computing resources leasing is an emerging service model designed to provide clients with flexible and efficient computing resources.[2] It is also better equipped to meet the increasingly diverse demands of the current internet industry. Under the computing resources leasing model, clients can lease computing resources from third-party providers based on their specific needs to carry out computational tasks, without establishing extensive computing infrastructure locally.

However, existing computing resources leasing models, exhibit a common issue of low computing resources utilization efficiency. The issue has become more critical in the current era of massive demand for computing resources. According to IDC data, the utilization rate of general-purpose computing centers in China, which primarily serve enterprises, is only 10% to 15%[3]. This indicates that a significant portion of computing resources remains idle due to a mismatch between supply and actual demand. Meanwhile, some existing clients—though actively consuming computing resources—allocate more resources than necessary to non-essential tasks, obscuring the true demand. It is a challenge for computing resources providers to accurately detect the actual effective utilization of their resources. Bridging these gaps through traditional infrastructure expansion often requires significant capital expenditures, extended implementation timelines, and increased energy demands, resulting in systemic inefficiencies that ultimately leads to substantial waste of resources on a social economical scale.

One economical and effective approach to address this problem is to reclaim idle resources of current clients through economic incentives. To this end, we focus on designing contract mechanisms that encourage clients to sell back their idle computing resources to the provider, who can then reallocate these recovered resources to other clients in need. The fundamental idea behind our contract

---

[1] https://www.ieisystem.com/global/file/2023-12-01/17014097286402c975afc8bfb91fe59018c23ec288049fd.pdf.
[2] https://www.21jingji.com/article/20231212/herald/4a5f93fbee91a636d7f324ea9ea69efd.html.
[3] https://news.qq.com/rain/a/20241029A06DME00.

design is to enable clients with idle resources to maximize their payoffs by carefully selecting a contract item that aligns with their true type. Specifically, we define a client's type as a tuple (v, c) that includes their capacity c of idle computing resources and their valuation v per unit of resource. On the other hand, the provider aims to maximize his utility, which is the difference between the revenue generated by reallocating the reclaimed resources to new clients and the cost incurred in repurchasing these resources. To achieve this goal, the provider designs customized contract items tailored to different client types and presents them as a contract menu. Clients then select a contract item from the menu (or choose not to participate) based on their type. Once a contract is chosen, clients return the resources as stipulated in the agreement and receive the corresponding compensation.

The main contribution of this paper is the design of an implementable contract that maximizes the provider's utility in a multi-dimensional private information environment (private capacity of idle resources and private valuation for one unit of resource) while simultaneously satisfying the requirements of individual rationality (IR) and incentive compatibility (IC) constraints. To address the challenges brought by the two-parameter setting, we assume that the provider knows the discrete probability distribution of clients' types (v, c). Such a setting can effectively help us reduce the computational complexity of the problem. Based on this assumption, we innovatively transform the contract design problem into a mathematical program. To be specific, the objective function of this program is formulated to maximize the expected utility of the provider, while the constraints are carefully designed to ensure IC, IR, and other feasibility requirements. Through this rigorous formulation, we are able to derive optimal contracts for the computing resources repurchasing problem.

2 Related Works

To meet the high demand of computing resources associated with the fast development of AI, large corporations build data centers, while small corporations that cannot afford such infrastructure choose to rely on leasing computing resources to address resource demand in AI research. [13]. Research on computing resources leasing remains limited, and even fewer studies focus on reclaiming idle resources. However, computing resources leasing is fundamentally very much similar to other leasing services, and reclaiming idle resources is comparable to utilizing idle virtual resources like CPU and bandwidth.

Mechanism Design for Idle Digital Resource Reutilization. Numerous studies have been conducted on the recycling or utilization of idle computer resources. In 2017, Quttoum et al., [11] proposed a fair resource allocation model (AFAM) aimed at optimizing resource allocation in Cloud Data Center Networks (CDNs) and improving resource utilization through the VCG auction mechanism. Later, in 2024 [12], they introduced a resource reclamation model (AMAD) based on the Stackelberg leadership framework to reclaim idle resources from users who have already leased resources by employing repeated leader-follower games, thereby further enhancing the actual resource utilization rate. Hosseini et al., [7] proposed a crowdsourced cloud infrastructure called Crowdcloud, which aims to repurpose users' idle computing resources through crowdsourcing methods, thereby creating a decentralized cloud computing platform. Muktadir et al. [6] proposed a leader-follower game-based mechanism for reclaiming unused virtual resources. The infrastructure provider acts as the leader and virtual network operators as followers, negotiating iteratively to set the compensation price for reclaimed resources. And in 2019 [10], the authors refined their original model by further categorizing the types of VNOs and introducing a strategy that adjusts resource compensation prices based on historical negotiation data. Liu et al., [8] approached the problem of reclaiming idle computing resources from organizations by modeling it as a non-cooperative game. The study proposed a dynamic pricing mechanism to integrate and redistribute these idle computing resources, making them available for other cloud users.

Contract-Based Mechanism Design. Contract-based incentive mechanism has been widely used in such fields as resource sharing and crowdsensing areas. Zhang et al., [15] developed a contract-theoretic incentive mechanism to encourage user participation in Device-to-Device communications within cellular networks. This mechanism models the interaction between the base station and users, offering performance-reward contracts that effectively boost user engagement and significantly increase network capacity compared to other methods. Ma et al., [9] proposed a contract-theory-based incentive mechanism to promote Wi-Fi resource sharing in crowdsourced wireless community networks. This mechanism aims to encourage users to share their private Wi-Fi access points, thereby expanding Wi-Fi coverage and reducing deployment costs for operators. Dai et al., [3] proposed a trust-driven contract incentive framework to address the issues of trust and incentives in mobile crowdsensing networks. This contract incentive scheme takes into account users' privacy preferences, to design a set of optimal contracts that maxi- mize the utility of both users and platforms. Zhao et al., [16] considered mobile users' varying privacy preferences and potential information asymmetry from users' passive disclosure of preferences, the study proposed a contract theory-driven incentive mechanism to optimize the balance between resource consumption and task completion under conditions of information asymmetry. Xie et al., [14] proposed an incentive mechanism design based on contract theory for resource trading in computational power networks. This mechanism takes into account the trust between resource providers and consumers, and promotes the effective transaction of resources by designing a credible incentive mechanism.

Although numerous studies have focused on the utilization of idle virtual resources, research on reclaiming leased computing resources remains scarce. In particular, there has been little discussion of computing resource reclamation from the perspective of contract theory. In practical computing resources leasing scenarios, clients not only conceal their true valuation of computing resources but also hide the actual scale of resources they are using, which increases the complexity of the problem beyond that of traditional studies. Given this, it is an innovative attempt to introduce contract theory to systematically explore the issue of computing resource reclamation.

3 Preliminaries

This section firstly introduces the problem of repurchasing computing resources and then formally define the contract, along with the desired properties of the contract design.

3.1 The Computing Resources Repurchasing Problem

In the computing resources repurchasing problem, a *provider* P offers computing resources, and there are n *clients* who have already rented these resources. Each current client may have idle resources. When new clients arrive seeking resources, the provider may not have enough available. Therefore, the provider P needs to recover idle resources from current clients and release them to the new clients. To ensure the minimum demand of the new coming clients is met, a lower bound on supply available for release (denoted by D) is set in advance. If the provider P fails to collect enough resources to meet D, he will face a penalty. This penalty is justified, as failing to satisfy the demand of new clients would force the provider to either lease computing resources from other providers to retain these clients or lose them, both of which would incur a loss. Therefore, provider P aims to maximize his utility by designing *contract* that incentivize clients to return their idle resources, which will be explained in detail in Sect. 3.2.

Let $N = \{*\}1, 2, \cdots, n$ denote the current client set. Each current client i has a *valuation* v_i for a unit of computing resource and a *capacity* c_i of her idle resources. We assume that the valuations of all clients are from a set, denoted by $V = \{v^1, v^2, \cdots, v^K\}$, which contains K distinct values with $v^1 < v^2 < \cdots < v^K$. Similarly, the capacities of idle resources are also assumed to be from a set, denoted by $C = \{c^1, c^2, \cdots, c^L\}$, with $c^1 < c^2 < \cdots < c^L$. Therefore, the private information of each client i can be characterized by a tuple $(v_i, c_i) \in V \times C$. We refer to the private information of client i as the *type* of client i. For simplicity, let $\Gamma = V \times C$ denote the set of all possible types, and let $\gamma_i = (v_i, c_i)$ represent the type of client i. The dual private nature of v_i and c_i induces a multi-parameter mechanism design problem. We further assume each client's type (v_i, c_i), $i \in N$, is independently drawn from a finite-support joint discrete distribution Δ_i, which is privately known to the provider. This is because the provider can infer these distributions from the previous historical transaction data. Let $\lambda_i^{l,k}$ denote the probability that client i's type is realized as (v^k, c^l). Therefore, we have $\sum_{k,l} \lambda_i^{l,k} = 1$ for all $i \in N$.

3.2 Contract

To facilitate the repurchasing of computing resources, we introduce a novel contract-based framework that empowers providers to incentivize clients to contribute their idle resources through tailored economic rewards. Informally, a contract specifies a set of contract items, each containing a recommended repurchasing amount and a payment price. Clients have the option to sign the contract and select one of items based on their types. Alternatively, they may choose not to sign the contract. In this case, they are not required to return any resources and will not receive any payment.

Definition 1 (Contract). *A contract is defined as a pair of functions* (x, p), *where* $x : \Gamma \to \mathbb{R}_+$ *is an allocation rule that maps each type to a recommended repurchasing amount, and* $p : \Gamma \to \mathbb{R}_+$ *is a payment rule that maps each type to a non-negative payment.*

For convenience, we slightly abuse the notation and use $x_{v^k}^{c^l}$ and $p_{v^k}^{c^l}$ to denote the recommended repurchasing amount and payment for type (v^k, c^l), respectively. That is, $x_{v^k}^{c^l} := x(v^k, c^l)$ and $p_{v^k}^{c^l} := p(v^k, c^l)$. Since the type set Γ is the domain of the allocation and payment rules of the contract, we refer to each type $\gamma \in \Gamma$ as a *contract item* in this context. In contract design, we only consider contracts where all payments are non-negative, i.e., $p_{v^k}^{c^l} \geq 0$ for all $(v^k, c^l) \in \Gamma$, as these serve as compensation for clients who return resources.

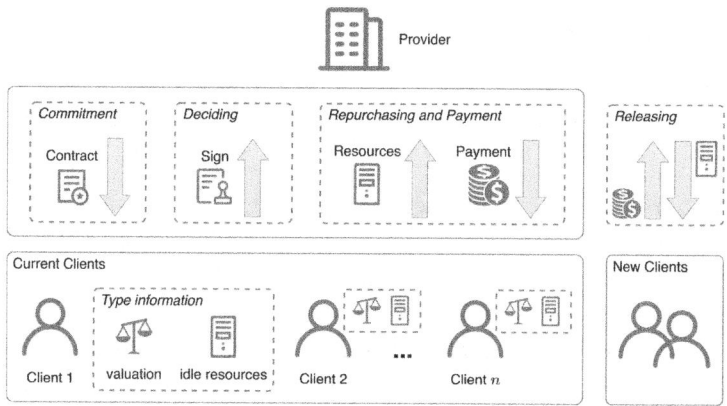

Fig. 1. The interactions between the provider and the clients.

The procedure of the interactions between the provider and the clients is as follows (as shows in Fig. 1):

- *Commitment.* The provider broadcasts the contract (x, p) to the current clients $\{*\}1, 2, \cdots, n$.
- *Deciding.* Upon receiving the proposed contract, clients evaluate it based on their true types $\gamma_i = (v_i, c_i)$ and decide whether to sign the contract with the provider. If they choose to sign, they will strategically select one contract item $\gamma_i' = (v_i', c_i')$ to optimize their utility $U_i(\gamma_i'; \gamma_i)$, defined as:

$$U_i(\gamma_i'; \gamma_i) = p(\gamma_i') - v_i x(\gamma_i').$$

The chosen item γ_i' does not need to be the client's true type, but it must satisfy $x(\gamma_i') \leq c_i$, meaning the selected recommended repurchasing amount cannot exceed the client's idle resource capacity. Once a contract is signed and an item is chosen, the provider and the client are obligated to strictly adhere to the terms specified in the contract item.
- *Repurchasing and Payment.* The client return $x(\gamma_i')$ amount of resource to the provider, and the provider pays $p(\gamma_i')$ to the client.
- *Releasing.* After collecting resources from all current clients, the provider shall release them to the new coming clients, and thus obtain corresponding utility, which is then defined as:

$$U_P(\gamma_1', \gamma_2', \cdots, \gamma_n')$$
$$= \alpha \cdot \sum_{i \in N} x(\gamma_i') - \sum_{i \in N} p(\gamma_i') + M \cdot \min\left\{0, \sum_{i \in N} x(\gamma_i') - D\right\}$$

where α is an exogenous variable, representing the rental price per unit of resource, and M denotes the penalty coefficient applied to each unit of shortfall in resources below the supply lower bound D.

3.3 Problem Formulation

The provider aims to design a contract (**x,p**) that maximizes her expected utility while ensuring several desired properties, including resource feasibility (Definition 2), incentive compatibility (Definition 4), and individual rationality (Definition 5).

The resource feasibility condition ensures that the recommended repurchasing amount does not exceed the client's idle resource capacity when the client selects the contract item that matches her true type.

Definition 2 (Resource Feasibility). *A contract is* resource feasible *if, for every contract item (v^k, c^l), the repurchased amount $x_{v^k}^{c^l}$ does not exceed the available capacity c^l. Specifically, the following condition must hold:*

$$x_{v^k}^{c^l} \leq c^l, \quad \forall c^l \in C, \quad v^k \in V. \tag{1}$$

The resource greedy condition ensures that, for the same private valuation, types with greater capacity are recommended more resources. Additionally, to encourage the return of more resources, the mechanism requires that as many resources as possible be repurchased, for each type γ.

Definition 3 (Resource Greedy). *A contract is* resource greedy *if given a valuation $v^k \in V$, the following two conditions hold for all types (v^k, c^l), $\forall c^l \in C$:*

- Monotonicity in Capacity. *For any two capacity $c^l > c^{l'}$, the repurchased amount from the higher-capacity type must be at least as large as that from the lower-capacity type:*

$$x_{v^k}^{c^l} \geq x_{v^k}^{c^{l'}}, \quad \forall c^l > c^{l'}, v^k \in V. \tag{2}$$

- **Maximal Recycling for Dominated Types.** *If a higher-capacity type c^l is repurchased strictly more than a lower-capacity type $c^{l'}$, i.e., $x_{v^k}^{c^l} > x_{v^k}^{c^{l'}}$, then the repurchased amount from the lower-capacity type must equal its full capacity $c^{l'}$:*

$$x_{v^k}^{c^{l'}} = c^{l'}, \quad \forall c^l, c^{l'} \in C, v^k \in V, \text{ s.t. } x_{v^k}^{c^l} > x_{v^k}^{c^{l'}}. \tag{3}$$

The incentive compatibility condition ensures that clients prefer the contract items specifically designed for their true types.

Definition 4 (Incentive Compatibility). *A contract is incentive compatible if it is resource feasible, and clients achieve the maximum utility by selecting the contract item that matches their true type, i.e.,*

$$p_{v^k}^{c^l} - v^k x_{v^k}^{c^l} \geq p_{v^{k'}}^{c^{l'}} - v^k x_{v^{k'}}^{c^{l'}}, \quad \forall v^k, v^{k'} \in V, c^l, c^{l'} \in C, \text{ s.t. } x_{v^{k'}}^{c^{l'}} \leq c^l. \tag{4}$$

The inequality in Definition 4 simultaneously accounts for the misreporting of both valuation and capacity. Lemma 1 shows that this can be decoupled into two separate inequalities, each focusing on a single parameter. We thus respectively name them as *the incentive compatibility w.r.t the valuation and the capacity.*

Lemma 1 (Equivalence of Incentive Compatibility). *A contract is incentive compatible if and only if it is resource feasible, resource greedy and satisfies the following conditions:*

$$p_{v^k}^{c^l} - v^k x_{v^k}^{c^l} \geq p_{v^{k'}}^{c^l} - v^k x_{v^{k'}}^{c^l}, \qquad \forall v^k, v^{k'} \in V, c^l \in C. \tag{5}$$

$$p_{v^k}^{c^l} - v^k x_{v^k}^{c^l} \geq p_{v^k}^{c^{l'}} - v^k x_{v^k}^{c^{l'}}, \qquad \forall v^k \in V, c^l, c^{l'} \in C \text{ s.t. } x_{v^k}^{c^{l'}} \leq c^l. \tag{6}$$

The full proof for Lemma 1 is left in full version in arXiv.

Definition 5 (Individual Rationality). *A contract is individually rational if it satisfies the following condition: each client achieves a non-negative utility when she signs the contract corresponding to her true types, that is*

$$p_{v^k}^{c^l} - v^k x_{v^k}^{c^l} \geq 0, \qquad \forall v^k \in V, c^l \in C. \tag{7}$$

In this paper, our goal is to design a contract which simultaneously satisfies the properties of resource feasibility and resource greediness, as well as incentive compatibility and individual rationality. For the sake of convenience, we refer to such a contract as a *feasible* contract. Therefore, given a feasible contract $(\mathbf{x},\mathbf{p})$, the expected utility of the provider can be expressed as

$$\mathbb{E}_{\forall i, \gamma_i \sim \Delta_i}[U_P(\gamma_1, \gamma_2, \cdots, \gamma_n)]$$
$$= \sum_{i=1}^{n} \sum_{l=1}^{L} \sum_{k=1}^{K} \lambda_i^{l,k} \left(\alpha x_{v^k}^{c^l} - p_{v^k}^{c^l} \right) + M \cdot \min \left\{ 0, \left(\sum_{i=1}^{n} \sum_{l=1}^{L} \sum_{k=1}^{K} \lambda_i^{l,k} x_{v^k}^{c^l} - D \right) \right\}. \tag{8}$$

Hence, the optimal contract design can be formulated as the following optimization problem:

$$\max_{(x,p)} \mathbb{E}_{\forall i, \gamma_i \sim \Delta_i} [U_P(\gamma_1, \gamma_2, \cdots, \gamma_n)] \quad \text{(Expected utility)} \quad (9)$$

$$\text{s.t.} \quad (1), (2), (3), (4), (7). \quad \text{(Feasibility constraints)}$$

4 Optimal Contract Design for Repurchasing Computing Resources Problem

4.1 Characterization of Feasible Contracts

In this section, we temporarily ignore the objective function of the optimization problem and characterize the feasible contracts in advance. Theorem 1 provides a comprehensive characterization of a feasible contract.

Theorem 1 (Feasible Contract). *A contract is feasible if and only if the following conditions hold:*

(P1) Resource feasibility: $x_{v^k}^{c^l} \leq c^l$ for all $v^k \in V$ and $c^l \in C$.

(P2) The recommended repurchasing amount decreases as the valuation increases, i.e., $c^l \geq x_{v^p}^{c^l} \geq x_{v^q}^{c^l} \geq 0$ for all $v^q \geq v^p$ and $c^l \in C$.

(P3) The payment satisfies the squeeze inequality: $v^p(x_{v^p}^{c^l} - x_{v^q}^{c^l}) \leq p_{v^p}^{c^l} - p_{v^q}^{c^l} \leq v^q(x_{v^p}^{c^l} - x_{v^q}^{c^l})$ for all $v^p, v^q \in V$ and $c^l \in C$.

(P4) Incentive compatibility w.r.t. the capacity: $p_{v^k}^{c^l} - v^k x_{v^k}^{c^l} \geq p_{v^k}^{c^{l'}} - v^k x_{v^k}^{c^{l'}}$ for all $v^k \in V, c^l, c^{l'} \in C$ subject to $x_{v^k}^{c^{l'}} \leq c^l$.

(P5) The utility of the client having the highest valuation is non-negative: $p_{v^K}^{c^l} - v^K x_{v^K}^{c^l} \geq 0$ for all $c^l \in C$.

(P6) The recommended repurchasing amount increases as the capacity increases (i.e., $\forall c^l > c^{l'}$, $x_{v^k}^{c^l} \geq x_{v^k}^{c^{l'}}$), and if $x_{v^k}^{c^l} > x_{v^k}^{c^{l'}}$, $x_{v^k}^{c^{l'}} = c^{l'}$.

Before proving Theorem 1, the following two lemmas are necessary.

Lemma 2. *A contract is incentive compatible w.r.t the valuation ((5) is satisfied), if and only if it satisfies properties (P2) and (P3) in Theorem 1.*

The full proof for Lemma 2 is left in full version.

Lemma 3. *For an incentive compatible contract, if (P5) in Theorem 1 is satisfied, then this contract must be individual rational.*

The full proof for Lemma 3 is left in full version.
Combining Lemma 2, Lemma 3 and Lemma 1, directly leads to Theorem 1.

4.2 Optimal Contract Design

The goal of this paper is to design a feasible contract that maximizes the provider's utility. We solve this problem in two steps. In the first step, we first propose the optimal payment rule $p^*(x) = (p_k^*(x))$ for any $x = (x_{v^1}, \cdots, x_{v^K})$, which is a function of a given allocation x. In the second step, we substitute the payment rule $p^*(x)$ into programming (9), resulting in a new programming whose variables are x. By solving this programming, we ultimately obtained the optimal contract, which includes the optimal allocation rule x^*, and, consequently, the optimal payment rule $p^*(x^*)$.

Based on this approach, we first consider a simpler scenario in which all clients have the same capacity c, and prove program (9) is equivalent to a linear programming problem. However, for the more complicated scenario where there are multiple capacity values, non-convex complementary constraints are additionally needed to ensure the property of resource greedyness. To handle these nonlinear constraints, we introduce slack variables to transform them into tractable smooth constraints, and then gradually approximate the original program.

Scenario: $|C| = 1$. In this case, all current clients have the same capacity c, and then we omit the superscript for convenience. The optimal payment rule is proposed in Proposition 1.

Proposition 1. *Suppose all clients have the same capacity. Then under a feasible contract, if its allocation profile is $x = (x_{v^1}, \cdots, x_{v^K})$, the corresponding optimal payment $p_k^*(x)$ is*

$$p_k^*(x) = \begin{cases} v^K x_{v^K} - \sum_{j=k}^{K-1} v^j (x_{v^{j+1}} - x_{v^j}), & k < K \\ v^K x_{v^K}, & k = K \end{cases} \tag{10}$$

Proof. Because the contract is feasible, we have $p_{v^K} \geq v^K x_{v^K}$, by individual rationality constraint. In addition, the squeeze inequality in (P3) of Theorem 1 implies $p_{v^k} \geq p_{v^{k+1}} - v^k(x_{v^{k+1}} - x_{v^k})$. Because the objective function in programming (9) is linear in $\{p_{v^k}\}$, we can obtain the optimal payment rule as (10) by backward induction. □

Next, to derive the optimal allocation rule, we substitute equation (10) into equation (8), which yields

$$\max_{(x,p)} \mathbb{E}_{\forall i, \gamma_i \sim \Delta_i}[U_P]$$

$$= \sum_{k=2}^{K} \sum_{i=1}^{n} \left[\lambda_i^k (\alpha - v^k) - \sum_{j=1}^{k-1} \lambda_i^j (v^k - v^{k-1}) \right] x_{v^k} + \sum_{i=1}^{n} \lambda_i^1 (\alpha - v^1) x_{v^1}$$

$$+ M \cdot \min\left\{ 0, \left(\sum_{i=1}^{n} \sum_{k=1}^{K} \lambda_i^k x_{v^k} - D \right) \right\}. \tag{11}$$

s.t. $c \geq x_{v^1} \geq \cdots \geq x_{v^k} \cdots \geq x_{v^K} \geq 0$

To transform this optimization problem into a linear programming problem (11), we need to linearize the nonlinear term involving the min function in the objective function. Here's the step-by-step reformulation:

Step 1: Introduce an Auxiliary Variable. Define a new variable t to replace the nonlinear term:

$$t = \min\left\{0, \left(\sum_{i=1}^{n}\sum_{k=1}^{K}\lambda_i^k x_{v^k} - D\right)\right\}.$$

Step 2: Linearize the min Function. The min function can be expressed with two linear inequalities:

$$t \leq 0; \text{ and, } t \leq \sum_{i=1}^{n}\sum_{k=1}^{K}\lambda_i^k x_{v^k} - D.$$

Thus, the reformulated linear program is:

$$\max_{(x,p)} \mathbb{E}_{\forall i, \gamma_i \sim \Delta_i}[U_P] \tag{12}$$

$$\text{s.t. } c \geq x_{v^1} \geq \cdots \geq x_{v^k} \cdots \geq x_{v^K} \geq 0$$

$$t \leq \sum_{k=1}^{K}\sum_{i=1}^{n}\lambda_i^k x_{v^k} - D$$

$$t \leq 0$$

By solving the above program (12), the optimal allocations of $\{x_{v^k}^*\}$ are achieved.

Scenario: $|C| > 1$. This case is significantly more complex than the one where $|C| = 1$, because the conditions of resource greedyness can not be expressed as linear constraints. Without loss of generality, assume $C = \{c^1, \cdots, c^L\}$. Similar to Proposition 1, we can construct an optimal payment rule under the case where $|C| > 1$.

Proposition 2. *Under a feasible contract, if its allocation profile is $x = (x_{v^k}^{c^l})_{c^l \in C, v^k \in V}$, then the corresponding optimal payment $p_{v^k}^{c^l*}(x)$ is*

$$p_{v^k}^{c^l*} = \begin{cases} v^K x_{v^K}^{c^l} - \sum_{j=k}^{K-1} v^j(x_{v^{j+1}}^{c^l} - x_{v^j}^{c^l}), & \forall k \leq K \\ v^K x_{v^K}^{c^l}, & k = K \end{cases}, \forall c_l \in C. \tag{13}$$

The full proof for Proposition 2 is left in ??. Similarly to the case of $|C| = 1$, by substituting (13) into (8), and replacing the penalty term as $t = \min\left\{0, \left(\sum_{i=1}^{n}\sum_{k=1}^{K}\sum_{l=1}^{L}\lambda_i^{l,k} x_{v^k}^{c^l} - D\right)\right\}$, we have the below program.

$$\max_{(x,p)} \mathbb{E}_{\forall i, \gamma_i \sim \Delta_i}[U_P] \tag{14}$$

$$= \sum_{l=1}^{L}\sum_{k=2}^{K}\sum_{i=1}^{n}\left[\lambda_i^{l,k}(\alpha - v^k) - \sum_{j=1}^{k-1}\lambda_i^{l,j}(v^k - v^{k-1})\right]x_{v^k}^{c^l} + \sum_{l=1}^{L}\sum_{i=1}^{n}\lambda_i^{l,1}(\alpha - v^1)x_{v^1}^{c^l} + Mt$$

$$\text{s.t.} \quad c^l \geq x_{v^1}^{c^l} \geq \cdots \geq x_{v^k}^{c^l} \cdots \geq x_{v^K}^{c^l} \geq 0 \quad \forall c^l \in C \tag{15}$$

$$x_{v^k}^{c^L} \geq \cdots \geq x_{v^k}^{c^l} \cdots \geq x_{v^k}^{c^1} \geq 0 \quad \forall v^k \in V \tag{16}$$

$$(x_{v^k}^{c^{l'}} - x_{v^k}^{c^l})(x_{v^k}^{c^l} - c^l) = 0 \quad \forall v^k \in V, \quad \forall c^l, c^{l'} \in C \tag{17}$$

$$t \leq \sum_{l=1}^{L}\sum_{k=1}^{K}\sum_{i=1}^{n}\lambda_i^k x_{v^k}^{c^l} - D \tag{18}$$

$$t \leq 0 \tag{19}$$

in which (18) and (19) are the constraints for the penalty term, (15) and (16) ensure the monotonic allocation, and (17) ensure the resources greedy in Definition 3. Clearly, the complementarity constraint of (17) brings an obstacle to solve program (14) Therefore, we adopt the method proposed by Fletcher and Leyffer [5] to transform the complementarity constraints into nonlinear inequality constraints.

Firstly, (17) is equivalent to $(x_{v^k}^{c^{l+1}} - x_{v^k}^{c^l})(c^l - x_{v^k}^{c^l}) \leq 0$ and $x_{v^k}^{c^{l+1}} - x_{v^k}^{c^l} \geq 0$, $c^l - x_{v^k}^{c^l} \geq 0$. The latter two have already been guaranteed in the linear constraints. This relaxation preserves the core characteristic of the complementarity constraint, which requires that at least one of the two factors is non-positive, while avoiding the restriction that the product must be strictly zero.

Next, we relax (17) as $(x_{v^k}^{c^{l+1}} - x_{v^k}^{c^l})(c^l - x_{v^k}^{c^l}) \leq \epsilon$, and then the program is transformed into (20):

$$\max_{(x,p)} \mathbb{E}_{\forall i, \gamma_i \sim \Delta_i}[U_P] \tag{20}$$

$$\text{s.t.} \quad c^l \geq x_{v^1}^{c^l} \geq \cdots \geq x_{v^k}^{c^l} \cdots \geq x_{v^K}^{c^l} \geq 0 \quad \forall c^l \in C \tag{21}$$

$$x_{v^k}^{c^L} \geq \cdots \geq x_{v^k}^{c^l} \cdots \geq x_{v^k}^{c^1} \geq 0 \quad \forall v^k \in V \tag{22}$$

$$(x_{v^k}^{c^{l'}} - x_{v^k}^{c^l})(c^l - x_{v^k}^{c^l}) \leq \epsilon \quad \forall c^{l'} \geq c^l, \quad c^{l'}, c^l \in C \tag{23}$$

$$t \leq \sum_{l=1}^{L}\sum_{k=1}^{K}\sum_{i=1}^{n}\lambda_i^k x_{v^k}^{c^l} - D$$

$$t \leq 0$$

We can solve this problem by invoking an existing NLP solver (such as filtermpec in [5] or knitro in [1]).

Notice that when solving the programming problem, we relax (17) to (23) by introducing a small positive tolerance parameter ϵ. This relaxation may violate incentive compatibility (Definition 4). We use *regret* to measure how closely the solution of the relaxed programming problem approximates incentive compatibility, which is a common metric for characterizing the extent of approximation to equilibria [2,4]. It represents the maximum excess utility a client can obtain by choosing a contract item that does not match their true type. Specifically, the regret of the solution of the relaxed programming problem is defined as

$$\max_{(v^k,c^l)\in V\times C} \max_{(v^{k'},c^{l'})\in \Omega(c^l)} \left(p^{c^{l'}}_{v^{k'}} - v^k x^{c^{l'}}_{v^{k'}}\right) - \left(p^{c^l}_{v^k} - v^k x^{c^l}_{v^k}\right),$$

$$\Omega\left(c^l\right) := \{*\}(v^{k'},c^{l'})\in V\times C\mid x^{c^{l'}}_{v^{k'}} \leq c^l.$$

Lemma 4 shows that the regret is at most $O(\sqrt{\epsilon})$.

Lemma 4 (Bounded Regret).
The regret of the solution to the optimization problem (20) is at most $O(\sqrt{\epsilon})$. Specifically, $p^{c^{l'}}_{v^{k'}} - v^k x^{c^{l'}}_{v^{k'}} - (p^{c^l}_{v^k} - v^k x^{c^l}_{v^k}) \leq \sum_{k=1}^{K} v^k \cdot \sqrt{\epsilon}$ for all $v^k, v^{k'} \in V, c^l, c^{l'} \in C$ subject to $x^{c^{l'}}_{v^{k'}} \leq c^l$.

Due to the space limitation, the proof of Lemma 4 can be found in full version.

5 Conclusion

In the context of low utilization of leased GPU computing power resources, this study/research proposes a contract-based incentive mechanism to encourage current clients to return their idle resources. Considering that clients may strategically conceal their true idle computing capacity and its actual valuation, we design a contract framework that incentivizes truthful resource reporting. The contract design problem is formulated as an optimization problem aimed at maximizing the utility of resource providers while ensuring IC, IR, and clients' maximum resource capacities. In this work, we assume finite discrete customer types, which is realistic and reasonable. In practice, resource providers can extract typical customer types from historical transaction data through cluster analysis. And in the process of computing resource rental, resource providers often provide users with limited types of packages, which is essentially a discrete type division. Such setting also effectively avoids the additional computational complexity brought about by continuous types. However, we still have to face the more complex IC and IR constraints caused by double information uncertainty. Given the non-convex nature of the optimization problem, we first characterize the feasibility of the proposed contract and transform the constraint conditions accordingly. Due to the presence of two-dimensional decision variables, we derived the optimal strategies in an iterative manner, that is, we initially derive the optimal payment strategy, and we then incorporate the payment strategy into the original problem and derive the optimal allocation strategy. Future work includes extending the current framework to consider a continuous client-type distribution rather than the discrete setting.

References

1. Byrd, R.H., Hribar, M.E., Nocedal, J.: An interior point algorithm for large-scale nonlinear programming. SIAM J. Optim. **9**(4), 877–900 (1999)
2. Corley, H.W.: A regret-based algorithm for computing all pure Nash equilibria for noncooperative games in normal form. Theoret. Econ. Lett. **10**(06), 1253–1259 (2020). https://doi.org/10.4236/tel.2020.106076, https://www.scirp.org/journal/doi.aspx?doi=10.4236/tel.2020.106076
3. Dai, M., Su, Z., Xu, Q., Wang, Y., Lu, N.: A trust-driven contract incentive scheme for mobile crowd-sensing networks. IEEE Trans. Veh. Technol. **71**(2), 1794–1806 (2022). https://doi.org/10.1109/TVT.2021.3117696
4. Erez, L., Lancewicki, T., Sherman, U., Koren, T., Mansour, Y.: Regret minimization and convergence to equilibria in general-sum Markov games. In: Proceedings of the 40th International Conference on Machine Learning, pp. 9343–9373. PMLR (2023). https://proceedings.mlr.press/v202/erez23a.html
5. Fletcher, R., Leyffer, S.: Solving mathematical programs with complementarity constraints as nonlinear programs. Optim. Methods Softw. **19**(1), 15–40 (2004). https://doi.org/10.1080/10556780410001654241, http://www.tandfonline.com/doi/abs/10.1080/10556780410001654241
6. Hena Al Muktadir, A., Jibiki, M., Martinez-Julia, P., Kafle, V.P.: Resource negotiation game for cloud networks with limited resources. In: 2018 IEEE 7th International Conference on Cloud Networking (CloudNet), pp. 1–4, October 2018. https://doi.org/10.1109/CloudNet.2018.8549557
7. Hosseini, M., Angelopoulos, C.M., Chai, W.K., Kundig, S.: Crowdcloud: a crowd-sourced system for cloud infrastructure. Clust. Comput. **22**(2), 455–470 (2019). https://doi.org/10.1007/s10586-018-2843-2
8. Liu, G., Xiao, Z., Tan, G., Li, K., Chronopoulos, A.T.: Game theory-based optimization of distributed idle computing resources in cloud environments. Theoret. Comput. Sci. **806**, 468–488 (2020). https://doi.org/10.1016/j.tcs.2019.08.019
9. Ma, Q., Gao, L., Liu, Y.F., Huang, J.: Incentivizing Wi-Fi network crowdsourcing: a contract theoretic approach. IEEE/ACM Trans. Netw. **26**(3), 1035–1048 (2018). https://doi.org/10.1109/TNET.2018.2812785
10. Muktadir, A., Jibiki, M., Martinez-Julia, P., Kafle, V.P.: Repeated leader follower game for managing cloud networks with limited resources. IEEE Access **7**, 108174–108188 (2019). https://doi.org/10.1109/ACCESS.2019.2933031
11. Quttoum, A.N.: AFAM: a fair allocation model for cloud-datacenter networks. In: Proceedings of the 2017 International Conference on Cloud and Big Data Computing, pp. 67–72. ACM, London United Kingdom, September 2017. https://doi.org/10.1145/3141128.3141147
12. Quttoum, A.N., Alshammari, M.: AMAD: adaptive mapping approach for datacenter networks, an energy-friend resource allocation framework via repeated leader follower game. Comput. Mater. Continua **80**(3), 4577–4601 (2024). https://doi.org/10.32604/cmc.2024.054102
13. Sun, W., Tohirovich Dedahanov, A., Li, W.P., Young Shin, H.: Sanctions and opportunities: factors affecting China's high-tech SMEs adoption of artificial intelligence computing leasing business. Heliyon **10**(16), e36620 (2024). https://doi.org/10.1016/j.heliyon.2024.e36620
14. Xie, R., et al.: Incentive mechanism design for trust-driven resources trading in computing force networks: contract theory approach. IEEE Trans. Netw. Serv. Manage. **22**(1), 618–634. https://doi.org/10.1109/TNSM.2024.3490734, https://ieeexplore.ieee.org/abstract/document/10756785/keywords#keywords

15. Zhang, Y., Song, L., Saad, W., Dawy, Z., Han, Z.: Contract-based incentive mechanisms for device-to-device communications in cellular networks. IEEE J. Sel. Areas Commun. **33**(10), 2144–2155 (2015). https://doi.org/10.1109/JSAC.2015.2435356, http://ieeexplore.ieee.org/document/7110552/
16. Zhao, N., Zhu, H., Sun, Y., Pei, Y., Niyato, D.: A contract-based incentive mechanism for joint data sensing and communication in mobile crowdsourcing networks. IEEE Trans. Veh. Technol. **73**(11), 17929–17934 (2024). https://doi.org/10.1109/TVT.2024.3429394

Characterizing Strategyproofness Through Score Functions in Voting Mechanisms

Felipe V. Furquim[1]([✉]) , Valentin Dardilhac[2] , Daniel Cordeiro[1] , and Johanne Cohen[2]

[1] Escola de Artes, Ciências e Humanidades, Universidade de São Paulo, São Paulo, Brazil
{fvfurq,daniel.cordeiro}@usp.br
[2] LISN CNRS, Université Paris-Saclay, Gif-sur-Yvette, France
{valentin.dardilhac,johanne.cohen}@lisn.fr

Abstract. This work focuses on the strategyproofness of voting systems in which voters select multiple options from a set of possibilities. These systems include those that are used for Participatory Budgeting, where elections are held to determine the allocation of a community budget (e.g., city or regional level) for funding various projects.

We present a model for analyzing voting mechanisms and the *Constrained Change Property (CCP)*, which serves as a criterion for designing strategyproof voting mechanisms. Additionally, we define a new concept of a social choice function and leverage it to develop a new class of utilitarian voting mechanisms, referred to as *score voting*. We prove that the mechanisms designed with core voting with a neutral score function are equivalent to knapsack voting on the same instance. Furthermore, we demonstrate that a score voting mechanism based on a total score function is strategyproof if and only if its score function satisfies CCP.

Building on these findings, we propose an algorithm capable of identifying the closest total score function that ensures the strategyproofness of any given score voting mechanism.

Keywords: strategyproofness · voting mechanism · algorithm

1 Introduction

In a decision-making process, individual preferences are aggregated in a collective choice. Those preferences are from people who are either affected by or interested in a decision; and they can be influenced by multiple criteria such as personal, social, economic, regional, and cultural interests. When in a collective group, those opinions and preferences will inevitably differ [14, 18].

How these individual preferences can be aggregated into a collective choice is studied by a branch of science called "social choice theory" [4]. The social choice theory has been applied to study applications in several domains, in particular, to study what is known in the literature as *Knapsack voting*, adapted from Cabannes' idea of Participatory Budgeting [5].

Knapsack Voting (KP) is a voting system that invites citizens to participate in the process of deciding how public money is spent. This form of participatory democracy was first employed by the city of Porto Alegre, Brazil, in 1989. Since then, different cities around the world have used it, such as Madrid, Seoul, Bogota, New York, and Paris. For example, in 2016, Paris applied KP to allow citizens to vote on how to allocate a budget of 100 million euros [6].

In Knapsack Voting (KP), citizens independently select a subset of projects based on multiple criteria (project cost, beneficiaries, etc.). Their individual preferences are then used to reach a joint decision in a fair and principled manner.

As we will explore further in this paper, voting systems are vulnerable to strategic manipulation when voters possess knowledge of the rules, voting patterns, or preferences of others. In such cases, individuals can strategically misrepresent their true preferences to influence the outcome in their favor, affecting the outcome that could be beneficial to the whole collective group participating in the ballot.

In this work, we focus on studying the properties of voting mechanisms—algorithms that select a solution by taking into account the opinion of the voters—that enhance resilience to manipulation. We study the concept of *strategyproofness* of these mechanisms, which is the idea that the best voting strategy for a voter is to be "sincere", i.e., the player has no incentive to strategically change their preferred vote to increase their outcome.

Designing a strategyproof voting mechanism is hard due to several impossibilities results resulting from the Gibbard-Satterthwaite theorem [11,17] and its extensions. The main contributions of this work are the following.

The first contribution of this work is the introduction of a new model for voting mechanisms that enables the analysis of voting systems independent of their type, whether utilitarian or fair. We called this model the *common Choice Mechanism* (CM). Using this model, we describe conditions that make non-dictatorial mechanisms non-strategyproof and show how the social choice functions respecting the "Constrained Change Property" (CCP) can be used to design CMs that are strategyproof for the unitary case.

The second contribution of this work is the notion of *score functions*. We use this notion to design a new class of utilitarian voting mechanisms that we call *score voting*. We show that the mechanisms designed with score voting with the "neutrality property" are equivalent to knapsack voting. We also show that any score voting designed with a *total* score function is strategyproof if and only if its score function satisfies CCP. We present an algorithm that uses this result to find the closest total score function that makes a score voting strategyproof.

The remainder of this document is organized as follows: Sect. 2 presents work on computational social choice and Participatory Budgeting. In Sect. 3, we present a model for studying choice mechanisms (CMs) and the notations used in this document; we also formally define the notion of strategyproofness. Section 4 studies the strategyproofness of different mechanisms and shows how the CCP property can be used to design CMs that are strategyproof. Section 5 studies several *score voting* mechanisms, shows how they can be used to design strategyproof mechanisms, and presents an algorithm to compute the closest strategyproof total score function. Proofs omitted due to space constraints can be found in [10].

2 Related Work

Before defining a strategyproof voting mechanism, we must first define some key concepts studied in previous works and understand why this is a difficult mechanism to build.

Participatory budgeting (PB) [5,6] has been applied by different municipalities as a democratic tool to allow citizens to prioritize investments on several projects given a limited budget. The idea was first applied in 1989 in the city of Porto Alegre, Brazil, and has been used in several cities, notably in Latin America and Europe [1].

Different electoral systems and their properties have been studied by *Computational social choice* [4], a branch of science that studies the computational aspects of collective decision-making. It covers problems regarding voting theory (including mechanisms design, the computational complexity of choosing a winner, strategic voting, fairness in allocations, coalition formation, etc.).

In particular, there is a great deal of interest in studying the manipulation of decisions by decision-makers. Voters can strategically change their true preferences to obtain a better outcome. Decision-making mechanisms that are immune to strategic voting are called *strategyproof*.

A common electoral system used in participatory budgeting is the k-approval voting. Each voter chooses ("approves") up to k projects, and the projects with the highest number of approvals are funded (with respect to budget constraints).

Goel et al. [12] introduced the *Knapsack Voting* scheme. The idea comes from the fact that the application of PB is conceptually similar to solving the classic Knapsack problem, with the set of chosen budget items fitting a limited budget B while maximizing social value [7]. In this scheme, each citizen votes for a subset of the objects such that the sum of the costs of the objects satisfies the budget constraint. They showed that this schema is strategyproof and welfare-maximizing when the outcome for the voter is given by the ℓ^1 distance from the outcome and its true preference and partially strategyproof under additive concave utilities.

Aggregating budget division schemes that maximize the utilitarian social welfare of voters have a tendency to overprioritize majority preferences, resulting in unfairness problems [9]. The seminal work by Moulin [15] shows that ℓ^1 preferences are a particular case of *single-peaked* preferences and presents a family of voting schemes that are both incentive compatible and proportional by adding some fixed ("phantom") ballots to the voter's ballots and choosing the median of the larger set. This result was later generalized by several works [2,3,8,9,16].

Voting schemes that maximize utilitarian social welfare are possible because *single-peaked* preferences assume that there exists an ordering of the alternatives. More general mechanisms may not be strategyproof due to an important impossibility result independently proved by Gibbard [11] and Satterthwaite [17]. The Gibbard-Satterthwaite theorem states that every resolute, non-imposed, and non-dictatorial social choice function for three or more alternatives is susceptible to strategic manipulation [4].

3 Preliminaries for Collective Decision-Making Problems

The notations used in this paper follows the standard conventions given by Brandt et al. [4].

A *common Choice Mechanism* (CM) is a 6-tuple $(\mathcal{V}, \mathcal{O}, (u_v)_{v \in \mathcal{V}}, \mathcal{S}, \mathcal{A}lgo)$, defined as follows. We consider a set of voters $\mathcal{V} = [n]$ (where $[i] = 1, \ldots, i$ for $i \in \mathbb{N}$) and a finite set $\mathcal{O}$ of m alternatives (or objects), which provide answers to the set of questions Q posed to the voters. The set of all valid collective decisions is denoted by $\mathcal{S} \subseteq \mathcal{O}$. Each voter $v \in \mathcal{V}$ has a private preference over all possible solutions, represented by a utility function[1] $u_v : \mathcal{P}(\mathcal{O}) \to \mathbb{R}$. A set of questions Q defines the format of the ballots. In response, each voter v submits a ballot b_v corresponding to their answers to Q. A *ballot profile* $\mathbf{B} = (b_1, \ldots, b_n)$ consists of a ballot b_v for each voter $v \in \mathcal{V}$. In the remainder of the document, we use the notation $(b_v; b_{-v})$ as a shorthand for the full profile $(b_1, \ldots, b_n)$, where the emphasis is placed on the ballot of voter v relative to those of all other voters.

Voting can take various forms, depending on the structure of the ballot and the nature of the collective decision. For instance, a ballot may consist of a subset $e \in \mathcal{P}(\mathcal{O})$, a linear ordering over the objects, or a valuation function defined on the set of objects. The *social choice function* $\mathcal{A}lgo$ aggregates a ballot profile $\mathbf{B}$ into a collective decision by returning a *solution*, that is, a *winning set* of objects, denoted by $\mathcal{A}lgo(\mathbf{B})$.

Our study focuses on the design of social choice functions that incentivize individual voters to vote sincerely. We denote by b_v^* the *sincere* ballot of voter v, i.e., the ballot that truthfully reflects her preferences according to her utility function (i.e., utility-maximizing).

Definition 1. *A Common Choice Mechanism* $(\mathcal{V}, \mathcal{O}, (u_v)_{v \in \mathcal{V}}, \mathcal{S}, \mathcal{A}lgo)$ *is said to be* strategyproof *if, for every ballot profile* $\mathbf{B}$ *and for every voter* $v \in \mathcal{V}$*, the following condition holds:*

$$u_v\left(\mathcal{A}lgo(b_v^*; \mathbf{B}_{-v})\right) \geq u_v\left(\mathcal{A}lgo(b_v; \mathbf{B}_{-v})\right),$$

that is, a voter cannot improve her utility by submitting any ballot other than her sincere one.

Our work focuses on mechanisms designed to *maximize social welfare*, that is, to select outcomes that best reflect the preferences of the majority of voters, under the assumption that each voter has a utility function over the set of alternatives.

---
[1] The power set of S, i.e., the set of all subsets of S, is denoted by $\mathcal{P}(S)$.

Definition 2. *If each ballot encodes a valuation over the objects, a social welfare-maximizing function (SWF) is an optimization function* $SWF : \mathcal{P}(\mathcal{O}) \to \mathbb{R}$ *that assigns to each feasible subset of objects a real-valued score representing its aggregate utility. The function selects the subset that maximizes total social welfare.*

We focus exclusively on the *utilitarian* social welfare function, denoted $f_{\text{utilitarian}}$, which selects the outcome $s \in \mathcal{S}$ that maximizes the total declared value across all voters: $f_{\text{utilitarian}}(s) = \sum_{v \in \mathcal{V}} \sum_{o \in s} b_v(o)$.

An algorithm $\mathcal{A}lgo$ that computes the optimal outcome of a utilitarian social welfare function is denoted by $\mathcal{A}lgo_{\text{utilitarian}}$.

To study different types of ballots, we consider a specific class of CMs, referred to as *Simple CMs*. These mechanisms incorporate object weights and evaluate the winning set based on the voter's sincere ballot. A *Simple CM* is defined as a 5-tuple $(\mathcal{V}, \mathcal{O}, W, (u_v)_{v \in \mathcal{V}}, \mathcal{A}lgo)$, characterized as follows. Each object $o \in \mathcal{O}$ is associated with a weight given by a function $w : \mathcal{O} \to \mathbb{R}$. A global weight constraint $W \in \mathbb{R}$ restricts the set of feasible solutions to $\mathcal{S} = \{e \in \mathcal{P}(\mathcal{O}) : \sum_{o \in e} w(o) \leq W\}$.

In this work, we focus on a single type of *Simple CM*, which we classify based on the form of the ballot. Specifically, we study the case where the mechanism is an *approval voting* mechanism, meaning that each ballot b_v is a function $b_v : \mathcal{O} \to \{0,1\}$. Equivalently, the ballot can be represented as a binary vector in $\{0,1\}^{|\mathcal{O}|}$, indicating which objects are approved by the voter.

Example of Participatory Budgeting Problem: Given a set of projects $\mathcal{O}$ (the alternatives), a set of voters $\mathcal{V}$ needs to select a set of projects they have identified to be interested in. Each project $o \in \mathcal{O}$ is associated with a cost $w(o)$, and the total available budget is fixed at W. The question Q arises as to which projects should be funded. Each voter v submits a ballot in the form of a subset $b_v \subseteq \mathcal{O}$, subject to the budget constraint, i.e., $\sum_{o \in b_v} w(o) \leq W$. Since the ballot consists of a subset of $\mathcal{O}$, this setting can be modeled as a Simple CM, denoted $(\mathcal{V}, \mathcal{O}, W, (u_v)_{v \in \mathcal{V}}, \mathcal{A}lgo_{\text{utilitarian}})$. The participatory budgeting problem aligns with a utilitarian social welfare framework, as the objective is to select the set of projects that maximizes the total number of approvals-equivalently, the total support expressed by voters.

Finally, *Knapsack voting* refers to the approval voting in the special case where all objects have identical weights. In this setting, the social choice function $\mathcal{A}lgo_{\text{utilitarian}}$ returns the optimal solution with respect to the utilitarian social welfare function.

4 Properties on Different Votes

We study the strategyproofness property on simple CM when its social choice function returns the optimal solution of the social welfare-maximizing function.

The Gibbard-Satterthwaite theorem[2] [11,17] gives some impossibility results about strategyproofness.

Theorem 1 (Gibbard-Satterthwaite Theorem[11,17]). *Whenever the utility u_v is represented by a ranking of the objects, one of the following propositions is true:*

- *The ballot only considers two possible outcomes (ex: a yes/no question);*
- *The social choice function is dictatorial; a voter can choose the outcome;*
- *The CM is not strategyproof.*

This theorem is powerful because it can be used for many existing voting mechanisms. Goel *et al.* [12] shows a surprising result:

Theorem 2 ([12]). *Unitary approval voting is strategyproof.*

We want to generalize Goel et al.'s result to other CM. Hence, we study the strategyproofness property on simple CM when its social choice function returns the optimal solution of the social welfare-maximizing function. Unfortunately, we obtain only results of impossibilities (by giving some counterexample or by applying the Gibbard-Satterthwaite theorem).

We find that a property seems essential to characterize strategyproofness. From now on, any voting mechanism that we study will be used on at least 3 objects. We focus on a compelling property:

Definition 3 (Constrained Change Property (CCP)). *Consider the approval vote $(\mathcal{V}, \mathcal{O}, W, (u_v)_{v \in \mathcal{V}}, \mathcal{A}lgo)$. The social choice function $\mathcal{A}lgo$ respects the Constrained Change Property (CCP) if a voter v switches from an object α to an object β in a ballot, the social choice function can either:*

- *Eject α of the output solution and have it replaced by another object;*
- *Get β to be chosen and eject a previously chosen object (and only one);*
- *Not change the output solution.*

For the first condition, it is enough to consider approval voting with a single voter. The social choice functions should select only the objects in her ballot: If the voter does not vote for the object α, it must no longer be in the output solution. For the last two conditions, modifying a ballot by replacing an object with another does not have to alter the whole solution because the projects have the same cost. Hence, there should be a relative symmetry between the objects.

This property on the social choice function implies the strategyproofness (corresponding to the following theorem):

Theorem 3. *Consider an approval voting CM $(\mathcal{V}, \mathcal{O}, W, (u_v)_{v \in \mathcal{V}}, \mathcal{A}lgo)$, in the unitary case. If the social choice function $\mathcal{A}lgo$ respects CCP, then the CM is strategyproof.*

In the next section, we present results regarding the CCP property. In order to study the conditions on which the converse of Theorem 3 does hold, we introduce the concept of *score functions*.

---

[2] Theorem 1 is a rephrasing of this theorem using our notation.

5 Score Voting

We design an extension of the simple CM $(\mathcal{V}, \mathcal{O}, W, (u_v)_{v \in \mathcal{V}}, \mathcal{A}lgo_{utilitarian})$ model for the case where objects have unitary weight, which is always strategyproof. We introduce the notion of *score functions*, which allows us to introduce a correlation between objects: selecting an object will impact the selection of another object. Each time an object is taken, it can favor another object via a scoring process.

Definition 4 (Score functions). *A score function on m objects is defined by a couple $(M, \mathcal{A}lgo)$ with $m \times m$ real matrix M and an associated social choice function $\mathcal{A}lgo$. Given an integer W, and a ballot profile $\mathbf{B} = (b_1, \ldots, b_n)$, with $b_1 + \dotplus, b_n$ seen as a column vector of m elements, the social choice function $\mathcal{A}lgo$ taking $\mathbf{B}$ as input returns the winning set of W objects which maximizes the inner product $M \cdot \mathbf{e}$ where the ith element of vector $\mathbf{e}$ represents the number of times where o_i is in $b_1 + \cdots + b_n$. If such objects are not clearly defined, we use a tie break, a strict order given by the social choice function $\mathcal{A}lgo$.*

Definition 5 (Score Voting). *A simple CM $(\mathcal{V}, \mathcal{O}, W, (u_v)_{v \in \mathcal{V}}, \mathcal{A}lgo)$ in which the social choice function $\mathcal{A}lgo$ is a score function, is called a Score Voting.*

Score Voting can be used to model utilitarian election systems where the voter must choose several options among a set of possible choices. The model is similar to the Knapsack Voting [12], where every ballot that contains an object o gives one point to o, and the winner is the object with the most points. The idea of a score function is to generalize the votes, allowing voters to give any amount of points to any given object.

Example: The city council would like to propose four projects to the residents, such as "renovating a library" (o_1), "creating a bike path" (o_2), "funding a soccer team" (o_3), or "funding a basketball team" (o_4). However, it can only fund two of the four, and she wants to avoid funding two sports-related projects. Thus, she can construct a score function with $M = \begin{pmatrix} 3 & 0 & 0 & 0 \\ 0 & 3 & 0 & 0 \\ 0 & 0 & 3 & -2 \\ 0 & 0 & -2 & 3 \end{pmatrix}$.

In this case, each column and row of the matrix M represents a project that needs funding. Each position of the matrix M represents the weight of influence between the column and row project. In this example, both objects o_3 and o_4 negatively influence each other, as the ballot organizer wants to avoid funding both projects.

The ballot distribution is the following: four voters want to fund only the sports projects, two voters want to fund only the non-sports projects, and two voters want to fund the "creating a bike path", and "funding a basketball team". The following vector is summarized by $\begin{pmatrix} 2 \\ 4 \\ 4 \\ 6 \end{pmatrix}$. Since $\begin{pmatrix} 3 & 0 & 0 & 0 \\ 0 & 3 & 0 & 0 \\ 0 & 0 & 3 & -2 \\ 0 & 0 & -2 & 3 \end{pmatrix} \begin{pmatrix} 2 \\ 4 \\ 4 \\ 6 \end{pmatrix} =$

in $\begin{pmatrix} 6 \\ 12 \\ 4 = (12-8) \\ 10 = (18-8) \end{pmatrix}$, the winner set is {"creating a bike path", "funding a basketball team" } because the scores of these objects are the highest.

5.1 Neutrality Property Over the Score Voting

In this section, we show how to use score voting to design an election system where objects receive equal treatment (*neutrality property*[3]). We use the standard matrix notations where $M_{i,j}$ is the coordinate (i,j) of matrix M, and $\mathbf{U}_i$ the coordinate i of vector $\mathbf{U}$. The vector such that $(\delta_i)_i = 1$ and $\forall j \neq i, (\delta_i)_j = 0$, is denoted by $\delta_{\mathbf{j}}$.

We start defining the *neutrality property* for election systems that treat objects equally:

Definition 6. *A score function* $(M, \mathcal{A}lgo)$ *is neutral if score matrix* M *has the following form:* $\forall k \in [m], i_1 \neq \cdots \neq i_k \in [m]$,

- $M(\delta_{\mathbf{i_1}} + \cdots + \delta_{\mathbf{i_k}})_{i_1} = \cdots = M(\delta_{\mathbf{i_1}} + \cdots + \delta_{\mathbf{i_k}})_{i_k}$;
- *For all other coordinates* $j \neq i_1, \ldots, i_k$, $M(\delta_{\mathbf{i_1}} + \cdots + \delta_{\mathbf{i_k}})_j$ *is strictly less than* $M(\delta_{\mathbf{i_1}} + \cdots + \delta_{\mathbf{i_k}})_{i_1}$.

For example, knapsack voting is a score voting respecting neutral property. Indeed, it is enough to notice that the identity matrix of size m

$$\begin{pmatrix} 1 & 0 & \cdots & 0 \\ 0 & 1 & \cdots & 0 \\ \vdots & \vdots & \ddots & \vdots \\ 0 & 0 & \cdots & 1 \end{pmatrix}.$$

with a tiebreak algorithm, $\mathcal{A}lgo$ corresponds to their score function.

We will now establish the set of score functions that are *neutral*. The proofs for the following results are described in details in [10].

Proposition 1. *If* $(M, \mathcal{A}lgo)$ *is a neutral score function, then* $M = D + \lambda_1 A^{(1)} + \cdots + \lambda_m A^{(m)}$ *with* $\lambda_1, \ldots, \lambda_m \in \mathbb{R}$, D *is a positive constant diagonal matrix and* $\forall i, j, k \in [m], i \neq j$ $A^{(j)}$ *is such that* $A^{(j)}_{k,j} = 1$ *and* $A^{(j)}_{k,i} = 0$.

We can now compute the winning set of all scoring votes according to knapsack voting.

Theorem 4. *If* $(M, \mathcal{A}lgo)$ *is a neutral score voting, then the outcome of the algorithm is the same as the one of a knapsack voting.*

---

[3] A voting function is *neutral* if it treats all candidates equally, ensuring that any systematic modification of votes—such as interchanging the names of two candidates—results in a correspondingly adjusted outcome without altering the overall decision structure.

5.2 Total Score Functions

The previous results show that neutrality is a hypothesis that is too restrictive for election systems. If neutrality is satisfied, then there is only one possible winning set: the one of knapsack voting.

We use score functions to study a class of mechanisms that are strategyproof under a less restrictive hypothesis. For that, we introduce the notion of the totality of score functions:

Definition 7. *A score function $(M, \mathcal{A}lgo)$ is called* total *if for any $s \in P(\mathcal{O})$, and for any $|s| \leq W < |\mathcal{O}|$, there exists a set of n voters and a ballot profile $\mathbf{B} = (b_1, \ldots, b_n)$ of ballots -that contain k objects each- such that for every objects $i \in s$ and $j \notin s$, $M(\mathbf{B})_i > M(\mathbf{B})_j$.*

We then note $M(\mathbf{B}) = s^+$ to indicate that the scores of the objects of s^+ are strictly greater than the scores of any other object.

Observe that Knapsack voting is not equivalent to total score functions (see Proposition 2).

Proposition 2. *There exists a score voting such that the associated score function $(M, \mathcal{A}lgo)$ is total and verifies CCP property. It does not have the same winning set as knapsack voting.*

From that, and from the fact that knapsack voting is total, we conclude that the set of total score functions strictly contains and is greater the set of neutral score functions. We now characterize the strategyproofness of total score functions. To do so, we will define another set of constraints Δ on the matrix of score functions. We will prove for the restricted case of $m = 3$ (3 objects) that strategyproofness is equal to CCP and to this new set Δ restricted to total functions.

Definition 8. *We define Δ as the following set of constraints on a score function $(M, \mathcal{A}lgo)$:*

$$\forall i,j,k, \quad M_{i,i} - M_{i,j} \geq M_{k,i} - M_{k,j}$$
$$\forall i,j, j \neq i, \quad M_{i,j} < M_{j,j} \text{ if } i > j$$
$$M_{i,j} \leq M_{j,j} \text{ otherwise}$$

We denote by T_Δ the set of all total score functions satisfy the constraints in Δ.

Proposition 3. *The set T_Δ is not equal to the set of score functions that respect the set of constraints Δ.*

CCP and strategyproofness are equivalent when $m = 3$ due to the following prepositions:

- Any strategyproof total score function satisfies Δ.

For any total score function $(M, \mathcal{A}lgo)$ that satisfies Δ, $\mathcal{A}lgo$ satisfies the CCP.

We will first establish in Proposition 4 that if $(M, \mathcal{A}lgo)$ is a strategy-proof total score function, then the inequalities $M_{i,j} < M_{j,j}$ hold for all i,j with $j \neq i$. Next, using Lemma 1, we will demonstrate in Proposition 5 that if $(M, \mathcal{A}lgo)$ is a strategy-proof total score function, then the inequalities $M_{j,k} - M_{j,i} \geq M_{k,k} - M_{k,i}$ are also satisfied.

Proposition 4. *If $(M, \mathcal{A}lgo)$ is a strategyproof total score function, then, for all i,j, $j \neq i$:*

- *if i is chosen over j in the tie break of $\mathcal{A}lgo$, then, $M_{i,j} < M_{j,j}$;*
- *otherwise, $M_{i,j} \leq M_{j,j}$.*

The next lemma highlights the strength of the total score functions. We show that given objects x, y, z and $W' \in [\![1, |\mathcal{O}| - 1]\!]$, there exists a set of ballots such that z gets the best score out of all the winning sets and y gets a higher score than x.

We use this result to prove the Propositions 5 and 9 by contraposition.

Lemma 1. *If $(M, \mathcal{A}lgo)$ is a total score function with a matrix of size at least 3×3 and x, y, z are three objects, then there exists a set of voters and a ballot profile $\mathbf{B} = k_1 \delta_1 + \cdots + k_x \delta_x + k_y \delta_y + k_z \delta_z + \cdots + k_m \delta_m$ such that $M(\mathbf{B}) = \begin{pmatrix} t_1 \\ \cdots \\ t_x \\ t_y \\ t_z \\ \cdots \end{pmatrix}$ with $t_z > t_y > t_x$, $t_z > t_i$ for every other object i and $k_1, k_2, k_3, \cdots \neq 0$ and $k_1, k_2, k_3, \cdots \neq |\mathbf{B}|$.*

Proposition 5. *For the total score functions, respecting the CCP property implies respecting the set of constraints Δ.*

Proposition 6. *For the total score functions, with 3 objects, respecting the set of constraints Δ implies respecting property CCP.*

The results indicate that, for total score functions, the set of functions satisfying CCP is a subset of those satisfying the set of constraints Δ. Moreover, these two sets are identical when $m = 3$.

Also, we can state the following important result regarding strategyproofness:

Proposition 7. *Strategyproof total score functions satisfy the set of constraints Δ.*

Combining this proposition, Theorem 3 and Proposition 5, we can show that using a total score function with $m = 3$, respecting CCP or being strategyproof are equivalent properties. Moreover, they show that voting mechanisms designed with a total score function respect the set of constraints Δ if and only if they are strategyproof.

Unfortunately, this result cannot be extended for $m > 3$ objects.

Proposition 8. *With strictly more than 3 objects, there exist some score functions that respect the set of constraints Δ and is not strategyproof.*

5.3 Total Score Functions with $m > 3$

In the previous section, we have defined the set of constraints Δ which helped us prove that CCP and strategyproofness are equivalent for total score functions whenever $m = 3$. When $m > 3$, the set of constraints Δ does not hold, even though the set T_Δ of total score functions that respect Δ is larger than the set of total score functions that are strategyproof. We study additional constraints for Δ that allow us to show the strategyproofness of some score voting systems with $m > 3$. We denote this extended set of restrictions Δ^+.

Definition 9. *Given a score function $(M, \mathcal{A}lgo)$, we define Δ^+ as the set of constraints that includes all constraints from Δ and additionally enforces the condition:* $\forall a, b, c, d \in \mathcal{O}$, $\quad M_{c,a} - M_{c,b} = M_{d,a} - M_{d,b}$.

The set of total score functions that satisfy Δ^+ is denoted as T_Δ^+.

Proposition 9. *If $m > 3$, every strategyproof total score function satisfies the set of constraints Δ^+.*

Proposition 10. *If $m > 3$, every total score function that satisfies Δ^+ satisfies CCP.*

From the result, we deduce that when $m = 3$ or $m > 3$:

Theorem 5. *A total score function is strategyproof iff it satisfies CCP.*

5.4 Finding the Closest Strategyproof Total Score Function

The set of total score functions T_Δ^+ can be utilized to develop an algorithm for computing the closest strategy-proof total score function—measured using the Frobenius norm $\|A\|_F = \sqrt{\sum_{i,j=1}^n |a_{ij}|^2} = \sqrt{\text{Tr}(A^*A)}$ [13]—to a given score function, provided such a function exists. Since T_Δ^+ forms a convex polyhedron, quadratic optimization techniques can be employed to project the given score function onto T_Δ^+, ensuring the closest approximation while preserving strategyproofness.

An issue is that some of the inequations of Δ are strict, but the projection is done on the closed set $\overline{T_\Delta^+}$. This leads to two possibilities:

1. the projection gives a strategyproof total score function;
2. the notion of the closest strategyproof total score function does not exist, and the projection gives a non-strategyproof score function.

In the first case, the projection is the result of the algorithm. We now show a way to handle the second case.

We project a matrix M onto $\overline{T_\Delta^+}$, which gives a matrix M'. If M' is not strategyproof, we cover the sphere centered in M', of radius δ (a small constant) with a set of points, so that every point in the sphere is at most at a distance ϵ to a point in the cover. To cover such a sphere, in dimension n^2, we consider the points $(k_1\epsilon, k_2\epsilon, \ldots, k_{n^2-1}\epsilon, \sqrt{1 - k_1^2\epsilon^2 - \cdots - k_{n^2-1}^2\epsilon^2})\ \forall k_1, \ldots, k_{n^2-1} \in [-\frac{1}{\epsilon}, \frac{1}{\epsilon}]$ whenever the square root is defined. With this cover, when ϵ is close to 0, the distance between a point on the cover and a point on the sphere is at most $n^2 * \sqrt{2\epsilon}$.

Using the inequalities of Δ^+, we can check on every point of the cover whether they are strategyproof or not and output a "close" strategyproof matrix. This procedure gives an algorithm that outputs the closest strategyproof total score function if it exists. It outputs a strategyproof total score function at distance δ from the optimal if the closest matrix does not exist.

Given a score function $(M, \mathcal{A}lgo)$, the algorithm can be described as follows:

1. Solve the quadratic optimization problem to project M on Δ^+;
2. Verify if the projection $(M', \mathcal{A}lgo)$ respects the constraints of Δ^+ and if so, return $(M', \mathcal{A}lgo)$;
3. Otherwise, find and return a score function that respects the constraints of Δ^+ on the sphere centered in M', of radius ϵ.

Theorem 6. *Given a score function $(M, \mathcal{A}lgo)$, there exists an algorithm that returns the closest strategyproof total score function.*

6 Conclusion and Perspectives

This paper investigates the design of strategyproof election systems, focusing on voting settings like Participatory Budgeting, where a fixed number of items (e.g., projects) must be selected under constraints.

Using a new model for describing choice mechanisms, we presented the Constrained Change Property (CCP), a powerful tool to characterize the strategyproofness of voting mechanisms. Using CCP, we showed how to design choice mechanisms that are strategyproof for the unitary case.

We further propose the notion of score functions and a new class of utilitarian mechanisms called Score Voting. Score functions offer a flexible framework for analyzing strategyproofness. We prove that score voting with a neutral score function coincides with knapsack voting, and that CCP-compliant score functions yield strategyproof mechanisms. These insights culminate in an algorithm that computes the closest total score function ensuring strategyproofness for any given score voting rule.

In future work, we intend to further improve the notion of score voting (and its score functions) to better characterize the relation between the CCP property and strategyproofness. We are also interested in quantifying the strategyproofness of a vote and creating approximation algorithms for the problem.

Acknowledgments. This work was partially funded by grants #2019/26702-8, #2021/06867-2, #2023/00811-0, and #2024/01115-0, São Paulo Research Foundation (FAPESP) and grant #444766/2024-3, National Council of Scientific and Technological Development (CNPq). This work was also developed within the 'Transitions' International Research Center (IRC), a joint initiative of CNRS and the University of São Paulo, with partial support from the FAPESP CIP Program (grant #2025/01171-0). It forms part of the THUS (Techno-Human Systems of the Future) research group, specifically within the IRP "Sustainable Megalopolises for Humans".

Disclosure of Interests. The authors have no competing interests to declare.

References

1. Aziz, H., Shah, N.: Participatory budgeting: models and approaches. In: Rudas, T., Péli, G. (eds.) Pathways Between Social Science and Computational Social Science. CSS, pp. 215–236. Springer, Cham (2021). https://doi.org/10.1007/978-3-030-54936-7_10
2. Barberà, S., Jackson, M.: A characterization of strategy-proof social choice functions for economies with pure public goods. Soc. Choice Welfare **11**(3), 241–252 (1994). https://doi.org/10.1007/BF00193809
3. Border, K.C., Jordan, J.S.: Straightforward elections, unanimity and phantom voters. Rev. Econ. Stud. **50**(1), 153–170 (1983). https://doi.org/10.2307/2296962
4. Brandt, F., Conitzer, V., Endriss, U., Lang, J., Procaccia, A.D.: Handbook of Computational Social Choice. Cambridge University Press (2016). https://doi.org/10.1017/CBO9781107446984
5. Cabannes, Y.: Participatory budgeting: a significant contribution to participatory democracy. Environ. Urban. **16**(1), 27–46 (2004). https://doi.org/10.1177/095624780401600104
6. Cabannes, Y.: Another City is Possible with Participatory Budgeting, 2 edn. University of Chicago Press (2023)
7. Chen, Y., Ghosh, A., Kearns, M., Roughgarden, T., Vaughan, J.W.: Mathematical foundations for social computing. Commun. ACM **59**(12), 102–108 (2016). https://doi.org/10.1145/2960403
8. Duddy, C.: Fair sharing under dichotomous preferences. Math. Soc. Sci. **73**, 1–5 (2015). https://doi.org/10.1016/j.mathsocsci.2014.10.005
9. Freeman, R., Pennock, D.M., Peters, D., Wortman Vaughan, J.: Truthful aggregation of budget proposals. J. Econ. Theory **193** (2021). https://doi.org/10.1016/j.jet.2021.105234
10. Furquim, F.V., Dardilhac, V., Cordeiro, D., Cohen, J.: Characterizing strategyproofness through score functions in voting mechanisms (2025). https://hal.science/hal-05040764
11. Gibbard, A.: Manipulation of voting schemes: a general result. Econometrica J. Econom. Soc. 587–601 (1973). https://doi.org/10.2307/1914083

12. Goel, A., Krishnaswamy, A.K., Sakshuwong, S., Aitamurto, T.: Knapsack voting for participatory budgeting. ACM Trans. Econ. Comput. (TEAC) **7**(2), 1–27 (2019)
13. Golub, G.H., Van Loan, C.F.: Matrix Computations. JHU Press (2013)
14. Maguire, L.A.: What can decision analysis do for invasive species management? Risk Anal. **24**, 859–868 (2004)
15. Moulin, H.: On strategy-proofness and single peakedness. Public Choice **35**(4), 437–455 (1980). https://doi.org/10.1007/BF00128122
16. Peters, H., van der Stel, H., Storcken, T.: Pareto optimality, anonymity, and strategy-proofness in location problems. Internat. J. Game Theory **21**(3), 221–235 (1992). https://doi.org/10.1007/BF01258276
17. Satterthwaite, M.A.: Strategy-proofness and arrow's conditions: existence and correspondence theorems for voting procedures and social welfare functions. J. Econ. Theory **10**(2), 187–217 (1975)
18. Slovic, P.: Trust, emotion, sex, politics, and science: surveying the risk-assessment battlefield. Risk Anal. **19**, 689–701 (1999)

Minimizing Blocking Agents for Stable Matching with Partial Approval Information

Yitian Gao, Jiaxue Li, Junjie Luo(✉) , and Yiheng Zhang

School of Mathematics and Statistics, Beijing Jiaotong University, Beijing, China
{22271087,22271092,jjluo1,22271080}@bjtu.edu.cn

Abstract. We study the stable matching problem with partial information, where agents submit only partial approval preferences, and the goal is to find a matching that is as stable as possible in the worst-case scenario. Unlike previous studies that focus solely on the Stable Marriage setting and measure stability by the number of blocking pairs, we explore another well-explored stability measure: the number of blocking agents. Additionally, we extend our analysis to both the Stable Roommates and Hospital/Residents problems. Our findings offer a comprehensive view of the computational complexity across these problem variants, highlighting interesting contrasts between blocking agents and blocking pairs, as well as among the three stable matching settings.

Keywords: Stable matching · Partial information · Blocking agents · Approval preference · Computational complexity · Nonlinear matching

1 Introduction

Matching under preferences is a prominent research topic that lies at the intersection of operations research, economics, and computer science. In the classical STABLE MARRIAGE problem, two disjoint sets of agents are given, each with a strict preference ordering over the agents in the opposite set. The objective is to find a matching between the two sets of agents such that no *blocking pair* exists, meaning no two agents would both prefer each other over their assigned partners in the matching. Such a matching is called *stable*. The seminal work of Gale and Shapley [10] established that a stable matching always exists and can be found efficiently. This foundational result has spurred extensive research within and beyond the STABLE MARRIAGE framework [12,15].

Traditional stable matching studies typically assume that participants can provide complete preferences over all other agents. While this assumption might be feasible in smaller applications, it has become increasingly impractical in larger markets, where participants often lack the information or cognitive ability to rank all others comprehensively. To address this challenge, several approaches have been proposed to deal with the setting where agents submit only partial preference information. Among these, Menon and Larson [17] introduced a worst-case model, which seeks to minimize the number of blocking pairs in the worst-case scenario across all potential linear extensions of the partial orders. This model offers a middle ground between other methods, such as selecting a stable matching for any arbitrary linear extension (which may poorly reflect the true preferences) and identifying a matching stable for all possible extensions (which is often unattainable). Furthermore, unlike probabilistic models [1], this worst-case approach does not depend on assumptions about underlying preference distributions.

Despite its many advantages, the worst-case model analyzed by Menon and Larson [17] admits no polynomial-time algorithm with a constant-factor approximation unless P = NP. This hardness barrier has prompted the exploration of simpler preference structures where positive results may still be achievable. A leading example is the approval-preference framework, where agents simply approve a subset of potential partners, an approach that has received much attention in the stable matching literature [2,5,6,14]. Chu et al. [7] provided a comprehensive analysis of the computational complexity for this setting, examining three well-established notions for blocking pairs (weak, strong, and super blocking pairs [13]) under two different partial information models: one-sided (one side submits full information while the other side provides partial information) and two-sided (both sides submit partial information).

Building on this line of research, we continue to focus on approval preferences but shift attention to another well-studied measure of stability: the number of agents involved in blocking pairs, known as *blocking agents*. The concept of blocking agents was first introduced by Roth and Xing [18] and has since attracted substantial attention in the literature [4,8,9,16]. It is worth noting that two matchings may have the same number of blocking pairs yet differ significantly in the number of blocking agents, and vice versa. For example, given k blocking pairs, if blocking agents are evenly distributed in both sides, the number of blocking agents might be as low as $2\sqrt{k} = \sqrt{k} + \sqrt{k}$, whereas in an uneven distribution, it could be as high as $k+1$. The same discrepancy can also arise in the partial-information setting (see a concrete example in the full version), which motivates studying blocking agents (rather than merely blocking pairs) in such scenarios. Furthermore, we extend this investigation to two well-known generalizations of STABLE MARRIAGE: namely STABLE ROOMMATES (where agents form pairings among a single set rather than two separate sets) and HOSPITALS/RESIDENTS (where each "hospital" is matched to possibly multiple residents).

Our Contributions. We study the stable matching problem with partial approval preferences, focusing on minimizing the maximum number of blocking agents in the worst case. Following Chu et al. [7], we consider three types of blocking agents–weak, strong, and super (see formal definitions in Sect. 2)–within two information settings: one-sided (1-SM) and two-sided (2-SM) for STABLE MARRIAGE. Additionally, we extend our analysis to STABLE ROOMMATES (SR) and HOSPITALS/RESIDENTS (HR), where we examine *both* blocking pairs and block-

Table 1. Results overview of stable marriage (SM), stable roommates (SR) and Hospital/Residents (HR). RP represents randomized polynomial time.

		weak	strong	super
Pair	2-SM	RP [7]	RP [7]	P [7]
	1-SM	NP-hard [7]	NP-hard [7]	P [7]
	SR	P (Thm 9)	RP (Thm 9)	P (Obs 8)
	HR	NP-hard (Thm 12)	NP-hard (Thm 12)	P (Obs 11)
Agent	2-SM	P (Thm 4)	P (Thm 5)	P (Obs 3)
	1-SM	NP-hard (Cor 1)	NP-hard (Thm 12)	P (Thm 7)
	SR	P (Thm 10)	P (Thm 10)	P (Obs 8)
	HR	NP-hard (Thm 12)	NP-hard (Thm 12)	P (Obs 11)

ing agents. We provide a comprehensive characterization of the computational complexity for all these problems, ranging from deterministic and randomized polynomial-time solvability to NP-hardness; see Table 1 for a summary.

We first consider 2-SM in the context of blocking agents (Sect. 3.1). Chu et al. [7] showed that for weak and strong blocking *pairs*, 2-SM is solvable in *randomized* polynomial time (via reductions to NONLINEAR MATCHING [3]) but it is challenging to solve them in *deterministic* polynomial time. In contrast, we show that for weak and strong blocking *agents*, 2-SM can be reduced to a polynomial number of MAXIMUM (WEIGHTED) MATCHING instances, and is thus solvable in *deterministic* polynomial time. This discrepancy illustrates how targeting blocking agents versus blocking pairs can lead to different computational complexities. Meanwhile, for the one-sided setting 1-SM (Sect. 3.2), our results for blocking agents mirror those for blocking pairs, though our algorithm for super blocking agents is more intricate than that of Chu et al. [7].

We then turn to SR (Sect. 4.1) and HR (Sect. 4.2). Similar to 2-SM, we show that SR with both weak and strong blocking pairs can be reduced to NONLINEAR MATCHING, making them solvable in randomized polynomial time. Surprisingly, for weak blocking pairs, the reduction yields a *convex* objective in NONLINEAR MATCHING, thus enabling *deterministic* polynomial-time solvability, which contrasts with the 2-SM case, where no deterministic polynomial-time algorithm is known. This result is surprising because STABLE ROOMMATES is typically viewed as more general (and often more difficult) than STABLE MARRIAGE. For SR with blocking *agents*, the problems become much simpler and can be easily solved in polynomial time. Thus, most problems for SR are solvable in deterministic polynomial time. For HOSPITALS/RESIDENTS, we find a sharp contrast: except for super blocking agents (pairs), the remaining four problems are all NP-hard via a single reduction from EXACT COVER BY 3-SETS.

Overall, our results highlight two main points: 1) blocking-agent problems are in general easier to solve than blocking-pair problems, and 2) While STABLE ROOMMATES shares closer similarities with 2-SM from a complexity perspective and is generally easier to solve, HOSPITALS/RESIDENTS exhibits parallels to 1-SM and thus more challenging. Due to space constraints, proofs for certain results (marked by ⋆) are deferred to the full version.

2 Preliminaries

Partial Approval Preferences. We adopt the notations from Chu et al. [7]. Let $U = \{m_1, m_2, \ldots, m_n\}$ and $W = \{w_1, w_2, \ldots, w_n\}$ be two disjoint sets of agents. Each agent $a \in U \cup W$ reveals a *partial* approval set, denoted by P_a, towards agents in the other set, where "partial" means that P_a is a subset of agents approved by agent a. Denote $C(P_a)$ the family of all possible complete approval sets for a, i.e., $C(P_a) = \{S \mid P_a \subseteq S \subseteq W\}$ if $a \in U$ and $C(P_a) = \{S \mid P_a \subseteq S \subseteq U\}$ if $a \in W$. Denote $P_{U \cup W} = \times_{a \in U \cup W} P_a$ the given partial preference profile, and denote $C = \times_{a \in U \cup W} C(P_a)$ the set of all possible completions. For every completion $c \in C$, denote P_a^c the corresponding complete approval set of agent a under c. We say that a *prefers* b_1 to b_2 under c if $b_1 \in P_a^c$ and $b_2 \notin P_a^c$, and a *weakly prefers* b_1 to b_2 under c if $b_2 \in P_a^c$ implies $b_1 \in P_a^c$.

Matching, Blocking Pair, and Blocking Agent. A matching $M \subseteq U \times W$ is a set of n disjoint agent pairs (m_i, w_j) with $m_i \in U$ and $w_j \in W$. Denote $\mathcal{M}$ the set of all possible matchings. For any $a \in U \cup W$, denote $M(a)$ the matched agent for a. Following Chu et al. [7], we consider three types of blocking pairs, which we refer to as weak, strong, and super blocking pairs. Because stability and blocking pairs are inherently opposing notions, this naming convention may seem counterintuitive. Nevertheless, we adopt these terms to maintain a clear correspondence with the classical framework, where each blocking concept directly complements its stability counterpart.

Definition 1 (Weak/Strong/Super Blocking Pair). *Given a matching and a completion $c \in C$, an agent pair $(m, w) \in U \times W \setminus M$ is called a*

- *weak blocking pair under c, if m and w prefer each other to $M(m)$ and $M(w)$ under c;*
- *strong blocking pair under c, if m prefers w to $M(m)$ and w weakly prefers m to $M(w)$, or m weakly prefers w to $M(m)$ and w prefers m to $M(w)$ under c;*
- *super blocking pair under c, if m and w weakly prefer each other to $M(m)$ and $M(w)$ under c.*

Similarly, we define three corresponding types of blocking agents.

Definition 2 (Weak/Strong/Super Blocking Agent). *Given a matching and a completion $c \in C$, an agent is called a* weak/strong/super blocking agent *if it is involved in at least one weak/strong/super blocking pair under c.*

By definition, weak blocking implies strong blocking, which in turn implies super blocking. For each $T \in \{\text{weak, strong, super}\}$, denote $bp_T(M, c)$ (resp. $ba_T(M, c)$) the set of T blocking pairs (resp. agents) for matching M under completion c.

Worst Completion. Our goal is to find a matching that has the minimum number of blocking agents in the worst case among all possible completions. Chu et al. [7] observe that for any matching there exists a worst completion that maximizes the number of blocking pairs.

Definition 3 (Worst Completion [7]). *For any matching M, define a completion $c_M \in C$ as follows:*

$$\forall m \in U : P_m^{c_M} = \begin{cases} W & \text{if } M(m) \in P_m; \\ W \setminus \{M(m)\} & \text{if } M(m) \notin P_m; \end{cases}$$

$$\forall w \in W : P_w^{c_M} = \begin{cases} U & \text{if } M(w) \in P_w; \\ U \setminus \{M(w)\} & \text{if } M(w) \notin P_w. \end{cases}$$

In words, under c_M, every agent approves all agents in the other set if its partner is already in its partial approval set. Otherwise, it approves all agents in the other set except for its partner. In fact, this completion is also the worst one for blocking agents. Indeed, for any $T \in \{\text{weak, strong, super}\}$, if there exists a completion such that an agent is a T blocking agent, which means a is involved in a T blocking pair, then this pair must also be a T blocking pair under c_M, implying that the agent must also be a blocking agent under c_M.

Observation 1. *For any matching M and any $T \in \{\text{weak, strong, super}\}$, the maximum number of T blocking pairs (resp. agents) among all completions equals the number of T blocking pairs (resp. agents) under c_M, i.e.,*
$\max_{c \in C} |bp_T(M, c)| = |bp_T(M, c_M)|$ *and* $\max_{c \in C} |ba_T(M, c)| = |ba_T(M, c_M)|.$

(Un)satisfied Agent. An agent is called *satisfied* if it is matched with an agent from its partial approval set; otherwise, it is *unsatisfied*. Note that while every weak blocking agent under the worst completion must be unsatisfied, the reverse is not always true. For instance, consider an unsatisfied agent $m \in U$. Two scenarios can prevent m from becoming a weak blocking agent. First, there may be no unsatisfied agent in W and hence there is no weak blocking pair. Second, even if there exists an unsatisfied agent $w \in W$, it could be that m is already matched to w, so (m, w) would not be a weak blocking pair. Algorithms must account for these possibilities.

However, if there are at least two unsatisfied agents on both sides, then every unsatisfied agent can form a weak blocking pair with at least one unsatisfied agent on the other side under the worst completion, thereby becoming a weak blocking agent. This leads to a sufficient condition for equivalence between being unsatisfied and being a weak blocking agent.

Observation 2. *If there are at least two unsatisfied agents on both sides, then an agent is unsatisfied if and only if it is a weak blocking agent under the worst completion.*

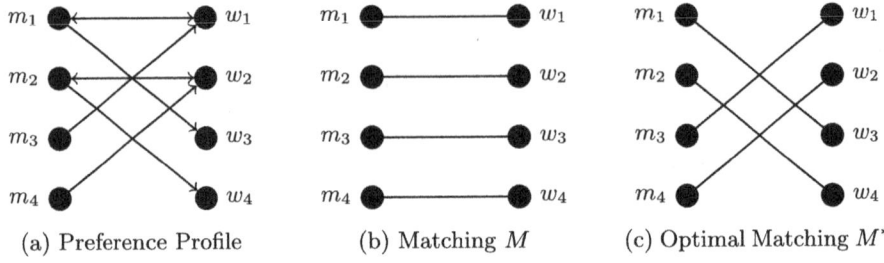

Fig. 1. An example with 8 agents and two matching solutions from [7].

Problem Definition. For each $T \in \{\text{weak, strong, super}\}$, define:

T-MINIMAX BLOCKING AGENTS (T-MBA)

Input: Two disjoint sets of agents U and W, and a partial approval preference profile $P_{U \cup W}$.

Task: Find a matching M^* that has the minimum number of T blocking agents with respect to all possible completions, i.e.,

$$M^* = \arg \min_{M \in \mathcal{M}} \max_{c \in C} |ba_T(M, c)| = \arg \min_{M \in \mathcal{M}} |ba_T(M, c_M)|.$$

The corresponding problem, T-MINIMAX BLOCKING PAIRS (T-MBP), which aims to minimize the number of *blocking pairs*, has been studied by Chu et al. [7][1]. We also consider the variant with one-sided partial information where agents in U already provide their full approval sets (i.e., $C(P_m) = \{P_m\}$ for each $m \in U$) whereas agents in W still provide partial approval sets. The worst completion is defined accordingly. We call this variant ONE-SIDED T-MINIMAX BLOCKING AGENTS (1-T-MBA). We will define and study the corresponding problems for STABLE ROOMMATES and HOSPITALS/RESIDENTS in Sect. 4.

Example. We take the example from Chu et al. [7] (see Fig. 1) to illustrate the above definitions and notations. The example contains 8 agents with partial approval sets, as shown in Fig. 1a; for instance, $P_{m_1} = \{w_1, w_3\}$ and $P_{w_1} = \{m_1\}$. A completion for P_{m_1} must include w_1 and w_3, and may also include w_2 and/or w_4, resulting in 4 different completions for P_{m_1}. Figure 1b displays a matching M where $M(m_i) = w_i$. In the worst completion c_M for M, m_1, m_2, w_1, w_2 approve all agents in the other set, while the remaining agents approve all except their partners. Under c_M, (m_3, w_4) forms a weak blocking pair since they prefer each other than w_3 and m_4. Similarly, (m_4, w_3) forms a weak blocking pair under c_M. Therefore, $\max_{c \in C} |ba_{\text{weak}}(M, c)| = |ba_{\text{weak}}(M, c_M)| = 4$. The optimal matching

---

[1] The problem was called T-MINIMAX APPROVAL MATCHING (T-MAM) in [7]. Here we use a different name to emphasize the two different optimization goals.

for WEAK-MBP is M^*, as shown in Fig. 1c, where $\max_{c \in C} |ba_{\text{weak}}(M^*, c)| = 0$ because all agents on the left side are already satisfied (although all agents on the right side are unsatisfied). For 1-WEAK-MBA, where the approval sets for m_i are already complete, both M and M^* are optimal with no weak blocking agents in the worst case. Note that in the one-sided setting m_3 only approves w_1 and will not prefer w_4 over $w_3 = M(m_3)$.

3 Blocking Agents in Stable Marriage

In this section, we consider blocking agents in the stable marriage setting.

3.1 Two-Sided Partial Information

We start with the two-sided information setting. First, observe that when every agent approves all other agents, all agents could participate in super blocking pairs as no one will get worse off, thus SUPER-MBA is trivial.

Observation 3. *For any instance of* SUPER-MBA, *every matching has $2n$ super blocking agents under the worst completion.*

It remains to consider weak and strong blocking agents. Chu et al. [7] proved that both WEAK-MBP and STRONG-MBP are solvable in *randomized* polynomial time. In contrast, we will show that the corresponding problems for blocking agents can be solved in *deterministic* polynomial time.

For WEAK-MBA, recall that any weak blocking agent must be unsatisfied. We first show that minimizing the number of unsatisfied agents can be done efficiently by a reduction to the maximum weighted matching problem.

Lemma 1. *Finding a matching that minimizes the number of unsatisfied agents can be done in $O(n^3)$ time.*

Proof. Given an instance $(U, W, P_{U \cup W})$ of WEAK-MBP, we construct a complete bipartite graph G whose two vertex sets are U and W. For each pair $(m, w) \in U \times W$, we assign weight 2 if m and w approve each other, weight 1 if exactly one approves the other, and weight 0 otherwise. Any perfect matching M in G then corresponds to a matching M' in WEAK-MBP, and the total edge weight of M is precisely the number of satisfied agents under M'. Thus, minimizing the the number of unsatisfied agents in WEAK-MBP reduces to finding a maximum weighted matching in G, which can be done in $O(n^3)$ time (e.g., via the Hungarian algorithm). □

However, not every unsatisfied agent is necessarily a weak blocking agent. By Observation 2, solving WEAK-MBA involves handling several special cases. One such case, where one side is fully satisfied, is addressed in the following lemma.

Lemma 2. *One can check in $O(n^{2.5})$ time whether there is a matching that satisfies all agents on one side. Moreover, if such a matching exists, then among all matchings that satisfy that side, one that minimizes the number of unsatisfied agents on the other side can be found in $O(n^3)$ time.*

Proof. Without loss of generality, we consider the case where all agents in U are satisfied. Given an instance $(U, W, P_{U \cup W})$ of WEAK-MBP, construct a bipartite graph G whose two vertex sets are U and W. For each pair (m, w), add an edge if m approves w. Then U can be fully satisfied if and only if G has a perfect matching. This can be tested in $O(n^{2.5})$ time (via the HopkroftKarp algorithm).

Next, suppose such a matching exists. To find one that minimizes the number of unsatisfied agents in W, we transform the above graph G to a weighted graph G', by assigning a weight of $n+1$ to each edge (m, w) if w approves m, and a weight of n otherwise. Compute a maximum weighted matching M^* in G', which can be done in $O(n^3)$ time. Because we already know all agents in U can be satisfied, M^* must be a perfect matching, so its total weight $w(M^*)$ is at least n^2. Furthermore, the difference $w(M^*) - n^2$ equals the maximum possible number of satisfied agents in W, which consequently minimizes the number of unsatisfied agents in W. □

Using Lemma 1 and 2, we show WEAK-MBA is solvable in polynomial time.

Theorem 4. WEAK-MBA *is solvable in $O(n^5)$ time.*

Proof. Let M be any matching, and let t_1 and t_2 be the number of unsatisfied agents from U and W, respectively, under M. Lemma 1 already handles the case with $ba_{\text{weak}}(M, c_M) = t_1 + t_2$. Here, we consider three additional special cases where $ba_{\text{weak}}(M, c_M) < t_1 + t_2$, which implies $\min\{t_1, t_2\} \le 1$ by Observation 2. We solve each case separately and then pick the best solution among them.

Case 1: $ba_{\text{weak}}(M, c_M) < t_1 + t_2$ **and** $t_1 = 0$ **(or symmetrically $t_2 = 0$).** For this case, there is no weak blocking pair and hence $ba_{\text{weak}}(M, c_M) = 0$. We can check this case in $O(n^{2.5})$ time according to Lemma 2.

Case 2: $ba_{\text{weak}}(M, c_M) < t_1 + t_2$ **and** $t_1 = t_2 = 1$. Let m^* and w^* be the two unsatisfied agents. Since $ba_{\text{weak}}(M, c_M) < t_1 + t_2$, it must be the case that m^* and w^* are matched, and hence $ba_{\text{weak}}(M, c_M) = 0$. To check such cases, we brute force over all pairs $(m^*, w^*) \in U \times W$ that disapprove each other. For each pair, remove m^* and w^* from the instance, and see whether the resulting problem admits a matching that has no unsatisfied agents. According to Lemma 1, each pair can be checked in $O(n^3)$ time and the overall time is $O(n^5)$.

Case 3: $ba_{\text{weak}}(M, c_M) < t_1 + t_2$ **and** $t_2 > t_1 = 1$ **(or symmetrically $t_1 > t_2 = 1$).** Let m^* be the unique unsatisfied agent in U. Since $t_2 > 1$, there exists at least one unsatisfied agent $w' \in W$ not matched with m^*. Hence, m^* must be a weak blocking agent. Since $ba_{\text{weak}}(M, c_M) < t_1 + t_2$, there is exactly one unsatisfied agent w^* in W that is matched with m^* but is not a weak blocking agent. Consequently, we have $ba_{\text{weak}}(M, c_M) = t_1 + t_2 - 1$. To check such cases, again brute force over all pairs (m^*, w^*) who disapprove each other. For each pair, remove m^* and w^* from the instance, and see whether the resulting problem admits a matching that satisfies all agents in $U \setminus \{m^*\}$. If no such matching exists, then this case cannot occur. Otherwise, we proceed to find a matching that minimizes the number of unsatisfied agents in $W \setminus \{w^*\}$. By Lemma 2, this can be done in $O(n^3)$ time for each pair, so the overall time is $O(n^5)$. □

We proceed to consider STRONG-MBA and show that it can also be solved in polynomial time using arguments similar to those for the weak blocking case.

Theorem 5 ($\star$). STRONG-MBA *is solvable in* $O(n^{6.5})$ *time.*

3.2 One-Sided Partial Information

We move on to the one-sided information setting, where we assume that agents in U submit full preferences while agents in W submit partial preferences. For weak blocking pairs, Chu et al. [7] proved that deciding whether there exists a matching with no weak blocking pairs in the worst case is NP-hard for 1-WEAK-MBP. Since for any matching M we have that $|bp_{\text{weak}}(M, c_M)| > 0$ if and only if $|ba_{\text{weak}}(M, c_M)| > 0$, so we get the NP-hardness of 1-WEAK-MBP as a corollary.

Corollary 1. *Deciding whether* $\arg\min_{M \in \mathcal{M}} |ba_{weak}(M, c_M)| = 0$ *is NP-hard for* 1-WEAK-MBP.

For strong blocking pairs, Chu et al. [7] proved that 1-STRONG-MBP is NP-hard but the target value for $|bp_{\text{weak}}(M, c_M)|$ is not 0 anymore, so it does not imply the NP-hardness of 1-STRONG-MBA. Nevertheless, we can show that 1-STRONG-MBA is also NP-hard.

Theorem 6. *The decision version of* 1-STRONG-MBA *is NP-hard.*

Proof. We present a reduction from the CLIQUE problem on regular graphs (all vertices have the same degree), which remains NP-hard. Consider an instance $I = (G = (V, E), k)$ of CLIQUE with $|V| = n$, $|E| = p$ and vertex degree d. Assume without loss of generality that $k^2 \leq n$. We construct an instance $I' = ((U, W, P_{U \cup W})$ of 1-STRONG-MBP as follows (See also Fig. 2). Let $U = E \cup S_c \cup S_r$ with $|S_c| = k$ and $|S_r| = n - k$. Let $W = V \cup T_c \cup T_r$ with $|T_c| = \binom{k}{2}$ and $|T_r| = p - \binom{k}{2}$. For convenience, we use E and V to refer the set of the corresponding agents. The approval preferences are: all agents from U approve all agents from T_r, and vice versa; all agents from $S_c \cup S_r$ approve all agents from V, but all agents from V only approve agents from S_r; for each edge $e = (v_i, v_j) \in E$, the corresponding agent e approves the two corresponding agents v_i and v_j from V.

In the full version of this paper, we show that G has a clique Q of size k if and only if there is a matching M for I' with $|ba_{\text{strong}}(M, c_M)| = B := n + p + k + \binom{k}{2}$. The idea is as follows. Suppose G has a clique Q of size k. Denote by $V_c \subseteq V$ and

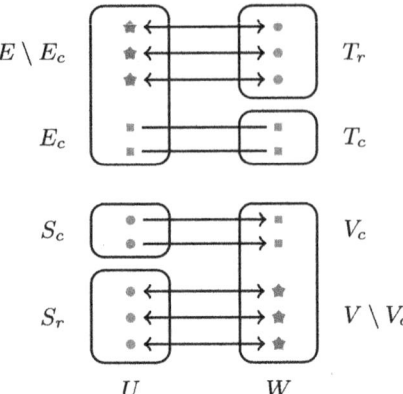

Fig. 2. Illustration of the matching in the proof of Theorem 6. The approval preferences are: U approves T_r and vice versa; $S_c \cup S_r$ approves V and V approves S_r; $e = (v_i, v_j) \in E$ approves $v_i, v_j \in V$. We use blue rectangular points to indicate unsatisfied agents and red stars to denote non-strong-blocking agents. (Color figure online)

$E_c \subseteq E$ the agents corresponding to the vertices and edges of Q, respectively. We construct a matching M according to Q by matching $E \setminus E_c$ to T_r, E_c to T_c, S_c to V_c, and S_r to $V \setminus V_c$. Then we can show that the unsatisfied agents are those in E_c and $T_c \cup V_c$, and there is no strong blocking agent in $(E \setminus E_c) \cup (V \setminus V_c)$. Particularly, since every edge in E_c has its endpoints in V (reflecting that Q is a clique), there are no strong blocking pairs between E_c and $V \setminus V_c$, ensuring that all agents in $V \setminus V_c$ are non-strong-blocking. □

It remains to consider super blocking agents.

Theorem 7. 1-SUPER-MBA *is solvable in* $O(n^{6.5})$ *time.*

Proof. Let M be any matching, and let t_1 be the number of unsatisfied agents in U under M. We consider three cases based on t_1. In each case, we provide a polynomial-time procedure to find an optimal solution for that case. Finally, we pick the best solution among the three.

Case 1: $t_1 = 0$. We can check whether there is a matching that fully satisfies U in $O(n^{2.5})$ time (by Lemma 2). Suppose such a matching M exists. In that case, each $m \in U$ has at least one approved partner ($|P_m| \geq 1$), and each $w \in W$ must lie in at least one P_m. Observe that a pair $(m, w) \notin M$ is a super blocking pair if and only if $w \in P_m$. Consequently, an agent $m \in U$ is a super blocking agent if and only if m approves at least two agents in W (its partner in M plus at least one additional agent), i.e., $|P_m| \geq 2$. An agent $w \in W$ is a super blocking agent if and only if w is approved by at least two agents in U (its matched partner in M plus at least one additional agent). Hence, the set of super blocking agents depends only on the partial approval sets $\{P_m\}$, namely, all agents $m \in U$ with $|P_m| \geq 2$ and all agents $w \in W$ approved by at least two agents in U are super blocking agents.

Case 2: $t_1 = 1$. Suppose exactly one agent m^* in U is unsatisfied. We guess both m^* and its partner $w^* \in W$ with $w^* \notin P_{m^*}$. For each pair (m^*, w^*), remove them from the instance and check, in $O(n^{2.5})$ time (by Lemma 2), whether the remaining agents admit a matching that fully satisfies $U \setminus \{m^*\}$. Since there could be $O(n^2)$ such guesses, this brute force takes $O(n^{4.5})$ overall. Next, assume M is a matching with m^* and w^* paired, and all other agents in U are satisfied. Then the set of super blocking agents depends only on $\{P_m\}$ and can be computed as follows. Every agent in $W \setminus \{w^*\}$ can form a super blocking pair with m^*, since m^* is unsatisfied. The remaining agent w^* is a super blocking pair if and only if it is approved by at least one agent from U. Any $m \in U \setminus \{m^*\}$ is a super blocking agent unless $|P_m| = 1$, in which case m has no alternative partner. Finally, m^* is a super blocking agent since it is unsatisfied.

Case 3: $t_1 \geq 2$. Since $t_1 \geq 2$, we first guess two agents m_1, m_2 that are unsatisfied under M and then guess their partners w_1, w_2 with $w_i \notin P_{m_i}$. In this case, every agent in W can form a super blocking pair with at least one of m_1 and m_2, so all of $W \cup \{m_1, m_2\}$ are super blocking agents. For every $m \in U \setminus \{m_1, m_2\}$, it is not a super blocking agent if and only if it is matched with its unique approved agent, i.e., $M(m) \in P_m$ and $|P_m| = 1$. Thus, minimizing the number of super blocking agents is equivalent to maximizing the number of agents $U \setminus \{m_1, m_2\}$ who are matched with their unique approved agent. We do this by building a bipartite graph G whose vertex sets are $U \setminus \{m_1, m_2\}$ and $W \setminus \{w_1, w_2\}$, and including an edge (m, w) if $w \in P_m$ and $|P_m| = 1$. We can then find a maximum matching in G in $O(n^{2.5})$ time. As there are $O(n^4)$ ways to guess (m_1, m_2, w_1, w_2), the total running time is $O(n^{6.5})$. □

4 Stable Roommates and Hospitals/Residents

In this section, we study two generalizations of the STABLE MARRIAGE setting: STABLE ROOMMATES (SR) and HOSPITALS/RESIDENTS (HR). We show that most problem variants for SR can be solved in deterministic polynomial time (similar to two-sided SM), whereas most problem variants for HR are NP-hard.

4.1 Stable Roommates

In SR, we consider a set V of $2n$ agents, where each agent $v \in V$ has a partial approval set P_v towards the remaining agents in $V \setminus \{v\}$. All the notations defined in Sect. 2 for the STABLE MARRIAGE setting can be defined analogously here. For each $T \in \{\text{weak, strong, super}\}$, define ROOMMATES T-MINIMAX BLOCKING AGENTS (R-T-MBA) as follows: Given a set V of agents and a partial approval preference profile P_V, the task is to find a matching M^* to minimize the number of T blocking agents under the worst completion. We also consider the variant where the goal is to minimize the number of blocking pairs, referred to as ROOMMATES T-MINIMAX BLOCKING PAIRS (R-T-MBP).

Observe that for any matching, every agent pair not in the matching is a super blocking pair under the worst completion. Thus, the number of super blocking pairs is always $n^2 - n$ and the number of super blocking agents is always $2n$.

Observation 8. *For* R-SUPER-MBP *(*R-SUPER-MBA*), every matching has the same number of super blocking pairs (agents) under the worst completion.*

It remains to consider weak and strong blocking agents/pairs. Let us define some notations we partition the set $V \times V$ of all agent pairs into three sets according to whether they are included in each other's partial approval sets: $A_1 := \{(v, v') \mid v' \in P_v \wedge v \in P_{v'}\}$, $A_2 := \{(v, v') \mid (v' \in P_v \wedge v \notin P_{v'}) \vee (v' \notin P_v \wedge v \in P_{v'})\}$, and $A_3 := \{(v, v') \mid v' \notin P_v \wedge v \notin P_{v'}\}$. Given a matching M, we partition the agent set V into three subsets $V_1 \cup V_2 \cup V_3$ as $V_i := \{v \in V \mid (v, M(v)) \in A_i\}, \forall i \in \{1, 2, 3\}$.

Finally, denote $k_i := |M \cap A_i| = |V_i|/2$ the number of matched pairs from A_i. Then the number of weak/strong blocking pairs for any matching in the worst case can be computed as follows.

Lemma 3 (⋆). *For any matching M, we have*

$$|bp_{weak}(M, c_M)| = f_1(k_1, k_2, k_3) := \binom{k_2 + 2k_3}{2} - k_3.$$

$$|bp_{strong}(M, c_M)| = f_2(k_1, k_2, k_3) := \left(2k_1 + \frac{3}{2}k_2 + k_3 - \frac{1}{2}\right)(2k_3 + k_2) - k_2 - k_3.$$

According to Lemma 3, we need to find a matching in a weighted graph that minimizes a nonlinear objective function. This is closely related to the NONLINEAR MATCHING studied by Berstein and Onn [3], which is known to be solvable in randomized polynomial time in general and solvable in deterministic polynomial time if the function f is convex [3]. We show that we can reduce both R-WEAK-MBP and R-STRONG-MBP to NM, proving that they can be solved in randomized polynomial time. Moreover, we show that R-WEAK-MBA can be solved in *deterministic* polynomial time since f_1 is convex. This is in contrast to WEAK-MBP, which can also be reduced to NM but the objective function is not convex [7].

Theorem 9 (⋆). R-WEAK-MBP *is solvable in polynomial time and* R-STRONG-MBP *is solvable in randomized polynomial time.*

Finally, we show the objective functions of R-WEAK-MBA and R-STRONG-MBA are linear functions of k_1, k_2, k_3, so they are polynomial-time solvable.

Theorem 10 (⋆). R-WEAK-MBA *and* R-STRONG-MBA *are solvable in polynomial time.*

4.2 Hospitals/Residents

In HR, we consider a set $H = \{h_1, h_2, \ldots, h_{n_1}\}$ of hospitals and a set $R = \{r_1, r_2, \ldots, r_{n_2}\}$ of residents. Each hospital $h_j \in H$ has a capacity c_j, representing the maximum number of residents it can accept. We assume the total capacity of all hospitals is at least n_2, i.e., $\sum_{h_j \in H} c_j \geq n_2$. A matching $M \subseteq H \times R$ is a set of n_2 agents pairs where each hospital appears at most c_j times and each resident appears exactly once. All the notations defined in Sect. 2 can be defined analogously here, except for blocking pairs. Given a matching and a completion $c \in C$, an agent pair $(h, r) \in (H \times R) \setminus M$ is called a *weak blocking pair* under c, if h prefers r than one of its matched residents or h still has unfilled capacity, and c prefers h than its matched hospital. Strong and super blocking pairs can be defined analogously. For each $T \in \{\text{weak, strong, super}\}$, define HOSPITALS/RESIDENTS T-MINIMAX BLOCKING AGENTS (HR-T-MBA) as follows: Given a set H of hospitals, a set R of residents, and a partial approval preference profile $P_{H \cup R}$, the task is to find a matching M^* to minimize the number of T blocking agents under the worst completion. The variant to minimize the number of blocking pairs is referred to as HOSPITALS/RESIDENTS T-MINIMAX BLOCKING PAIRS (HR-T-MBP).

Observe again that the problems for super blocking pairs are trivial since every unmatched agent pair is a super blocking pair under the worst completion.

Observation 11. *For HR-SUPER-MBP (HR-SUPER-MBA), every matching has the same number of super blocking pairs (agents) under the worst completion.*

Finally, we show the remaining four problems are NP-hard via a reduction from EXACT COVER BY 3-SETS (X3C) [11]. In X3C, we are given a set S and a family F of 3-element subsets of S. The task is to find a subset $F' \subseteq F$ such that each element of S is contained in exactly one subset from F'. The idea is to build an instance of our matching problem such that each element in S corresponds to a resident, each subset in F corresponds to a hospital, and each hospital approves exactly those 3 residents corresponding to the 3-elements subset. Consequently, a matching that satisfies $|S|/3$ hospitals and hence minimizes the number of blocking pairs (agents) aligns with an exact cover in X3C.

Theorem 12 (⋆). *HR-WEAK-MBP, HR-WEAK-MBA, HR-STRONG-MBP, and HR-STRONG-MBA are NP-hard.*

5 Conclusion

We study the problem of minimizing blocking agents in three stable matching settings with partial approval preferences. Our work provides a complete picture of the computational complexity for all problem variants and reveals interesting contrasts to previous studies. For future work, it would be valuable to improve the running times of our positive results to enhance their practical applicability. In the NP-hard cases, one could also consider imposing restrictions on the amount of missing information [17]. Finally, it is interesting to explore additional preference restrictions, such as single-peakedness.

Acknowledgments. We thank the reviewers for their constructive suggestions. This research is supported by the National Natural Science Foundation of China (No. 12301412) and the Talent Fund of Beijing Jiaotong University (No. 2023XKRC007).

References

1. Aziz, H., Biró, P., Gaspers, S., de Haan, R., Mattei, N., Rastegari, B.: Stable matching with uncertain linear preferences. Algorithmica **82**(5), 1410–1433 (2020)
2. Bentert, M., Boehmer, N., Heeger, K., Koana, T.: Stable matching with multilayer approval preferences: approvals can be harder than strict preferences. Games Econ. Behav. **142**, 508–526 (2023)
3. Berstein, Y., Onn, S.: Nonlinear bipartite matching. Discret. Optim. **5**(1), 53–65 (2008)
4. Biró, P., Manlove, D.F., Mittal, S.: Size versus stability in the marriage problem. Theoret. Comput. Sci. **411**(16–18), 1828–1841 (2010)
5. Boehmer, N., Brill, M., Schmidt-Kraepelin, U.: Proportional representation in matching markets: selecting multiple matchings under dichotomous preferences. In: AAMAS 2022, pp. 136–144 (2022)
6. Bogomolnaia, A., Moulin, H.: Random matching under dichotomous preferences. Econometrica **72**(1), 257–279 (2004)
7. Chu, Y., Luo, J., Zheng, T.: Stable matching with approval preferences under partial information. In: AAIM 2024, pp. 64–75 (2024)
8. Cseh, Á., Escamocher, G., Quesada, L.: Computing relaxations for the three-dimensional stable matching problem with cyclic preferences. Constraints **28**(2), 138–165 (2023)
9. Eriksson, K., Häggström, O.: Instability of matchings in decentralized markets with various preference structures. Internat. J. Game Theory **36**, 409–420 (2008)
10. Gale, D., Shapley, L.S.: College admissions and the stability of marriage. Am. Math. Mon. **120**(5), 386–391 (2013)
11. Garey, M.R., Johnson, D.S.: Computers and Intractability: A Guide to the Theory of NP-Completeness. W. H. Freeman (1979)
12. Gusfield, D., Irving, R.W.: The Stable Marriage Problem: Structure and Algorithms. MIT Press (1989)
13. Irving, R.W.: Stable marriage and indifference. Discret. Appl. Math. **48**(3), 261–272 (1994)
14. Knittel, M., Dooley, S., Dickerson, J.P.: The dichotomous affiliate stable matching problem: approval-based matching with applicant-employer relations. In: IJCAI 2022, pp. 356–362 (2022)
15. Manlove, D.: Algorithmics of Matching Under Preferences, vol. 2. World Scientific (2013)
16. Meeks, K., Rastegari, B.: Solving hard stable matching problems involving groups of similar agents. Theoret. Comput. Sci. **844**, 171–194 (2020)
17. Menon, V., Larson, K.: Robust and approximately stable marriages under partial information. In: WINE 2018, pp. 341–355 (2018)
18. Roth, A.E., Xing, X.: Turnaround time and bottlenecks in market clearing: decentralized matching in the market for clinical psychologists. J. Polit. Econ. **105**(2), 284–329 (1997)

The Capacity-Constrained Facility Location Problem with Ordinal Preferences: Algorithmic and Mechanism Design Perspectives

Zifan Gong[1], Alexander Lam[1], Momcilo Mrkaic[1], Yachao Yan[1](✉), and Yingchao Zhao[2]

[1] City University of Hong Kong, Hong Kong, Hong Kong
{zifangong2-c,mmrkaic2-c,yachaoyan2-c}@my.cityu.edu.hk,
alexlam@cityu.edu.hk
[2] Saint Francis University, Hong Kong, Hong Kong
yczhao@sfu.edu.hk

Abstract. We study a variation of the facility location problem that involves finding ideal locations for capacitated facilities and assigning agents to these facilities. Additionally, each agent has an ordinal ranking over the facilities and incurs a cost related to both the ranking and the distance from their assigned facility. Our work focuses on minimizing the maximum cost and total cost. For these objectives, we show that computing an optimal solution is intractable in general, but we provide exact algorithms that run in polynomial time when the number of facilities is constant. We then move to the mechanism design setting, where the agents' preferences are private information, and design strategy-proof mechanisms which have a bounded approximation for our objectives.

Keywords: Facility location · Capacity constraints · Ordinal preferences

1 Introduction

Facility location problems are prevalent throughout society, and have been widely studied in computer science, microeconomic theory, and operations research. The problem fundamentally involves determining the ideal locations for a set of facilities, which are being built to serve a group of agents. In this work, we study a variation of the facility location problem which is particularly applicable to school placement. In our model, each facility has a limit to the number of agents that it can serve, and each agent has a complete ordering indicating

---

The authors are ordered alphabetically. Yingchao Zhao is supported by the Research Grants Council of the Hong Kong Special Administrative Region, China, under Project UGC/FDS11/E03/21 and the Institutional Develop Grant of Saint Francis University, under Project IDGC240206.

their personal ranking of the facilities. Due to these nuances, our problem also inherently involves the simultaneous problem of determining the allocation of agents to facilities.

We assume that each agent incurs a cost equal to their distance to the facility they are allocated to, multiplied by a constant factor relating to their ranking of the facility. Our focus is on the objectives of maximum cost, an egalitarian fairness measure which considers the worst-off agent, and total cost, a utilitarian measure of efficiency. We first address these objectives from an algorithmic perspective, assuming that the agents' preferences and locations are public information, and aim to compute solutions which are optimal with respect to our objectives. We then take a mechanism design perspective, assuming that the agents' preferences are instead private information, and may be misreported. Our focus is therefore to design strategy-proof mechanisms which approximate the optimal objective values.

For this work, we restrict our attention to the real-line domain, which is standard in many related papers [2,11,14,15,20]. A one-dimensional domain is a natural starting point for more general metrics such as tree or network graphs, or higher dimensions. In particular, problem instances under an L1-norm can be decomposed into multiple unidimensional problems, under which our results are applicable.

1.1 Our Contributions

We introduce a novel capacity-constrained facility location model with ordinal preferences, where each facility has limited capacity and agents rank their preferences across all facilities. Our study approaches the problem from both algorithmic and mechanism design perspectives.

From an algorithmic perspective, we have the following results in Sect. 4:

- Minimizing either the maximum or total cost is NP-hard.
- We propose optimal algorithms that are polynomial when the number of facilities is bounded for both maximum cost and total cost.

In Sect. 5, we focus on mechanism design, developing strategy-proof mechanisms to minimize both the maximum and total cost under private ordinal preferences. We first examine the setting with m facilities and no spare capacity in Sect. 5.1, where we propose mechanisms with an approximation ratio of α_m. Additionally, we establish the lower bounds of $\frac{\alpha_m+1}{2}$ for maximum cost and $\max\{1, \frac{\alpha_m+2m-2}{2m}\}$ for total cost. We then explore the case of two facilities with spare capacities, where $c_1, c_2 \geq \frac{n}{2}$, with c_1 and c_2 representing the capacities of the two facilities. Finally, we extend our model to more general case of m facilities with arbitrary capacities, showing that the results from Sect. 5.1 continue to hold. A summary of our contributions is provided in Table 1. All the missing proofs can be found in the Appendix.

Table 1. Summary of results. The first row in the mechanisms section refers to m-facility setting, in which each facility has identical capacity k, and there is no spare capacity. In the second row in the mechanisms section, c_1 and c_2 refer to the capacity of the two facilities. The lower bound $\max\{1, \frac{\alpha_m+1}{4}\}$ in Theorem 4 of [5] still holds in our model. The last row is the setting where there are m facilities, each with an arbitrary capacity c_j.

Algorithmic Results	Maximum Cost		Total Cost	
Optimization	NP-hard		NP-hard	
Algorithm	$O(\log(mn^2)mn^{m+2})$		$O(m^3 n^{m+3})$	
Mechanisms	LB	UB	LB	UB
m facilities, capacity k	$\frac{\alpha_m+1}{2}$	α_m	$\max\{1, \frac{\alpha_m+2m-2}{2m}\}$	α_m
two facilities, $c_1+c_2 > n$	$\frac{\alpha_m+1}{2}$	α_m	$\max\{1, \frac{\alpha_m+1}{4}\}$ [5]	α_m
m facilities, capacity c_j	$\frac{\alpha_m+1}{2}$	α_m	$\max\{1, \frac{\alpha_m+2m-2}{2m}\}$	α_m

2 Related Work

The classic facility location problem has been studied in many fields such as operations research and social choice [7,13,18]. In social choice literature, Moulin [12] proposed that strategy-proof mechanisms for locating one facility on the line coincide with the median voter schemes.

In the mechanism design field, the facility location problem has become popular since Procaccia and Tennenholtz [15] introduced approximate mechanisms. In their work, strategy-proof mechanisms without payments were proposed and then extended to the setting of two facilities. After that, Fotakis and Tzamos [9] proposed that no deterministic, anonymous, strategy-proof mechanism with a bounded approximation ratio exists when there are more than three facilities.

There are many variants of the problem. One research direction focuses on studying models with preferences, including the optional preferences model. In that model, there are two facilities and an agent "likes" either one or both facilities. The cost is the sum of distances from his position to the facilities he likes. Serafino and Ventre [16,17] and Fong et al. [8] also proposed the facility location problem with fractional preferences, where the preference of each agent for a facility is a number between 0 and 1. Another direction is the obnoxious facility location problem with preferences, where the agents want to be as far away as possible from the facility [6,21]. In most existing models that involve preferences, the agent only cares about its closest/furthest facility. These models cannot model agents' interests effectively. In our model, the agents can express preferences over the set of m facilities through the coefficients.

Another research direction involves more constraints. The capacitated facility location is one important direction, where each facility can serve a limited number of agents, which is close to many real-life scenarios [1,4]. Aziz et al. [3] initiated the mechanism design approach for the capacitated facility location problem. From a computational perspective, they showed that the objective of minimizing either the total cost or the maximum cost is NP-hard. They also proposed algorithms that run in polynomial time when the number of facilities is bounded. For the mechanism design approach, they proposed many strategy-proof mechanisms including Extended Endpoint Mechanism and InnerPoint Mechanism. The InnerPoint Mechanism is for two facilities with equal capacity, and achieves an approximation ratio of $\frac{n-1}{2}$ for total cost and 2 for maximum cost. Recently, Auricchio et al. [2] presented strategy-proof mechanisms for k facilities with equal capacities. They proposed two strategy-proof and anonymous mechanisms with bounded approximation ratio. On the line of ordinal preferences, agents report linear ordinal preferences on the set of potential facility sites to be opened [10, 19]. We adopt the model introduced by Chan et al. [5]. They considered mechanism design perspective for the problem of locating two facilities on a bounded interval, where agents express ordinal preferences on two facilities. We extend their framework by involving more than two facilities and capacity constraints.

3 Preliminaries

Let $N = \{1, 2, \ldots, n\}$ denote the set of agents on a real-line. The location profile is $\boldsymbol{x} = (x_1, x_2, \ldots x_n)$. Without loss of generality, we suppose $x_1 \leq x_2 \leq \cdots \leq x_n$. Let $F_1, F_2, \ldots, F_m$ be the m facilities to be located. The capacity of each facility F_j is c_j, $j \in [m]$. We denote $\boldsymbol{c} = (c_1, c_2, \ldots, c_m)$. We assume that all agents can be served, that is $\Sigma_{i=1}^{m} c_i \geq n$.

A mechanism f is a function that maps the location profile $\boldsymbol{x}$ and preference profile $\boldsymbol{p}$ to an output $\boldsymbol{y} = (y_1, y_2, \ldots, y_m)$ that locates facility F_j at y_j. The mechanism f also outputs an assignment: $\boldsymbol{N} = \{N_1, N_2, \ldots, N_m\}$. Given an agent i, let $a_i \in \{1, 2, \ldots, m\}$ be the facility that agent i is assigned to. Let N_j denote the set of agents that are assigned to facility j, i.e. $N_j = \{i | a_i = j\}$.

Let $d(x_i, y_j) = |x_i - y_j|$ be the distance between x_i and y_j. Let $1 = \alpha_1 \leq \alpha_2 \leq \cdots \leq \alpha_m$ be constant coefficients. We use these coefficients to characterize the agent's preference over different facilities. Each agent $i \in N$ has an ordinal preference over the m facilities. Agent i's ordinal preference is denoted by a ranking σ_i over the m facilities. The agents' preference profile is denoted by $\boldsymbol{p} = (p_1, p_2, \ldots, p_n)$. Let $p_i = (\sigma_1, \sigma_2, \ldots, \sigma_m)$, which indicates that facility F_{σ_k} is the agent's k-th most preferred facility. Therefore, we define the cost that agent i has when assigned to facility F_j as

$$c_i(y_j) = \alpha_k d(x_i, y_j),$$

F_j is agent i's k-th preferred facility and $k \in [m]$. The total (social) cost of all agents is defined as $SC(\boldsymbol{x}, \boldsymbol{y}) = \sum_{i \in N} c_i(y_j)$. The maximum cost is defined as $MC(\boldsymbol{x}, \boldsymbol{y}) = \max_{i \in N} c_i(y_j)$. The optimal maximum cost and total cost are denoted as OPT_{MC} and OPT_{TC} respectively.

4 Computational Result

In this section, we consider the capacity-constrained facility location problem with ordinal preferences from the algorithmic perspective. We first show that the problem is NP-hard and design an algorithm that runs in polynomial time when m is a given constant.

4.1 Maximum Cost

We first consider the objective of minimizing the maximum cost. In this section, we prove that achieving this objective is NP-hard and propose a polynomial time algorithm given a fixed number of facilities.

Proposition 1. *Computing a solution that minimizes the maximum cost is NP-hard even when there is no spare capacity.*

Proof. Aziz et al. in [3] proved that computing a solution that minimizes the maximum cost is NP-hard for facility location problem with capacities. Since their problem is a special case of our minimizing the maximum cost for facility location problem with capacities and ordinal preferences with all the $\alpha_j = 1, \forall j \in [m]$, our problem is also NP-hard.

Before going deeper, we have the following observation. For facility F_j, knowing all agents assigned to it, we can calculate the minimum $MC(\boldsymbol{x}, \boldsymbol{y})$ of agents from N_j. Thus, it is possible to determine all feasible agent assignments to facilities and to derive the minimum total cost among them. However, the number of ways of assigning agents to facilities is an exponential function in terms of the number of agents. We want to find an algorithm such that it is possible to compute the optimal maximum cost in polynomial time for a bounded number of facilities. Lemma 1 provides a direction for us, which narrows down the possible values of minimum maximum cost to $O(mn^2)$ numbers. For each number, we can check its feasibility in polynomial time for bounded number of facilities.

Lemma 1. *In the optimal solution, there are at least two agents assigned to the same facility that have cost of OPT_{MC} or the maximum cost is 0.*

We design an algorithm for finding the optimal solution for maximum cost, given subroutine COMPUTEFEASIBLESOLUTION corresponding to the problem of checking feasibility and computing a solution for given maximum cost. With the help of Lemma 1, we have the set of candidate values. Among them, there are values for which there is no solution with smaller maximum cost, thus we use COMPUTEFEASIBLESOLUTION to determine OPT_{MC}.

Theorem 1. OPTIMALMAXIMUMCOST *computes a solution that minimizes the maximum cost in $O(log(mn^2)T(n,m))$ time, given* COMPUTEFEASIBLESOLUTION *which runs in $T(n,m)$.*

Then, we design an algorithm for computing a feasible solution given a maximum cost. First, for each facility j and given max cost r, each agent i has a feasible interval of radius $R_{i,j} = \frac{r}{\alpha_k}$ around the agent's location where we can place facility F_j, where k is the index of facility F_j in p_i. We are interested in

Algorithm 1. Computing Optimal Solution for Maximum Cost

1: **procedure** OPTIMALMAXIMUMCOST(n, m, x, c, α, p)
2:     $q \leftarrow []$
3:     **for** $i \leftarrow 1$ to $n-1$ **do**
4:         **for** $j \leftarrow i+1$ to n **do**
5:             **for** $k \leftarrow 1$ to m **do**
6:                 $a \leftarrow$ index of k in p_i
7:                 $b \leftarrow$ index of k in p_j
8:                 $y \leftarrow \frac{\alpha[a] \cdot x[i] + \alpha[b] \cdot x[j]}{\alpha[a] + \alpha[b]}$
9:                 Add $\alpha[a](y - x[i])$ to q
10:             **end for**
11:         **end for**
12:     **end for**
13:     Sort q in ascending order
14:     $l \leftarrow 1, r \leftarrow size(q)$
15:     **while** $l < r$ **do**
16:         $mid \leftarrow \frac{l+r}{2}$
17:         **if** ComputeFeasibleSolution$(q[mid], n, m, x, c, \alpha, p) \neq$ false **then**
18:             $r \leftarrow mid$
19:         **else**
20:             $l \leftarrow mid + 1$
21:         **end if**
22:     **end while**
23:     **return** ComputeFeasibleSolution$(q[l], n, m, x, c, \alpha, p)$
24: **end procedure**

the subsets of agents for whom it is feasible to assign them to one facility simultaneously, which means the intersection of their intervals is not empty so we place the facility at that intersection. Finding such a subset for each facility, so that subsets are disjoint and their union is the whole set of agents, is a feasible solution for a given maximum cost.

We will find all feasible subsets for which no other agent can be added without violating the feasibility of the subset. In other words, it is a feasible subset of agents that is not a subset of any larger feasible subset (i.e., a maximal subset). Each of these subsets corresponds to the end point of some interval, where we place a facility, and there are $O(n)$ subsets. In Algorithm 2, we use A to represent intervals. We make use of data structure q that can extract element with the minimum rightEnd, and we store all feasible subsets in Q.

In total, there are $O(n^m)$ combinations of feasible subsets, given that there are $O(n)$ feasible subsets for each facility. Now taking one combination of such subsets, one subset for each facility, we need to check whether the solution is feasible, i.e. there is an assignment of agents to the facilities that meets capacity constraints, and their union is the whole set of agents. This can be done with maximum flow algorithm. We construct a flow network where the source points to each agent with capacity 1. Agents from feasible subsets in the chosen combination point to a vertex that represents a corresponding facility, also with capacity 1. Finally, each facility points to the sink with its capacity constraint. If the max flow in this network is n, we find a feasible solution. Moreover, if $f(v, w)$ represents the flow from vertex v to vertex w, we can reconstruct the solution. If $f(i, F_j) = 1$, we assign agent i to facility j. For example, in Fig. 1, we see that the maximum flow is 5, and we assign agent 1 to F_1, agent 2 to F_2, and the remaining agents to F_3.

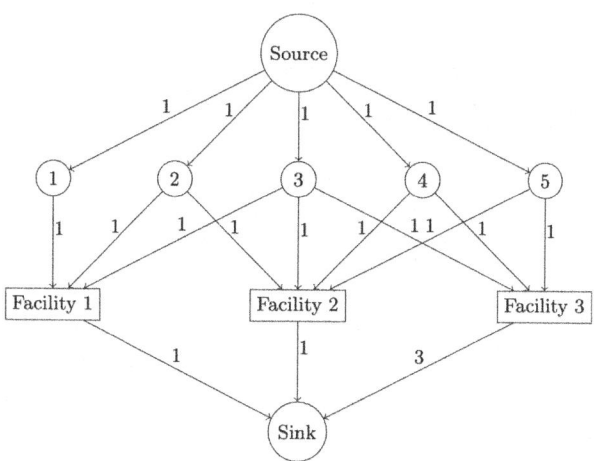

Fig. 1. Checking the feasibility of given subsets with flow. Here, we see that the maximum flow is 5, and we assign agent 1 to F_1, agent 2 to F_2, and the remaining agents to F_3.

Algorithm 2. Finding Feasible Subsets for Facility j

1: **procedure** SWEEPLINE(r, n, j, x, α, p)
2:     $A \leftarrow [], Q \leftarrow [], q \leftarrow []$
3:     **for** $i \leftarrow 1$ to n **do**
4:         $a \leftarrow$ index of j in p_i
5:         $A[i].\text{leftEnd} \leftarrow x[i] - \frac{r}{\alpha[a]}$
6:         $A[i].\text{rightEnd} \leftarrow x[i] + \frac{r}{\alpha[a]}$
7:     **end for**
8:     Sort A by leftEnd in ascending order, breaking ties arbitrarily
9:     **for** $i \leftarrow 1$ to n **do**
10:        **while** q is not empty and Extract-Min(q).rightEnd $< A[i].\text{leftEnd}$ **do**
11:            Remove Extract-Min(q) from q
12:        **end while**
13:        Add i to q, and then add (q) to Q
14:     **end for**
15:     **return** Q
16: **end procedure**

Theorem 2. *Given maximum cost r, COMPUTEFEASIBLESOLUTION computes a feasible solution with the maximum cost no greater than r in $O(mn^{m+2})$. Hence, it runs in polynomial time for a fixed number of facilities.*

Algorithm 3. Checking Feasibility of Given Max Cost

1: **procedure** COMPUTEFEASIBLESOLUTION$(r, n, m, x, c, \alpha, p)$
2:   $Q \leftarrow []$
3:   **for** $j \leftarrow 1$ to m **do**
4:     $Q[j] \leftarrow$ SweepLine(r, n, j, x, α, p)
5:   **end for**
6:   **return** CHECKCOMBINATIONS$(n, m, c, Q, [], 1)$
7: **end procedure**
8: **procedure** CHECKCOMBINATIONS$(n, m, c, Q,$ combination, index$)$
9:   **if** index $<= m$ **then**
10:     **for** num $\leftarrow 1$ to size$(Q[$ index $])$ **do**
11:       combination[index] $\leftarrow$ num
12:       solution $\leftarrow$ CHECKCOMBINATIONS$(n, m, c, Q,$ combination, index $+ 1)$
13:       **if** solution $\neq$ false **then return** solution
14:     **end if**
15:     **end for**
16:     **return** false
17:   **else**
18:     Initialize a graph G with vertices for source, sink, n agents, and m facilities
19:     **for** $i \leftarrow 1$ to n **do**
20:       Add edge (source, agent i, 1) to G
21:     **end for**
22:     **for** $j \leftarrow 1$ to m **do**
23:       Add edge (facility j, sink, $c[j]$) to G
24:       **for** agent in $Q[j][$combination$[j]]$ **do**
25:         Add edge (agent, facility j, 1) to G
26:       **end for**
27:     **end for**
28:     $f \leftarrow$ MaxFlow$(G,$ source, sink$)$
29:     **if** $|f| == n$ **then**
30:       assignment $\leftarrow []$
31:       **for** $i \leftarrow 1$ to $n, j \leftarrow 1$ to m **do**
32:         **if** $f($agent $i,$ facility $j) == 1$ **then**
33:           Add i to assignment $[j]$
34:         **end if**
35:       **end for**
36:       **return** assignment
37:     **else**
38:       **return** false
39:     **end if**
40:   **end if**
41: **end procedure**

Theorem 3. *Algorithm 1 computes a solution that minimizes the maximum cost in $O(log(mn^2))mn^{m+2})$, which is polynomial for a fixed number of facilities.*

Proof. Using Algorithm 3 as the subroutine in Algorithm 1 enables us to solve the problem with the maximum cost objective in polynomial time for fixed number of facilities. From Theorem 1, we conclude that the running time is $O(log(mn^2)mn^{m+2})$.

4.2 Total Cost

In this section, we consider the objective of minimizing the total cost. We prove that achieving this objective is NP-hard and propose an algorithm which can solve it in polynomial time for a fixed number of facilities.

Proposition 2. *Computing a solution that minimizes the total cost is NP-hard even when there is no spare capacity.*

As discussed above, if we approach the problem by determining all feasible agent assignments to facilities first and to deriving the minimum total cost among them, the time complexity will be exponential in terms of number of agents. We want to find an algorithm such that it is possible to compute optimal maximum cost in polynomial time for bounded number of facilities. Thus, we approach the problem from the other side. Instead of first deciding on the assignment of agents to facilities, we locate each facility first and then we figure out the optimal assignment for that facility locations profile. By Lemma 2, we can see that for each facilities, there are need $O(n)$ possible locations given the objective of minimizing total cost.

Lemma 2. *There exists an optimal solution where each facility has the same location with at least one agent.*

Algorithm 4 uses backtracking to construct all facility location profiles such that each facility is placed at location of some agent. For each profile we construct sets of jobs and workers. For each facility location profile, we use Hungarian algorithm to find the optimal assignment. In this setting, we set $sumC := \sum_{j \in [m]} min(n, c_j)$ jobs and workers. A job is an agent that needs to be served by some facility, while free capacities in facilities represent the workers. Thus, facility F_j provides $min(n, c_j)$ workers. We also have the following cost function: if job i corresponding to agent i is assigned to a worker provided by the facility F_j, it produces a cost of $\alpha_a|x_i - y_j|$, where a is the index of facility F_j in p_i. If there are some spare capacity, which means $sumC > n$, the extra jobs than n will lead to 0 cost no matter which facility it is served by. We can see that this setting will not influence the capacity constraints and value of total cost. Therefore, we transform the problem to the assignment problem which can be solved in polynomial time with Hungarian algorithm.

F stores the information which facility provides the workers. We call the OPTIMALTOTAL COST subroutine to retrieve the solution.

Theorem 4. *Algorithm 4 computes a solution that minimizes total cost in $O(m^3n^{m+3})$ time, which is polynomial for a fixed number of facilities.*

Algorithm 4. Computing Optimal Solution for Total Cost

1. Initialize: set best assignment and minimum total cost to empty and infinity, respectively.
2. Recursively try all combinations of m agent locations as candidate facility sites.
    2.1. For each combination, construct a multiset of facility slots: facility j appears $c[j]$ times.
    2.2. Define cost of assigning agent i to facility j as the product of the distance $|x[i] - y[j]|$ and the weight corresponding to j's rank in i's preference list.
    2.3. Use the Hungarian algorithm to compute an assignment minimizing total cost under this cost function. A job represents an agent and a facility's free capacity corresponds to workers.
    2.4. If the resulting total cost is lower than the current minimum, update the best assignment and cost.
3. After all combinations are evaluated, return the assignment with the lowest total cost.

5 Mechanism Design

In this section, we consider the mechanism design setting, in which each agent i's preference p_i is private information, and thus an agent may misreport his preference to p_i'.

A mechanism f is strategy-proof (SP) if by reporting the information truthfully, each agent gains at least as much as that when misreporting, regardless of what others report, i.e., for any agent i and p_i' we have $c_i(f(p_i, \boldsymbol{P}_{-i}, \boldsymbol{x})) \leq c_i(f(p_i', \boldsymbol{P}_{-i}, \boldsymbol{x}))$. We say a mechanism f has an approximation ratio $r \geq 1$ under the objective of minimizing the maximum (resp. total) cost, if for any instance $I = (\boldsymbol{P}, \boldsymbol{x})$, the output satisfies:

$$r = \sup_I \frac{MC(f(I))}{OPT_{MC}(I)} \quad (\text{resp.} \quad r = \sup_I \frac{SC(f(I))}{OPT_{TC}(I)}),$$

where $OPT_{MC}(I)$ (resp. $OPT_{TC}(I)$) is the optimal maximum (resp. total) cost of the instance I.

In this section, we explore three different scenarios. First, we analyze the case where there are m facilities, each with an equal capacity of k, and no excess capacity. Next, we investigate a setting with two facilities that have spare capacity. Finally, we extend the model to a scenario with m facilities, where each facility F_i has an arbitrary capacity c_i. For all three settings, we propose strategy-proof mechanisms for minimizing the maximum cost and the total cost. We also provide corresponding lower bounds for each objective.

Let Φ^m denote the mechanism design problem for optimizing a specific system objective and Φ^1 denote the special setting when $\alpha_1 = \alpha_2 = \cdots = \alpha_m$.

We extend the Proposition 1 introduced by Chan et al. [5] to the m-facility setting, characterizing the relationship between Φ^1 and Φ^m to demonstrate its applicability to our problem.

Lemma 3. *Under the maximum cost and total cost objectives, if a mechanism f has an approximation ratio of β for Φ^1, then it has an approximation ratio of at most $\beta \alpha_m$ for Φ^m.*

5.1 M Facilities, No Spare Capacity

Mechanism 1. *Let k be the capacity of each facility and $n = km$. Let $X_j = \{x_{k(j-1)+i} | i \in [k]\}$. Place facility F_j at the location of $\frac{x_{k(j-1)+1} + x_{kj}}{2}$. Then assign the agents in X_j to facility F_j.*

Theorem 5. *Mechanism 1 is strategy-proof and optimal for minimizing the maximum cost in the Φ^1 setting, and achieves an approximation ratio of α_m for the maximum cost in the Φ^m setting.*

Theorem 6. *For Φ^m with private preferences, no SP mechanism has an approximation ratio less than $\frac{\alpha_m+1}{2}$ under the maximum cost objective.*

Mechanism 2. *Set $\alpha_1 = \alpha_2 = \cdots = \alpha_m = 1$, then run Algorithm 4 and get the location $\mathbf{y}$ and assignment $\mathbf{N}$.*

Theorem 7. *Mechanism 2 is a strategy-proof mechanism with an approximation ratio of α_m for total cost in Φ^m setting.*

Theorem 8. *For Φ^m with private preferences, no SP mechanism has an approximation ratio less than $\max\{1, \frac{\alpha_m + 2m - 2}{2m}\}$ under the total cost objective.*

5.2 Two Facilities, Spare Capacity

In this section, we consider the scenario that there are two facilities, and the capacity $c_1, c_2 \geq \frac{n}{2}$. We propose strategy-proof mechanisms with an approximation ratio of α_m for maximum cost and total cost respectively. We also prove that for maximum cost, no SP mechanism has an approximation ratio less than $\max\{1, \frac{\alpha_m+1}{2}\}$.

We first consider the problem when $\alpha_m = 1$, where the two facilities are indifferent to the agents. To minimize the maximum cost, the optimal solution can be computed in $O(n^2)$ time in the following way: For $i \in [2, n-1]$, let $y_{1i} = \frac{x_1+x_i}{2}$, and $y_{2i} = \frac{x_{i+1}+x_n}{2}$. Then, place F_1 at y_{1i} and F_2 at y_{2i}, and assign agents in $[x_1, x_i]$ to F_1 and agents in $[x_{i+1}, x_n]$ to F_2. Finally, return the solution with the smallest maximum cost among the $n-1$ solutions (y_{1i}, y_{2i}), while also ensuring that $|N_j| \leq c_j$, for $j \in \{1, 2\}$. Based on this optimal mechanism (denoted by f_{MC^*}), we have the following theorem.

Theorem 9. *Mechanism 1 is a strategy-proof mechanism with an approximation ratio of α_m for the maximum cost.*

Theorem 10. *For two facilities with spare capacity, no strategy-proof mechanism has an approximation ratio less than $\frac{\alpha_m+1}{2}$ under the maximum cost objective.*

For the total cost objective, we also consider the case when $\alpha = 1$, where the two facilities are indifferent to the agents. To minimize the total cost, the optimal solution can be computed in $O(n^2)$ time as follows: For $i \in [2, n-1]$, let y_{1i} denote the median of $x_1, x_2, \ldots, x_i$, and y_{2i} denote the median of $x_{i+1}, x_{i+2}, \ldots, x_n$. Then, place F_1 at y_{1i} and F_2 at y_{2i}, and assign agents in $[x_1, x_i]$ to F_1 and agents in $[x_{i+1}, x_n]$ to F_2. Finally, return the solution with the smallest maximum cost among the $n-1$ solutions (y_{1i}, y_{2i}), while also ensuring that $|N_j| \leq c_j$, for $j \in \{1, 2\}$. Based on this optimal mechanism (denoted by f_{TC^*}), we have the following theorem.

Theorem 11. *Mechanism f_{TC*} is a strategy-proof mechanism with an approximation ratio of α_m for the total cost objective.*

5.3 m Facilities, Arbitrary Capacity

In this section, we consider the case with m different facilities. Each facility F_j has a capacity of c_j, where $c_j \geq 1$.

Theorem 12. *There is a strategy-proof mechanism with an approximation ratio of α_m for the max cost and total-cost objective, when there are m facilities.*

Theorem 13. *For m facilities with arbitrary capacity, no SP mechanism has an approximation ratio less than $\frac{\alpha_m+1}{2}$ under the maximum cost objective.*

Theorem 14. *For m facilities with arbitrary capacity, no SP mechanism has an approximation ratio less than $\max\{1, \frac{\alpha_m + 2m - 2}{2m}\}$ under the total cost objective.*

6 Discussion and Conclusion

In this paper, we introduce the one-dimensional capacity-constrained facility location model with ordinal preferences, where each facility has limited capacity, and agents rank their preferences across all facilities. Our study approaches the problem from both algorithmic and mechanism design perspectives. From the algorithmic perspective, we show that minimizing either the maximum or total cost is NP-hard. We also design optimal algorithms that are polynomial when the number of facilities is bounded. For the mechanism design approach, we consider three different scenarios and propose strategy-proof mechanisms with bounded approximation ratios.

There are many possible directions for future work. An interesting direction could be the setting where the agents can misreport their locations. The existence of a mechanism that achieves both strategyproofness and anonymity is also an open question.

References

1. Aardal, K., van den Berg, P.L., Gijswijt, D., Li, S.: Approximation algorithms for hard capacitated k-facility location problems. Eur. J. Oper. Res. **242**(2), 358–368 (2015)
2. Auricchio, G., Wang, Z., Zhang, J.: Facility location problems with capacity constraints: two facilities and beyond. In: Larson, K. (ed.) Proceedings of the Thirty-Third International Joint Conference on Artificial Intelligence, IJCAI 2024, pp. 2651–2659. International Joint Conferences on Artificial Intelligence Organization (2024). https://doi.org/10.24963/ijcai.2024/293
3. Aziz, H., Chan, H., Lee, B., Li, B., Walsh, T.: Facility location problem with capacity constraints: algorithmic and mechanism design perspectives. In: Proceedings of the AAAI Conference on Artificial Intelligence, vol. 34, pp. 1806–1813 (2020)
4. Brimberg, J., Korach, E., Eben-Chaim, M., Mehrez, A.: The capacitated p-facility location problem on the real line. Int. Trans. Oper. Res. **8**(6), 727–738 (2001)
5. Chan, H., Gong, Z., Li, M., Wang, C., Zhao, Y.: Facility location games with ordinal preferences. Theoret. Comput. Sci. **979**, 114208 (2023)
6. Cheng, Y., Yu, W., Zhang, G.: Strategy-proof approximation mechanisms for an obnoxious facility game on networks. Theoret. Comput. Sci. **497**, 154–163 (2013)
7. Farahani, R.Z., Hekmatfar, M.: Facility Location: Concepts, Models, Algorithms and Case Studies. Springer (2009)
8. Fong, C.K.K., Li, M., Lu, P., Todo, T., Yokoo, M.: Facility location games with fractional preferences. In: Proceedings of the AAAI Conference on Artificial Intelligence, vol. 32 (2018)
9. Fotakis, D., Tzamos, C.: On the power of deterministic mechanisms for facility location games. ACM Trans. Econ. Comput. (TEAC) **2**(4), 1–37 (2014)
10. Hanjoul, P., Peeters, D.: A facility location problem with clients' preference orderings. Reg. Sci. Urban Econ. **17**(3), 451–473 (1987)
11. Lu, P., Wang, Y., Zhou, Y.: Tighter bounds for facility games. In: Leonardi, S. (ed.) WINE 2009. LNCS, vol. 5929, pp. 137–148. Springer, Heidelberg (2009). https://doi.org/10.1007/978-3-642-10841-9_14
12. Moulin, H.: On strategy-proofness and single peakedness. Public Choice **35**(4), 437–455 (1980)
13. Owen, S.H., Daskin, M.S.: Strategic facility location: a review. Eur. J. Oper. Res. **111**(3), 423–447 (1998)
14. Pal, M., Tardos, T., Wexler, T.: Facility location with nonuniform hard capacities. In: Proceedings 42nd IEEE Symposium on Foundations of Computer Science, pp. 329–338. IEEE (2001)
15. Procaccia, A.D., Tennenholtz, M.: Approximate mechanism design without money. In: Proceedings of the 10th ACM Conference on Electronic Commerce, pp. 177–186 (2009)
16. Serafino, P., Ventre, C.: Heterogeneous facility location without money on the line. In: ECAI 2014, pp. 807–812. IOS Press (2014)
17. Serafino, P., Ventre, C.: Truthful mechanisms without money for non-utilitarian heterogeneous facility location. In: Proceedings of the AAAI Conference on Artificial Intelligence, vol. 29 (2015)
18. Snyder, L.V.: Facility location under uncertainty: a review. IIE Trans. **38**(7), 547–564 (2006)
19. Vasil'ev, I.L., Klimentova, K.B., Kochetov, Y.A.: New lower bounds for the facility location problem with clients' preferences. Comput. Math. Math. Phys. **49**, 1010–1020 (2009)

20. Walsh, T.: Strategy proof mechanisms for facility location with capacity limits. In: Raedt, L.D. (ed.) Proceedings of the Thirty-First International Joint Conference on Artificial Intelligence, IJCAI 2022, pp. 527–533. International Joint Conferences on Artificial Intelligence Organization (2022). https://doi.org/10.24963/ijcai.2022/75
21. Ye, D., Mei, L., Zhang, Y.: Strategy-proof mechanism for obnoxious facility location on a line. In: International Computing and Combinatorics Conference, pp. 45–56. Springer (2015)

Regularized Minimax-V Learning for Solving Randomly Terminating Two-Player Zero-Sum Markov Games

Xinxiang Guo[1,2] and Yifen Mu[2,3(✉)]

[1] School of Mathematical Sciences, University of Chinese Academy of Sciences, Beijing, China
[2] Key Laboratory of Systems and Control, Institute of Systems Science, Academy of Mathematics and Systems Science, Chinese Academy of Sciences, Beijing, China
mu@amss.ac.cn
[3] State Key Laboratory of Mathematical Sciences, Academy of Mathematics and Systems Science, Chinese Academy of Sciences, Beijing, China

Abstract. In this paper, we investigate randomly terminating two-player zero-sum Markov games. This game model differs from infinite-horizon discounted zero-sum Markov games and can be used to model many practical scenarios; however, related work on such games is still insufficient. We propose the regularized minimax-V learning algorithm and prove that the value sequence generated by this algorithm converges to the minimax value function with an appropriate choice of regularization parameter sequence. This algorithm applies the Euclidean regularization technique to accelerate the convergence in a different way compared with previous literature. Through simulation experiments, we demonstrate the convergence of regularized minimax-V learning algorithm in the ratio game and a randomly generated Markov game. To the best of our knowledge, for randomly terminating two-player zero-sum Markov games, this paper presents the first accelerated NE-solving algorithm.

Keywords: Markov game · Euclidean regularization · V-value iteration · Accelerating technique

1 Introduction

In recent years, multi-agent reinforcement learning [3] has found widespread applications in modeling interactions among multiple agents or between agents and the environment. Specific applications include real-time strategy games [29], chess and card games [21,26], robot control [17], energy networks [10] and so on. In game theory, multi-agent reinforcement learning can be framed as stochastic games [20,25]. Depending on the relationship among agents, stochastic games are categorized into cooperative [30], competitive [20], and mixed settings [34]. Competitive settings in stochastic games often refer to two-player zero-sum Markov games (MGs). Specifically, one variety is the randomly terminating MG, which can be used to model numerous scenarios, such as robot sports, gaming, racing, and so on. However, most literature focuses on infinite-horizon MGs with a discount parameter, and research on randomly terminating MGs is relatively scarce [25]. This paper considers the randomly terminating MGs and investigates methods of finding Nash equilibria for such games.

Previous equilibrium-solving algorithms can be categorized into two types: value-based methods and policy-based methods. Policy-based methods [1,35] typically parameterize policies and update these parameters directly using gradient information. Such methods include REINFORCE [32], actor-critic [18], extra-gradient [19] algorithms, and so on. Despite their advantage of scalability to large action spaces through function approximation, policy-based methods often face challenges in directly obtaining gradients.

Value-based methods maintain a value sequence to estimate the value of a state or state-action pair and iteratively update these estimates. Most value-based methods iterate over the value of state-action pairs (i.e., Q-values) and generate the current policy using the updated Q-values [20,24]. Beyond value-based methods that iteratively update Q-value functions, there are also value-based methods that iteratively update the state value function (i.e., V-values) [16,25]. Compared to iteratively updating the Q-value function, iteratively updating the V-value function is simpler and can still be theoretically guaranteed to learn a Nash equilibrium under certain conditions. The algorithms proposed in this paper fall into this class.

Recently, research has extended beyond focusing solely on the convergence of algorithms to Nash equilibria to also include considerations of convergence efficiency. This encompasses the convergence conditions [4,14], the convergence rate [6,9], and the convergence mode (i.e., time-average or last-iterate) [13]. Regularization techniques are simple yet effective methods that balance exploitation and exploration by introducing a regularization term into the objective function. This technique finds wide applications in fields such as machine learning, statistical learning, online learning, game theory, and multi-agent reinforcement learning.

The main motivation of this paper is to study the randomly terminating two-player zero-sum MGs and propose an accelerated algorithm using regularization techniques without losing convergence. We propose a regularized minimax-V learning algorithm to solve the NE and introduce Euclidean regularization to accelerate the process of solving a saddle point problem. We prove the convergence of the value sequence generated by this algorithm to the minimax value function with an appropriate choice of the regularization parameter sequence. Through simulation experiments, we demonstrate the convergence of the regularized minimax-V learning algorithm in the ratio game and a randomly generated Markov game.

1.1 Related Work

Two-Player Zero-Sum Markov Games (MGs). According to the time horizon, two-player zero-sum MGs can be classified into three categories: infinite-horizon [4,31,33], episodic finite-horizon [2,16], and randomly terminating horizon. Compared with the first two categories, research on randomly terminating MGs is relatively lacking. Shapley [25] first introduced this game model and proposed an algorithm analogous to value iteration. Daskalakis et al. [8] showed that if both players run policy gradient methods in tandem, their policies will converge to a min-max equilibrium of the game, as long as their learning rates follow a necessary two-timescale rule. Policy-based learning algorithms usually require additional conditions on the game model to achieve convergence. This paper also considers the randomly terminating version and proposes two algorithms that converge to the NE without additional conditions.

The regularized minimax-V learning algorithm is inspired by the work of Littman [20]. Compared with the V-learning algorithm [16], which also updates V-values at each iteration, this paper does so in a completely different manner. The V-learning algorithm is designed for finite-horizon episodic MGs, where the value sequence can be updated from the last step to the first step within each episode. However, since the horizon of randomly terminating MGs is potentially infinite, V-learning is no longer applicable.

Regularization Technique for Learning in Games. Generally speaking, there are two main approaches to applying regularization techniques in game learning. One approach is to incorporate the regularization term into the objective function of the learning algorithm, which can improve its performance [5,27]. The other approach is to integrate the regularization term into the payoff information [23]. Cen et al. [7] use entropy regularization to modify the original matrix game, resulting in an algorithm that can find an approximate Nash equilibrium at a sublinear rate. In our work, we utilize the Euclidean norm as the regularization term to attain a faster convergence rate, and the entropy regularization does not apply here because the entropy of a probability distribution is unbounded.

Paper Structure: In Sect. 2, we introduce the randomly terminating two-player zero-sum Markov game model and the definition of norm used later; in Sect. 3, we propose the regularized minimax-V learning algorithm, prove its convergence to the minimax-value function; in Sect. 4, we implement several simulation experiments including ratio game and randomly generated Markov game; in Sect. 5, we conclude this paper and introduce some possible future directions.

2 Problem Formulation

2.1 Game Model

The randomly terminating two-player zero-sum Markov game (MG) was proposed by Shapley [25] and can be represented by a tuple $(\mathcal{N}, \mathcal{S}, \mathcal{A}, \mathcal{B}, r, p, \zeta)$ with each element defined as follows:

- $\mathcal{N} := \{X, Y\}$ is the set of players, i.e., the participants of the game are player X and Y.
- $\mathcal{S}$ is the finite state space.
- $\mathcal{A}$ is the finite action set for player X and $\mathcal{B}$ is the finite action set for player Y.
- $r : \mathcal{S} \times \mathcal{A} \times \mathcal{B} \to [-1, 1]$ is the payoff function for player Y and the cost function for player X.
- $p : \mathcal{S} \times \mathcal{A} \times \mathcal{B} \to \mathcal{S}$ is the state transition function, i.e., $p(s'|s, a, b)$ gives the probability of transmitting from state s to state s' if player X takes action a and player Y takes action b.
- $\zeta := \min_{(s,a,b) \in \mathcal{S} \times \mathcal{A} \times \mathcal{B}} (1 - \sum_{s'} p(s'|s, a, b)) > 0$ is the lower bound on the probability that the MG will terminate at some state after both players take some actions.

A stationary policy (aka. Markovian policy) for player X refers to a set of mappings denoted by $\Delta(\mathcal{A})^{|\mathcal{S}|} \ni \mathbf{x} := \{x_s\}_{s \in \mathcal{S}}$ where $x_s : \mathcal{S} \to \Delta(\mathcal{A})$ maps each state to an action distribution. The stationary policy $\mathbf{y}$ and mapping y_s for player Y can be similarly defined.

For a pair of stationary policies $(\mathbf{x}, \mathbf{y})$ and an initial state s, the value that player Y gains and player X pays can be represented as

$$V^{\mathbf{x},\mathbf{y}}(s) = \mathbb{E}[\sum_{t=0}^{T} r(s_t, a_t, b_t)|s_0 = s, a_t \sim x_{s_t}(\cdot), \quad (1)$$
$$b_t \sim y_{s_t}(\cdot), s_{t+1} \sim p(\cdot|s_t, a_t, b_t), t \geq 0],$$

where the expectation above is taken with respect to the trajectory induced by the joint policy $(\mathbf{x}, \mathbf{y})$. We call $V^{\mathbf{x},\mathbf{y}} = (V^{\mathbf{x},\mathbf{y}}(s))_{s \in \mathcal{S}}$ the *V-value function*. The minimax game value on state s is then defined as

$$V^*(s) = \min_{x \in \Delta(\mathcal{A})^{|\mathcal{S}|}} \max_{y \in \Delta(\mathcal{B})^{|\mathcal{S}|}} V^{\mathbf{x},\mathbf{y}}(s). \quad (2)$$

We call $V^* = (V^*(s))_{s \in \mathcal{S}}$ the *minimax value function*.

A pair of stationary policies $(\mathbf{x}^*, \mathbf{y}^*)$ is called a *Nash equilibrium* (NE) of the MGs if for any stationary policy $\mathbf{x}$ and $\mathbf{y}$,

$$V^{\mathbf{x}^*,\mathbf{y}^*}(s) \leq V^{\mathbf{x},\mathbf{y}^*}(s), \quad \forall s \in \mathcal{S},$$
$$V^{\mathbf{x}^*,\mathbf{y}^*}(s) \geq V^{\mathbf{x}^*,\mathbf{y}}(s), \quad \forall s \in \mathcal{S}. \quad (3)$$

From the definition (2) of minimax game value, it can be easily proven that if a pair of stationary policies $(\mathbf{x}, \mathbf{y})$ satisfies $V^{\mathbf{x},\mathbf{y}}(s) = V^*(s)$ for all s, then the stationary policies $(\mathbf{x}, \mathbf{y})$ is a Nash equilibrium. Moreover, it is known that a pair of stationary policies attaining the minimax value on state s is necessarily attaining the minimax value on all states [11].

Given an initial state s and a pair of stationary policies $(\mathbf{x}, \mathbf{y})$, the *Q-value function* (with respect to player Y) is defined as

$$Q_s^{\mathbf{x},\mathbf{y}}(a,b) = \mathbb{E}[\sum_{t=0}^{T} r(s_t, a_t, b_t)|s_0 = s, a_0 = a, b_0 = b, \\ s_{t+1} \sim p(\cdot|s_t, a_t, b_t), a_{t+1} \sim x_{s_{t+1}}(\cdot), \\ b_{t+1} \sim y_{s_{t+1}}(\cdot), t \geq 0] \quad (4)$$

for all $a \in \mathcal{A}$ and $b \in \mathcal{B}$, where the expectation above is taken with respect to the trajectory induced by the joint policy $(\mathbf{x}, \mathbf{y})$. From their definitions, we know that the V-value function and the Q-value function satisfy

$$Q_s^{\mathbf{x},\mathbf{y}}(a,b) = r(s,a,b) + \sum_{s' \in \mathcal{S}} p(s'|s,a,b) V^{\mathbf{x},\mathbf{y}}(s') \quad (5)$$

and

$$V^{\mathbf{x},\mathbf{y}}(s) = \sum_{a \in \mathcal{A}} \sum_{b \in \mathcal{B}} x_s(a) y_s(b) Q_s^{\mathbf{x},\mathbf{y}}(a,b)$$

for all $s \in \mathcal{S}$, $a \in \mathcal{A}$ and $b \in \mathcal{B}$.

Remark 1. The game defined above is analogous to the infinite-horizon discounted MGs introduced by Littman [20]. In the latter model, the game continues infinitely with a discount parameter γ that diminishes the importance of future rewards. From the perspective of accounting for future reward uncertainty, these two game models share similarities. The randomly terminating version handles uncertainty by allowing the game to potentially terminate with certain probabilities, whereas the infinite-horizon version uses a discount parameter γ less than 1 to address such uncertainties.

For the infinite-horizon discounted MGs, the Eq. (5) turns to be

$$Q_s^{\mathbf{x},\mathbf{y}}(a,b) = r(s,a,b) + \gamma \sum_{s' \in \mathcal{S}} p(s'|s,a,b) V^{\mathbf{x},\mathbf{y}}(s'), \quad (6)$$

where $\sum_{s' \in \mathcal{S}} p(s'|s,a,b) = 1$. From the aspect of the relation between the V-value and the Q-value function, the infinite-horizon model can be seen as a special case of the randomly terminating model because when we let the state transition probability p in (5) satisfy $\sum_{s' \in \mathcal{S}} p(s'|s,a,b) = \gamma < 1$ for all s, a, b, the relation (6) is obtained.

2.2 Notations

For a matrix $Q^{|\mathcal{A}| \times |\mathcal{B}|}$, its norm is defined to be

$$||Q|| := \max_{a \in \mathcal{A}} \max_{b \in \mathcal{B}} |Q(a,b)| \quad (7)$$

where $Q(a,b)$ is the according element given action profile (a,b). Hence, given two matrices Q_1 and Q_2, their distance is

$$||Q_1 - Q_2|| := \max_{a \in \mathcal{A}} \max_{b \in \mathcal{B}} |Q_1(a,b) - Q_2(a,b)|.$$

Given two V-value functions V_1 and V_2, their distance is defined to be

$$||V_1 - V_2|| := \max_{s \in \mathcal{S}} |V_1(s) - V_2(s)|. \qquad (8)$$

3 Regularized Minimax-V Learning

3.1 Contraction Operator

In this part, we outline the structure of the regularized minimax-V learning algorithm without introducing the regularization technique.

Let V be a vector in $\mathbb{R}^{|\mathcal{S}|}$ and then $V(s)$ represents the value estimate of state s. Given any vector V, a payoff matrix can be obtained by the formula

$$Q_s(a,b) = r(s,a,b) + \sum_{s' \in \mathcal{S}} p(s'|s,a,b) V(s'), \qquad (9)$$

similar to (5). We call the game with payoff matrix Q_s the auxiliary game at state s.

Given the payoff matrix Q_s at state s, denote the value of the according two-player zero-sum game by $f_v(Q_s)$, i.e.,

$$f_v(Q_s) = \min_{x \in \Delta(\mathcal{A})} \max_{y \in \Delta(\mathcal{B})} x^T Q_s y. \qquad (10)$$

Then, consider the mapping $M : \mathbb{R}^{|\mathcal{S}|} \to \mathbb{R}^{|\mathcal{S}|}$ defined by

$$MV(s) = (1-\alpha)V(s) + \alpha f_v(Q_s) \qquad (11)$$

where $\alpha \in (0,1)$. For the mapping M, it is easy to prove the following lemma.

Lemma 1. *The mapping $M : \mathbb{R}^{|\mathcal{S}|} \to \mathbb{R}^{|\mathcal{S}|}$ is a contraction operator and thus has one unique fixed point. Moreover, the fixed point is the minimax value function of the Markov game.*

Once the fixed point V^* of M is obtained, by solving the NE (x_s^*, y_s^*) under payoff matrix Q_s^* for all state s, we obtain the NE $(\mathbf{x}^*, \mathbf{y}^*)$ of the MGs because the strategy profile $\{(x_s^*, y_s^*)\}_{s \in \mathcal{S}}$ forms the NE strategy of the MGs.

Remark 2. For the fixed point V^*, we have

$$V^* = f_v(Q_s^*), \quad \forall s \in \mathcal{S}. \qquad (12)$$

By Lemma 1, the fixed point V^* can be obtained iteratively as below. Starting from any initial point $V_0 \in \mathbb{R}^{|\mathcal{S}|}$ (e.g., $V_0(s) = 0$, $\forall s \in \mathcal{S}$), define

$$V_t = MV_{t-1}, \; t \geq 1. \qquad (13)$$

Then, the sequence $\{V_t\}_{t=0}^{\infty}$ converges to V^*. When the learning rate α is set to be 0, the formula (11) is just the value iteration format in [25].

3.2 Regularized Minimax-V Learning Algorithm

In this part, we focus on solving the problem (10) and introduce the regularization technique to accelerate the process.

To solve the saddle point problem (10), an approach is to use learning algorithms such as Fictitious Play [12] and Hedge [15]. In the following simulations, we employ the Optimistic Multiplicative Weights Update (OMWU) algorithm [31] as the learning algorithm.

However, the original problem (10) is only convex-concave. Our idea is that by adding an appropriate regularization term, we can transform this problem into a strongly-convex-strongly-concave saddle point problem, and thus accelerate the convergence. Meanwhile, by controlling the regularization parameter, the deviation from the original minimax value is tolerable, and the overall convergence to the minimax value function can be retained.

To be specific, consider the regularized function $f_v^\tau : \mathbb{R}^{|\mathcal{A}| \times |\mathcal{B}|} \to \mathbb{R}$, defined as

$$f_v^\tau(Q) := \min_{x \in \Delta(\mathcal{A})} \max_{y \in \Delta(\mathcal{B})} \left(x^T Q y + \frac{\tau}{2}||x||^2 - \frac{\tau}{2}||y||^2 \right), \tag{14}$$

where $\tau > 0$ is called the regularization coefficient. It is easy to see that the problem (14) is a strongly-convex-strongly-concave saddle point problem.

The difference between $f_v^\tau(Q)$ and $f_v(Q)$ can be proven to be bounded by a constant times of the regularization coefficient. However, this does not apply to entropy regularization since the entropy of a probability distribution is unbounded. Therefore, if we use entropy regularization like [7], the difference between the solutions of the regularized problem and the original problem becomes uncontrollable.

By replacing f_v in (11) with f_v^τ, we obtain the regularized operator $M^\tau : \mathbb{R}^{|\mathcal{S}|} \to \mathbb{R}^{|\mathcal{S}|}$, i.e.,

$$M^\tau V(s) = (1-\alpha)V(s) + \alpha f_v^\tau(Q_s), \quad \forall s \in \mathcal{S}, \tag{15}$$

where f_v^τ is defined as (14) and Q_s is defined as (9). For any two value functions V_1 and V_2, by simple calculation, we have

$$||M^\tau V_1 - M^\tau V_2|| \leq (1-\alpha\zeta)||V_1 - V_2|| + \alpha\tau C,$$

where $C = \max\{\frac{|\mathcal{A}|-1}{|\mathcal{A}|}, \frac{|\mathcal{B}|-1}{|\mathcal{B}|}\}$. Hence, the operator M^τ is not a contraction operator anymore and we cannot get the minimax value function V^* by directly iteratively letting $V_{t+1} = M^\tau V_t$. Therefore, we make a little modification to (15), the minimax value function can still be reached via the method of value iteration.

Rather than use a constant regularization coefficient, we use a time-varying regularization coefficient τ_t, i.e., for $t = 0, 1, 2, \ldots$,

$$V_{t+1}(s) = (1-\alpha)V_t(s) + \alpha f_v^{\tau_t}(Q_{t,s}), \quad \forall s \in \mathcal{S}, \tag{16}$$

where the initial condition V_0 can be arbitrarily set. Usually, we would set $V_0(s) = 0$ for all $s \in \mathcal{S}$.

Algorithm 1: Regularized Minimax-V Learning Algorithm

Input: Learning rate $\alpha \in (0,1)$, regularization coefficient sequence $\{\tau_t\}_{t=0}^{\infty}$ satisfying (17)
Initialization: $V_0(s) \leftarrow 0, \forall s \in \mathcal{S}, t \leftarrow 0$;
repeat
$\quad Q_{t,s}(a,b) \leftarrow r(s,a,b) + \sum_{s'} p(s'|s,a,b) V_t(s')$;
$\quad f_v^{\tau_t}(Q_{t,s}) := \min_{x \in \Delta(\mathcal{A})} \max_{y \in \Delta(\mathcal{B})} \left(x^T Q_{t,s} y + \frac{\tau_t}{2} \|x\|^2 - \frac{\tau_t}{2} \|y\|^2 \right)$;
$\quad V_{t+1}(s) \leftarrow (1-\alpha) V_t(s) + \alpha f_v^{\tau_t}(Q_{t,s}), \forall s \in \mathcal{S}$;
$\quad t \leftarrow t + 1$;
until *stopping condition is met*;
Output: $\{V_t(s)\}_{s \in \mathcal{S}}$

Next, we present regularized minimax-V learning algorithm and its pseudocode is given in Algorithm 1. Then, we prove that the generated value sequence converges to the minimax value function V^* when the time-varying regularization coefficient sequence $\{\tau_t\}_{t=0}^{\infty}$ in (16) satisfies given condition, as demonstrated by the following theorem.

Theorem 1. *For $t = 0, 1, \cdots$, let*

$$V_{t+1}(s) = (1-\alpha) V_t(s) + \alpha f_v^{\tau_t}(Q_{t,s})$$
$$Q_{t,s}(a,b) = r(s,a,b) + \sum_{s'} p(s'|s,a,b) V_t(s'),$$

where function $f_v^{\tau_t}$ is defined by (14). If the regularization coefficient sequence $\{\tau_t\}_{t=0}^{\infty}$ satisfies $\lim_{t \to 0} \tau_t = 0$ and

$$\lim_{t \to \infty} \sum_{k=0}^{t} (1-\alpha)^k \tau_{t-k} = 0, \tag{17}$$

then from any initial condition V_0, the function sequence $\{V_t\}_{t=0}^{\infty}$ converges to the minimax value function V^*, i.e., $\lim_{t \to \infty} \|V_t - V^*\| = 0$.

The proof of Theorem 1 relies on the following result established in [28].

Lemma 2 (Corollary 5 in [28]). *Consider the process generated by the iteration of Eq. (16) where $0 < \alpha < 1$. Assume that the process defined by*

$$U_{t+1}(s) = (1-\alpha) U_t(s) + \alpha f_v^{\tau}(Q_s^*)$$

converges to V^*. Assume further that there exist number $0 < \gamma < 1$ and a sequence $\lambda_t \geq 0$ converging to zero such that

$$|f_v^{\tau_t}(Q_s) - f_v^{\tau_t}(Q_s^*)| \leq \gamma \|V - V^*\| + \lambda_t$$

holds for all V. Then, the iteration defined by Eq. (16) converge to V^*.

Then, we can prove Theorem 1.

Proof. By Lemma 2, we only need to verify that the two required assumptions hold here. First, we prove that the process defined by

$$U_{t+1}(s) = (1-\alpha)U_t(s) + \alpha f_v^{\tau_t}(Q_s^*)$$

converges to V^*.

$$\begin{aligned}
\|U_{t+1} - V^*\| &= \max_s |U_{t+1}(s) - V^*(s)| \\
&= \max_s |(1-\alpha)(U_t(s) - V^*(s)) + \alpha(f_v^{\tau_t}(Q_s^*) - V^*(s))| \\
&\leq \max_s [(1-\alpha)|U_t(s) - V^*(s)| + \alpha|f_v^{\tau_t}(Q_s^*) - V^*(s)|] \\
&\leq (1-\alpha)\max_s |U_t(s) - V^*(s)| + \alpha \max_s |f_v^{\tau_t}(Q_s^*) - V^*(s)| \\
&= (1-\alpha)\|U_t - V^*\| + \alpha \max_s |f_v^{\tau_t}(Q_s^*) - f_v(Q_s^*)| \\
&\leq (1-\alpha)\|U_t - V^*\| + \frac{\alpha \tau_t}{2} \max\left\{\frac{|\mathcal{A}|-1}{|\mathcal{A}|}, \frac{|\mathcal{B}|-1}{|\mathcal{B}|}\right\},
\end{aligned}$$

where the last equality holds by Eq. (12). Let $C = \max\{\frac{|\mathcal{A}|-1}{|\mathcal{A}|}, \frac{|\mathcal{B}|-1}{|\mathcal{B}|}\}$. Further, for the similar reason, we have that for all $k = 0, 1, \cdots, t$,

$$\|U_{k+1} - V^*\| \leq (1-\alpha)\|U_k - V^*\| + \frac{\alpha C}{2}\tau_k.$$

Therefore,

$$\begin{aligned}
\|U_{t+1} - V^*\| &\leq (1-\alpha)\|U_t - V^*\| + \frac{\alpha C}{2}\tau_t \\
&\leq (1-\alpha)((1-\alpha)\|U_{t-1} - V^*\| + \frac{\alpha C}{2}\tau_{t-1}) + \frac{\alpha C}{2}\tau_t \\
&= (1-\alpha)^2\|U_{t-1} - V^*\| + \frac{\alpha C}{2}((1-\alpha)\tau_{t-1} + \tau_t) \\
&\vdots \\
&\leq (1-\alpha)^{t+1}\|U_0 - V^*\| + \frac{\alpha C}{2}\sum_{k=0}^{t}(1-\alpha)^k \tau_{t-k}.
\end{aligned}$$

Since $0 < \alpha < 1$ and $\lim_{t \to \infty} \sum_{k=0}^{t}(1-\alpha)^k \tau_{t-k} = 0$, we have that $\lim_{t \to \infty}\|U_t - V^*\| = 0$, i.e., the sequence $\{U_t\}_{t=0}^{\infty}$ converges to V^* given any initial condition U_0.

Second, we show that $f_v^{\tau_t}$ is a pseudo-contraction operator, i.e., there exist number $0 < \gamma < 1$ and a sequence $\lambda_t \geq 0$ converging to zero such that for any V_1 and V_2,

$$|f_v^{\tau_t}(Q_{1,s} - f_v^{\tau_t}(Q_{2,s})| \leq \gamma \|V_1 - V_2\| + \lambda_t$$

where $Q_{1,s}$ and $Q_{2,s}$ is defined as (9). For all $s \in \mathcal{S}$, we have

$$\begin{aligned}
&|f_v^{\tau_t}(Q_{1,s}) - f_v^{\tau_t}(Q_{2,s})| \\
&= |f_v^{\tau_t}(Q_{1,s}) - f_v(Q_{1,s}) + f_v(Q_{1,s}) - f_v(Q_{2,s}) + f_v(Q_{2,s}) - f_v^{\tau_t}(Q_{2,s})| \\
&\leq |f_v^{\tau_t}(Q_{1,s}) - f_v(Q_{1,s})| + |f_v(Q_{1,s}) - f_v(Q_{2,s})| + |f_v(Q_{2,s}) - f_v^{\tau_t}(Q_{2,s})| \\
&\leq \frac{C}{2}\tau_t + (1-\zeta)\|V_1 - V_2\| + \frac{C}{2}\tau_t \\
&= (1-\zeta)\|V_1 - V_2\| + C\tau_t,
\end{aligned} \qquad (18)$$

where $C = \max\{\frac{|\mathcal{A}|-1}{|\mathcal{A}|}, \frac{|\mathcal{B}|-1}{|\mathcal{B}|}\}$. Since the above relation holds for any two V_1 and V_2, take $V_2 = V^*$ and write V_1 as V, then we obtain that

$$|f_v^{\tau_t}(Q_s) - f_v^{\tau_t}(Q_s^*)| \leq (1-\zeta)\|V - V^*\| + \alpha C\tau_t.$$

Since $\lim_{t \to 0} \tau_t = 0$ and $0 < \zeta < 1$, the second required assumption in Lemma 2 holds.

Therefore, by Lemma 2, the sequence $\{V_t\}_{t=0}^{\infty}$ generated by

$$V_{t+1}(s) = (1-\alpha)V_t(s) + \alpha f_v^{\tau_t}(Q_{t,s})$$
$$Q_{t,s}(a,b) = r(s,a,b) + \sum_{s'} p(s'|s,a,b)V_t(s'),$$

converges to V^*.

4 Simulation Results

4.1 Ratio Game

First, we consider ratio game proposed by von Neumann [22]. In ratio game, the state set contains only one state and each player has two actions. Its state transition relation and reward information can be shown by

$$R = \begin{pmatrix} -1 & \epsilon \\ -\epsilon & 0 \end{pmatrix} \quad \text{and} \quad S = \begin{pmatrix} w & w \\ 1 & 1 \end{pmatrix}$$

respectively, where $\epsilon, w \in (0,1)$ with $\epsilon < \frac{1-w}{2w}$. This means that the immediate reward of player Y for selecting actions (a,b) is $R_{a,b}$ and the probability of stopping in each round is $S_{a,b}$. By calculating, given any strategy $x \in \Delta(\mathcal{A})$ and $y \in \Delta(\mathcal{B})$, the value function is

$$V^{x,y} = \frac{x^T R y}{x^T S y}.$$

For the ratio game, there is a unique Nash equilibrium, which is $x^* = y^* = (0,1)$ and thus the minimax value function $V^* = 0$.

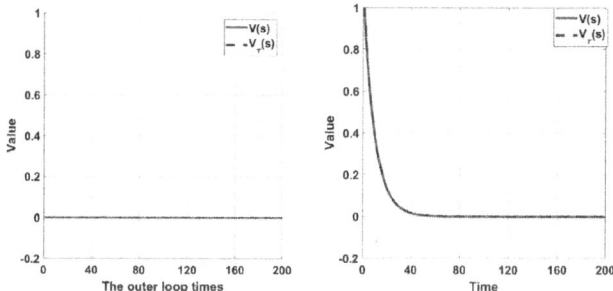

(a) The value sequence when initial condition is $V_0 = 0$.   (b) The value sequence when initial condition is $V_0 = 1$.

Fig. 1. The value sequence for ratio game.

We apply regularized minimax-V learning algorithm to solve this game. For comparison, we also consider the performance of the algorithm without introducing the regularization technique. For this algorithm, we call it minimax-V learning algorithm. We set regularization parameter to $\tau_t = (1-\alpha)^t$. For both algorithms, we let them proceed 2000 outer iterations, which are sufficient to ensure convergence.

The induced value sequences with different initial condition are illustrated in Fig. 1. Figure 1a shows the value sequence when the initial condition is $V_0 = 0$ while Fig. 1b shows the value sequence when the initial condition is $V_0 = 1$. We can see that no matter what the initial condition is, both algorithms converge to the minimax value function. When the initial condition is set to $V_1 = 1$, both algorithms converge in approximately 40 outer iterations, a notably small number compared to the preset 2000 outer iterations.

4.2 Randomly Generated Markov Game

In this part, we consider a Markov game with randomly generated reward function and state transition function. This Markov game contains 3 states and each player has 3 actions.

For regularized minimax-V learning algorithm, we still set the regularization parameter sequence $\tau_t = (1-\alpha)^t$ and let both algorithms proceed 2000 outer iterations. The value sequences generated by both algorithms are illustrated in Fig. 2.

In the figure, the first sub-figure, the second sub-figure and the third sub-figure depict the value sequence of the first, second and third state, respectively. The last sub-figure integrate the first three sub-figures.

This figure shows that the value sequences generated by both learning algorithms gradually converge to the minimax value function and both algorithms only need nearly 200 outer loop rounds to converge.

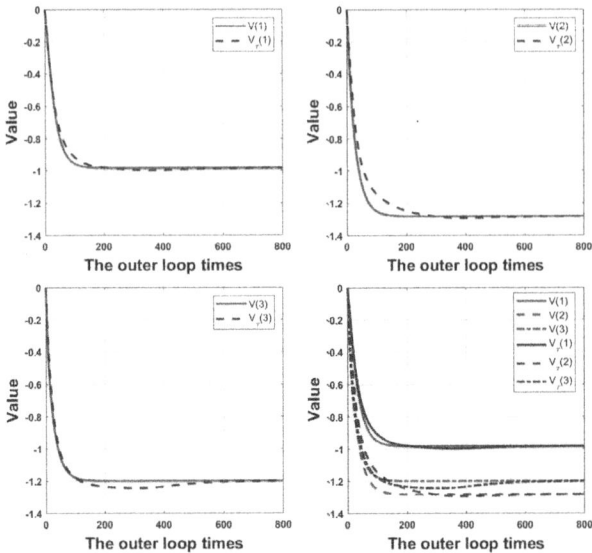

Fig. 2. The value sequence for the randomly generated Markov game: 2000 outer iterations and 5000 inner iterations for both algorithms.

5 Conclusion and Future Work

In this paper, we investigate randomly terminating two-player zero-sum Markov games and propose regularized minimax-V learning algorithm. We introduce the regularization technique to accelerate the process of solving a saddle point problem in the inner loop. The value sequence generated by this algorithm is proven to converge to the minimax value function with appropriate choice of the regularization parameter sequence. Through simulations, we demonstrate that the effectiveness of regularized minimax-V learning algorithm in the ratio game and a randomly generated Markov game.

This paper has some limitations that need to be addressed. Both algorithms require complete information about the reward function and state transition function. In the future, we could consider designing a model-free learning algorithm by replacing the constant learning parameter α with a time-varying learning parameter. Besides, it would be better to provide a theoretical guarantee that regularized minimax-V algorithm converges faster than the minimax-V algorithm. We leave these questions as future work.

Acknowledgements. This work was supported by the Strategic Priority Research Program of Chinese Academy of Sciences under Grant XDA27030201, the Natural Science Foundation of China under Grant T2293770 and the National Key Research and Development Program of China under grant No.2022YFA1004600.

References

1. Agarwal, A., Kakade, S.M., Lee, J.D., Mahajan, G.: Optimality and approximation with policy gradient methods in Markov decision processes. In: Conference on Learning Theory, pp. 64–66. PMLR (2020)
2. Bai, Y., Jin, C., Yu, T.: Near-optimal reinforcement learning with self-play. Adv. Neural. Inf. Process. Syst. **33**, 2159–2170 (2020)
3. Buşoniu, L., Babuška, R., De Schutter, B.: Multi-agent reinforcement learning: an overview. In: Innovations in Multi-Agent Systems and Applications-1, pp. 183–221 (2010)
4. Cai, Y., Oikonomou, A., Zheng, W.: Tight last-iterate convergence of the extragradient and the optimistic gradient descent-ascent algorithm for constrained monotone variational inequalities. arXiv preprint arXiv:2204.09228 (2022)
5. Cen, S., Cheng, C., Chen, Y., Wei, Y., Chi, Y.: Fast global convergence of natural policy gradient methods with entropy regularization. Oper. Res. **70**(4), 2563–2578 (2022)
6. Cen, S., Chi, Y., Du, S., Xiao, L.: Faster last-iterate convergence of policy optimization in zero-sum Markov games. In: International Conference on Learning Representations (ICLR) (2023)
7. Cen, S., Wei, Y., Chi, Y.: Fast policy extragradient methods for competitive games with entropy regularization. Adv. Neural. Inf. Process. Syst. **34**, 27952–27964 (2021)
8. Daskalakis, C., Foster, D.J., Golowich, N.: Independent policy gradient methods for competitive reinforcement learning. Adv. Neural. Inf. Process. Syst. **33**, 5527–5540 (2020)
9. Daskalakis, C., Panageas, I.: Last-iterate convergence: zero-sum games and constrained min-max optimization. arXiv preprint arXiv:1807.04252 (2018)
10. Fang, X., Zhao, Q., Wang, J., Han, Y., Li, Y.: Multi-agent deep reinforcement learning for distributed energy management and strategy optimization of microgrid market. Sustain. Urban Areas **74**, 103163 (2021)
11. Filar, J., Vrieze, K.: Competitive Markov Decision Processes. Springer (2012)
12. Fudenberg, D., Levine, D.K.: The Theory of Learning in Games, vol. 2. MIT Press (1998)
13. Golowich, N., Pattathil, S., Daskalakis, C., Ozdaglar, A.: Last iterate is slower than averaged iterate in smooth convex-concave saddle point problems. In: Conference on Learning Theory, pp. 1758–1784. PMLR (2020)
14. Gorbunov, E., Loizou, N., Gidel, G.: Extragradient method: $O(1/k)$ last-iterate convergence for monotone variational inequalities and connections with cocoercivity. In: International Conference on Artificial Intelligence and Statistics, pp. 366–402. PMLR (2022)
15. Guo, X., Mu, Y., Yang, X.: Periodicity in hedge-myopic system and an asymmetric ne-solving paradigm for two-player zero-sum games. arXiv preprint arXiv:2403.04336 (2024)
16. Jin, C., Liu, Q., Wang, Y., Yu, T.: V-learning–a simple, efficient, decentralized algorithm for multiagent RL. arXiv preprint arXiv:2110.14555 (2021)
17. Kober, J., Bagnell, J.A., Peters, J.: Reinforcement learning in robotics: a survey. Int. J. Robot. Res. **32**(11), 1238–1274 (2013)
18. Konda, V., Tsitsiklis, J.: Actor-critic algorithms. In: Advances in Neural Information Processing Systems, vol. 12 (1999)

19. Korpelevich, G.M.: The extragradient method for finding saddle points and other problems. Matecon **12**, 747–756 (1976)
20. Littman, M.L.: Markov games as a framework for multi-agent reinforcement learning. In: Machine Learning Proceedings 1994, pp. 157–163. Elsevier (1994)
21. Moravčík, M., et al.: Deepstack: expert-level artificial intelligence in heads-up no-limit poker. Science **356**(6337), 508–513 (2017)
22. Neumann, J.V.: A model of general economic equilibrium. Rev. Econ. Stud. **13**(1), 1–9 (1945)
23. Perolat, J., et al.: From poincaré recurrence to convergence in imperfect information games: finding equilibrium via regularization. In: International Conference on Machine Learning, pp. 8525–8535. PMLR (2021)
24. Sayin, M., Zhang, K., Leslie, D., Basar, T., Ozdaglar, A.: Decentralized q-learning in zero-sum Markov games. Adv. Neural. Inf. Process. Syst. **34**, 18320–18334 (2021)
25. Shapley, L.S.: Stochastic games. Proc. Natl. Acad. Sci. **39**(10), 1095–1100 (1953)
26. Silver, D., et al.: Mastering the game of go with deep neural networks and tree search. Nature **529**(7587), 484–489 (2016)
27. Syrgkanis, V., Agarwal, A., Luo, H., Schapire, R.E.: Fast convergence of regularized learning in games. In: Advances in Neural Information Processing Systems, vol. 28 (2015)
28. Szepesvári, C., Littman, M.L.: A unified analysis of value-function-based reinforcement-learning algorithms. Neural Comput. **11**(8), 2017–2060 (1999)
29. Vinyals, O., et al.: Grandmaster level in starcraft ii using multi-agent reinforcement learning. Nature **575**(7782), 350–354 (2019)
30. Wang, X., Sandholm, T.: Reinforcement learning to play an optimal nash equilibrium in team Markov games. In: Advances in Neural Information Processing Systems, vol. 15 (2002)
31. Wei, C.Y., Lee, C.W., Zhang, M., Luo, H.: Linear last-iterate convergence in constrained saddle-point optimization. arXiv preprint arXiv:2006.09517 (2020)
32. Williams, R.J.: Simple statistical gradient-following algorithms for connectionist reinforcement learning. Mach. Learn. **8**, 229–256 (1992)
33. Xie, Q., Chen, Y., Wang, Z., Yang, Z.: Learning zero-sum simultaneous-move Markov games using function approximation and correlated equilibrium. In: Conference on Learning Theory, pp. 3674–3682. PMLR (2020)
34. Zhang, K., Yang, Z., Başar, T.: Multi-agent reinforcement learning: a selective overview of theories and algorithms. In: Handbook of Reinforcement Learning and Control, pp. 321–384 (2021)
35. Zhao, Y., Tian, Y., Lee, J., Du, S.: Provably efficient policy optimization for two-player zero-sum Markov games. In: International Conference on Artificial Intelligence and Statistics, pp. 2736–2761. PMLR (2022)

Improved Approximation of Maximin Share Fair Allocation Under Generalized Assignment Constraints

Bo Li, Md Habibur Rahman Sifat, and Ankang Sun[✉]

Department of Computing, The Hong Kong Polytechnic University, Hong Kong, China
{comp-bo.li,ankang.sun}@polyu.edu.hk, habib.sifat@connect.polyu.hk

Abstract. Generalized assignment is one of the constraints that has been widely studied in fair allocation problems. Existing research has mostly focused on (approximate) envy-freeness. In this work, we consider the maximin share (MMS) fairness. It has been observed that under generalized assignment constraints, the agents' valuations are fractionally subadditive and thus a $\frac{3}{13}(\approx 0.230)$-approximate MMS fair allocation exists, as shown by Akrami et al. [NeurIPS 2023]. In this paper, we improve the approximation ratio to $\frac{4}{15} \approx 0.266$. For the case of two agents, we improve the approximation to $\frac{2}{3}$, and prove that this bound is tight.

Keywords: Fair division · Maximin share · Generalized assignment constraint

1 Introduction

Fair resource allocation is an important problem originating in mathematics and economics [21]. It has attracted increasing attention in operations research and computer science due to its game-theoretic nature and agent-based behaviors. Two golden standard fairness criteria are the *envy-freeness* (EF) and the *proportionality* (PROP). An EF allocation requires that every agent prefers her bundle to the bundle of any other agents, while a PROP one requires that every agent's utility for her bundle is no smaller than $\frac{1}{n}$ fraction of her utility for all the resources, where n is the number of agents. When the resources can be arbitrarily divided (aka cake-cutting problem), an EF or PROP allocation can be found easily [10,23]. However, the problem becomes more complex when the resources cannot be divided (aka discrete fair allocation problem). For instance, consider a scenario where there is a single valuable and indivisible resource to be allocated to two agents. No matter who receives the resource, the other agent will not be satisfied according to either EF or PROP.

The authors are ordered alphabetically.

There is a lengthy and heated debate about the definition of fairness for discrete fair allocation problems. One of the most widely accepted notions is the maximin share (MMS) fairness, introduced by Budish [6]. MMS defines a benchmark share for each agent, which represents the optimal utility they can ensure by dividing the items into n bundles but receiving the worst bundle. Although MMS is a weaker fairness requirement than EF and PROP, it is not guaranteed to be satisfiable [16,17]. The good news is that there are good approximations, where constant fractions of every agent's MMS value can be guaranteed simultaneously, if the agents have additive valuations; see, e.g., [1,12,17]. So far the best-known approximation is $\frac{3}{4} + \frac{3}{3836} \approx 0.751$ [1].

One common assumption in all the above-mentioned works is that no matter how the items are allocated among the agents, the allocation is feasible. However, in some practical applications, the allocation faces constraints. Consider an example where each agent i has a knapsack with capacity B, and every item e_j has size s_j. A set of items is *feasible* if the total size of the items is no greater than B. Such a setting is known as the budget-feasible constraint, which has been studied in the fair division literature. For example, Wu et al. [25], Gan et al. [11], and Barman et al. [3], respectively, considered approximate envy-freeness. Deng and Li [7] considered the MMS fairness in a broader setting where the agents may have different budgets and proved the existence of $\frac{n}{3n-2}$-approximate MMS allocations. Barman et al. [4] further extended this line of study to the *generalized assignment* constraints, where item sizes and agents are both personalized; that is, the size of an item for different agents can also be different. This is a broad setting that not only contains budget-feasible constraints but also other combinatorial problems, such as weighted bipartite matching. Barman et al. [4] focused on envy-free fairness and proved the existence of a feasibly envy-free up to any item under charity (FEFX) allocation. However, MMS has been overlooked in this generalized setting, which motivates our work.

1.1 Our Contribution

In this work, we study the MMS fair allocation of indivisible items under generalized assignment constraints, which has been overlooked in prior research. It can be verified that under generalized assignment constraints, agents' valuations belong to the class of fractionally subadditive (XOS) functions. The state-of-the-art approximation ratio of MMS for XOS valuations is $\frac{3}{13} \approx 0.231$ proved by Akrami et al. [2]. This directly implies the existence of $\frac{3}{13}$-MMS allocations under generalized assignment constraints.

To improve the known approximation ratio, we first consider the setting of two agents as a warm-up. We present a simple instance for which no better than $\frac{2}{3}$-MMS allocations exist. This shows a sharp contrast to the setting with no constraints, where an exact MMS allocation always exists (for two agents). We then prove that a divide-and-choose algorithm guarantees $\frac{2}{3}$-MMS and thus achieves the best possible approximation.

We then consider the general case with an arbitrary number of agents, and prove the existence of a $\frac{4}{15}(\approx 0.266)$-MMS allocation for generalized assignment constraints, which improves the best-known result implied by Akrami et al. [2]. Our algorithm is simple and has been widely studied in the literature. The algorithm runs in rounds. In each round, we allocate a bundle with a minimum number of items to satisfy an agent's $\frac{4}{15}$-MMS. The algorithm has been proven to ensure $\frac{4}{11} \approx 0.3636$ MMS for the fair division of hereditary set systems (where all agents face the same set of feasible allocations). Generalized assignment constraints do not satisfy hereditary set systems. Though the algorithm is simple, the analysis is not straightforward.

1.2 Related Work

Fair division under non-additive valuations has been widely studied in the literature. For XOS valuation functions, Ghodsi et al. [13] first proved that the social welfare maximizing allocation of $\frac{1}{5}$-capped valuations[1] guarantees $\frac{1}{5}$-approximate MMS. This approximation ratio was later improved to $\frac{1}{4.6} \approx 0.219$ [20]. Very recently, Akrami et al. [2] proved that combining the algorithm of [19] and the capped social welfare maximizing technique of [13], the approximation can be further enhanced to $\frac{3}{13} \approx 0.231$ [2]. Their results directly imply the existence of $\frac{3}{13}$-MMS allocations for the generalized assignment constraints, currently best-known. Besides XOS, submodular and subadditive valuations are also studied [8,22]. There is also a line of research studying fair division under combinatorial constraints. Besides budget-feasible and generalized assignment constraints, connectivity constraints [5,14], vertex cover constraints [24], and scheduling constraints [15,18] have also been studied in fair division models.

2 Model and Definitions

Denote by $[k] = \{1, \ldots, k\}$ for any positive integer k. For the model of allocating indivisible items, we have a set $E = \{e_1, \ldots, e_m\}$ of m items to be allocated to a set $N = [n]$ of n agents. Each agent i has a valuation function $v_i : 2^E \to \mathbb{R}_{\geq 0}$, expressing the cardinal preference of agents over items. For simplicity, we write $v_i(e)$ to denote $v_i(\{e\})$ for all i and e. If there are no constraints, we assume the valuations are additive, i.e., $v_i(S) = \sum_{e \in S} v_i(e)$. An allocation $\mathbf{A} = (A_0, A_1, \ldots, A_n)$ is an $(n+1)$-partition of E, i.e., $A_i \cap A_j = \emptyset$ for $i \neq j$ and $\bigcup_{i=0}^{n} A_i = E$, where A_i is the set of items allocated to agent i for $i \in N$ and A_0 is the set of unallocated items. When all items are required to be allocated, $A_i = \emptyset$ and an allocation is also denoted as $(A_1, \ldots, A_n)$, an n-partition of E. In this work, we are concerned with two allocation constraints, namely, *generalized assignment constraints* and *budget-feasible constraints*.

Generalized Assignment Constraints. For generalized assignment constraints, items have personalized sizes, and agents have personalized budgets. Formally, for any $i \in N$ and $e_j \in E$, $s_i(e_j)$ is the size of item e_j for agent i and B_i is

---

[1] The α-capped valuation of $v(\cdot)$, denoted by $v^\alpha(\cdot)$, is that $v^\alpha(S) = \min\{\alpha, v(S)\}$ for all S.

the budget of agent i. Note that the size of an item can be different to different agents in this general setting. The sizes are additive, i.e., $s_i(T) = \sum_{e \in T} s_i(e)$ for any $T \subseteq E$. A set of items T is *feasible* for agent i if the total size of T is at most the budget of agent i, i.e., $s_i(T) \leq B_i$. Throughout the paper, denote by $\mathcal{F}_i$ the set of feasible sets for agent i for all $i \in [n]$. Therefore, the agents' valuations are additive only for feasible sets of items. Although our algorithms do not allocate an infeasible set to an agent, it is convenient to extend the valuations to all sets of items. In particular, an agent i's value of an *infeasible* set S is the maximum value it has for any feasible subset of S, i.e., $v_i(S) = \max_{T \in \mathcal{F}_i : T \subset S} \sum_{e \in T} v_i(e)$. An instance under generalized assignment constraints is denoted by $\mathcal{I} = \langle N, E, \{v_i\}, \{s_i\}, \{B_i\} \rangle$.

Budget-Feasible Constraints. This is a special setting of generalized assignment constraints, where $\mathcal{F}_i = \mathcal{F}_j$ for all $i, j \in N$. That is, agents have the same budget and an item has the same size to every agent. Formally, the size of item e_j is denoted by $s(e_j)$ for all agents, and every agent has a common budget B. Denote by $\mathcal{F}$ the set of feasible sets (for all agents). Note that the valuation functions are not identical in the budget-feasible setting. In the following of this paper, the homogeneous model represents the budget-feasible constraints, and the heterogeneous model represents generalized assignment constraints.

We next introduce our solution concept, *maximin share* (MMS) fairness. The maximin share of agent i, denoted by MMS_i, is the least value she can guarantee in the partition that maximizes the least value of a bundle. Let $\Pi_n(E)$ be the set of n-partitions of E, then

$$\mathsf{MMS}_i(n, E) = \max_{(P_1,\ldots,P_n) \in \Pi_n(E)} \min_{j \in [n]} v_i(P_j).$$

When n and E are clear from the context, instead of $\mathsf{MMS}_i(n, E)$, we write MMS_i. An n-partition $\{P_1, \ldots, P_n\}$ of E is said to be an MMS_i-defining partition if $v_i(P_j) \geq \mathsf{MMS}_i$ for all $j \in [n]$. Throughout the paper, for any i, let $\{P_1^i, P_2^i, \ldots, P_n^i\}$ be an MMS_i-defining partition; note that some part P_l^i can be infeasible for agent i.

We pursue the allocation that assigns every agent a bundle with a value of at least her maximin share, and unfortunately, such an allocation does not always exist [9]. Consequently, we search for approximate solutions; given an $\alpha \in (0, 1]$, allocation $\mathbf{A} = (A_1, \ldots, A_n)$ is α-MMS if $v_i(A_i) \geq \alpha \cdot \mathsf{MMS}_i$ for all $i \in [n]$.

Connections Among Valuation Functions. We now discuss the connections between underlying valuations in this paper and valuations that are widely studied in the literature. First, for the homogeneous model, valuation functions are reduced to additive valuation functions when set E is feasible for all agents, and therefore, the former generalizes the latter. Moreover, valuation functions in the homogeneous model are a subclass of valuation functions regarding *hereditary set system* where feasible sets are independent sets in a set system, and any subset of a feasible set is also feasible (downward-closed). To the best of our knowledge, the current best-known bound on the approximation of maximin share in the hereditary set system is $\frac{11}{31}$ [19], which also implies the existence of $\frac{11}{31}$-MMS allocations in the homogeneous budget-feasible model. Thus, our objective in the homogeneous setting is to improve upon the $\frac{11}{30}$-MMS performance guarantee.

In the heterogeneous model, valuation functions belong to the class of *fractionally subadditive* functions, also known as XOS. The XOS function is defined as the maximum over a collection of additive set functions. The current best approximation of maximin share under XOS valuations is $\frac{3}{13}$ presented in [2], and this result naturally implies the existence of $\frac{3}{13}$-MMS allocations in heterogeneous model. Thus, we aim at improving upon the performance guarantee of $\frac{3}{13}$-MMS.

When computing the performance guarantee under generalized assignment constraints, one can focus on the instance where, for any agent i, every bundle in MMS_i-defining partition is feasible. In particular, for an arbitrary instance $\mathcal{I} = \langle N, E, \{v_i\}, \{s_i\}, \{B_i\}\rangle$ and an arbitrary agent i, let $\{P_1, \ldots, P_n\}$ be an MMS_i-defining partition. Moreover for any $j \in [n]$, suppose $T_j \subseteq P_j$ is the feasible set with the maximum value for agent i, i.e., $v_i(T_j) = v_i(P_j)$. We construct another instance $\mathcal{I}' = \langle N, E, \{v_i'\}, \{s_i'\}, \{B_i\}\rangle$ as follows; for any agent i and item e, if $e \in \bigcup_{j \in [n]} T_j$, define $v_i'(e) = v_i(e)$ and $s_i'(e) = s_i(e)$; otherwise, $v_i'(e) = 0$ and $s_i'(e) = 0$. One can verify that, the maximin share of every agent i in $\mathcal{I}'$ is equal to that in $\mathcal{I}$. Moreover, for any agent i and $S \subseteq E$, $v_i(S) \geq v_i'(S)$ holds, and thus, an α-MMS allocation of instance $\mathcal{I}'$ is also an α-MMS allocation of $\mathcal{I}$. Therefore, we can assume without loss of generality that for any $i \in [n]$, every part in MMS_i-defining partition is feasible for agent i when computing the performance guarantee.

3 The Case of Two Agents

We first present an instance with two agents where no allocation can have a performance guarantee better than $\frac{2}{3}$-MMS.

Theorem 1. *For any $\epsilon > 0$, no allocation can be $(\frac{2}{3} + \epsilon)$-MMS, even for two agents.*

Proof. Let us consider an instance with two agents and four items. Agents have an identical budget of 1, i.e., $B_1 = B_2 = 1$. The valuations and sizes are shown in tables where the left (resp., the right) one presents the valuation and size functions of agent 1 (resp., agent 2).

Items	e_1	e_2	e_3	e_4
$v_1(\cdot)$	$\frac{2}{3}$	$\frac{1}{3}$	$\frac{1}{3}$	$\frac{2}{3}$
$s_1(\cdot)$	$\frac{3}{10}$	$\frac{7}{10}$	$\frac{1}{5}$	$\frac{4}{5}$

Items	e_1	e_2	e_3	e_4
$v_2(\cdot)$	$\frac{2}{3}$	$\frac{2}{3}$	$\frac{1}{3}$	$\frac{1}{3}$
$s_2(\cdot)$	$\frac{3}{10}$	$\frac{4}{5}$	$\frac{1}{5}$	$\frac{7}{10}$

One can check that an MMS_1-defining partition is $\{\{e_1,e_2\},\{e_3,e_4\}\}$ and MMS_2-defining partition is $\{\{e_1,e_4\},\{e_2,e_3\}\}$, and thus $\mathsf{MMS}_1 = \mathsf{MMS}_2 = 1$. Suppose $\mathbf{A} = (A_1, A_2)$ is an allocation assigning every agent a value of at least $\frac{2}{3} + \epsilon$. It is not hard to verify that A_1 contains at least either $\{e_1, e_2\}$ or $\{e_3, e_4\}$, and in both cases, $v_2(A_2) \leq \frac{2}{3}$, leading to a contradiction.

For two agents, the above result is tight; that is, there always exist $\frac{2}{3}$-MMS allocations for two agents. Due to space limit, the proof of Theorem 2 is omitted.

Theorem 2. *For two agents, there always exists an allocation guaranteeing $\frac{2}{3}$-MMS for both agents.*

4 The Existence of $\frac{4}{15}$-MMS Allocations for n Agents

The main result of this section is the existence of $\frac{4}{15}$-MMS allocations. When analyzing the performance guarantee, without loss of generality, we can assume $\mathsf{MMS}_i > 0$ for all $i \in [n]$ as the agent with zero maximin share is happy even when no item is assigned to her. Moreover, we can scale the valuation so that for any $i \in [n]$, $\mathsf{MMS}_i = 1$.

To find the allocation with the desired performance guarantee, we adapt an algorithm that has been studied in [19] to the context of the heterogeneous model. Intuitively, given a target $\alpha \in (0,1]$, the algorithm searches for the minimal cardinality feasible set of at least a value α for some agent; note that we have scaled the maximin share of every agent to 1. Once such a set is found, assign the set to the corresponding agent and then recurse on the remaining items and agents. We formally present the procedure in Algorithm 1. Due to space limit, the proof of Theorem 3 is omitted.

Algorithm 1. $\frac{4}{15}$-MMS Allocation under Generalized Assignment Constraints

---
Input: An instance $\mathcal{I} = \langle N, E, \{v_i\}, \{s_i\}, \{B_i\}\rangle$ and a target $\alpha > 0$.
1: **for** $\tau = 1$ to m **do**
2:     **while** there exists set a $T \subseteq E$ with $|T| = \tau$ and an agent $i \in [n]$ (ties break arbitrarily) with $v_i(T) \geq \alpha$ and $T \in \mathcal{F}_i$ **do**
3:         Allocate bundle T to agent i.
4:         $N \leftarrow N \setminus \{i\}$ and $E \leftarrow E \setminus T$.
5:     **end while**
6: **end for**

Theorem 3. *Algorithm 1 computes a $\frac{4}{15}$-MMS allocation in the model with generalized assignment constraints.*

Acknowledgments. This work is funded by the Hong Kong SAR Research Grants Council (No. PolyU 15224823) and the Guangdong Basic and Applied Basic Research Foundation (No. 2024A1515011524).

References

1. Akrami, H., Garg, J.: Breaking the 3/4 barrier for approximate maximin share. In: Woodruff, D.P. (ed.) SODA 2024, Alexandria, VA, USA, January 7–10, 2024, pp. 74–91. SIAM (2024)
2. Akrami, H., Mehlhorn, K., Seddighin, M., Shahkarami, G.: Randomized and deterministic maximin-share approximations for fractionally subadditive valuations. In: NeurIPS (2023)
3. Barman, S., Khan, A., Shyam, S., Sreenivas, K.V.N.: Finding fair allocations under budget constraints. In: AAAI, pp. 5481–5489. AAAI Press (2023)
4. Barman, S., Khan, A., Shyam, S., Sreenivas, K.V.N.: Guaranteeing envy-freeness under generalized assignment constraints. In: EC, pp. 242–269. ACM (2023)
5. Bouveret, S., Cechlárová, K., Elkind, E., Igarashi, A., Peters, D.: Fair division of a graph. In: Sierra, C. (ed.) IJCAI 2017, Melbourne, Australia, August 19–25, 2017, pp. 135–141. ijcai.org (2017)
6. Budish, E.: The combinatorial assignment problem: approximate competitive equilibrium from equal incomes. In: Dror, M., Sosic, G. (eds.) BQGT '10, Newport Beach, California, USA, May 14–16, 2010, pp. 74:1. ACM (2010)
7. Deng, B., Li, W.: The budgeted maximin share allocation problem. Optim. Lett. 1–14 (2024)
8. Feige, U., Huang, S.: Concentration and maximin fair allocations for subadditive valuations. CoRR arXiv:2502.13541 (2025)
9. Feige, U., Sapir, A., Tauber, L.: A tight negative example for MMS fair allocations. In: WINE, volume 13112 of Lecture Notes in Computer Science, pp. 355–372. Springer (2021)
10. Foley, D.K.. Resource Allocation and the Public Sector. Yale University (1966)
11. Gan, J., Li, B., Wu, X.: Approximation algorithm for computing budget-feasible EF1 allocations. In: AAMAS, pp. 170–178. ACM (2023)
12. Garg, J., Taki, S.: An improved approximation algorithm for maximin shares. Artif. Intell. **300**, 103547 (2021)
13. Ghodsi, M., Hajiaghayi, M.T., Seddighin, M., Seddighin, S., Yami, H.: Fair allocation of indivisible goods: beyond additive valuations. Artif. Intell. **303**, 103633 (2022)
14. Igarashi, A.: How to cut a discrete cake fairly. In: Williams, B., Chen, Y., Neville, J. (eds.) AAAI 2023, IAAI 2023, EAAI 2023, Washington, DC, USA, February 7–14, 2023, pp. 5681–5688. AAAI Press (2023)
15. Kumar, Y., Equbal, S., Gurjar, R., Nath, S., Vaish, R.: Fair scheduling of indivisible chores. In: Dastani, M., Sichman, J.S., Alechina, N., Dignum, V. (eds.) AAMAS 2024, Auckland, New Zealand, May 6–10, 2024, pp. 2345–2347. International Foundation for Autonomous Agents and Multiagent Systems / ACM (2024)
16. Kurokawa, D., Procaccia, A.D., Wang, J.: When can the maximin share guarantee be guaranteed? In: AAAI, pp. 523–529. AAAI Press (2016)
17. Kurokawa, D., Procaccia, A.D., Wang, J.: Fair enough: guaranteeing approximate maximin shares. J. ACM **65**(2), 8:1-8:27 (2018)
18. Li, B., Li, M., Zhang, R.: Fair scheduling for time-dependent resources. In: Ranzato, M., Beygelzimer, A., Dauphin, Y.N., Liang, P., Vaughan, J.W. (eds.) NeurIPS 2021, December 6–14, 2021, virtual, pp. 21744–21756 (2021)
19. Li, Z., Vetta, A.: The fair division of hereditary set systems. ACM Trans. Econ. Comput. **9**(2), 12:1-12:19 (2021)

20. Seddighin, M., Seddighin, S.: Improved maximin guarantees for subadditive and fractionally subadditive fair allocation problem. Artif. Intell. **327**, 104049 (2024)
21. Steinhaus, H.: The problem of fair division. Econometrica **16**(1), 101–104 (1948)
22. Uziahu, G.B., Feige, U.: On fair allocation of indivisible goods to submodular agents. CoRR arXiv:2303.12444 (2023)
23. Varian, H.R.: Equity, envy, and efficiency (1973)
24. Wang, F., Li, B.: Fair surveillance assignment problem. In: Chua, T.-S., Ngo, C.-W., Kumar, R., Lauw, H.W., Lee, R.K. (eds.) WWW 2024, Singapore, May 13–17, 2024, pp. 178–186. ACM (2024)
25. Wu, X., Li, B., Gan, J.: Budget-feasible maximum nash social welfare is almost envy-free. In: IJCAI, pp. 465–471. ijcai.org (2021)

Optimal Hiring Strategy in Auction-Based Crowdsourcing Systems

Hongtao Liu[1], Weiran Shen[1(✉)], and Yiheng Shen[2]

[1] Gaoling School of Artificial Intelligence, Renmin University of China, Beijing, China
{ht6,shenweiran}@ruc.edu.cn
[2] Computer Science, Duke University, Durham, NC, USA
ys341@duke.edu

Abstract. We consider an auction-based crowdsourcing system. A requester is faced with a binary choice question and decides to hire workers to answer the question. The workers can ask prices for answering the question and the requester can choose to hire which workers based on their skills and ask prices. We model the problem as a mechanism design problem and characterize the optimal hiring policy. We show that the problem of computing the accuracy of a given set of workers is #P-hard. However, we prove that choosing at most k workers into committee can achieve at least $1/\lceil n/k \rceil$ of the optimal utility. Finally, we also provide a polynomial algorithm for computing the optimal hiring strategy when the number of workers' skill levels is constant.

Keywords: Optimal strategy · Auction · Crowdsourcing system

1 Introduction

Crowdsourcing is a way of hiring a group of participants to finish a certain task together. In a crowdsourcing task, there is a requester and a set of workers. The requester first posts a task advertisement on a certain crowdsourcing platform. A typical advertisement usually contains a brief description of the task and a reward for finishing the task. Then interested workers participate to finish the task and get paid accordingly.

In recent years, crowdsourcing applications have significantly expanded, covering areas such as data labeling for machine learning tasks, information collection, and human subject studies. The requester typically breaks the task into smaller sub-tasks and offers a uniform price for each sub-task. This outsourcing model has both advantages and disadvantages. It can quickly attract many workers, speeding up task completion, but it may also increase task completion costs.

Quality control is another issue that can arise in such a procedure. The workers may have different skill levels and thus may complete each sub-task with varying qualities. To solve the problem, the requester can send each sub-task to multiple workers and then aggregate their solutions to increase the solution quality. However, workers with higher skill levels often cost more to provide higher-quality solutions, while the reward usually remains the same. This hiring policy could lead to an adverse selection problem and deter workers with higher skill levels.

An auction-based crowdsourcing system can reduce the adverse selection problem. In such a system, each worker sets an asking price, which is the minimum reward the worker must receive if hired. After collecting the asking prices, the requester decides which workers to hire.

In this paper, we consider such an auction-based crowdsourcing system and study the optimal hiring policy of the requester. Our model combines both mechanism design and solution aggregation. We show that computing the optimal set of workers to hire is #P-hard. The difficulty is directly inherited from the hardness of computing the best accuracy of a set of workers. We also put forward an efficient algorithm for computing the optimal hiring strategy when there are only a constant number of different skill levels.

1.1 Related Works

Our paper is related to the crowdsourcing literature. A popular model for crowdsourcing task assignment market is to find a matching between the workers and the tasks [3,6,7,16]. Most of these works model the problem as a bipartite graph, and assign each task to one worker. Different from them, we need to assign the task to an unknown number of workers and infer the true answer from the workers' reported answers. We refer interested readers to [18] for a more comprehensive survey of the history works on truth inference in crowdsourcing. In their setting, the principal needs to learn the quality of the workers while assigning them tasks.

There is another line of work that combines crowdsourcing and mechanism design. Dominic [4] models the crowdsourcing system as an all-pay auction. In their work, each agent exerts effort into each contest with a certain cost and the agent with the highest effort wins. As a follow-up work, Luo [10] considers a similar model, where each agent contributes a certain amount, and the agent with the most contribution wins and gets all the other contributions as their utility. In this paper, however, multiple agents can obtain positive utility by answering the question. There are other works applying peer prediction techniques in designing mechanisms for crowdsourcing platforms [8,14]. Unlike this paper, they do not consider the incentive issues. More recently, Wang [15] applies the auction design into mobile crowdsourcing systems.

In the voting theory, Nitzan [12] first proposes the optimal method for aggregating a set of independent answers with varying accuracies, known as the optimal Nitzan-Paroush weighted majority rule. Later, Ben-Yashar and Nitzan [1] extend their results by adding the prior differences and the dependency of experts' skills on the state of nature. More recently, Berend [2] provides a constant approximation for the error rate of the aggregation rule. In our work, we mainly focus on the "accuracy gain" (accuracy minus 1/2) of the aggregated answer since there is a penalty if the answer is wrong.

In machine learning, the Nitzan-Paroush rule is also applied as the well-known Naive Bayes approach. There are both empirical studies [5,9,13] and theoretical analyses [5,17] showing that this approach performs well on aggregating answers to a binary problem.

2 Model

A requester (she) is faced with a binary choice question. If she answers the question correctly, she gets a reward of R. If she answers incorrectly, her utility is $-R$. Assume that the requester has no knowledge about the question and can only take a random guess (with a 0.5 probability being correct, giving her an expected utility of 0). However, she can hire skillful workers to answer the question through a crowdsourcing system. Assume that the crowdsourcing system uses an auction-based mechanism so that the requester can decide who to hire to answer the question.

Let $N = \{1, 2, \ldots, n\}$ represent the set of all workers. Suppose that each worker i has a skill level that completely determines the probability a_i of his answer being correct. We assume that crowdsourcing system can estimate a_i for the requester based on the worker's past performances, i.e., a_i is known to the requester. Clearly, we have that $a_i \in [0.5, 1]$ since a random guess could already give an accuracy of 0.5. For simplicity, we assume that the workers will try their best to answer the question and that their answers are independent from one another. Therefore, the best accuracy that the requester can get from the set C of hired workers is:

$$A(C) = \sum_{S \subseteq H} \left(\prod_{i \in S} a_i \prod_{j \in H \setminus S} (1 - a_i) \right) \qquad (1)$$

We call Eq. (1) the accuracy function.

Each worker also has a cost $c_i \in \mathbb{R}_+$ for answering the question. Suppose that c_i is a random variable that is drawn from cumulative distribution function $F_i(c_i)$ with a density function $f_i(c_i)$. We consider a Bayesian environment and assume that c_i is worker i's private information, whereas the distribution function $F_i(c_i)$ is common knowledge. The workers are asked to report their costs so that the requester can decide who to hire.

Denote by $H = (x, p)$ the requester's hiring policy, where:

- $x : \mathbb{R}_+^n \mapsto \{0, 1\}^n$ is the hiring function that maps the reported costs to a hiring decision, where $x_i = 1$ if and only if worker i is hired;
- $p : \mathbb{R}_+^n \mapsto \mathbb{R}_+^n$ is the payment function that determines the amount of money that the requester pays to each worker.

Worker i's utility u_i is defined as the money he is paid for participation less his cost for answering the question, i.e.,

$$u_i = p_i - c_i x_i, \qquad (2)$$

Denote the committee with candidates selected in vector x as $C_x = \{i \in N : x_i = 1\}$. The requester's utility is the expected reward she gets from aggregating the hired workers' answers minus the payments she distributed to the workers:

$$u_0 = R \cdot A(C_x) - R(1 - A(C_x)) - \sum_{i \in N} p_i \qquad (3)$$

$$= 2R \cdot A(C_x) - R - \sum_{i \in N} p_i. \qquad (4)$$

The complete crowdsourcing procedure is as follows:

1. The requester announces the hiring policy $H = (x, p)$;
2. The workers report their costs c' to the requester (note that it is possible that $c' \neq c$ if misreporting benefits a worker);
3. The requester decides who to hire and the payments by computing $x(c')$ and $p(c')$, and sends the question to the hired workers;
4. Each worker submits his answer to the requester, and gets paid $p_i(c')$.

Following game-theoretic conventions, we use $a = (a_1, a_2, \ldots, a_n)$ to denote the accuracy profile of all workers and $a_{-i} = (a_1, a_2, \ldots, a_{i-1}, a_{i+1}, \ldots, a_n)$ to denote the accuracy profile of all workers except i. Similarly, we can also define c and c_{-i}, respectively.

The workers' goal is to maximize their utilities. Thus they will choose not to participate in the procedure if their utilities are less than 0. Therefore, the requester needs to ensure that the workers' utilities are non-negative if they report truthfully.

Definition 1 (Individual Rationality). *A hiring policy $H = (x, p)$ is individually rational, if $\forall c, i$,*

$$u_i = p_i(c) - c_i x_i(c) \geq 0 \tag{5}$$

And to incentivize the workers to report truthfully, the hiring policy should also satisfy the following incentive compatibility constraint.

Definition 2 (Incentive Compatibility (Truthfulness)). *A hiring policy is said to be incentive-compatible (or truthful), if reporting the cost truthfully is an optimal strategy for each worker. That is, $\forall c_i, c'_i, \forall c_{-i}$ and $\forall i$,*

$$p_i(c) - c_i x_i(c) \geq p_i(c'_i, c_{-i}) - c_i x_i(c'_i, c_{-i}) \tag{6}$$

Thanks to the celebrated revelation principle [11], we can, without loss of generality, focus on the set of truthful hiring policies.

Theorem 1 (Revelation Principle [11]). *For any hiring policy, there exists a truthful hiring policy that implements the same hiring and payment function.*

3 Problem Analysis

In this section, we derive the optimal hiring policy and show the hardness of computing the optimal policy. Note that our setting is very similar to a reverse auction or procurement setting, where a buyer wants to buy an item from multiple sellers. Before we present the optimal hiring policy, we define the following virtual cost function, which is commonly used in the reverse auction design literature.

Definition 3 (Virtual Cost Function). *Suppose worker i's cost is drawn from cumulative distribution function $F_i(c_i)$ with density function $f_i(c_i)$. Then the virtual cost function of worker i is:*

$$\varphi_i(c_i) = c_i + \frac{F_i(c_i)}{f_i(c_i)} \tag{7}$$

A virtual cost function is called regular *if it is monotone increasing.*

Following the standard Myersonian approach, we have the following two results:

Lemma 1. *A hiring policy $H = (x, p)$ is incentive compatible, if and only if the following two conditions hold:*

- *The hiring function $x_i(c_i, c_{-i})$ is monotone decreasing in c_i for all c_{-i};*
- *The payment function $p_i(c_i, c_{-i})$ satisfies:*

$$p_i(c_i, c_{-i}) = \int_0^{c_i} s \, dx_i(s, c_{-i}), \tag{8}$$

where we fix c_{-i} and view the hiring function $x_i(s, c_{-i})$ as a function of s.

Lemma 2. *For any truthful hiring policy, the requester's utility can be written as:*

$$u_0 = 2RA(C_x) - R - \sum_{i \in N} x_i \varphi_i(c_i). \tag{9}$$

We omit the proofs for the above two lemmas here, as they can be easily derived using standard mechanism design techniques. To optimize the requester's utility, we need to choose a monotone decreasing hiring function x to maximize Eq. (9). The last term in Eq. (9) is easy to maximize if for each i, $\varphi_i(c_i)$ is regular. Even if $\varphi_i(c_i)$ is irregular, we can still apply the so-called "ironing" trick [11] to obtain a monotone increasing version. Define the quantile of a cost c_i as $q_i(c_i) = 1 - F_i(c_i)$ and a function $H_i(q) = (1 - q) \cdot c_i(q)$, where $c_i(q)$ is the cost at quantile q. The ironed virtual cost is the following:

Definition 4 (Ironed Virtual Cost). *The ironed virtual cost of an agent i is defined as:*

$$\overline{\varphi}_i(c_i) = G'_i(q_i(c_i)),$$

where $G'_i(q)$ is the first order derivative (with respect to q) of the convex hull of function H_i:

$$G_i(q) = \max_{\substack{\omega \in [0,1] \\ \alpha \cdot \omega + \beta \cdot (1-\omega) = q}} \{\omega H_i(\alpha) + (1 - \omega) H_i(\beta)\}.$$

Since G_i is convex, the ironed virtual cost is monotone decreasing. It can also be proved that Eq. (9) still holds if we substitute φ_i with $\overline{\varphi}_i$. Based on the previous two lemmas, we present our main mechanism as follows:

Mechanism 1 (Mechanism for Hiring Workers).

1. *Collect the reported cost c_i from all workers.*
2. *Compute each worker's ironed virtual cost $\overline{\varphi}_i(c_i)$.*
3. *Select x that maximizes*

$$u_0 = 2RA(C_x) - \sum_{i \in N} x_i \overline{\varphi}_i(c_i).$$

4. Set the payment rule as:

$$p_i(c_i, c_{-i}) = \int_0^{c_i} s \, dx_i(s, c_{-i}).$$

Theorem 2. *Mechanism 1 is incentive-compatible, individually rational, and maximizes the requestor's utility.*

Proof. The proof directly follows from Lemma 1 and Lemma 2. □

The major challenge for implementing the mechanism is in maximizing the first term in Eq. (9), or equivalently, maximizing the accuracy function. We present the difficult results starting from some simple cases showing that the function is not submodular.

3.1 Simple Case for $n \leq 2$

Let's first consider the easy case with $n \leq 2$. In this case, optimizing Eq. (9) is easy since we can simply enumerate all possible hiring outcomes. There are only 2 outcomes ($x = 0$ and $x = 1$) for $n = 1$, and 4 outcomes ($x \in \{(0,0), (0,1), (1,0), (1,1)\}$) for $n = 2$. A closer investigation into this case, however, shows that the optimal hiring policy is always to hire at most a single worker.

Observation 1. *When $n \leq 2$, the optimal hiring strategy always hires at most 1 worker.*

Proof. The statement is true for $n = 1$. For $n = 2$, we claim that hiring a single worker can achieve the same accuracy as hiring both workers. This option is cheaper and therefore a better strategy. To prove the claim, simply note that the requester will choose to use the answer provided by the worker with a higher accuracy, regardless of whether the two workers' answers agree with each other. Thus, the aggregated answer is always the same as the better worker.

Remark. Since $\ln\left(\frac{0.5+x}{0.5-x}\right)$ is a convex function, similar proof can be extended to show that, if a committee C contains a "dominating high-skill" worker with accuracy a_{big} and other k workers with accuracy $a_1, a_2, \ldots, a_k$ satisfying

$$a_{big} - 0.5 \geq \sum_{i=1}^{k}(a_i - 0.5),$$

we have $A(C) = a_{big}$. Moreover, Observation 1 can be generalized to the following result:

Lemma 3. *When all workers have the same accuracy, i.e., $a_i = a, \forall i$, for any nonnegative integer k, hiring $2k + 1$ workers yields the same overall accuracy as that of hiring $2k + 2$ workers.*

The proof of the Lemma 3 was also deferred to the full version.
We can directly obtain non-submodularity of the accuracy function from Lemma 3.

Corollary 1. *The function A is not submodular.*

4 Hardness of Computing Accuracy

Despite the fact that the accuracy function is not submodular, we try to characterize it in this section since it help us select the workers. When the problem is turned to $n \geq 3$ with different accuracies, the first natural question that comes to our mind is: given the answers from all the workers, how do we optimally aggregate them? Fortunately, the answer to this question is known. We present the result here for completeness.

Lemma 4 (Optimal Weighted Voting Scheme). *The best strategy to aggregate hired workers' answers is equivalent to carrying out weighted voting among the workers with the following voting weight scheme:*

$$w_i \propto \ln \frac{a_i}{1-a_i}.$$

Proof. See Theorem 1 in [12]. □

The lemma above leads to our next result. Given a binary vector x, it is already a difficult problem to compute the accuracy function $A(C_x)$. Optimizing it is even more challenging.

Theorem 3. *Take each worker's accuracy a_i in a committee C as input, computing $A(C)$ is #P-hard.*

To prove Theorem 3, we reduce the following subset sum problem, which is known to be #P-hard, to our problem.

Lemma 5 (#Subset Sum is #P-hard). *Given a non-negative integer set $I = \{y_1, y_2, \cdots, y_n\}$ and a constant boundary $B = \frac{1}{2}\sum_{i=1}^{n} y_i$, computing $|\{I' \subseteq I \mid \sum_{e \in I'} e > B\}|$ is #P-hard.*

Before proving Theorem 3, we first consider the following lemma:

Lemma 6 (Random Variable Correspondance for Workers). *Suppose in the set C of m hired workers, each worker has accuracies $a_{k_1}, a_{k_2}, \cdots, a_{k_m}$ respectively. Then we can find m random variables, denoted by $X_1, X_2, \cdots, X_m$ satisfying the following:*

$$\Pr\left[X_i = -\ln \frac{a_{k_i}}{1-a_{k_i}}\right] = 1 - a_{k_i};$$

$$\Pr\left[X_i = \ln \frac{a_{k_i}}{1-a_{k_i}}\right] = a_{k_i}.$$

If we denote the sum of these variables by $S = \sum_{i=1}^{m} X_i$, we have

$$A(C) = \Pr[S > 0] + \frac{1}{2}\Pr[S = 0].$$

The proof of Lemma 6 was deferred to the full version. Now we are ready to prove Theorem 3.

Proof for Theorem 3. We reduce the #Subset Sum problem to our problem of computing the accuracy. For each instance of #Subset Sum with $B = (y_1 + y_2 + \cdots + y_m)/2$, we assume the answer to the #Subset Sum is K. We construct the following instance for a group of selected workers C.

$$C = \{i_1, i_2, \cdots, i_m\},$$

$$a_{i_j} = 0.5\left[1 + \left(\frac{e^{y_j/2z} - 1}{e^{y_j/2z} + 1}\right)\right], \forall j \in [m]$$

$$z = 1 + \frac{\sum_{j \in [m]} y_j}{\ln\left(1 + \frac{1}{4m \cdot 2^m}\right)}.$$

By construction, we have $\ln\left(\frac{a_{i_j}}{1 - a_{i_j}}\right) = \frac{y_j}{2}$. Thus the final accuracy $A(C)$ of the committee C is:

$$\sum_{\substack{C' \subset C, \\ \sum_{i_u \in C'} y_u > \sum_{i_v \notin C'} y_v}} \left(\prod_{i_j \in C'} a_{i_j} \cdot \prod_{i_k \notin C'} (1 - a_{i_k})\right) +$$

$$\sum_{\substack{C' \subset C, \\ \sum_{i_u \in C'} y_u = \sum_{i_v \notin C'} y_v}} \left(\prod_{i_j \in C'} a_{i_j} \cdot \prod_{i_k \notin C'} (1 - a_{i_k})\right) \cdot \frac{1}{2}.$$

We can divide the previous formula into two parts, the first part is the probability that the sum of all the workers' corresponding random variables is larger than 0; the second part is half of the probability that the sum of all the workers' corresponding random variables is 0. Our goal is to only keep the first part (otherwise the reduction will be affected), so we consider adding a new worker i_{m+1} to form a new committee. We let the accuracy of the new worker be the following:

$$a_{i_{m+1}} = 0.5\left[1 + \left(\frac{e^{\delta/2z} - 1}{e^{\delta/2z} + 1}\right)\right], 0 < \delta \ll 1.$$

Since $\ln\left(\frac{a_{i_{m+1}}}{1 - a_{i_{m+1}}}\right) = \frac{\delta}{2}$, the random variable corresponding to the new worker X_{m+1} is δ with probability $a_{i_{m+1}}$, and $-\delta$ with probability $1 - a_{i_{m+1}}$. The realized value of all other random variables are all integers. Therefore, when the outcome of X_{m+1} is $-\delta$, the final sum can only be positive if the number of positive variables is at least one more than that of the negative variables. So the final accuracy $A(C \cup \{i_{m+1}\})$ would turn into

$$\sum_{\substack{C' \subset C, \\ \sum_{i_u \in C'} y_u > \sum_{i_v \notin C'} y_v}} \left(\prod_{i_j \in C'} a_{i_j} \cdot \prod_{i_k \notin C'} (1 - a_{i_k}) \right) +$$

$$a_{i_{m+1}} \cdot \sum_{\substack{C' \subset C, \\ \sum_{i_u \in C'} y_u = \sum_{i_v \notin C'} y_v}} \left(\prod_{i_j \in C'} a_{i_j} \cdot \prod_{i_k \notin C'} (1 - a_{i_k}) \right).$$

Thus we have

$$\sum_{\substack{C' \subset C, \\ \sum_{i_u \in C'} y_u > \sum_{i_v \notin C'} y_v}} \left(\prod_{i_j \in C'} a_{i_j} \cdot \prod_{i_k \notin C'} (1 - a_{i_k}) \right) = \frac{a_{i_{m+1}} \cdot a_f(C) - \frac{A(C \cup \{i_{m+1}\})}{2}}{a_{i_{m+1}} - \frac{1}{2}}.$$

By letting $\epsilon_j = 0.5 \cdot \left(\frac{e^{y_j/2z} - 1}{e^{y_j/2z} + 1} \right)$, $\epsilon_{\max} = \max_j \{\epsilon_j\}$. We have the following inequality:

$$\sum_{\substack{C' \subset C, \\ \sum_{i_u \in C'} y_u > \sum_{i_v \notin C'} y_v}} \left(\prod_{i_j \in C'} a_{i_j} \cdot \prod_{i_k \notin C'} (1 - a_{i_k}) \right)$$

$$= \sum_{\substack{C' \subset C, \\ \sum_{i_u \in C'} y_u > B}} \left(\prod_{i_j \in C'} (0.5 + \epsilon_j) \cdot \prod_{i_k \notin C'} (0.5 - \epsilon_k) \right)$$

$$\leq K \cdot \prod_{i_j \in C'} (0.5 + \epsilon_j) \cdot \prod_{i_k \notin C'} (0.5 + \epsilon_k)$$

$$\leq K \cdot (0.5 + \epsilon_{\max})^m \leq \frac{K}{2^m} \cdot (1 + 4m \cdot \epsilon_{\max}).$$

Similarly, we can get

$$\sum_{\substack{C' \subset C, \\ \sum_{i_u \in C'} y_u > \sum_{i_v \notin C'} y_v}} \left(\prod_{i_j \in C'} a_{i_j} \cdot \prod_{i_k \notin C'} (1 - a_{i_k}) \right) \geq \frac{K}{2^m} \cdot (1 - 4m \cdot \epsilon_{\max}).$$

Therefore, we have $z > \frac{\sum_{j \in [m]} y_j}{\ln\left(1 + \frac{1}{4m \cdot 2^m}\right)}$, suppose $\epsilon_{\max}$ is achieved by the worker i_j, we have

$$\epsilon_{\max} = 0.5 \left(\frac{e^{y_j/2z} - 1}{e^{y_j/2z} + 1} \right) < 0.5 \left(e^{y_j/2z} - 1 \right) \leq \frac{1}{8m \cdot 2^m}.$$

So the lower and upper bounds are:

$$\frac{K}{2^m} \cdot (1 - 4m \cdot \epsilon_{\max}) > \frac{K - \frac{1}{2}}{2^m},$$

$$\frac{K}{2^m} \cdot (1 + 4m \cdot \epsilon_{\max}) > \frac{K + \frac{1}{2}}{2^m}.$$

Thus we have

$$K - \frac{1}{2} < 2^m \cdot \frac{a_{i_{m+1}} \cdot A(C) - \frac{A(C \cup \{i_{m+1}\})}{2}}{a_{i_{m+1}} - \frac{1}{2}} < K + \frac{1}{2}.$$

Therefore after we computed $A(C)$ and $A(C \cup \{i_{m+1}\})$, we can get K by finding the integer nearest to

$$2^m \cdot \frac{a_{i_{m+1}} \cdot A(C) - \frac{A(C \cup \{i_{m+1}\})}{2}}{a_{i_{m+1}} - \frac{1}{2}}.$$

Thus we have completed the reduction and proved that calculating the accuracy of committee C is #P hard. □

5 Revenue Approximation with Limited-Size Committee

Since it is difficult to compute the utility, we now focus on the case when we only have limited power to compute the optimal strategy with a limited committee size. We studied the utility that this type of committee can achieve to the optimal utility without the size limit. Before presenting our main approximation results, we provide an observation on the corresponding random variable S (in Lemma 6) of a committee. This observation may help readers gain more insights into the accuracy function A.

Observation 2. *The distribution of S is a discrete distribution whose supports are symmetric points:*

$$\{-s_1, -s_2, \ldots, -s_\ell, 0, +s_\ell, +s_{\ell-1}, +s_{\ell-2}, \ldots, +s_1\},$$

where $s_i > 0, \forall i \in [\ell]$. Moreover, we have $\Pr[S = +s_j] > \Pr[S = -s_j], \forall j \in [\ell]$.

The proof of Observation 2 was deferred to full version.

Theorem 4. *The optimal hiring strategy with k workers gives the requester a utility at least $1/\lceil \frac{n}{k} \rceil \cdot U^*$, where U^* is the maximum utility with unlimited committee size.*

Before proving this theorem, we first consider a useful lemma:

Lemma 7. *For any two groups of workers C_1 and C_2, $A(C_1 \cup C_2) - 1/2 \leq A(C_1) + A(C_2) - 1$.*

The proof of Lemma 7 was also deferred to the full version. With the above lemma, we are ready to prove Theorem 4.

Proof of Theorem 4. For simplicity, we define the "accuracy gain" of a worker group C as $A(C) - 1/2$. Suppose that the optimal hiring strategy contains n^* workers, we can divide the whole committee into $q = \lceil \frac{n^*}{k} \rceil$ disjoint groups, each group contains at most k workers. Denote these groups by $G_1, G_2, \ldots, G_q$. Suppose that the maximum utility with a committee of size at most k is U'_k. Then the utility of hiring each group is at most U'_k. Then we have

$$U^* = \left[(2A\left(\bigcup_{i=1}^{q} G_i\right) - 1 \right] \cdot R - \sum_{i=1}^{q} \bar{\varphi}(G_i)$$

$$= R \cdot \sum_{i=1}^{q} (2A(G_i) - 1) - \sum_{i=1}^{q} \bar{\varphi}(G_i)$$

$$\leq \sum_{i=1}^{q} (R \cdot (2A(G_i) - 1) - \bar{\varphi}(G_i))$$

$$\leq U'_k \cdot q = U'_k \left\lceil \frac{n^*}{k} \right\rceil \leq U'_k \left\lceil \frac{n}{k} \right\rceil,$$

where the second equation follows from Lemma 7, and the second inequality is due to the fact that $|G_i| \leq k$. Therefore, we have

$$U'_k \geq \frac{U^*}{\left\lceil \frac{n}{k} \right\rceil},$$

completing the proof. □

6 Algorithm for Selecting Workers with L Possible Accuracies

In this section, we consider a specific case where there are only a constant number of possible accuracies and propose an algorithm for selecting the optimal committee. Let L be the number of possible accuracies. We can categorize the workers into L groups with each group containing workers with the same accuracy. Denote the accuracies by $a^1, a^2, \ldots, a^L$, and the number of workers in each group by $n^1, n^2, \ldots, n^L$, we give an algorithm that returns the optimal committee in polynomial time. The algorithm is as follows:

1. Sort workers in each group according to their ironed virtual costs in ascending order. Let $p_{i,j}$ be the sum of the first j workers in group i.
2. Compute the accuracy $A_{w_1, w_2, \ldots, w_L}$ of a committee with w_i workers in each group i.
3. Compute $(w_1^*, w_2^*, \ldots, w_L^*) =$

$$\arg\max\nolimits_{w_1, \ldots, w_L} \left\{ (2A_{w_1, w_2, \ldots, w_L} - 1) \cdot R - \sum_{i=1}^{L} p_{i, w_i} \right\}.$$

If the maximum value is positive, select the first w_i workers in group i into the committee. Otherwise, do not select any worker.

Note that in the second step, the w_i workers in group i has the same accuracy. Thus in the third step, we will clearly choose the w_i workers with the lowest virtual costs.

Theorem 5. *The above algorithm runs in polynomial time and induces a truthful mechanism.*

Proof. There are only $\prod_{i=1}^{L}(n^i + 1)$ different combinations in step 3. The only missing point is how to compute the $A_{w_1,w_2,\ldots,w_L}$ in the second step for any combination of $\{w_i\}$ in polynomial time.

We construct an L-dimension "matrix" M with $n^1 \times n^2 \times \cdots \times n^L$ elements. Each element $m_{w_1,w_2,\ldots,w_L}$ stores the distribution of the random variable corresponding to (by Lemma 6) the committee with w_i workers in layer i. For any random variable, each group of workers can contribute a value from the following set to the sum:

$$\left\{-n^i \cdot \ln \frac{a^i}{1-a^i}, -(n^i-1) \cdot \ln \frac{a^i}{1-a^i}, \ldots, (n^i-1) \cdot \ln \frac{a^i}{1-a^i}, n^i \cdot \ln \frac{a^i}{1-al^i}\right\},$$

thus there are at most $\prod_{i=1}^{L}(2n^i + 1)$ possible values of the random variable.

We compute the distribution for $m_{w_1,w_2,\ldots,w_L}$ using the distribution for elements with smaller indices. For example, when computing the distribution for $m_{w_1,w_2,\ldots,w_L}$, we choose an arbitrary $i \in [L]$ where $w_i \geq 1$. We begin with the distribution for the element $m' = m_{w_1,w_2,\ldots,w_i-1,\ldots,w_L}$, where the i-th dimension index is decreased by 1. The support of m' is on at most $G = \prod_{i=1}^{L}(2n^i + 1)$ values. We enumerate all its support x with probability $p(x)$. We then add two resulting points $x + \ln \frac{a^i}{1-a^i}$ with probability $p(x) \cdot a^i$ and $x - \ln \frac{a^i}{1-a^i}$ with probability $p(x) \cdot (1-a^i)$ to the distribution of $m_{w_1,w_2,\ldots,w_L}$.

Therefore, each distribution takes $O\left(\prod_{i=1}^{L}(2n^i + 1)\right)$ computations and there are in total $\prod_{i=1}^{L}(n^i + 1)$ accuracies to compute. Therefore, the total running time of the algorithm is $O((2n)^L \cdot n^L) = O(2^L \cdot n^{2L})$, which is polynomial in n since L is a constant.

To show that the induced mechanism is truthful, it suffices to show that the probability of a worker being chosen decreases as their cost increases. For any worker i, the ironed virtual cost function $\bar{\varphi}(c_i)$ is monotone increasing in c_i. So if c_i increases, worker i is more likely to be placed at the front, and all $p_{i,j}$ are weakly increasing. As a result, the committee containing worker i is less likely to be chosen according to step 3 of our algorithm. □

7 Conclusion

In this paper, we consider the mechanism design problem where a platform needs to hire a group of workers to answer a binary question. We characterize truthful mechanisms and give an optimal mechanism. We also show that computing the overall accuracy for a given group of workers is #P-hard, and give approximation results when the committee size is capped by a constant. Finally, when the workers' skill levels fall in constant layers, we give a polynomial algorithm for computing the optimal hiring strategy.

Since our paper only considers a dichotomous choice question for the requestor, a natural extension of our work is to consider a non-binary question with more than 2 answers. It is also interesting to consider the interdependency among the workers' answers and the accuracy of workers are affected by the state of nature (true answer).

References

1. Ben-Yashar, R.C., Nitzan, S.I.: The optimal decision rule for fixed-size committees in dichotomous choice situations: the general result. Int. Econ. Rev. 175–186 (1997)
2. Berend, D., Kontorovich, A.: Consistency of weighted majority votes. In: Advances in Neural Information Processing Systems, vol. 27 (2014)
3. Dickerson, J.P., Sankararaman, K.A., Srinivasan, A., Xu, P.: Assigning tasks to workers based on historical data: online task assignment with two-sided arrivals. In: International Conference on Autonomous Agents and Multiagent Systems (AAMAS) (2018)
4. DiPalantino, D., Vojnovic, M.: Crowdsourcing and all-pay auctions. In: Proceedings of the 10th ACM Conference on Electronic Commerce, pp. 119–128 (2009)
5. Domingos, P., Pazzani, M.: Beyond independence: conditions for the optimality of the simple bayesian classifier. In: Proc. 13th International Conference Machine Learning, pp. 105–112. Citeseer (1996)
6. Goel, G., Nikzad, A., Singla, A.: Mechanism design for crowdsourcing markets with heterogeneous tasks. In: Proceedings of the AAAI Conference on Human Computation and Crowdsourcing, vol. 2 (2014)
7. Hassan, U.U., Curry, E.: A multi-armed bandit approach to online spatial task assignment. In: 2014 IEEE 11th International Conference on Ubiquitous Intelligence and Computing and 2014 IEEE 11th International Conference on Autonomic and Trusted Computing and 2014 IEEE 14th International Conference on Scalable Computing and Communications and Its Associated Workshops, pp. 212–219. IEEE (2014)
8. Kamar, E., Horvitz, E.: Incentives for truthful reporting in crowdsourcing. In: AAMAS, vol. 12, pp. 1329–1330 (2012)
9. Langley, P., Iba, W., Thompson, K., et al.: An analysis of bayesian classifiers. In: AAAI, vol. 90, pp. 223–228. Citeseer (1992)
10. Luo, T., Das, S.K., Tan, H.P., Xia, L.: Incentive mechanism design for crowdsourcing: an all-pay auction approach. ACM Trans. Intell. Syst. Technol. (TIST) **7**(3), 1–26 (2016)
11. Myerson, R.B.: Optimal auction design. Math. Oper. Res. **6**(1), 58–73 (1981)
12. Nitzan, S., Paroush, J.: Optimal decision rules in uncertain dichotomous choice situations. Int. Econ. Rev. 289–297 (1982)
13. Pazzani, M.J.: Searching for dependencies in bayesian classifiers. In: Pre-proceedings of the Fifth International Workshop on Artificial Intelligence and Statistics, pp. 424–429. PMLR (1995)
14. Satzger, B., Psaier, H., Schall, D., Dustdar, S.: Auction-based crowdsourcing supporting skill management. Inf. Syst. **38**(4), 547–560 (2013)
15. Wang, Y., Cai, Z., Zhan, Z.H., Gong, Y.J., Tong, X.: An optimization and auction-based incentive mechanism to maximize social welfare for mobile crowdsourcing. IEEE Trans. Comput. Soc. Syst. **6**(3), 414–429 (2019)
16. Xu, P., Srinivasan, A., Sarpatwar, K.K., Wu, K.L.: Budgeted online assignment in crowdsourcing markets: theory and practice. In: AAMAS, pp. 1763–1765 (2017)
17. Zhang, H.: The optimality of naive bayes. Aa **1**(2), 3 (2004)
18. Zheng, Y., Li, G., Li, Y., Shan, C., Cheng, R.: Truth inference in crowdsourcing: is the problem solved? Proc. VLDB Endowment **10**(5), 541–552 (2017)

Large-Scale Contextual Market Equilibrium Computation Through Deep Learning

Yunxuan Ma[1,2], Yide Bian[3], Hao Xu[4], Weitao Yang[4], Jingshu Zhao[4], Zhijian Duan[2], Feng Wang[4], and Xiaotie Deng[2,5(✉)]

[1] PKU-WUHAN Institute for Artificial Intelligence, Wuhan, China
[2] CFCS, School of Computer Science, Peking University, Beijing, China
{yunxuanma,zjduan,xiaotie}@pku.edu.cn
[3] Yuanpei College, Peking University, Beijing, China
bian1d@stu.pku.edu.cn
[4] School of Computer Science, Wuhan University, Wuhan, China
{xuhao2002,yangweitao,candyzhao,fengwang}@whu.edu.cn
[5] CMAR, Institute for AI, Peking University, Beijing, China

Abstract. Market equilibrium is one of the most fundamental solution concepts in economics and social optimization analysis. Existing works on market equilibrium computation primarily focus on settings with relatively few buyers. Motivated by this, our paper investigates the computation of market equilibrium in scenarios with a large-scale buyer population, where buyers and goods are represented by their contexts. Building on this realistic and generalized contextual market model, we introduce MarketFCNet, a deep learning-based method for approximating market equilibrium. We start by parameterizing the allocation of each good to each buyer using a neural network, which depends solely on the context of the buyer and the good. Next, we propose an efficient method to unbiasedly estimate the loss function of the training algorithm, enabling us to optimize the network parameters through gradient. To evaluate the approximated solution, we propose a metric called Nash Gap, which quantifies the deviation of the given allocation and price pair from the market equilibrium. Experimental results indicate that MarketFCNet delivers competitive performance and significantly lower running times compared to existing methods as the market scale expands, demonstrating the potential of deep learning-based methods to accelerate the approximation of large-scale contextual market equilibrium.

Keywords: market equilibrium · contextual market · equilibrium measure · neural networks · machine learning

---

Full version of this research can be found at https://arxiv.org/abs/2406.15459.

1 Introduction

Market equilibrium is a solution concept in microeconomics theory, which studies how *individuals* amongst groups will exchange their *goods* to get each one better off [31]. The importance of market equilibrium is evidenced by the 1972 Nobel Prize awarded to John R. Hicks and Kenneth J. Arrow "for their pioneering contributions to general economic equilibrium theory and welfare theory" [36]. Market equilibrium has wide application in fair allocation [22], as a few examples, fairly assigning course seats to students [8] or dividing estates, rent, fares, and others [25]. Besides, market equilibrium is also considered for ad auctions with budget constraints where money has real value [12,13].

Existing works often use traditional optimization methods or online learning techniques to solve market equilibrium, which can tackle one market with around 400 buyers and goods in experiments [21,32]. However, in realistic scenarios, there might be millions of buyers in one market (*e.g.* job market, online shopping market). In these scenarios, the description complexity for the market is $\Omega(nm)$ and existing algorithms need $\Omega(nm)$ computational costs to do one iteration step on the market, if there are n buyers and m goods, which is unacceptable when n is extremely large and potentially infinite.

Contextual models come to the rescue. The success of contextual auctions [4,17] demonstrates the power of contextual models, in which each bidder and item can be described with low-dimensional representations, which consist of the information about bidders and items and further determines the value (or its distribution) of items to bidders. In this way, auctions, as well as other economic problems, can be described more memory-efficiently, making it possible to accelerate the computation of these problems. Inspired by the models of contextual auctions, we propose the concept of contextual markets in a similar way. We argue that contextual markets can be useful to model large-scale markets aforementioned, since prior works have assumed the real market to be within some low-dimension space, and the values of goods to buyers are often not hard to speculate given the knowledge of goods and buyers [28,29]. Besides, contextual models never lose expressive power compared with raw models in the worst case [6].

This paper initiates the study of *deep learning* for *contextual* market equilibrium computation with a large (and potential infinite) number of buyers. Following the framework of differentiable economics [14,18,40], we propose a deep-learning-based approach, *MarketFCNet*, that takes the representations of one buyer and one good as input, and outputs the allocation of the good to the buyer. The training on MarketFCNet targets at an unbiased estimator of the objective function in EG-convex program, which can be achieved with constant buyer samples. Through learning an allocation function, we decrease the network iteration costs to $O(m)$ compared with $O(nm)$ in traditional buyer-wise methods, greatly accelerating the learning of market equilibrium. In addition, the allocation of any item to any buyer, as well as the price of any item, can be accessed through one network call and thus have $O(1)$ query complexity.

In this way, we optimize the allocation function on "buyer space" implicitly, rather than optimizing the allocation to each buyer separately. Therefore, the performance of MarketFCNet becomes independent of the number of buyers n. As a consequence, MarketFCNet shows its efficiency when n becomes large.

The effectiveness of MarketFCNet is demonstrated by our experimental results. As the market scale expands, MarketFCNet delivers competitive performance and significantly lower running times compared to existing methods in different experimental settings, demonstrating the potential of deep learning-based methods to accelerate the approximation of large-scale contextual market equilibrium.

To conclude, the contributions of this paper consist of three parts,

- We propose a method, MarketFCNet, to approximate the contextual market equilibrium in which the number of buyers is large.
- We propose a measure, Nash Gap, to quantify the deviation of the given allocation and price pair from the market equilibrium.
- We conduct extensive experiments, demonstrating promising performance on the approximation measure and running time compared with existing methods.

2 Related Works

Market Equilibriums. The history of market equilibrium arises from microeconomics theory, where the concept of competitive equilibrium [31, §10] was proposed, and the existence of market equilibrium is guaranteed in a general setting [2,39]. [20] first considered the linear market case, and proved that the solution of the EG-convex program constitutes a market equilibrium, which lays the polynomial-time algorithmic foundations for market equilibrium computation. [19] later showed that the EG program also works for a class of CCNH utility functions. Shmyrev program later is also proposed to solve market equilibrium with linear utility with a perspective shift from allocation to price [35], while [11] later found that Shmyrev program is the dual problem of EG program with a change of variables. There is also a branch of literature that consider computational perspective in more general settings such as indivisible goods [15,16,34] and piece-wise linear utility [23,24,38].

Algorithms of Solving Market Equilibriums. There are abundant works that present algorithms to solve the market equilibrium and show the convergence results theoretically [10]. [21] discusses the convergence rates of first-order algorithms for EG convex program under linear, quasi-linear, and Leontief utilities. [32] later designs stochastic optimization algorithms for the EG convex program and Shmyrev program with convergence guarantee and show some economic insight. [27] proposes an ADMM algorithm for CCNH utilities and shows linear convergence results. Besides, researchers are more engaged in designing dynamics that possess more economic insight. For example, PACE dynamic [22,30,42] and proportional response dynamic [9,41,43], though the original idea of PACE arise from auction design [12,13].[1]

3 Contextual Market Modelling

In this section, we focus on the model of contextual market equilibrium in which goods are assumed to be divisible. Let the market consist of n buyers, denoted as $1, ..., n$, and m goods, denoted as $1, ..., m$. We denote $[k]$ as the abbreviation of the set $\{1, 2, ..., k\}$. Each buyer $i \in [n]$ has a representation b_i, and each good $j \in [m]$ has a representation g_j. We assume that b_i belongs to the buyer representation space $\mathcal{B}$, and g_j belongs to the good representation space $\mathcal{G}$. For a buyer with representation $b \in \mathcal{B}$, she has a budget $B(b) > 0$. Denote $Y(g) > 0$ as the supply of good with representation g. Although many existing works [21] assume that each good j has *unit* supply (i.e. $Y(g) \equiv 1$ for all $g \in \mathcal{G}$) without loss of generality, their models can be easily generalized to our settings.

An *allocation* is a matrix $\boldsymbol{x} = (x_{ij})_{i \in [n], j \in [m]} \in \mathbb{R}_+^{n \times m}$, where x_{ij} is the amount of good j allocated to buyer i. We denote $\boldsymbol{x}_i = (x_{i1}, ..., x_{im})$ as the vector of the bundle of goods that is allocated to buyer i. The buyers' utility function is denoted as $u : \mathcal{B} \times \mathbb{R}_+^m \to \mathbb{R}_+$, here $u(b_i; \boldsymbol{x}_i)$ denotes the utility of buyer i with representation b_i when she chooses to buy $\boldsymbol{x}_i$. We denote $u_i(\boldsymbol{x}_i)$ as an equivalent form of $u(b_i; \boldsymbol{x}_i)$ and often refer to them as the same thing. Similarly, $B(b_i), Y(g_j)$ and B_i, Y_j are often referred to as the same thing, respectively.

Let $\boldsymbol{p} = (p_1, ..., p_m) \in \mathbb{R}_+^m$ be the prices of the goods, the *demand set* of buyer with representation b_i is defined as the set of utility-maximizing allocations within budget constraint.

$$D(b_i; \boldsymbol{p}) := \arg\max_{\boldsymbol{x}_i} \left\{ u(b_i; \boldsymbol{x}_i) \mid \boldsymbol{x}_i \in \mathbb{R}_+^m, \langle \boldsymbol{p}, \boldsymbol{x}_i \rangle \leq B(b_i) \right\}. \quad (1)$$

A *contextual market* is a 4-tuple: $\mathcal{M} = \langle n, m, (b_i)_{i \in [n]}, (g_j)_{j \in [m]} \rangle$, where buyer utility $u(b_i; \boldsymbol{x}_i)$ is known given the information of the market. We also assume budget function $B : \mathcal{B} \to \mathbb{R}_+$ represents the budget of buyers and capacity function $Y : \mathcal{G} \to \mathbb{R}_+$ represents the supply of goods. All of u, B, and Y are assumed to be public knowledge and excluded from a market representation. This assumption mainly comes from two aspects: (1) these functions can be learned from historical data and (2) budgets and supplies can be either encoded in b and g in some way.

The *market equilibrium* is represented as a pair $(\boldsymbol{x}, \boldsymbol{p})$, $\boldsymbol{x} \in \mathbb{R}_+^{n \times m}$, $\boldsymbol{p} \in \mathbb{R}_+^m$, which satisfies the following conditions.

- *Buyer optimality*: $\boldsymbol{x}_i \in D(b_i, \boldsymbol{p})$ for all $i \in [n]$,
- *Market clearance*: $\sum_{i=1}^n x_{ij} \leq Y(g_j)$ for all $j \in [m]$, and equality must hold if $p_j > 0$.

---

[1] For more related works, please refer to full version.

We say that u_i is *homogeneous* (with degree 1) if it satisfies $u_i(\alpha \boldsymbol{x}_i) = \alpha u_i(\boldsymbol{x}_i)$ for any $\boldsymbol{x}_i \geq 0$ and $\alpha > 0$ [33, §6.2]. Following existing works, we assume that u_is are CCNH utilities, where CCNH represents concave, continuous, non-negative, and homogeneous functions [21]. For CCNH utilities, a market equilibrium can be computed using the following *Eisenberg-Gale convex program* (EG):

$$\max \sum_{i=1}^{n} B_i \log u_i(\boldsymbol{x}_i) \quad \text{s.t.} \quad \sum_{i=1}^{n} x_{ij} \leq Y_j, \ \boldsymbol{x} \geq 0. \tag{EG}$$

Theorem 1 shows that the market equilibrium can be represented as the optimal solution of (EG).

Theorem 1 (Gao and Kroer[21]). *Let u_i be concave, continuous, non-negative, and homogeneous (CCNH). Assume $u_i(\mathbf{1}) > 0$ for all i. Then, (i) (EG) has an optimal solution and (ii) any optimal solution $\boldsymbol{x}$ to (EG) together with its optimal Lagrangian multipliers $\boldsymbol{p}^* \in \mathbb{R}_+^m$ constitute a market equilibrium, up to arbitrary assignment of zero-price items. Furthermore, $\langle \boldsymbol{p}^*, \boldsymbol{x}_i^* \rangle = B_i$ for all i.*

Based on Theorem 1, it's easy to find that we can always assume $\sum_{i \in [n]} x_{ij} = Y_j$ while preserving the existence of market equilibrium, which states as follows.

Proposition 1. *Following the assumptions in Theorem 1. For the following EG convex program with equality constraints,*

$$\max \sum_{i=1}^{n} B_i \log u_i(\boldsymbol{x}_i) \quad \text{s.t.} \quad \sum_{i=1}^{n} x_{ij} = Y_j, \ \boldsymbol{x} \geq 0. \tag{2}$$

Then, an optimal solution $\boldsymbol{x}^*$ together with its Lagrangian multipliers $\boldsymbol{p}^* \in \mathbb{R}_+^m$ constitute a market equilibrium. Moreover, assume more that for each good j, there is some buyer i such that $\frac{\partial u_i}{\partial x_{ij}} > 0$ always hold whenever $u_i(\boldsymbol{x}_i) > 0$, then all prices are strictly positive in market equilibrium. As a consequence, Eqs. (EG) and (2) derive the same solution.

All proofs can be found in full version. Since the additional assumption in Proposition 1 is fairly weak, without further clarification, we always assume the conditions in Proposition 1 hold and the market clearance condition becomes $\sum_{i \in [n]} x_{ij} = Y(g_j), \ \forall j \in [m]$.

4 MarketFCNet

In this section, we introduce the MarketFCNet (denoted as Market Fully-Connected Network) approach to solve the contextual market equilibrium in the previous section. The key point of MarketFCNet is to design an unbiased estimator of an optimization program whose solution coincides with the market equilibrium. MarketFCNet does not rely on the number of buyers, making it an advantage to fit the infinite-buyer case without scaling on computational complexity.

4.1 Problem Reformulation

Following the idea of differentiable economics [18], we utilize parameterized models to represent the allocation of good j to buyer i, denoted as $x_\theta(b_i, g_j)$, and denote it as the allocation network, where θ is the network parameter. Given buyer i and good j, the network can automatically compute the allocation $x_{ij} = x_\theta(b_i, g_j)$. The allocation to buyer i is represented as $\boldsymbol{x}_i = \boldsymbol{x}_\theta(b_i, \boldsymbol{g})$ and the allocation matrix is represented as $\boldsymbol{x} = \boldsymbol{x}_\theta(\boldsymbol{b}, \boldsymbol{g})$. Then the market clearance constraint can be reformulated as $\sum_{i \in [n]} x_\theta(b_i, g_j) = Y(g_j), \forall j \in [m]$ and the price constraint can be reformulated as $\boldsymbol{x}_\theta(\boldsymbol{b}, \boldsymbol{g}) \geq 0$. Let $\mathcal{U}(\mathcal{B})$ be the uniform distribution on the discrete set $\mathcal{B} = \{b_i : i \in [n]\}$, then the EG program (EG) becomes,

$$\max_{x_\theta} \quad \text{OBJ}(x_\theta) = \mathbb{E}_{b \sim \mathcal{U}(\mathcal{B})}[B(b) \log u(b; \boldsymbol{x}_\theta(\boldsymbol{b}, \boldsymbol{g}))]$$
$$\text{s.t.} \quad \mathbb{E}_{b \sim \mathcal{U}(\mathcal{B})}[x_\theta(b, g_j)] = Y(g_j)/n, \forall j \in [m] \quad \text{(EG-FC)}$$
$$\boldsymbol{x}_\theta(\boldsymbol{b}, \boldsymbol{g}) \geq 0$$

For simplicity, we take $Y(g_j)/n \equiv 1$ for all g_j and omit $\mathbb{E}_{b \sim \mathcal{U}(\mathcal{B})}$ as $\mathbb{E}_b$ when the context is clear. By this reformulation, the number of buyers, n, disappears from the program (EG-FC). As a consequence, MarketFCNet can potentially capture the scenarios even buyers are infinite or follow some distribution $\mathcal{F} \in \Delta(\mathcal{B})$.

4.2 Optimization

The second constraint in (EG-FC) can be hardcoded with network architecture (for example, a post-processed element-wise softplus function $\sigma(x) = \log(1 + \exp(x))$ that maps the set of real numbers to the set of positive numbers). To address the first constraint, notice that the prices of goods are simply the Lagrangian multipliers of the first constraint in (EG-FC). Therefore, we employ the Augmented Lagrange Multiplier Method (ALMM) to explicitly extract the multipliers and solve the optimization problem (EG-FC). We define $\mathcal{L}_\rho(x_\theta, \boldsymbol{\lambda})$ as the Lagrangian where $\boldsymbol{\lambda} \in \mathbb{R}^m$ is the multipliers and ρ is the quadratic penalty term. The Lagrangian has the form as follows:

$$\mathcal{L}_\rho(x_\theta; \boldsymbol{\lambda}) = -\text{OBJ}(x_\theta) + \sum_{j=1}^m \lambda_j \left(\mathbb{E}_b[x_\theta(b, g_j)] - 1\right) + \frac{\rho}{2} \sum_{j=1}^m \left(\mathbb{E}_b[x_\theta(b, g_j)] - 1\right)^2 \quad (3)$$

Exactly computing the objective function seems intractable due to the extremely large and potentially infinite buyer size. Therefore, we follow the framework in learning theory culture that many first-order algorithms can be applied (*e.g.*, SGD, Adam) with only an unbiased gradient of the objective function [1,7]. Given such unbiased gradients as oracle, we provide our pseudo-codes in Algorithm 1. An illustration of the training procedure is also provided in Fig. 1.

Let $X(w)$ be a deterministic function where w is a (potentially high-dimensional) random variable with distribution $\mathcal{P}(w)$ and x be a real number. We call $X(w)$ is an unbiased estimator of x if $\mathbb{E}_{w \sim \mathcal{P}(w)}[X(w)] = x$.

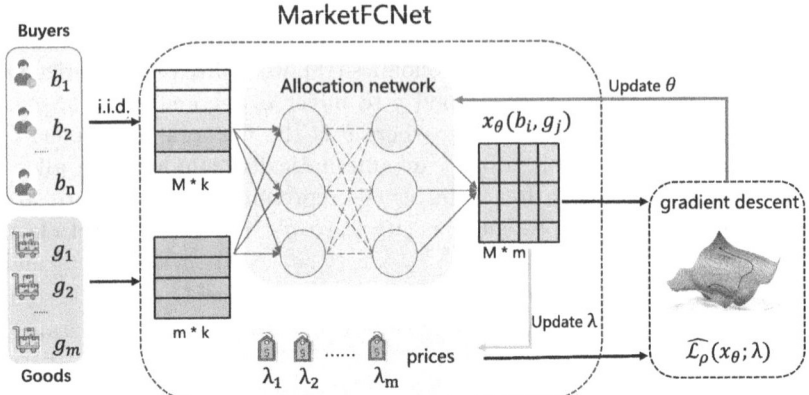

Fig. 1. Training process of MarketFCNet. On each iteration, the batch of M independent buyers are drawn. Each buyer and each good are represented as k-dimension context. The (i,j)'th element in the allocation matrix represents the allocation computed from i'th buyer and j'th good. MarketFCNet training process alternates between the training of allocation network and prices. The training of allocation network needs to achieve an unbiased estimator $\widehat{\mathcal{L}}_\rho(x_\theta;\lambda)$ of the loss function $\mathcal{L}_\rho(x_\theta;\lambda)$, followed by gradient descent. The training of prices need to get an unbiased estimator $\widehat{\Delta\lambda}_j$ of $\Delta\lambda_j$, followed by ALMM updating rule $\lambda_j \leftarrow \lambda_j + \beta_t \widehat{\Delta\lambda}_j$.

From Algorithm 1, it's clear that in order to finish the ALMM algorithm, we need to obtain unbiased estimators of the following two expressions.

- An unbiased estimator of $\mathcal{L}_\rho(x_\theta;\boldsymbol{\lambda})$.
- An unbiased estimator of $\Delta\lambda_j$, where $\Delta\lambda_j$ is given by $\Delta\lambda_j = \rho\left(\mathbb{E}_b[x_\theta(b,g_j)] - 1\right)$.

4.3 Extract Unbiased Estimators

Unbiased Estimator of $\Delta\lambda_j$. To get an unbiased estimator of $\Delta\lambda_j$, it's enough to obtain an unbiased estimator of $\mathbb{E}_b[x_\theta(b,g_j)]$. By applying Monte Carlo method, we can choose batch size M and sample $b_1, b_2, ..., b_M \sim \mathcal{U}(\mathcal{B})$, then $\frac{1}{M}\sum_{i=1}^{M} x_\theta(b_i, g_j)$ forms an unbiased estimator.

Define $\hat{\Delta}_j(b_1,\ldots,b_M) = \rho(\frac{1}{M}\sum_{i=1}^{M} x_\theta(b_i, g_j) - 1)$. It's clear to see that $\mathbb{E}_{b_i \stackrel{\text{i.i.d.}}{\sim} \mathcal{U}(\mathcal{B})}[\hat{\Delta}_j(b_1,\ldots,b_M)] = \Delta_j$.

Unbiased Estimator of $\mathcal{L}_p(x_\theta;\boldsymbol{\lambda})$. Notice that $\mathcal{L}_p(x_\theta;\boldsymbol{\lambda})$ is the sum of three terms: the objective term $\text{OBJ}(x_\theta)$, the multiplier term $\sum_{j=1}^{m} \lambda_j \left(\mathbb{E}_b[x_\theta(b,g_j)] - 1\right)$ as well as the quadratic term $\frac{\rho}{2}\sum_{j=1}^{m}\left(\mathbb{E}_b[x_\theta(b,g_j)] - 1\right)^2$. We only need to get an unbiased estimator for each term. For the first and the second term, the technique to achieve an unbiased estimator is similar to those in $\Delta\lambda_j$ and thus omitted.

Algorithm 1: MarketFCNet

Input: An oracle of buyer sampler $b \sim \mathcal{U}(\mathcal{B})$, goods $g_1, ..., g_m$, batch size M_1 and M_2, quadratic penalty term ρ, iteration K for optimizing Lagrangian, step size $(\beta_t)_{t=1}^{\infty}$ for optimizing multipliers.
Output: Allocation network $x_\theta(b, g)$, price p_j for each good g_j.
1 Initialize an allocation network $x_\theta(b,g)$ and multipliers $\{\lambda_j\}_{j \in [m]}$.
2 **for** $t = 1, 2, ...$ until converged **do**
3   **for** $k = 1, 2, ...K$ **do**
4   Get an unbiased estimator $\hat{\mathcal{L}}_\rho(x_\theta; \boldsymbol{\lambda})$ with batch size M_1 such that $\mathbb{E}[\hat{\mathcal{L}}_\rho(x_\theta; \boldsymbol{\lambda})] = \mathcal{L}_\rho(x_\theta; \boldsymbol{\lambda})$.
5   Call the optimizer to optimize θ on $\hat{\mathcal{L}}_\rho(x_\theta; \boldsymbol{\lambda})$ for one step.
6   **end**
7   **for** $j = 1, ..., m$ **do**
8   Get an unbiased estimator $\hat{\Delta}\lambda_j$ with batch size M_2 such that $\mathbb{E}[\hat{\Delta}\lambda_j] = \Delta\lambda_j := \rho(\mathbb{E}_b[x_\theta(b,g_j)] - 1)$.
9   Updates λ_j with $\lambda_j \leftarrow \lambda_j + \beta_t \Delta \lambda_j$.
10   **end**
11 **end**
12 **return** $x_\theta(b,g), \{\lambda_j\}_{j \in [m]}$.

For the last term, the quadratic dependency on expectation requires additional attention. Notice that

$$(\mathbb{E}_b[x_\theta(b, g_j)] - 1)^2 = (\mathbb{E}_b[x_\theta(b, g_j)] - 1) \cdot (\mathbb{E}_{b'}[x_\theta(b', g_j)] - 1) \tag{4}$$

Therefore, we can sample $b_1, ..., b_M, b'_1, ..., b'_M \sim U(\mathcal{B})$ and compute

$$\frac{\rho}{2} \cdot \frac{1}{M} \sum_{i=1}^{M} \sum_{j=1}^{m} (x_\theta(b_i, g_j) - 1) \cdot (x_\theta(b'_i, g_j) - 1) \tag{5}$$

which provides an unbiased estimator for the last term.

5 Performance Measures of Market Equilibrium

In this section, we propose *Nash Gap* to measure the performance of an approximated market equilibrium and show that Nash Gap preserves the economic interpretation for market equilibrium. To introduce Nash Gap, we first introduce two types of welfare, Log Nash Welfare and Log Fixed-price Welfare in Definition 1 and Definition 2, respectively.

Definition 1 (Log Nash Welfare). *The Log Nash Welfare (abbreviated as LNW) is defined as*

$$\text{LNW}(\boldsymbol{x}) = \frac{1}{B_{\text{total}}} \sum_{i \in [n]} B_i \log u_i(\boldsymbol{x}_i), \tag{6}$$

where $B_{\text{total}} = \sum_{i \in [n]} B_i$ is the total budget for buyers.

Log Nash Welfare is fundamentally the logarithm of Nash Welfare with an economic interpretation of the geometric mean of buyers' utilities, which has been widely studied in literature [5,26].

Definition 2 (Fixed-Price Utility and Log Fixed-Price Welfare). *We define the fixed-price utility for buyer i as,*

$$\tilde{u}(b_i; \boldsymbol{p}) = \max_{\boldsymbol{x}_i}\{u(b_i; \boldsymbol{x}_i) \mid \boldsymbol{x}_i \in \mathbb{R}^m_+, \langle \boldsymbol{p}, \boldsymbol{x}_i \rangle \leq B(b_i)\} \tag{7}$$

which represents the optimal utility that buyer i can obtain at the price level $\boldsymbol{p}$, regardless of the market clearance constraints. The Log Fixed-price Welfare (abbreviated as LFW) is defined as the weighted sum of the logarithm of Fixed-price utility,

$$\mathrm{LFW}(\boldsymbol{p}) = \frac{1}{B_{\text{total}}} \sum_{i \in [n]} B_i \log \tilde{u}_i(\boldsymbol{p}) \tag{8}$$

Based on these definitions, we present the definition of Nash Gap.

Definition 3 (Nash Gap). *We define Nash Gap (abbreviated as NG) as the difference between Log Nash Welfare and Log Fixed-price Welfare, i.e.*

$$\mathrm{NG}(\boldsymbol{x}, \boldsymbol{p}) = \mathrm{LFW}(\boldsymbol{p}) - \mathrm{LNW}(\boldsymbol{x}) \tag{9}$$

5.1 Properties of Nash Gap

To show why NG is useful in the measure of market equilibrium, we first observe that,

Proposition 2 (Price Constraints). *If $(\boldsymbol{x}, \boldsymbol{p})$ constitutes a market equilibrium, the following identity always holds,*

$$\sum_{j \in [m]} p_j Y_j = \sum_{i \in [n]} B_i \tag{10}$$

Below, we state the most important theorem in this paper.

Theorem 2. *Let $(\boldsymbol{x}, \boldsymbol{p})$ be a pair of allocation and price. Assuming the allocation satisfies market clearance and the price meets price constraint, then we have* $\mathrm{NG}(\boldsymbol{x}, \boldsymbol{p}) \geq 0$.

Moreover, $\mathrm{NG}(\boldsymbol{x}, \boldsymbol{p}) = 0$ if and only if $(\boldsymbol{x}, \boldsymbol{p})$ is a market equilibrium.

Theorem 2 shows that Nash Gap is an ideal measure of the solution concept of market equilibrium since it holds the following properties,

- $\mathrm{NG}(\boldsymbol{x}, \boldsymbol{p})$ is continuous on the inputs $(\boldsymbol{x}, \boldsymbol{p})$.
- $\mathrm{NG}(\boldsymbol{x}, \boldsymbol{p}) \geq 0$ always hold. (under conditions in Theorem 2)
- $\mathrm{NG}(\boldsymbol{x}, \boldsymbol{p}) = 0$ if and only if $(\boldsymbol{x}, \boldsymbol{p})$ meets the solution concept.

- The computation of NG does not require the knowledge of an equilibrium point $(\boldsymbol{x}^*, \boldsymbol{p}^*)$

Since some may argue that $\text{NG}(\boldsymbol{x}, \boldsymbol{p})$ is not intuitive to understand, we consider some more intuitive measures, the Euclidean distance to the market equilibrium, i.e., $\|\boldsymbol{x} - \boldsymbol{x}^*\|$ and $\|\boldsymbol{p} - \boldsymbol{p}^*\|$, as well as the difference on Weighted Social Welfare, $|\text{WSW}(\boldsymbol{x}) - \text{WSW}(\boldsymbol{x}^*)|$, where $\text{WSW}(\boldsymbol{x}) := \sum_{i \in [n]} \frac{B_i}{B_{\text{total}}} u_i(\boldsymbol{x}_i)$, and show the connection between NG and these intuitive measures.

Proposition 3 (Informal). *Under some technical assumptions, if* $\text{NG}(\boldsymbol{x}, \boldsymbol{p}) = \varepsilon$, *we have:*

- $\|\boldsymbol{p} - \boldsymbol{p}^*\| = O(\sqrt{\varepsilon})$.
- $\|\boldsymbol{x}_i - \boldsymbol{x}_i^*\| = O(\sqrt{\varepsilon})$ *for all* i.
- $|\text{WSW}(\boldsymbol{x}) - \text{WSW}(\boldsymbol{x}^*)| = O(\varepsilon)$.

Proposition 3 means that intuitive measures (*e.g.*, Euclidean distance to the equilibrium point) are upper bounded by Nash Gap with a square root rate, serving as a certificate that Nash Gap is an ideal measure.

Finally, we give a saddle-point explanation for Nash Gap.

Corollary 1. *Within market clearance and price constraint, we have*

$$\min_{\boldsymbol{p}} \text{LFW}(\boldsymbol{p}) = \max_{\boldsymbol{x}} \text{LNW}(\boldsymbol{x}) \qquad (11)$$

Corollary 1 provides an economic interpretation for GAP. Market equilibrium can be seen as the saddle point over social welfare, and the social welfare for $\boldsymbol{x}$ can be actually implemented while the social welfare for $\boldsymbol{p}$ is virtual and desired by buyers. Nash Gap measures the gap between the "desired welfare" and the "implemented welfare" for buyers.

5.2 Measures in General Cases

Since NG only works for $(\boldsymbol{x}, \boldsymbol{p})$ that satisfies market clearance and price constraints, we generalize the measure of NG to the full outcome space in this section, which drives us to design some reasonable measures that can measure the performance of all positive $(\boldsymbol{x}, \boldsymbol{p})$.

We first notice that any equilibrium must satisfy the conditions of *market clearance* and *price constraint*, we first project arbitrary positive $(\boldsymbol{x}, \boldsymbol{p})$ to the space where these constraints hold. Specifically, if we let

$$\alpha_j = \frac{Y_j}{\sum_i x_{ij}}, \quad \tilde{x}_{ij} = x_{ij} \cdot \alpha_j \qquad \beta = \frac{\sum_i B_i}{\sum_j Y_j p_j}, \quad \tilde{p}_j = \beta \cdot p_j \qquad (12)$$

then $(\tilde{\boldsymbol{x}}, \tilde{\boldsymbol{p}})$ satisfies these constraints and we consider $\text{NG}(\tilde{\boldsymbol{x}}, \tilde{\boldsymbol{p}})$ as the equilibrium measure.

Besides, we also need to measure how far is the point $(\boldsymbol{x}, \boldsymbol{p})$ to the space within the conditions of *market clearance* and *price constraint*. We propose the following two measurements, called Violation of Allocation (abbreviated as VoA) and Violation of Price (abbreviated as VoP), respectively.

$$\text{VoA}(\boldsymbol{x}) := \frac{1}{m}\sum_j |\log \alpha_j|, \qquad \text{VoP}(\boldsymbol{p}) := |\log \beta| \qquad (13)$$

From the expressions of VoA and VoP, we know that these two constraints hold if and only if $\text{VoA}(\boldsymbol{x}) = 0$ and $\text{VoP}(\boldsymbol{p}) = 0$.

We argue that this projection is of economic meaning. If $(\boldsymbol{x}, \boldsymbol{p})$ constitutes a market equilibrium and we scale the budget with a factor of β, then $(\boldsymbol{x}, \beta \boldsymbol{p})$ constitutes a market equilibrium in the new market. Similarly, if we scale the value for each buyer with factor $\boldsymbol{\alpha}^{-1}$ (here $\boldsymbol{\alpha}$ is a vector in $\mathbb{R}_+^m$ and $\boldsymbol{\alpha}^{-1}$ is element-wise inverse of $\boldsymbol{\alpha}$) and capacity with factor $\boldsymbol{\alpha}$, then, $(\boldsymbol{\alpha}\cdot\boldsymbol{x}, \frac{\boldsymbol{p}}{\boldsymbol{\alpha}})$ constitute a market equilibrium in the new market. These instances are evidence that market equilibrium holds a linear structure over market parameters. Therefore, a linear projection can eliminate the effect from linear scaling, while preserving the effect from orthogonal errors.

Notice that $\boldsymbol{x} = \tilde{\boldsymbol{x}}$ and $\boldsymbol{p} = \tilde{\boldsymbol{p}}$ if and only if $\text{VoA}(\boldsymbol{x}) = 0$ and $\text{VoP}(\boldsymbol{p}) = 0$, respectively. From Theorem 2 We can easy derive following statement:

Proposition 4. *For arbitrary $\boldsymbol{x} \in \mathbb{R}_+^{n\times m}, \boldsymbol{p} \in \mathbb{R}_+^m$, we have $\text{VoA}(\boldsymbol{x}) \geq 0, \text{VoP}(\boldsymbol{p}) \geq 0, \text{NG}(\tilde{\boldsymbol{x}}, \tilde{\boldsymbol{p}}) \geq 0$ always hold. Moreover, $(\boldsymbol{x}, \boldsymbol{p})$ is a market equilibrium if and only if $\text{VoA}(\boldsymbol{x}) = \text{VoP}(\boldsymbol{p}) = \text{NG}(\tilde{\boldsymbol{x}}, \tilde{\boldsymbol{p}}) = 0$.*

Proposition 4 is a certificate that $\text{VoA}(\boldsymbol{x}), \text{VoP}(\boldsymbol{p}), \text{NG}(\tilde{\boldsymbol{x}}, \tilde{\boldsymbol{p}})$ together form a good measure for market equilibrium. Therefore, in our experiments, we compute these measures of solutions and prefer a lower measure without further clarification.

6 Experiments

In this section, we present empirical experiments that evaluate the effectiveness of MarketFCNet. Though briefly mentioned in this section, we leave the details of baselines, implementations, hyper-parameters, and experimental environments to full version.

6.1 Experimental Settings

In our experiments, all utilities are chosen as CES utilities, which satisfies CCNH conditions and captures a wide utility class including linear utilities, Cobb-Douglas utilities, and Leontief utilities [3,37]. CES utilities have the form of $u_i(x_i) = \left(\sum_{j\in[m]} v_{ij}^\alpha x_{ij}^\alpha\right)^{1/\alpha}$ with $\alpha \leq 1$. The fixed-price utilities for CES utility, which are necessary to measure the performance, are derived in full version.

Table 1. Comparison of MarketFCNet with baselines: $n = 1,048,576$ buyers and $m = 10$ goods. The GPU time for MarketFCNet represents the training time and testing time, respectively.

Methods	NG	VoA	VoP	GPU Time
Naïve	3.65e-1	0	0	3.57e-3
EG	2.17e-2	2.620e-1	7.031e-2	197
EG-m	**2.49e-4**	6.01e-2	9.77e-2	100
FC	1.63e-3	**1.416e-2**	**6.750e-3**	**43.6; 9.63e-2**

In order to evaluate the performance of MarketFCNet, we compare them mainly with a baseline that directly maximizes the objective in EG convex program with gradient ascent algorithm (abbreviated as *EG*), which is widely used in the field of market equilibrium computation. Besides, we also consider a momentum version of *EG* algorithm with momentum $\beta = 0.9$ (abbreviated as *EG-m*).

We also consider a naïve allocation and pricing rule (abbreviated as *Naïve*), which can be regarded as the benchmark of the experiments:

$$x_{ij} = 1, \quad p_j = \frac{\sum_{i \in [n]} B_i}{mV_j}, \quad \text{for all } i, j \tag{14}$$

In *Naïve*, the allocation is evenly distributed such that the market clearance holds. The price for each good is designated such that the price constraints also hold.

In the following experiments, MarketFCNet is abbreviated as *FC*. Notice that *Naïve* always gives an allocation that satisfies market clearance and price constraints, while *EG*, *EG-m* and *FC* do not.

6.2 Experiment Results

Comparing with Baselines. We choose the number of buyers $n = 1,048,576 = 2^{20}$, number of items $m = 10$, CES utility parameter $\alpha = 0.5$ and representation with standard normal distribution as the basic experimental environment of MarketFCNet; We consider $NG(\tilde{x}, \tilde{p})$, $VoA(x)$, $VoP(p)$ and the running time of algorithms as the measures. Without special specifications, these parameters are default settings among other experiments. Results are presented in Table 1. From these results, we can see that the approximations of MarketFCNet are competitive with *EG* and *EG-m* and far better than Naïve, which means that the solution of MarketFCNet is very close to market equilibrium. MarketFCNet also achieves a much lower running time compared with *EG* and *EG-m*, which indicates that these methods are more suitable for large-scale market equilibrium computation. In the following experiments, VoA and VoP measures are omitted and we only report NG and running time.

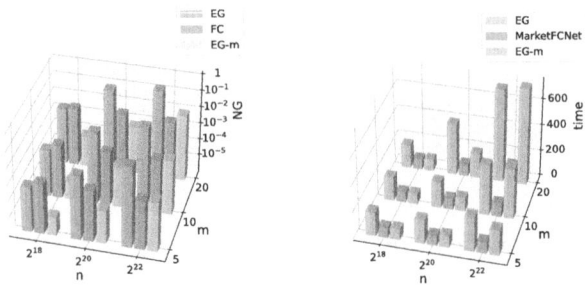

Fig. 2. The Nash Gap (left) and GPU time (right) for MarketFCNet, EG, and EG-m. Market size varies from $n = 2^{18}, 2^{20}, 2^{22}$ buyers and $m = 5, 10, 20$ goods.

Different Market Scale for MarketFCNet. A natural question is how market size (here n and m) will have an impact on the efficiency of MarketFCNet and traditional baselines. We take $m = 5, 10, 20$ and $n = 2^{18} = 262,114, 2^{20} = 1,048,576, 2^{22} = 4,194,304$ as the experimental settings. For each combination of n and m, we trained MarketFCNet and compared it with EG and EG-m. The Nash Gap and training time are provided in Fig. 2. Experimental results show that, as the market size varies, MarketFCNet has almost the same Nash Gap and running time. However, as the market size increases, both EG and EG-m have larger Nash Gaps and longer running times, demonstrating the scalability of MarketFCNet to solve large-scale contextual market equilibrium compared with traditional methods.

7 Conclusions and Future Work

This paper initiates the problem of large-scale contextual market equilibrium computation from a deep learning perspective. We believe that our approach will pioneer a promising direction for deep learning-based market equilibrium computation.

For future works, it would be promising to extend these methods to the case when only the number of goods is large, or both the numbers of goods and buyers are large, which stays a blank throughout our works.[2] Since many existing works proposed dynamics for online market equilibrium computation, it's also promising to extend our approaches to the online setting. Besides, both existing works and ours consider sure budgets and values for buyers, and it would be interesting to extend the fisher market and equilibrium concept when the budgets or values of buyers are stochastic or uncertain.

Acknowledgments. This work is supported by Wuhan East Lake High-Tech Development Zone (also known as the Optics Valley of China, or OVC) National Comprehensive

---

[2] The most ideal method is that the method shall work on the setting when the goods, or both the buyers and goods, potentially obey some certain distribution.

Experimental Base for Governance of Intelligent Society, and the National Natural Science Foundation of China (NSFC) under grant number [62172012]. The authors would like to thank Ningyuan Li, Yurong Chen, Shicheng Li, and many anonymous referees for their suggestions and help with this work.

References

1. Amari, S.: Backpropagation and stochastic gradient descent method. Neurocomputing 5(4–5), 185–196 (1993)
2. Arrow, K.J., Debreu, G.: Existence of an equilibrium for a competitive economy. Econometrica J. Econometric Soc. 265–290 (1954)
3. Arrow, K.J., Chenery, H.B., Minhas, B.S., Solow, R.M.: Capital-labor substitution and economic efficiency. Rev. Econ. Stat. 225–250 (1961)
4. Balseiro, S., Kroer, C., Kumar, R.: Contextual standard auctions with budgets: revenue equivalence and efficiency guarantees. Manage. Sci. **69**(11), 6837–6854 (2023)
5. Banerjee, S., Gkatzelis, V., Gorokh, A., Jin, B.: Online nash social welfare maximization with predictions. In: Proceedings of the 2022 Annual ACM-SIAM Symposium on Discrete Algorithms (SODA), pp. 1–19. SIAM (2022)
6. Bengio, Y., Louradour, J., Collobert, R., Weston, J.: Curriculum learning. In: Proceedings of the 26th Annual International Conference on Machine Learning, pp. 41–48, (2009)
7. Bottou, L.: Large-scale machine learning with stochastic gradient descent. In: Proceedings of COMPSTAT'2010: 19th International Conference on Computational StatisticsParis France, August 22–27, 2010 Keynote, Invited and Contributed Papers, pp. 177–186. Springer (2010)
8. Budish, E.: The combinatorial assignment problem: approximate competitive equilibrium from equal incomes. J. Polit. Econ. **119**(6), 1061–1103 (2011)
9. Cheung, Y.K., Cole, R., Tao, Y.: Dynamics of distributed updating in fisher markets. In: Proceedings of the 2018 ACM Conference on Economics and Computation, pp. 351–368 (2018)
10. Cole, R., Fleischer, L.: Fast-converging tatonnement algorithms for one-time and ongoing market problems. In: Proceedings of the Fortieth Annual ACM Symposium on Theory of Computing, pp. 315–324 (2008)
11. Cole, R., et al.: Convex program duality, fisher markets, and nash social welfare. In: Proceedings of the 2017 ACM Conference on Economics and Computation, pp. 459–460 (2017)
12. Conitzer, V., Kroer, C., Sodomka, E., Stier-Moses, N.E.: Multiplicative pacing equilibria in auction markets. Oper. Res. **70**(2), 963–989 (2022)
13. Conitzer, V., et al.: Pacing equilibrium in first price auction markets. Manage. Sci. **68**(12), 8515–8535 (2022)
14. Curry, M., Sandholm, T., Dickerson, J.: Differentiable economics for randomized affine maximizer auctions. arXiv preprint arXiv:2202.02872 (2022)
15. Deng, X., Papadimitriou, C., Safra, S.: On the complexity of equilibria. In: Proceedings of the Thiry-fourth Annual ACM Symposium on Theory of Computing, pp. 67–71 (2002)
16. Deng, X., Papadimitriou, C., Safra, S.: On the complexity of price equilibria. J. Comput. Syst. Sci. **67**(2), 311–324 (2003)

17. Duan, Z., et al.: A context-integrated transformer-based neural network for auction design. In: International Conference on Machine Learning, pp. 5609–5626. PMLR (2022)
18. Dütting, P., Feng, Z., Narasimhan, H., Parkes, D.C., Ravindranath, S.S.: Optimal auctions through deep learning: advances in differentiable economics. J. ACM (JACM) (2023)
19. Eisenberg, E.: Aggregation of utility functions. Manage. Sci. **7**(4), 337–350 (1961)
20. Eisenberg, E., Gale, D.: Consensus of subjective probabilities: the pari-mutuel method. Ann. Math. Stat. **30**(1), 165–168 (1959)
21. Gao, Y., Kroer, C.: First-order methods for large-scale market equilibrium computation. In: Advances in Neural Information Processing Systems, vol. 33, pp. 21738–21750 (2020)
22. Gao, Y., Peysakhovich, A., Kroer, C.: Online market equilibrium with application to fair division. In: Advances in Neural Information Processing Systems, vol. 34, pp. 27305–27318 (2021)
23. Garg, J., Tao, Y., A Végh, L.: Approximating equilibrium under constrained piecewise linear concave utilities with applications to matching markets. In: Proceedings of the 2022 Annual ACM-SIAM Symposium on Discrete Algorithms (SODA), pp. 2269–2284. SIAM (2022)
24. Garg, J., Mehta, R., Vazirani, V.V., Yazdanbod, S.: Settling the complexity of leontief and plc exchange markets under exact and approximate equilibria. In: Proceedings of the 49th Annual ACM SIGACT Symposium on Theory of Computing, pp. 890–901 (2017)
25. Goldman, J., Procaccia, A.D.: Spliddit: unleashing fair division algorithms. ACM SIGecom Exchanges **13**(2), 41–46 (2015)
26. Huang, Z., Li, M., Shu, X., Wei, T.: Online nash welfare maximization without predictions. In: International Conference on Web and Internet Economics, pp. 402–419. Springer (2023)
27. Jalota, D., Pavone, M., Qi, Q., Ye, Y.: Fisher markets with linear constraints: equilibrium properties and efficient distributed algorithms. Games Econom. Behav. **141**, 223–260 (2023)
28. Kroer, C., Peysakhovich, A.: Scalable fair division for'at most one'preferences. arXiv preprint arXiv:1909.10925 (2019)
29. Kroer, C., Peysakhovich, A., Sodomka, E., Stier-Moses, N.E.: Computing large market equilibria using abstractions. In: Proceedings of the 2019 ACM Conference on Economics and Computation, pp. 745–746 (2019)
30. Liao, L., Gao, Y., Kroer, C.: Nonstationary dual averaging and online fair allocation. In: Advances in Neural Information Processing Systems, vol. 35, pp. 37159–37172 (2022)
31. Mas-Colell, A., Whinston, M.D., Green,J.R., et al.: Microeconomic Theory, volume 1. Oxford University Press New York (1995)
32. Nan, T., Gao, Y., Kroer, C.: Fast and interpretable dynamics for fisher markets via block-coordinate updates. arXiv preprint arXiv:2303.00506 (2023)
33. Nisan, N., Roughgarden, T., Tardos, E., Vazirani, V.V.: Algorithmic Game Theory, 2007. Book available for free online (2007)
34. Papadimitriou, C.: Algorithms, games, and the internet. In: Proceedings of the Thirty-third Annual ACM Symposium on Theory of Computing, pp. 749–753 (2001)
35. Shmyrev, V.I.: An algorithm for finding equilibrium in the linear exchange model with fixed budgets. J. Appl. Ind. Math. **3**, 505–518 (2009)

36. The Sveriges Riksbank Prize in Economic Sciences in Memory of Alfred Nobel 1972. Nobelprize.org. Nobel Prize Outreach AB 2024, Sun (2024). https://www.nobelprize.org/prizes/economic-sciences/1972/summary/
37. Varian, H.R., Varian, H.R.: Microeconomic Analysis, vol. 3. Norton New York (1992)
38. Vazirani, V.V., Yannakakis, M.: Market equilibrium under separable, piecewise-linear, concave utilities. J. ACM (JACM) **58**(3), 1–25 (2011)
39. Walras, L.: Elements of Pure Economics. Routledge (2013)
40. Wang, T., Dütting, P., Ivanov, D., Talgam-Cohen, I., Parkes, D.C.: Deep contract design via discontinuous piecewise affine neural networks. arXiv preprint arXiv:2307.02318 (2023)
41. Wu, F., Zhang, L.: Proportional response dynamics leads to market equilibrium. In: Proceedings of the Thirty-ninth Annual ACM Symposium on Theory of Computing, pp. 354–363 (2007)
42. Yang, Z., Liao, L., Kroer, C.: Greedy-based online fair allocation with adversarial input: enabling best-of-many-worlds guarantees. arXiv preprint arXiv:2308.09277 (2023)
43. Zhang, L.: Proportional response dynamics in the fisher market. Theoret. Comput. Sci. **412**(24), 2691–2698 (2011)

Fair Value Distribution in Cooperative Committee Election

Ying Qin[1], Zeyu Ren[1], Zihe Wang[1(✉)], and Jie Zhang[2(✉)]

[1] Renmin University of China, 59 Zhongguancun Street, Beijing, China
[2] University of Bath, Claverton Down, Bath, UK

Abstract. We consider a scenario where a group of agents needs to elect a committee to lead them in accomplishing a project. They elect a committee to maximize social welfare, and the question is how to distribute the total value of the project to every agent. This scenario encodes a cooperative game setting where the reward of the chosen coalition must be distributed fairly.

First, we establish the axiomatic foundation of solution concepts in this cooperative committee election game. We show that a natural extension of Shapley value to this game does not meet the classical axioms when the values of different coalitions are binary. We then propose a value distribution rule that satisfies all the desired properties. Furthermore, we prove that this rule is unique in meeting these properties and also satisfies an additional monotonicity property.

When the values of the coalitions can take any general values, we decompose the game into a linear combination of simple games. This decomposition is unique, allowing us to extend our value distribution rule to solve this general class of games.

Keywords: Game theory · Cooperative game · Shapley value · Value distribution

1 Introduction

Distributing the worth of a coalition fairly to the agents forming the coalition is an important problem in cooperative game theory. This issue becomes particularly challenging when the agents involved have varying contributions. Ensuring that each member receives a fair share of the rewards, reflective of their input and importance to overall success, is crucial for maintaining cooperation and motivation within the group.

In our research, we explore a scenario where a group of agents needs to elect a committee from among themselves to lead them in accomplishing a project. They aim to choose a committee that can guide them to great success, but they also care about the value they can gain from the project's successful completion, including those who are not elected to the committee. This problem encompasses a wide range of real-life situations, such as:

Corporate board elections, where shareholders must choose a board that can maximize company profits while ensuring fair dividend distribution.

Community project teams, where residents elect leaders to oversee neighborhood improvements, ensuring that the benefits are distributed impartially among all participants.

Political coalition formations, where parties must form alliances to achieve legislative goals while reasonably sharing power and influence.

By examining a baseline model that resembles these scenarios, our research seeks to develop methods for fair coalition reward distribution that adhere to standard and practical axioms in cooperative game theory.

1.1 Our Contribution

Our model differs from classical cooperative games in that we do not specify a value for every possible coalition formed by n players. Instead, our model only knows the worth of coalitions of size $m < n$. Additionally, we do not assume that the players will form the grand coalition. Instead, they will elect a committee of size m, and we seek solutions that fairly distribute the worth of the elected committee to all n players. Because of these differences, we define several axioms that are appropriate in this context. We start with simple games, in which the worth of every committee is binary. We design a simple-to-implement rule that meets all axioms and demonstrate the uniqueness of this rule for simple games. We then generalize the rule to accommodate general games where the worth of a committee is not necessarily binary.

1.2 Related Work

The two typical solution concepts in cooperative game theory are the core and Shapley value. The *core*, which dates back to the last century [36], is the set of feasible allocations or imputations where no coalition of agents can benefit by breaking away from the grand coalition. The *Shapley value*, introduced by Shapley in 1951 [31,32], distributes the total gains to the players, assuming they all collaborate. It is considered a fair distribution because it is the only distribution with certain desirable properties: symmetry, dummy player, efficiency, and additivity.

Since then, many works consider other axioms. Myerson [25] obtains an axiomatization by balanced contributions, meaning that j's contribution to i always equals i's contribution to j. Young [38] introduces monotonicity, which states that if a player's contribution to all coalitions weakly increases, then the player's allocation should not decrease. Kalai and Samet [21] study the weighted Shapley value and introduce the partnership axiom. Hart and Mas-Colell [20] characterize Shapley value using the potential function. Chun [12] uses coalitional strategic equivalence to replace additivity. Haller [19] investigates collusion properties, where one player becomes a proxy player and holds the power of both players. Radzik [29] discusses several new axioms, including "amalgamating payoffs," which Lehrer [23] uses for axiomatizing the Banzhaf value. Besner [5] introduces the player splitting property, which states that players' payoffs do not change if another player splits into two new players. Besner [6] considers the disjointly productive player and stated that the payoffs of other players should not change when such a player leaves the game. For further investigations, please refer to the survey by Driessen [13] and the book by Algaba et al. [2].

There also exist many studies that consider classic cooperative games with some constraints. Aumann and Dreze [3], Owen [28], Winter [37] assume a partition of players so that cooperation within groups is easier than that between groups. Myerson [24], Owen [27], Borm et al. [10] use undirected graphs to describe communication possibilities between players, making players' ability to operate dependent on their positions in graphs. Gilles et al. [17] provide a full characterization of dividends in cooperative games with permission structures. Faigle and Kern [15] consider situations where only coalitions of players that respect a given precedence structure on the set of players are feasible. They propose three axioms and demonstrate their unique "Shapley value". Van Den Brink [34], Van Den Brink and Gilles [33] examine the conjunctive and disjunctive settings, respectively, providing axiomatic characterizations of Shapley value of modified games. Bilbao and Edelman [8] study cooperative games on convex geometry and identify two classes of axioms that give rise to a unique Shapley value. These constraints also extend to union stable systems [1], matroids [7], and augmenting systems [9]. Nowak and Radzik [26] first introduce cooperative games in generalized characteristic function form, where the utility of a coalition depends not only on its members but also on the order of formation. Sanchez and Bergantiños [30] study the same game and propose different axiomatic characterizations using marginality, balanced contributions, the potential, and the consistency property. Bergantiños and Sánchez [4] examine the family of weighted Shapley values for games in generalized characteristic function form. Van Den Brink et al. [35] introduce the axiom of order monotonicity, requiring that players who enter earlier should not get more. Zou et al. [39] introduce a new class of cooperative games that combines precedence constraints and the generalized characteristic function. Ge et al. [16] study cooperative games where players join sequentially and propose the desideratum that players should have incentives to join as early as possible. They present a natural rule called "Rewarding First Critical Player".

Committee election is a classical and well-understood situation. Several papers consider metric distortion for analyzing, comparing, and designing voting rules [11,18]. Ebadian et al. [14] use fairness both as a tool to derive novel low-distortion randomized voting rules. Kalayci et al. [22] consider proportional representation problem, a novel definition which is a strengthening of proportional fairness and core fairness. Our work is different from those. We consider value distribution problem in committee election.

2 The Committee Election Game and Axiomatic Characterization of the Solution

We consider a coalitional game consisting of a set of players $N = \{1, \ldots, n\}$. The players need to select a committee $C_m = \{c_1, c_2, \ldots, c_m\}$ from among themselves to lead the remaining players in accomplishing a project. We use $\mathcal{C}_m$ to denote the set of all potential committees. Let $v : \mathcal{C}_m \to \mathbb{R}_{\geq 0}$ be the characteristic function of the game, which determines the worth of committee C_m. Thus, a cooperative committee election game is represented as (v, N, m).

Let $\psi_i(v)$ denote the value distributed to the player $i \in N$. A socially desirable committee C is one whose worth $v(C)$ is maximized among all $C \in \mathcal{C}_m$. Without loss of generality, we assume that this socially desirable committee C is elected, and we study the problem of fairly distributing the committee's worth, $v(C)$, to all players in N. To this end, we aim to find a fair value distribution rule that meets the following axioms.

Definition 1. *(Anonymity) For any player $i \in N$, any committee $C = \{c_1, c_2, \ldots, c_m\}$, and any permutation $\pi : N \to N$, denote the committee $C_\pi = \{c_{\pi(c_1)}, c_{\pi(c_2)}, \ldots, c_{\pi(c_m)}\}$. When we have $v(C) = v'(C_\pi)$ for all committees C, it holds that $\psi_i(v) = \psi_{\pi(i)}(v'), \forall i \in N$.*

The *anonymity* property ensures that the value distributed to the players does not depend on identities of the players, so that every player is treated equally. It's worth noting that anonymity is typically stronger than the *symmetric* property [31,32] in most cooperative games, which requires that any player contributing equally to a coalition receive the same payoff.

A player i is called a *dummy player* if for any committee $C_{m-1} \subseteq N$ of size $m-1$ and $i \notin C_{m-1}$, the worth $v(C_{m-1} \cup \{i\}) - v(C_{m-1}) = 0$. In other words, player i does not contribute to any committee that has a positive worth.

A fair value distribution rule should not allocate any payoff to dummy players, i.e., those players who do not contribute to any committee formation.

Definition 2. *(Dummy player) For any cooperative committee election game (v, N, m), if a player i is a dummy player, then $\psi_i(v) = 0$.*

We also desire the property that the value of the selected committee is fully distributed to all players.

Definition 3. *(Efficiency) For any cooperative committee election game (v, N, m), $\sum_{i \in N} \psi_i(v) = \max_{C \in \mathcal{C}_m} v(C)$.*

Finally, consider a scenario where a coalition of players wishes to pool their abilities and act as a single entity to extract more value. Such merging behavior was first considered by Lehrer [23] for axiomatizing the Banzhaf value, and then further investigated by Radzik [29]. The following property ensures that no two agents can gain additional benefits or change the benefits of others by merging their identities. This maintains fair value distribution and prevents manipulation by stopping agents from combining their identities for a better outcome to prevent collusion attack.

Definition 4. *(Merge-proofness)* Consider a committee election game (v, N, m) and two players i and j that do not appear simultaneously in C such that $v(C) > 0$. Besides, there does not exist a committee C_{m-1} of size $m - 1$ such that $v(C_{m-1} \cup \{i\}) > 0$ and $v(C_{m-1} \cup \{j\}) > 0$. When they merge their identities into one player $\overline{ij}$, such that in the new game $(v', N \cup \{\overline{ij}\} \setminus \{i,j\}, m)$, $v'(C_{m-1} \cup \{\overline{ij}\}) = \max\{v(C_{m-1} \cup \{i\}), v(C_{m-1} \cup \{j\})\}$ for any committee C_{m-1}, $C_{m-1} \cap \{i,j\} = \emptyset$, and $v'(C_m) = v(C_m)$ for any committee C_m, $C_m \cap \{i,j\} = \emptyset$. A value distribution rule is merge-proof if $\psi_{\overline{ij}}(v') = \psi_i(v) + \psi_j(v)$ and $\psi_t(v') = \psi_t(v)$ for any other player $t \in N \setminus \{i,j\}$.

3 Simple Games

A cooperative committee election game is called *simple* if for each committee $C \in \mathcal{C}_m$, either $v(C) = 0$ or $v(C) = 1$. Shapley value is a classical solution in cooperative game theory. However, for simple cooperative committee election games, an adaptation of Shapley value fails to meet the specified axioms illustrated in Example 1. We then propose a value distribution rule that fulfills all of these properties. Additionally, we prove that our proposed solution is the only value distribution rule that satisfies all the desired properties in these games.

Example 1. Consider a game where a committee of size 2 is selected from 4 players, i.e., $N = \{1, 2, 3, 4\}$. The characteristic function is given as $v(\{1,2\}) = 1, v(\{3,4\}) = 1$ and the value of other committees of size 2 is 0. For convenience, we denote the value of these committees as $v(12) = 1, v(34) = 1$. To deploy Shapley value to determine the value distributed to each player, we need to know the value of every possible coalition. For any committee whose size is smaller than 2, the project cannot be finished. Thus, we have $v(1) = v(2) = v(3) = v(4) = v(\emptyset) = 0$. For any coalition whose size is larger than 2, we generalize the value as the maximum value of all committees that are subsets of the coalition. For example, $v(123) = \max\{v(12), v(13), v(23)\} = 1$. Similarly, we have $v(1234) = v(124) = v(134) = v(234) = 1$. Using Shapley value, we can calculate the value for each player as $\psi_1(v) = \psi_2(v) = \psi_3(v) = \psi_4(v) = \frac{1}{4}$.

We note that this solution is not merge-proof. To illustrate this, assume that player 1 and player 3 merge into a single player $\overline{13}$. We check that $v(13) = 0$. And for player 2, we have $v(12) = 1$ and $v(23) = 0$. For player 4, we have $v(14) = 0$ and $v(34) = 1$. Thus, we can construct the new valuation function $\bar{v}$, we have $\bar{v}(\overline{13}) = \bar{v}(2) = \bar{v}(4) = 0$, because the committee size is 1. We have $\bar{v}(\overline{13}2) = \max\{v(12), v(23)\} = 1, \bar{v}(\overline{13}4) = \max\{v(14), v(34)\} = 1, \bar{v}(24) = v(24) = 0$ by the merge manipulation. Also we have $\bar{v}(\overline{13}24) = 1$. Consequently, we consider all 6 permutations and find that player $\overline{13}$ gets value when players come in order $24\overline{13}, 2\overline{13}4, 4\overline{13}2$ and $4\overline{13}2$. Thus, player $\overline{13}$ should be allocated $\frac{2}{3}$ by Shapley value. Since $\frac{2}{3} > \frac{1}{4} + \frac{1}{4}$, player 1 and player 3 have an incentive to merge to obtain more value. Therefore, Shapley value is not merge-proof.

Thus, we propose the Proportional rule. We use k to denote the number of committees that achieve a value of 1 in the valuation function v. Note that we only discuss the case $k > 0$ in the following part. For case $k = 0$, it is no doubt that $\psi_i(v) = 0$ for any player i. Additionally, we use k_i to denote the number of committees containing player i that achieve a value of 1.

Rule 1 *(Proportional rule).* *Given a simple game (v, N, m), the value allocated to player i is $\psi_i(v) = \frac{k_i}{m \cdot k}$.*

Next, we demonstrate that the Proportional rule satisfies all four of the previously defined axioms.

Theorem 1. *The Proportional rule satisfies anonymity, dummy player property, efficiency, and merge-proofness.*

Proof. The rule is anonymous because the value distributed to player i depends solely on the number of committees that include player i and achieve value 1.

The rule does not distribute any value to a dummy player i since $\psi_i(v) = 0$ if $k_i = 0$.

The rule is efficient, as $\sum_{i=1}^{n} \frac{k_i}{m \cdot k} = 1$.

Finally, recalling the merge condition for player i and j, we know that i and j do not appear simultaneously in C such that $v(C) > 0$. Besides, there does not exist a committee C_{m-1} of size $m - 1$ such that $v(C_{m-1} \cup \{i\}) > 0$ and $v(C_{m-1} \cup \{j\}) > 0$. Thus, when player i and player j are merged into a single player $\overline{ij}$, k remains unchanged and $k_{\overline{ij}} = k_i + k_j$. Hence, the rule is merge-proof.

We then demonstrate that the Proportional rule is the only rule that satisfies anonymity, dummy player property, efficiency, and merge-proofness. Given m and k, we prove this by induction on t, the number of players that are not dummy. The key lies in how to derive the allocated value of each player when there are $t - 1$ players that are not dummy by assuming the Proportional rule is suitable for t players. We can always find a simple game of t players and get the simple game of $t - 1$ players after merging two players. By merge-proofness, we derive the allocated value.

Theorem 2. *For any simple game, the Proportional rule is the unique rule that satisfies anonymity, dummy player property, efficiency, and merge-proofness.*

Proof. Given m and k, there are at most km players that are not dummy, that is, $v(\{c_1, \cdots, c_m\}) = v(\{c_{m+1}, \cdots, c_{2m}\}) = \cdots = v(\{c_{(k-1)m+1}, \cdots, c_{km}\}) = 1$. Because of efficiency, the sum of the distributed value is 1. Due to anonymity, the value should be allocated equally. Thus, the allocated value should be $\frac{1}{mk}$ for each player. We can derive the same value by the Proportional rule.

Then, we use induction on the number of players that are not dummy. Assuming that the distributed value of each player can be derived by the Proportional rule when there are t players that are not dummy, we need to prove that the value of each player can still be derived by the Proportional rule when there are $t - 1$ players which are not dummy. Given a simple game (v, N, m) in which $t - 1$ players are not dummy. There must be a player i with $k_i > 1$. We use $\mathcal{C}_i$ to denote the set of all the committees C_{m-1} such that $v(C_{m-1} \cup \{i\}) = 1$. We know that $|\mathcal{C}_i| = k_i > 1$. Suppose that $\hat{C} \in \mathcal{C}_i$. Thus, we construct a simple game $(v', N \cup \{i_1, i_2\} \setminus \{i\}, m)$ in which t players are not dummy. The construction is as follows: $v'(\hat{C} \cup \{i_1\}) = 1$, $v'(C \cup \{i_2\}) = 1$ for any $C \in \mathcal{C}_i \setminus \{\hat{C}\}$ and $v'(C) = v(C)$ for any C that $i, i_1, i_2 \notin C$. Despite these committees, we set $v'(C) = 0$ for any other committee C. Because the number of players that are not dummy is t, we can get the allocated value for each player in $(v', N \cup \{i_1, i_2\} \setminus \{i\}, m)$ by assumption. We have $\psi_{i_1}(v') = \frac{k_{i_1}}{mk}$ and $\psi_{i_2}(v') = \frac{k_{i_2}}{mk}$. Next, we check the condition of the merge-proofness if we merge the player i_1 and i_2. Through our construction, we know that player i_1 and i_2 do not appear simultaneously in C such that $v'(C) > 0$. Besides, for any committee C_{m-1} of size $m-1$, we know that $v'(C_{m-1} \cup \{i_1\}) \cdot v'(C_{m-1} \cup \{i_2\}) = 0$. Thus, we can merge the player i_1, i_2 and get the simple game (v, N, m). According to our construction, we know that $k_{i_1} + k_{i_2} = k_i$. Therefore, we have $\psi_i(v) = \psi_{i_1}(v') + \psi_{i_2}(v') = \frac{k_i}{mk}$, and $\psi_{i'}(v) = \psi_{i'}(v') = \frac{k_{i'}}{mk}$ for any player $i' \neq i$ by merge-proofness. Consequently, we prove that the Proportional rule can derive the distributed value when there are $t-1$ players that are not dummy.

By applying induction, we know that given any m, k, and for any number of players that are not dummy, if there exists a rule satisfying anonymity, dummy player property, efficiency, and merge-proofness, then we can get the distributed value by the Proportional rule. That is, the Proportional rule is unique. We finish the proof.

Besides these desired properties, it is also important that as the worth of a committee C increases, the value distributed to players in C also increases.

Definition 5. *(Monotonicity). For any two games (v_1, N, m) and (v_2, N, m), if there exists a committee C such that $v_1(C) \geq v_2(C)$, and for any $C' \neq C$, $v_1(C') = v_2(C')$, then we have $\psi_i(v_1) \geq \psi_i(v_2)$ for any $i \in C$.*

We show that the Proportional rule also satisfies this property.

Theorem 3. *The Proportional rule is monotone.*

Proof. We consider the variation of k between two different simple games (v_1, N, m) and (v_2, N, m). If there exists a committee C such that $v_1(C) > v_2(C)$, and for any $C' \neq C$, $v_1(C') = v_2(C')$, then $k_{v_1} = k_{v_2} + 1$ where k_{v_1} and k_{v_2} denote the number of committees that achieve a value of 1 in v_1 and v_2 respectively. And for any player $i \in C$, we have $k_{v_1, i} = k_{v_2, i} + 1$. Under the Proportional rule, we consider the difference between $\frac{k_{v_1, i}}{mk_{v_1}}$ and $\frac{k_{v_2, i}}{mk_{v_2}}$. It is

$$\frac{1}{m} \cdot \frac{(k_{v_2, i} + 1)k_{v_2} - k_{v_2, i}(k_{v_2} + 1)}{k_{v_2}(k_{v_2} + 1)} = \frac{k_{v_2} - k_{v_2, i}}{mk_{v_2}(k_{v_2} + 1)}.$$

Because $k_{v_2} \geq k_{v_2, i}$, we have $\frac{k_{v_1, i}}{mk_{v_1}} - \frac{k_{v_2, i}}{mk_{v_2}} \geq 0$. Therefore, the Proportional rule satisfies monotonicity.

4 General Games

In this section, we extend our results from simple games to general games where the worth of a committee can take any value. We desire a property in order to decompose a general game into simple games, enabling us to establish our results based on previous findings. First, we demonstrate that additivity considered in Shapley value [32] cannot apply to our problem directly.

Example 2. Consider 3 games where a committee of size 2 is selected from 3 players. The characteristic functions are given as $v_1(12) = 2, v_1(13) = 1, v_1(23) = 1, v_2(12) = 3, v_2(13) = 0, v_2(23) = 0$ and $v_3(12) = 1, v_3(13) = 2, v_3(23) = 2$. Note that the mechanism should extract the maximum value, thus $\sum_{i \in N} \psi_i(v_1) = \max_{C \in \mathcal{C}_m} v_1(C) = 2$. Similarly, we have $\sum_{i \in N} \psi_i(v_2) = 3$ and $\sum_{i \in N} \psi_i(v_3) = 2$. We know that $v_1 + v_1 = v_2 + v_3$. However, $\sum_{i \in N} \psi_i(v_1) + \sum_{i \in N} \psi_i(v_1) \neq \sum_{i \in N} \psi_i(v_2) + \sum_{i \in N} \psi_i(v_3)$. Therefore, additivity is not satisfied in our problem. We then introduce order-preserving linearity.

Definition 6. *(Order-preserving linearity). Given two general games (v, N, m) and (u, N, m) that only differ in their value functions v and u, if for any two committees C_1 and C_2, the condition $(v(C_1) - v(C_2))(u(C_1) - u(C_2)) \geq 0$ holds, then in the combined game $(l_1 v + l_2 u, N, C)$ for $l_1, l_2 \in \mathbb{R}_+$, each player's value is given by $\psi_i(l_1 v + l_2 u) = l_1 \psi_i(u) + l_2 \psi_i(v)$ for any player $i \in N$.*

By order-preserving linearity, we can decompose any general valuation function v into a linear combination of binary valuation functions v_j, such that $v = \sum_j l_j \cdot v_j$, where $l_j \in \mathbb{R}_+$. We propose an algorithm similar to [16]. The algorithm constructs this decomposition in iterations. In each iteration, we first construct a binary valuation function that maintains the same order as v. Next, we determine the coefficient l_j. We then subtract $l_j \cdot v_j$ from v and proceed to the next iteration. This process continues until the decomposition is complete.

Observation 4. *Given a general valuation function v, the positive linear binary valuation function decomposition is unique.*

Now, in conjunction with the Proportional rule introduced for simple games, we can utilize the decomposition algorithm to break down a general game into a series of simple games. This allows us to derive a value distribution function for the general game. We illustrate this process with the following example.

Example 3. Given a general game (v, N, C), in which $N = \{1, 2, 3\}$ and $m = 2$. The general valuation function is $v(12) = 10, v(13) = 8$ and $v(23) = 5$. By applying the decomposition algorithm, we obtain the following three simple games, such that $v = 5v_1 + 3v_2 + 2v_3$.

1. $v_1(12) = 1, v_1(13) = 1$ and $v_1(23) = 1$.
2. $v_2(12) = 1, v_2(13) = 1$ and $v_2(23) = 0$.
3. $v_3(12) = 1, v_3(13) = 0$ and $v_3(23) = 0$.

Algorithm 1: Decomposition Algorithm

Input: A general valuation function v.
Output: The positive linear combination $\sum_j l_j \cdot v_j$.

Let $\tilde{v}$ be the copy of v;
Sort the values in $\tilde{v}$ in descending order;
$j \leftarrow 1$;
while $\max(\tilde{v}) > 0$ **do**
    Let k be the number of positive values in $\tilde{v}$;
    Construct binary valuation function v_j that the first k values are 1 and the others are 0;
    Let l_j be the minimum that makes the number of positive values in $\tilde{v} - l_j \cdot v_j$ different from k;
    $\tilde{v} \leftarrow \tilde{v} - l_j \cdot v_j$;
    $j \leftarrow j + 1$;
return $\sum_j l_j \cdot v_j$.

By employing the Proportional rule on these simple games, we obtain their value distribution functions.

For v_1, the allocated value should be $\psi_1(v_1) = \frac{1}{3}, \psi_2(v_1) = \frac{1}{3}$ and $\psi_3(v_1) = \frac{1}{3}$.
For v_2, the allocated value should be $\psi_1(v_2) = \frac{1}{2}, \psi_2(v_2) = \frac{1}{4}$ and $\psi_3(v_2) = \frac{1}{4}$.
For v_3, the allocated value should be $\psi_1(v_3) = \frac{1}{2}, \psi_2(v_3) = \frac{1}{2}$ and $\psi_3(v_3) = 0$.
Finally, by applying order-preserving linearity, we derive the value distribution function for the original general game. That is, $\psi_1(v) = \frac{5}{3} + \frac{3}{2} + 1 = \frac{25}{6}, \psi_2(v) = \frac{5}{3} + \frac{3}{4} + 1 = \frac{41}{12}$ and $\psi_3(v) = \frac{5}{3} + \frac{3}{4} = \frac{29}{12}$.

Then, we extend the Proportional rule to general games. We use k_j to denote the number of committees that achieve value 1 in the valuation function v_j. We also use k_{ji} to denote the number of committees containing player i that gets value 1 in the valuation function v_j. Combining the decomposition $v = \sum_j l_j \cdot v_j$, we get the following rule.

Rule 2 *(Generalized Proportional rule)*. *Given a general game whose characteristic function v can take any value, the allocated value to player i is given by* $\psi_i(v) = \sum_j l_j \cdot \frac{k_{ji}}{mk_j}$.

Given that the decomposition of general games into simple games is unique, and the Proportional rule is unique for simple games meeting all desired properties, we arrive at the following result.

Theorem 5. *The Generalized Proportional rule is the only rule that satisfies anonymity, dummy player, efficiency, merge-proofness and order-preserving linearity for general games.*

Proof. Generalized Proportional rule is anonymous because no player's identity information is used.

The rule does not distribute any value to a dummy player i since $\psi_i(v) = 0$ if $k_{ji} = 0$ for any simple game v_j.

As the summation of the distributed value, we have

$$\sum_{i=1}^n \psi_i(v) = \sum_{i=1}^n \sum_j l_j \cdot \frac{k_{ji}}{mk_j} = \sum_j l_j \cdot \frac{\sum_{i=1}^n k_{ji}}{mk_j} = \sum_j l_j = \max_{C \in \mathcal{C}} v(C).$$

Thus, the rule is efficient.

Next, considering the merge condition for players i_1 and i_2 in a general game (v, N, m), we know that i_1 and i_2 do not appear simultaneously in C such that $v(C) > 0$. Besides, there does not exist a committee C_{m-1} of size $m-1$ such that $v(C_{m-1} \cup \{i_1\}) > 0$ and $v(C_{m-1} \cup \{i_2\}) > 0$. Suppose the decomposition is $v = \sum_j l_j \cdot v_j$. For each binary function v_j, it also holds that i_1 and i_2 do not appear simultaneously in C such that $v_j(C) > 0$ and there does not exist a committee C_{m-1} of size $m-1$ such that $v_j(C_{m-1} \cup \{i_1\}) > 0$ and $v_j(C_{m-1} \cup \{i_2\}) > 0$. That is, the merge condition is also satisfied for each v_j. For each binary function v_j, when i_1, i_2 are merged into a single player, k_j remains unchanged and $k_{i_1, i_2} = k_{i_1} + k_{i_2}$. That is, the rule is merge-proof for every v_j. The new game $(v', N \cup \{\overline{i_1 i_2}\} \setminus \{i_1, i_2\}, m)$ can be obtained by the sum of the merged v_j. Hence, the rule satisfies merge-proofness.

If two general games (v, N, m) and (u, N, m) are order-preserving, then u and v can be represented by the binary functions v_j. We assume that $u = \sum_j l_{j,u} \cdot v_j$ and $v = \sum_j l_{j,v} \cdot v_j$. For any $\alpha_1, \alpha_2 \in \mathbb{R}_+$, we have $\alpha_1 u + \alpha_2 v = \sum_j (\alpha_1 l_{j,u} + \alpha_2 l_{j,v}) v_j$. Thus, for any player i, we have $\alpha_1 \psi_i(u) + \alpha_2 \psi_i(v) = \sum_j \alpha_1 l_{j,u} \cdot \frac{k_{ji}}{mk_j} + \sum_j \alpha_2 l_{j,v} \cdot \frac{k_{ji}}{mk_j} = \psi_i(\alpha_1 u + \alpha_2 v)$. Therefore, the rule is order-preserving linearity.

Consequently, the Generalized Proportional rule satisfies all axioms. We know that the decomposition of general games into simple games is unique, and Proportional rule is unique for simple games. Lastly, the Generalized Proportional rule is unique in meeting all desired properties.

Finally, we demonstrate that Generalized Proportional rule also satisfies monotonicity.

Theorem 6. *The Generalized Proportional rule is monotone.*

Proof. For two different general games (v, N, m) and (v', N, m), suppose that there exists a committee C such that $v(C) > v'(C)$, and for any $C' \neq C$, $v(C') = v'(C')$. We sort the values of v and v' in ascending order. Consider the different values between $v'(C)$ and $v(C)$ and set the values to $u_1 < \cdots < u_{t-1} \in (v'(C), v(C))$. We also set $u_0 := v'(C)$ and $u_t := v(C)$. Then, we construct a series of general games $(v_0, N, m), \cdots, (v_t, N, m)$. For any $v_s (0 \leq s \leq t)$, let $v_s(C) = u_s$, and for any $C' \neq C$, $v_s(C') = v(C')$. Thus, v_s and $v_{s+1} (0 \leq \mu \leq t-1)$ are order-preserving. We consider the value difference that players in C receive between v_{s+1} and v_s. If $v_{s+1}(C)$ is the maximum among the value of all the potential committees, assuming $\hat{v}$ is the binary function that only $\hat{v}(C) = 1$, we have $v_{s+1} = v_s + (u_t - u_{t-1}) \cdot \hat{v}$. Thus, we have $\psi_i(v_{s+1}) = \psi_i(v_s) + \frac{u_t - u_{t-1}}{m}$ for any player $i \in C$. If $v_{s+1}(C)$ is not the maximum, we construct two order-preserving binary functions $\hat{v}$ and $\tilde{v}$. For $\hat{v}$, we have $\hat{v}(C') = 0$ if $v_{s+1}(C') < u_{s+1}$ and $\hat{v}(C') = 1$ if $v_{s+1}(C') \geq u_{s+1}$. For $\tilde{v}$, we have $\tilde{v}(C') = 0$ if $v_s(C') < u_s$ and $\hat{v}(C') = 1$ if $v_s(C') \geq u_s$. We know that v_{s+1} and v_s are order-preserving. And the only difference between v_{s+1} and v_s is the worth of committee C, which are u_{s+1} and u_s respectively. Thus, we can use a general value function v^* to represent v_{s+1} and v_s combined with $\hat{v}$ and $\tilde{v}$. That is, $v_{s+1} = v^* + (u_{s+1} - u_s)\hat{v}$ and $v_s = v^* + (u_{s+1} - u_s)\tilde{v}$. Note that $\tilde{v}(C) = 0, \hat{v}(C) = 1$ and $\tilde{v}(C') = \hat{v}(C')$ for any other $C' \neq C$. We compare the distributed value of player $i \in C$ in $\tilde{v}$ and $\hat{v}$. Because Proportional rule is monotone, we have $\psi_i(\hat{v}) \geq \psi_i(\tilde{v})$ for any player $i \in C$. Consequently, we have $\psi_i(v_t) \geq \psi_i(v_{t-1}) \geq \cdots \geq \psi_i(v_0)$, that is $\psi_i(v) \geq \psi_i(v')$. Therefore, the Generalized Proportional rule satisfies monotonicity.

5 Discussion

We considered a maximum welfare committee election problem. For players to elect the maximum welfare committee, the value distributed to each player needs to be fair to maintain the stability of their decision. While dummy player property and efficiency are standard axioms in cooperative game theory, we require the solution to be anonymous, which is stronger than symmetry. Additionally, our approach incorporates monotonicity and merge-proofness. We showed that Shapley value is not merge-proof, and thus we proposed the Proportional rule, which meets all the above axioms. Furthermore, this is the only rule that satisfies these properties. We also generalized this series of results from simple games to general games.

Suppose that the value of a committee is not determined by the characteristic function, but rather reported by committee members. We assume that each player in a committee C can influence the entire report. By monotonicity, we know that Generalized Proportional rule can prevent under report. However, Generalized Proportional rule is not truthful. Players can obtain higher value through misreporting.

References

1. Algaba, E., Bilbao, J.M., Borm, P., López, J.: The position value for union stable systems. Math. Methods Oper. Res. **52**, 221–236 (2000)
2. Algaba, E., Fragnelli, V., Sánchez-Soriano, J.: Handbook of the Shapley value. CRC Press (2019)
3. Aumann, R.J., Dreze, J.H.: Cooperative games with coalition structures. Internat. J. Game Theory **3**, 217–237 (1974)
4. Bergantiños, G., Sánchez, E.: Weighted shapley values for games in generalized characteristic function form. TOP **9**, 55–67 (2001)
5. Besner, M.: Axiomatizations of the proportional shapley value. Theor. Decis. **86**(2), 161–183 (2019)

6. Besner, M.: Disjointly productive players and the shapley value. Games Econom. Behav. **133**, 109–114 (2022)
7. Bilbao, J.M., Driessen, T.S., Jiménez Losada, A., Lebrón, E.: The shapley value for games on matroids: The static model. Math. Methods Oper. Res. **53**, 333–348 (2001)
8. Bilbao, J.M., Edelman, P.H.: The shapley value on convex geometries. Discret. Appl. Math. **103**(1–3), 33–40 (2000)
9. Bilbao, J.M., Ordóñez, M.: Axiomatizations of the shapley value for games on augmenting systems. Eur. J. Oper. Res. **196**(3), 1008–1014 (2009)
10. Borm, P., Owen, G., Tijs, S.: On the position value for communication situations. SIAM J. Discret. Math. **5**(3), 305–320 (1992)
11. Caragiannis, I., Shah, N., Voudouris, A.A.: The metric distortion of multiwinner voting. Artif. Intell. **313**, 103802 (2022)
12. Chun, Y.: On the symmetric and weighted shapley values. Internat. J. Game Theory **20**, 183–190 (1991)
13. Driessen, T.S.: A survey of consistency properties in cooperative game theory. SIAM Rev. **33**(1), 43–59 (1991)
14. Ebadian, S., Kahng, A., Peters, D., Shah, N.: Optimized distortion and proportional fairness in voting. In: Proceedings of the 23rd ACM Conference on Economics and Computation, pp. 563–600 (2022)
15. Faigle, U., Kern, W.: The shapley value for cooperative games under precedence constraints. Internat. J. Game Theory **21**, 249–266 (1992)
16. Ge, Y., Zhang, Y., Zhao, D., Tang, Z.G., Fu, H., Lu, P.: Incentives for early arrival in cooperative games. In: Proceedings of the 23rd International Conference on Autonomous Agents and Multiagent Systems, pp. 651–659 (2024)
17. Gilles, R.P., Owen, G., Van Den Brink, R.: Games with permission structures: the conjunctive approach. Internat. J. Game Theory **20**(3), 277–293 (1992)
18. Goel, A., Hulett, R., Krishnaswamy, A.K.: Relating metric distortion and fairness of social choice rules. In: Proceedings of the 13th Workshop on Economics of Networks, Systems and Computation, pp. 1–1 (2018)
19. Haller, H.: Collusion properties of values. Internat. J. Game Theory **23**, 261–281 (1994)
20. Hart, S., Mas-Colell, A.: Potential, value, and consistency. Econometrica J. Econ. Soc., 589–614 (1989)
21. Kalai, E., Samet, D.: On weighted shapley values. Internat. J. Game Theory **16**(3), 205–222 (1987)
22. Kalayci, Y., Kempe, D., Kher, V.: Proportional representation in metric spaces and low-distortion committee selection. In: Proceedings of the AAAI Conference on Artificial Intelligence, vol. 38, pp. 9815–9823 (2024)
23. Lehrer, E.: An axiomatization of the Banzhaf value. Internat. J. Game Theory **17**, 89–99 (1988)
24. Myerson, R.B.: Graphs and cooperation in games. Math. Oper. Res. **2**(3), 225–229 (1977)
25. Myerson, R.B.: Conference structures and fair allocation rules. Internat. J. Game Theory **9**, 169–182 (1980)
26. Nowak, A.S., Radzik, T.: The shapley value for n-person games in generalized characteristic function form. Games Econom. Behav. **6**(1), 150–161 (1994)
27. Owen, G.: Values of graph-restricted games. SIAM J. Algebraic Discrete Meth. **7**(2), 210–220 (1986)
28. Owen, G.: Values of games with a priori unions. In: Mathematical economics and game theory: Essays in honor of Oskar Morgenstern, pp. 76–88. Springer (1977)

29. Radzik, T.: A new look at the role of players' weights in the weighted shapley value. Eur. J. Oper. Res. **223**(2), 407–416 (2012)
30. Sanchez, E., Bergantiños, G.: On values for generalized characteristic functions. Oper. Res. Spektrum **19**, 229–234 (1997)
31. Shapley, L.S.: Notes on the N-Person Game—ii: The Value of an N-person Game (1951)
32. Shapley, L.S., et al.: A Value for N-person Games (1953)
33. Van Den Brink, R.: An axiomatization of the disjunctive permission value for games with a permission structure. Internat. J. Game Theory **26**(1), 27–43 (1997)
34. Van Den Brink, R., Gilles, R.P.: Axiomatizations of the conjunctive permission value for games with permission structures. Games Econom. Behav. **12**(1), 113–126 (1996)
35. Van Den Brink, R., González-Arangüena, E., Manuel, C., del Pozo, M.: Order monotonic solutions for generalized characteristic functions. Eur. J. Oper. Res. **238**(3), 786–796 (2014)
36. Von Neumann, J., Morgenstern, O.: Theory of Games and Economic Behavior, 2nd Rev (1947)
37. Winter, E.: A value for cooperative games with levels structure of cooperation. Internat. J. Game Theory **18**, 227–240 (1989)
38. Young, H.P.: Monotonic solutions of cooperative games. Internat. J. Game Theory **14**(2), 65–72 (1985)
39. Zou, Z., Zhang, Q., Borkotokey, S., Yu, X.: The extended shapley value for generalized cooperative games under precedence constraints. Oper. Res. Int. J. **20**, 899–925 (2020)

A Payoff-Based Policy Gradient Method in Stochastic Games with Long-Run Average Payoffs

Junyue Zhang[1,2] and Yifen Mu[1(✉)]

[1] State Key Laboratory of Mathematical Sciences, Academy of Mathematics and Systems Science, Chinese Academy of Sciences, Beijing 100190, China
{zhangjunyue,mu}@amss.ac.cn
[2] School of Mathematical Sciences, University of Chinese Academy of Sciences, Beijing 100049, China

Abstract. Despite the significant potential for various applications, stochastic games with long-run average payoffs have received limited scholarly attention, particularly concerning the development of learning algorithms for them due to the challenges of mathematical analysis. In this paper, we study the stochastic games with long-run average payoffs and present an equivalent formulation for individual payoff gradients by defining advantage functions which will be proved to be bounded. This discovery allows us to demonstrate that the individual payoff gradient function is Lipschitz continuous with respect to the policy profile. Leveraging these insights, we devise a payoff-based gradient estimation approach and integrate it with the Regularized Robbins-Monro method from stochastic approximation theory to construct a bandit learning algorithm suited for stochastic games with long-run average payoffs. Additionally, we prove that if all players adopt our algorithm, the policy profile employed will asymptotically converge to a Nash equilibrium with probability one, provided that all Nash equilibria are globally neutrally stable and a globally variationally stable Nash equilibrium exists. This condition represents a wide class of games, including monotone games.

Keywords: Dynamic games · Multi-agent systems · Learning theory · Optimization under uncertainties

1 Introduction

Ever since they were proposed by Shapley [26] in the 1950's, stochastic games have been extensively studied with a large amount of applications in fields such as multi-agent reinforcement learning [33], robotics [15], autonomous driving [4]. Unlike static games, in stochastic game settings, the games will be played in many rounds. Before choosing their actions, players know the current state which determines the rules of the game for that stage in advance. After the players act in this round, they will receive their own instantaneous rewards. At the same time, the state of the game will move to the next state based on the transition probabilities induced by the current state and the players' actions. Therefore, in contrast to matrix games, in stochastic games, each player has to balance two objectives: maximizing the one-step gains or optimizing the long-run payoffs.

In general, there are four classic long-run payoff models in stochastic games [27]: total payoffs in finite horizon, total payoffs in random stopping time frameworks, discounted payoffs in infinite horizon, and average payoffs in infinite horizon. Regardless of the specific long-run payoff function utilized, these functions are significantly influenced by the transition probabilities that are a direct consequence of the players' strategic choices. As a result, they often exhibit a considerable degree of nonlinearity, even when the action space available to players in each state is finite. Therefore, some well-known algorithms such as fictitious play [3] and no-regret learning [17], which are widely used in matrix games, cannot be simply applied to learning in stochastic games.

However, as demonstrated in [10], calculating a Nash equilibrium for stochastic games with discounted payoffs is classified as PPAD-Complete, a complexity equivalent to that of matrix games [7]. This parallel suggests that it may be possible to adapt learning algorithms for finding Nash equilibria in some special stochastic games, or correlated equilibria in all such games, analogous to methods employed in matrix games. The literature on this topic is extensive; however, the majority of studies have concentrated on finite horizon payoffs [19,28] or discounted payoffs in infinite horizons [18,34], with less emphasis on random stopping time payoffs [14] and average payoffs over infinite horizons [11]. Nevertheless, as highlighted in [30], the first two models possess inherent limitations. In certain scenarios, the model of average payoffs in infinite horizon may more accurately reflect real-world conditions, particularly when players are required to take actions at a high frequency within a short time frame.

In our study, we will focus on stochastic games with long-run average payoffs, and the main contributions of our work are as follows:

1. We extend the concept of advantage functions from reinforcement learning [31] to stochastic games with long-run average payoffs, and prove that it is bounded and well-defined, thereby laying a solid foundation for further analysis.
2. We prove that the individual payoff gradients in stochastic games with long-run average payoffs are Lipschitz continuous, as researchers have done for the other three types of stochastic games [9,13,16] [34].
3. Capitalizing on our observations, we develop a payoff-based gradient estimation approach inspired by simultaneous perturbation stochastic approximation method [5] and integrate it with the Robbins-Monro method [23] and the mirror descent algorithm [2] to construct a bandit learning algorithm suited for stochastic games with long-run average payoffs. Our algorithm is distributed, relatively simple, and can be applied to lots of games. What the players only need is instantaneous rewards they receive in games.
4. We prove that using our algorithm, the learning process will converge to a Nash equilibrium with probability 1 if all Nash equilibria are globally neutrally stable and a globally variationally stable Nash equilibrium exists.

The paper is organized as follows. In Sect. 2, we provide an overview of the fundamental concepts associated with stochastic games. In Sect. 3, we analyse the properties of the value function and demonstrate that the individual payoff gradients are Lipschtiz continuous. In Sect. 4, we introduce a payoff-based methods for estimating the individual payoff gradients, and present a mirror descent algorithm for learning the Nash equilibrium. Our algorithm is derived based on these methodologies. In Sect. 5, we introduce the concept of stability in stochastic games including neutrally stable and variationally stable, and show the convergence of the progress induced by our algorithm. Discussions are given in Sect. 6. The details of the proofs in this paper can be found in the ArXiv version [32].

2 Problem Setup and Preliminaries

Throughout this paper, we focus on stochastic games involving a finite set of players, denoted by $N = \{1, 2, \ldots, n\}$, and a finite set of states S. Each player $i \in N$ possesses a finite set of actions A_i available in every state $s \in S$. The Cartesian product $A = \prod_{i \in N} A_i$ represents the set of all possible joint actions, while $A_{-i} = \prod_{j \neq i} A_j$ signifies the set of all joint actions except player i. Upon reaching a state s, player i chooses an action $a_i \in A_i$ and receives an immediate reward $r_i(a_i, a_{-i})$. Subsequently, the game transitions to a new state $s' \in S$ with probability $\mathbb{P}[s'|s, a]$, where a represents the joint action $(a_1, \ldots, a_n)$. Formally, we use a tuple $\mathcal{G} = (S, N, (A_i)_{i \in N}, \mathbb{P}, (r_i)_{i \in N})$ to denote such a stochastic game.

The game will be played repeatedly as follows. At any discrete time $t = 0, 1, 2, \ldots$, all players observe their current state s^t and choose her action a_i^t from A_i. Then they receive their immediate reward $r_i(s^t, a^t)$, denoting the joint action as $a^t = (a_i^t)_{i \in N}$. After that, the state of the game will move to s^{t+1} according to the transition probability $\mathbb{P}(s^{t+1}|s^t, a^t)$, and players choose their actions in the next state s^{t+1}.

At each discrete time step t, for any player i, we define the history $\mathcal{H}_i^t$ as the collection of all information available to player i, encompassing their realized states, actions, and rewards. This history is formally represented as the set

$$\mathcal{H}_i^t = \{s^l, a_i^l, r_i(s^l, a^l) : l = 0, 1, \ldots, t-1\} \cup \{s^t\}$$

Players will determine their actions at time t based on $\mathcal{H}_i^t$. A general policy for player i is characterized by a mapping $\pi_i : \mathcal{H} \to \Delta(A_i)$, where $\mathcal{H}$ is the set of all histories and $\Delta(A_i)$ represents the set of all probability distributions over the set A_i. Upon observing the history $\mathcal{H}_i^t$, player i will execute the mixed strategy $\pi_i(\mathcal{H}_i^t)$. However, employing a policy contingent on the history can be intricate. In practice, there is a preference for more straightforward policies, particularly stationary policies, which are the focus of this paper and are defined subsequently.

Definition 1 (Stationary policy). *A policy π_i for player i is stationary if it is solely dependent on the current state, i.e., for any history $\mathcal{H}_i^t$, we have $\pi_i(\mathcal{H}_i^t) =$*

$\pi_i(s^t)$. And we will use $\pi_i(a_i|s)$ to denote the probability of player i taking action a_i in state s. Concurrently, we use Π_i to denote the set of all stationary policies of player i. Furthermore, the set of all stationary policy profiles is denoted by $\Pi = \prod_{i \in N} \Pi_i$.

Here we note that the stationary policy set Π_i can be represented as a subset of $\mathbb{R}^{|S| \times |A_i|}$. More precisely, we have

$$\Pi_i = \{(\pi_i(a_i|s))_{(a_i,s) \in A_i \times S} : \sum_{a_i \in A_i} \pi_i(a_i|s) = 1, \forall s \in S, \pi_i(a_i|s) \geqslant 0, \forall (a_i, s) \in A_i \times S\}.$$

Given a initial state s^0 and a stationary policy profile π, a Markov chain on the set of states can be naturally induced. The (s, s')-element of the transition matrix P_π is

$$P_\pi(s'|s) = \sum_{a \in A} \mathbb{P}[s'|s, a] (\prod_{i=1}^{n} \pi_i(a_i|s)). \tag{1}$$

Thus, assuming that the Markov chain P_π is ergodic, if we let p_π be the unique stationary distribution of P_π, we have $\lim_{t \to \infty} \mathbb{P}(s^t = s|s^0 = s_0, \pi) = p_\pi(s)$. And for the convenience of our analysis, we will use the following standard assumption used in the MDP literature [12].

Remark 1. In this paper we assume that for any stationary policy profile π chosen by the players, the induced Markov chain P_π is ergodic, and its mixing time is uniformly bounded above by a constant $\tau > 0$, that is

$$\|(w - w')P_\pi\|_1 \leqslant e^{-\frac{1}{\tau}} \|w - w'\|_1, \quad \forall i \in N, \pi \in \Pi, w, w' \in \Delta(S).$$

This is an assumption that holds in a wide array of scenarios. For instance, if there exists a constant $\varepsilon > 0$ such that for all state s, s' and for all joint action a, the transition probability satisfies $\mathbb{P}[s'|s, a] > \varepsilon$. Under these conditions, for any policy profile π, the induced transition matrix P_π ensures that $P_\pi(s'|s) > \varepsilon$. It can be readily demonstrated that the assumption in Remark 1 is valid in this context.

The objective for each player i in stochastic games with long-run average payoffs is to choose a stationary policy π_i to maximize her expected long-run average payoffs, which is given by

$$V_i(\pi_i, \pi_{-i}) = \mathbb{E}\left[\lim_{T \to \infty} \frac{1}{T} \sum_{t=0}^{T} r_i(s^t, a^t)\right]. \tag{2}$$

With the assumption in Remark 1, we have

$$V_i(\pi_i, \pi_{-i}) = \sum_{s \in S} p_{\pi_i, \pi_{-i}}(s) \sum_{a} (\prod_{i=1}^{n} \pi_i(a_i|s)) r_i(s, a_i, a_{-i}),$$

which shows that $V_i(\pi_i, \pi_{-i})$ is well-defined, i.e., convergent if all players take stationary policies.

Now we define Nash equilibrium in stochastic games with long-run average payoff as follows.

Definition 2 (Nash equilibrium). *A policy profile $\pi^* = (\pi_i^*)_{i \in N}$ is a Nash equilibrium for the stochastic game $\mathcal{G}$ with long-run average payoffs if*

$$V_i(\pi_i^*, \pi_{-i}^*) \geq V_i(\pi_i, \pi_{-i}^*), \quad \forall i \in N, \forall \pi_i \in \Pi_i.$$

3 Properties of Individual Payoff Functions

In this section, we analyze the properties of the long-run average payoff function $V_i(\pi_i, \pi_{-i})$ and its associated individual payoff gradient $\nabla_i V_i(\pi_i, \pi_{-i})$. For notational simplicity, the individual payoff gradient may occasionally be represented by $v_i(\pi_i, \pi_{-i})$, and the vector of all players' individual payoff gradients by $v(\pi) = (v_i(\pi_i, \pi_{-i}))_{i \in N}$.

Inspired by [31], the advantage functions of a state-action pair (s, a) for player i given a policy π are defined as:

$$\mathrm{adv}_i^\pi(s, a) = \sum_{t=0}^{\infty} \mathbb{E}\big[r_i(s^t, a^t) - V_i(\pi_i, \pi_{-i}) | s_0 = s, a_0 = a, \pi\big], \quad (3)$$

which represents the sum of the differences between the rewards received by player i and the value function $V_i(\pi_i, \pi_{-i})$, given that the game begins at state s with action a and subsequently follows the stationary policy profile π.

We also define the average advantage functions as:

$$\overline{\mathrm{adv}}_i^\pi(s, a_i) := \mathbb{E}_{a_{-i} \sim \pi_{-i}(\cdot|s)}\big[\mathrm{adv}_i^\pi(s, (a_i, a_{-i}))\big]. \quad (4)$$

Before embarking on the subsequent analysis, we need to clarify that $\mathrm{adv}_i^\pi(s, a)$ and $\overline{\mathrm{adv}}_i^\pi(s, a_i)$ is well-defined, which means they will not be infinity.

Lemma 1. *The advantage functions $\mathrm{adv}_i^\pi(s, a)$ are bounded with respect to the policy profile π, and so are the average advantage functions $\overline{\mathrm{adv}}_i^\pi(s, a_i)$.*

We now provide an equivalent formulation of the individual payoff gradients $v_i(\pi)$ which will help us to analyse the boundedness and smoothness of the gradients. We start with the following versions of the policy gradient theorem for stochastic games with long-run average payoffs.

Theorem 1 (Policy gradient theorem). *For any player i and any stationary policy profile π, we have*

$$\nabla_i V_i(\pi) = \sum_{s \in S} p_\pi(s) \sum_{a_i \in A_i} (\nabla_i \pi_i(a_i|s)) \overline{\mathrm{adv}}_i^\pi(s, a_i). \quad (5)$$

Remark 2. As we have $v_i(\pi) = \nabla_i V_i(\pi) = \big(\frac{\partial}{\partial \pi_i(a_i|s)} V_i(\pi)\big)_{(a_i, s) \in A_i \times S}$, it can be obtained that

$$\frac{\partial}{\partial \pi_i(a_i|s)} V_i(\pi) = p_\pi(s) \overline{\mathrm{adv}}_i^\pi(s, a_i). \quad (6)$$

By Theorem 1 and Lemma 1, it can be immediately demonstrated that v_i is bounded with respect to policy profile π.

Lemma 2. *For any player i, the individual payoff gradient $v_i(\pi)$ is bounded with respect to the policy profile π.*

However, it is insufficient to apply the method that will be presented in the subsequent section to estimate $v_i(\pi)$. It is necessary to demonstrate that $v_i(\pi)$ is Lipschitz continuous with respect to π. This will be done by proving the following theorem based on some auxiliary lemmas, which will be provided in the appendix.

Theorem 2. *For any player i and state-action pair (s, a_i), the partial derivative of the value function $\frac{\partial}{\partial \pi_i(a_i|s)} V_i(\pi)$ is Lipschitz continuous with respect to π. Consequently, the individual payoff gradient $\nabla_i V_i(\pi)$ is also Lipschitz continuous.*

4 The Learning Framework

In this section, we present our algorithm in stochastic games. It is based on a fundamental learning framework known as the regularized Robbins-Monro template [21]. However, applying this method to stochastic games necessitates the estimation of the individual payoff gradients, i.e., $v(\pi) = (v_i(\pi))_{i \in N}$, which is challenging to compute directly. To overcome this difficulty, we employ the simultaneous perturbation stochastic approximation method [5], which requires only the individual payoffs for estimating $v_i(\pi)$. By integrating these two approaches, we devise an algorithm aimed at attaining the Nash equilibrium for stochastic games with long-run average payoffs. Notably, in our algorithm, players only need to observe their own actions and resulting payoffs to learn their individual payoff gradients, so it is a distributed bandit algorithm.

4.1 Regularized Robbins-Monro Process

Combining Robbins-Monro algorithm [23], a famous stochastic approximation method, and "follow-the-regularized-leader" family of algorithms of Shalev-Shwartz & Singer [25], we obtain the following learning framework called regularized Robbins-Monro template [21]:

$$Y^{t+1} = Y^t + \gamma^t \hat{v}^t,$$
$$\pi^{t+1} = Q(Y^{t+1}) \qquad (7)$$

where

1. $\pi^t = (\pi_i^t)_{i \in N}$ is the players' policy profile at time t.
2. $\hat{v}^t = (\hat{v}_i^t)_{i \in N}$ is an individual "gradient-like" signal that estimates the individual gradient $v(\pi^t) = (v_i(\pi^t))_{i \in N}$.
3. $Y^t = (Y_i^t)_{i \in N}$ is the weighted sum of $\{\hat{v}^s | s \leqslant t\}$ in dual space of Π.

4. $\gamma^t > 0$ is the step-size, and to guarantee the perpetual update of parameters throughout the iterative process, we assume that $\sum_t \gamma^t = \infty$.

The most important element is the function $Q : \mathcal{Y} \to \Pi \subseteq \mathbb{R}^{|S| \times |A|}$, where $\mathcal{Y}$ is the dual space of Π. It is a "general projection" map that mirrors gradient steps in $\mathcal{Y}$ to policy space Π, and we call it the players' mirror map, which is related to the mirror descent algorithm, as decribed in reference [8].

To define the mirror map more specifically, we decompose it into the product form $Q = (Q_i)_{i \in N}$, where $Q_i : \mathcal{Y}_i \to \Pi_i \subseteq \mathbb{R}^{|S| \times |A_i|}$ is the mirror map of a single player i. We begin with introducing the concept of "regularizer" on Π_i as follows:

Definition 3 (Regularizer). *For any player i, $h_i : \mathbb{R}^{|S| \times |A_i|} \to \mathbb{R} \cup \{\infty\}$ is a regularizer on Π_i if*

1. *h_i is supported on Π_i, i.e., $\{x_i \in \mathbb{R}^{|S| \times |A_i|} : h_i(x_i) < \infty\} = \Pi_i$.*
2. *h_i is continuous and K_i-strongly convex on Π_i, i.e., there exists a constant $K_i > 0$ such that for all $\pi_i, \pi'_i \in \Pi_i$ and all $\lambda \in [0, 1]$*

$$h_i(\lambda \pi_i + (1-\lambda)\pi'_i) \leq \lambda h_i(\pi_i) + (1-\lambda) h_i(\pi'_i) - \frac{1}{2} K_i \lambda (1-\lambda) \|\pi'_i - \pi_i\|_2^2.$$

And we define the mirror map $Q : \mathcal{Y}_i \to \Pi_i$ and convex conjugate $h_i^* : \mathcal{Y}_i \to \mathbb{R}$ induced by regularizer h_i:

$$Q_i(Y_i) = \operatorname{argmax}_{\pi_i \in \Pi_i} \{\langle Y_i, \pi_i \rangle - h_i(\pi_i)\}, \tag{8}$$

$$h_i^*(Y_i) = \max_{\pi_i \in \Pi_i} \{\langle Y_i, \pi_i \rangle - h_i(\pi_i)\}. \tag{9}$$

For convenience, we denote $h(\pi) = \sum_{i \in N} h_i(\pi_i)$ for the players' aggregate regularizer and $Q = (Q_i)_{i \in N}$ for the induced mirror map.

Example 1 (Entropic regularizaion). Let $h_i(\pi_i) = \sum_{a_i \in A_i, s \in S} \pi_i(a_i|s) \log \pi_i(a_i|s)$ be the (negative) Gibbs-Shannon entropy on Π_i. With straightforward calculations, the induced mirror map of each player i is the logit choice map, i.e.,

$$(Q_i(Y_i))_{a_i, s} = \frac{\exp(Y_i(a_i, s))}{\sum_{a_i \in A_i} \exp(Y_i(a_i, s))}, \tag{10}$$

which is Hedge algorithm in learning in finite games [6].

For the analysis of the learning process, we introduce the Fenchel coupling:

$$F(p, y) = h(p) + h^*(y) - \langle y, p \rangle, \tag{11}$$

which can be seen as the global energy function if the equilibrium point satisfies variational stability [21]. In more specific terms, it can be employed to measure the distance between a Nash equilibrium π^* and any policy profile π, as it has the following properties.

Lemma 3. ([22]) If h is a K-strongly convex regularizer on $\mathcal{X}$, let Q be the mirror map induced by h, fix some $p \in \mathcal{X}$, and for all $y, y' \in \mathcal{Y}$, the dual space of $\mathcal{X}$, we have

1. $F(p, y) \geq \frac{1}{2} K \|Q(y) - p\|^2$.
2. $F(p, y') \leq F(p, y) + \langle y' - y, Q(y) - p \rangle + \frac{1}{2K} \|y' - y\|^2$.

And we assume that

$$F(p, y_n) \to 0 \text{ whenever } Q(y_n) \to p, \tag{12}$$

which is called "reciprocity condition" in the theory of Bregman function [20].

4.2 Simultaneous Perturbation Stochastic Approximation

Now, what we need is to estimate the gradients. However, directly estimating the individual payoff gradients $v_i(\pi^t)$ from the historical information presents a formidable challenge for players. Nevertheless, under the assumption in Remark 1, players are able to efficiently evaluate their policies and approximate the payoff function $V_i(\pi_i, \pi_{-i})$ provided they adhere to a consistent strategy π_i over an extended sequence of moves.

Once we get the approximate value of $V_i(\pi)$, following [5,13,29], players can use the simultaneous perturbation stochastic approximation approach that allows them to estimate their individual payoff gradients based on their own payoffs only. In detail, this estimation process can be summarized in the following steps for each player $i \in N$:

1. Choose a query radius $\delta > 0$.
2. Determine the policy $\pi_i \in \Pi_i$ where player i estimate her payoff gradient.
3. Draw a vector z_i from the unit sphere $\mathbb{S}_i$ of $\mathbb{R}^{|S| \times |A_i|}$ uniformly and play $\hat{\pi}_i = \pi_i + \delta z_i$.
4. Get the approximate value $\hat{V}_i$ of value function $V_i(\pi_i, \pi_{-i})$ and obtain

$$\hat{v}_i = \frac{d_i}{\delta} \hat{V}_i z_i, \quad d_i \text{ is the dimension of } \Pi_i. \tag{13}$$

Using Stoke's theorem, we show that $\hat{v}_i$ is an unbiased estimator of the individual gradient of the δ-averaged payoff function:

$$V_i^\delta(\pi_i, \pi_{-i}) = \frac{1}{vol(\delta \mathbb{B}_i) \prod_{j \neq i} vol(\delta \mathbb{S}_j)} \int_{\delta \mathbb{B}_i} \int_{\prod_{j \neq i} \delta \mathbb{S}_j} u_i(\pi_i + w_i, \pi_{-i} + z_{-i}) dz_1 \cdots dw_i \cdots dz_N, \tag{14}$$

where $\mathbb{B}_i$ is the unit ball of $\mathbb{R}^{|S| \times |A_i|}$. And the Lipschitz continuity of v_i guarantee that $\|\nabla_i V_i - \nabla_i V_i^\delta\|_\infty = \mathcal{O}(\delta)$.

Lemma 4. The estimator $\hat{v} = (\hat{v}_i)_{i \in N}$ given by (13) satisfies

$$\mathbb{E}[\hat{v}_i] = \nabla_i V_i^\delta(\pi_i, \pi_{-i}),$$

with $V_i^\delta(\pi_i, \pi_{-i})$ as in (14). Moreover, we have $\|\nabla_i V_i - \nabla_i V_i^\delta\|_\infty = \mathcal{O}(\delta)$ if $\nabla_i V_i$ is Lipschitz continuous.

Before we use the method above, there are some issues to clarify. The first thing is that as the perturbation direction z_i is drawn from the unit sphere $\mathbb{S}_i$, it may fail to be tangent to Π_i, especially by our definition, all individual policies π_i are in the surface of Π_i. Thus we need to build a equivalent representation of the policy π_i which lie in a low-dimensional space and satisfy our need. Let

$$\mathcal{X}_i = \{x_i \in \mathbb{R}^{|S| \times (|A_i|-1)} : 0 \leqslant \sum_{a_i \in A_{i,last}} x_i(a_i, s) \leqslant 1, \forall s \in S, x_i \geqslant 0\}. \tag{15}$$

where $A_{i,last}$ means the set A_i deletes its last action. There is a natural bijection M between Π_i and $\mathcal{X}_i$, i.e., $M(\pi_i)(a_i, s) = x_i(a_i, s), \forall a_i \in A_{i,last}, \forall s \in S$. And it is easy to check that

$$\begin{aligned}\frac{\partial}{\partial \pi_i(a_i|s)} V_i(\pi_i, \pi_{-i}) &= \frac{\partial}{\partial x_i(a_i, s)} V_i(x_i, x_{-i}), \forall a_i \in A_{i,last}, \forall s \in S, \\ \frac{\partial}{\partial \pi_i(a_{i,last}|s)} V_i(\pi_i, \pi_{-i}) &= - \sum_{a_i \in A_{i,last}} \frac{\partial}{\partial x_i(a_i, s)} V_i(x_i, x_{-i}),\end{aligned} \tag{16}$$

where $a_{i,last}$ is the action deleted in the set $A_{i,last}$. So we can use $\mathcal{X}_i$ to represent the set of player i's stationary policies and the interior of $\mathcal{X}_i$ is a closed convex set. If we let x_i being in the interior of $\mathcal{X}_i$ and draw z_i from $\mathbb{S}_i \subseteq \mathbb{R}^{|S| \times (|A_i|-1)}$, we can guarantee that z_i is tangent to $\mathcal{X}_i$ and estimate the individual payoff gradient $\nabla_i V_i(x_i, x_{-i})$ using $\hat{x}_i = x_i + \delta z_i$. Subsequently, we derive the gradient $\nabla_i V_i(\pi_i, \pi_{-i})$ from $\nabla_i V_i(x_i, x_{-i})$.

On the other hand, even when z_i is a feasible direction of perturbation, the query point $\hat{x}_i = x_i + \delta z_i$ may not be in $\mathcal{X}_i$. To address this issue, we adopt the concept of a "safety net", as introduced in the work of [1]. Let $\mathbb{B}_{r_i}(p_i)$ be an r_i-ball centered at $p_i \in \mathcal{X}_i$ so that $\mathbb{B}_{r_i}(p_i) \subseteq \mathcal{X}_i$. Then instead of using the direction z_i, we consider the feasibility adjustment

$$w_i = z_i - r_i^{-1}(x_i - p_i), \tag{17}$$

and player i plays $\hat{x}_i = x_i + \delta w_i$. When $\delta/r_i < 1$, one can check that $\hat{x}_i \in \mathcal{X}_i$.

4.3 A Distributed Learning Algorithm in Stochastic Games

With the preparation above, we introduce the following learning algorithm, Algorithm 1, in stochastic games. In practice, each player's step size can vary, but for convenience, we assume that all players have the same step size in this paper.

The reason why players need a time threshold T^t in the algorithm is that from any initial state s, after the players play a fixed policy $\hat{x}^t$ in few rounds T^t, the instantaneous reward $\hat{V}_i^t$ for player i at time $T^t + 1$ will be close to the value function $V_i(\hat{x}^t)$ if the probability distribution over S is not far from the stationary distribution $p_{\hat{x}^t}$ at that time.

Lemma 5. *If the assumption in Remark 1 holds and players play games by the Algorithm 1, we have*

$$|\mathbb{E}[\hat{V}_i^t] - V_i(\hat{x}^t)| \leqslant |S|(\max_{s,a} r_i(s,a))e^{-\frac{T^t}{\tau}}. \tag{19}$$

Algorithm 1. Multi-agent Distributed Learning Algorithm for Player i

Require: step-size γ^t, query radius δ^t, safety ball $\mathbb{B}_{r_i}(p_i)$, time threshold T^t, mirror map Q
Choose initial policy $\pi_i^0 \in \Pi_i$ ($x_i^0 \in \mathcal{X}_i$ equivalently).
set $Y_i^0 = 0$.
for period $t = 0, 1, 2, \ldots$ **do**
  draw z_i^t uniformly from $\mathbb{S}^{|S|(|A_i|-1)}$
  let $w_i^t = z_i^t - r_i^{-1}(x_i^t - p_i)$
  play policy $\hat{x}_i^t = x_i^t + \delta^t w^t$ in the following T^t stage, and get the one-step reward $\hat{V}_i^t$ at time $T^t + 1$
  get the individual gradient $\hat{v}_i^t$ by the estimator of $\nabla_i V_i(x_i, x_{-i})$, in details

$$\hat{v}_i^t(a_i, s) = \begin{cases} \frac{|S|(|A_i|-1)}{\delta^t} \hat{V}_i^t z_i^t(a_i, s) & \text{if } a_i \neq a_{i, last}, \\ -\sum_{a_i' \neq a_i} \frac{|S|(|A_i|-1)}{\delta^t} \hat{V}_i^t z_i^t(a_i, s) & \text{if } a_i = a_{i, last}. \end{cases} \quad (18)$$

  update $Y_i^{t+1} = Y_i^t + \gamma^t \hat{v}_i^t$
  update $\pi_i^{t+1} = Q(Y_i^{t+1})$, and corresponding x_i^{t+1}
end for

5 Convergence Analysis and Results

If all players use the Algorithm 1 in stochastic games, we construct the following stochastic process to describe their behaviors:

$$\begin{aligned} \hat{x}^t &= M(\pi^t) + \delta^t w^t, \\ Y^{t+1} &= Y^t + \gamma^t \hat{v}^t, \\ \pi^{t+1} &= Q(Y^{t+1}). \end{aligned} \quad (20)$$

In the above, $w^t = (w_i^t)_{i \in N}$ and $\hat{v}^t = (\hat{v}_i^t)_{i \in N}$ have the following expressions:

$$w_i^t = z_i^t - r_i^{-1}(x_i^t - p_i), \quad \hat{v}_i^t = \frac{|S|(|A_i|-1)}{\delta^t} \hat{V}_i^t \cdot F_i z_i^t, \quad (21)$$

where

$$F_i = \begin{pmatrix} G_i & & \\ & \ddots & \\ & & G_i \end{pmatrix}_{(|S| \times |A_i|) \times (|S| \times (|A_i|-1))} \quad \text{and} \quad G_i = \begin{pmatrix} 1 & & \\ & \ddots & \\ & & 1 \\ -1 & \cdots & -1 \end{pmatrix}_{|A_i| \times (|A_i|-1)}.$$

For the estimator $\hat{v}_i^t$, as we have proved in Lemma 4

$$\mathbb{E}_{z_i^t}\left[\frac{|S|(|A_i|-1)}{\delta^t} \hat{V}_i^t \cdot z_i^t\right] = \nabla_i V_i^\delta(x_i, x_{-i}),$$

we obtain that

$$\mathbb{E}[\hat{v}_i^t] = \mathbb{E}_{z_i^t}\left[\frac{|S|(|A_i|-1)}{\delta^t} \hat{V}_i^t \cdot F_i z_i^t\right] = F_i \nabla_i V_i^\delta(x_i, x_{-i}). \quad (22)$$

Thus, we can write

$$\begin{aligned}\hat{v}_i^t =& \nabla_i V_i(\pi_i^t, \pi_{-i}^t) + (F_i \nabla_i V_i^{\delta^t}(x_{i,\delta^t}^t, x_{-i,\delta^t}^t) - \nabla_i V_i(\pi_i^t, \pi_{-i}^t)) \\
& + F_i(\frac{|S|(|A_i|-1)}{\delta^t} V_i(\hat{x}^t) \cdot z_i^t - \mathbb{E}_{z_i^t}[\frac{|S|(|A_i|-1)}{\delta^t} V_i(\hat{x}^t) \cdot z_i^t]) \\
& + F_i(\frac{|S|(|A_i|-1)}{\delta^t} \hat{V}_i^t \cdot z_i^t - \frac{|S|(|A_i|-1)}{\delta^t} V_i(\hat{x}^t) \cdot z_i^t) \\
\triangleq & \nabla_i V_i(\pi_i^t, \pi_{-i}^t) + b_i^t + U_i^t + \epsilon_i^t, \end{aligned} \qquad (23)$$

where in the stochastic approximation process, b_i^t and ϵ_i^t are bias terms, while U_i^t is the noise term.

Having completed the preceding preparations, we are now in a position to analyze the convergence of the process (20). As a learning process in games, it is expected to converge to a Nash equilibrium. However, as it is of PPAD-complete complexity to find a Nash equilibrium even in finite games [7] and stochastic games with discounted payoffs [10], it is therefore not to be expected that our algorithm will converge to some Nash equilibrium in all stochastic games with long-run average payoffs. However, the system will converge to the Nash equilibrium globally if we assume that the game have some great properties.

Definition 4 (Neutrally stable [21]). *A policy profile π^* is globally neutrally stable if*

$$\langle v(\pi), \pi - \pi^* \rangle \leqslant 0, \forall \pi \in \Pi, \qquad (24)$$

where $v(\pi) = (v_i(\pi))_{i \in N}$ is the individual payoff gradient. Furthermore, if the equality holds only when π is a Nash equilibrium, we say the policy profile π^* is globally variationally stable.

A classic type of games is monotone games [24] which can be defined as follows in stochastic games.

Example 2 (Monotone games). A stochastic game $\mathcal{G}$ is a monotone game if

$$\langle v(\pi) - v(\pi'), \pi - \pi' \rangle \leqslant 0, \forall \pi, \pi' \in \Pi_i. \qquad (25)$$

Based on the first-order stationary property of Nash equilibria, any Nash equilibrium in monotone games is globally neutrally stable.

The following convergence theorem holds when Nash equilibria possess the aforementioned stability conditions.

Theorem 3. *Suppose that in stochastic games, all Nash equilibria are globally neutrally stable and a globally variationally stable Nash equilibrium exists. If all players employ Algorithm 1 with parameters having the following property*

$$\lim_{t \to \infty} \gamma^t = \lim_{t \to \infty} \delta^t = 0, \quad \sum_{t=0}^{\infty} \gamma^t = \infty, \quad \sum_{t=0}^{\infty} \gamma^t \delta^t < \infty, \quad \sum_{t=0}^{\infty} (\frac{\gamma^t}{\delta^t})^2 < \infty, \quad \sum_{t=0}^{\infty} \frac{\gamma^t}{\delta^t} e^{-\frac{T^t}{\tau}} < \infty,$$

then the sequences of learned policies π^t and realized policies $\hat{x}^t$ each converge to a Nash equilibrium with probability 1.

It is shown [24] that, strict monotone games admit a unique Nash equilibrium and possess the globally variationally stability. So the following corollary holds.

Corollary 1. *The sequences of learned policies π^t and realized policies $\hat{x}^t$ converge to the Nash equilibrium with probability 1 in strict monotone games if the assumption of Theorem 3 holds.*

The proof of Theorem 3 is divided into three steps as follows. First, for any globally neutrally stable Nash equilibrium π^*, we will demonstrate that $F(\pi^*, Y^t)$ converges to a finite random variable F^∞. Then we will show that there exists a subsequence $\{\pi^{t_k}\}$ which will converges to some Nash equilibrium if there exists a globally stable Nash equilibrium. Then the proof of Theorem 3 can be completed with these two facts as well as the properties of the Fenchel coupling. The details of the proof are provided in [32].

It is necessary to choose the appropriate parameters when implementing the algorithm. We consider parameters of the following form $\gamma^t = \gamma/t^p$, $\delta^t = \delta/t^q$ with $\gamma, \delta > 0$ and $0 < p, q \leq 1$. In order to satisfy the assumptions of the Theorem 3, we need $p + q > 1$ and $p - q > 1/2$. For T^t, when τ is known to the players, we can let $T^t = [T \log t] + 1$ with $T > 0$ and $p - q + T/\tau > 1$. Otherwise, we can let $T^t = [t^T] + 1$ with $T > 0$. A suitable pair of parameters is $(\gamma^t, \delta^t, T^t) = (t^{-1}, t^{-\frac{1}{3}}, \sqrt{t})$.

6 Discussion

In this work, we demonstrate that the individual payoff gradients for stochastic games with long-run average payoffs exhibit Lipschitz continuity, a feature that is also observed in the context of discounted payoffs. Based on these facts observed, we have developed a promising learning algorithm predicated on policy gradient estimation. The algorithm ensures that, provided the games possess a globally variationally stable Nash equilibrium with all Nash equilibria globally neutrally stable, the system will converge to some Nash equilibrium with probability one, if all players employ this approach.

However, our current analysis only confirms the global asymptotic convergence of the algorithm for certain special cases within game theory. The investigation into non-asymptotic convergence properties is still pending. Additionally, the applicability of the algorithm to a broader spectrum of stochastic games, including zero-sum games, is an open question. Moreover, it is yet to be determined if the algorithm exhibits a high probability of converging to the Nash equilibrium when the initial policy profile is in close proximity to it.

Acknowledgments. This work was supported by the Strategic Priority Research Program of Chinese Academy of Sciences under Grant XDA27030201, the Natural Science Foundation of China under Grant T2293770, the National Key Research and Development Program of China under grant No.2022YFA1004600.

Disclosure of Interests. The authors have no competing interests to declare that are relevant to the content of this article.

References

1. Agarwal, A., Dekel, O., Xiao, L.: Optimal algorithms for online convex optimization with multi-point bandit feedback. In: Proceedings of the 23rd Annual Conference on Learning Theory (COLT) (2010). https://www.microsoft.com/en-us/research/publication/optimal-algorithms-for-online-convex-optimization-with-multi-point-bandit-feedback/
2. Beck, A., Teboulle, M.: Mirror descent and nonlinear projected subgradient methods for convex optimization. Oper. Res. Lett. **31**, 167–175 (2003). https://api.semanticscholar.org/CorpusID:7036108
3. Berger, U.: Brown's original fictitious play. J. Econ. Theory **135**(1), 572–578 (2007). https://doi.org/10.1016/j.jet.2005.12.010, https://www.sciencedirect.com/science/article/pii/S0022053106000299
4. Bhalla, S., Ganapathi Subramanian, S., Crowley, M.: Deep multi agent reinforcement learning for autonomous driving. In: Goutte, C., Zhu, X. (eds.) Advances in Artificial Intelligence, pp. 67–78. Springer International Publishing, Cham (2020)
5. Bravo, M., Leslie, D., Mertikopoulos, P.: Bandit learning in concave n-person games. In: Proceedings of the 32nd International Conference on Neural Information Processing Systems, pp. 5666–5676. NIPS'18, Curran Associates Inc., Red Hook, NY, USA (2018)
6. Cesa-Bianchi, N., Lugosi, G.: Prediction, Learning, and Games. Cambridge University Press (2006)
7. Chen, X., Deng, X.: Settling the complexity of two-player NASH equilibrium. In: 2006 47th Annual IEEE Symposium on Foundations of Computer Science (FOCS'06), pp. 261–272 (2006). https://doi.org/10.1109/FOCS.2006.69
8. Darzentas, J.: Problem complexity and method efficiency in optimization. J. Oper. Res. Soc. **35**(5), 455–455 (1984). https://doi.org/10.1057/jors.1984.92, https://doi.org/10.1057/jors.1984.92
9. Daskalakis, C., Foster, D.J., Golowich, N.: Independent policy gradient methods for competitive reinforcement learning. In: Proceedings of the 34th International Conference on Neural Information Processing Systems. NIPS '20, Curran Associates Inc., Red Hook, NY, USA (2020)
10. Deng, X., Li, N., Mguni, D., Wang, J., Yang, Y.: On the complexity of computing Markov perfect equilibrium in general-sum stochastic games. National Sci. Rev. **10**(1) (2022). https://doi.org/10.1093/nsr/nwac256, https://doi.org/10.1093/nsr/nwac256, _eprint: https://academic.oup.com/nsr/article-pdf/10/1/nwac256/49182748/nwac256.pdf
11. Etesami, S.R.: Learning stationary NASH equilibrium policies in n-player stochastic games with independent chains. SIAM J. Control Optimization **62**(2), 799–825 (2024). https://doi.org/10.1137/22M1512880, https://doi.org/10.1137/22M1512880
12. Even-Dar, E., Kakade, S.M., Mansour, Y.: Online Markov decision processes. Math. Oper. Res. **34**(3), 726–736 (2009). https://doi.org/10.1287/moor.1090.0396, https://doi.org/10.1287/moor.1090.0396
13. Flaxman, A.D., Kalai, A.T., McMahan, H.B.: Online convex optimization in the bandit setting: gradient descent without a gradient. In: Proceedings of the Sixteenth Annual ACM-SIAM Symposium on Discrete Algorithms, pp. 385–394. SODA '05, Society for Industrial and Applied Mathematics, USA (2005)
14. Giannou, A., Lotidis, K., Mertikopoulos, P., Vlatakis-Gkaragkounis, E.V.: On the convergence of policy gradient methods to NASH equilibria in general stochastic

games. In: Proceedings of the 36th International Conference on Neural Information Processing Systems. NIPS '22, Curran Associates Inc., Red Hook, NY, USA (2024)
15. Gu, S., et al.: Safe multi-agent reinforcement learning for multi-robot control. Artificial Intell. **319**, 103905 (2023). https://doi.org/10.1016/j.artint.2023.103905, https://www.sciencedirect.com/science/article/pii/S0004370223000516
16. Hambly, B., Xu, R., Yang, H.: Policy gradient methods find the NASH equilibrium in n-player general-sum linear-quadratic games. J. Mach. Learn. Res. **24**(1) (mar 2024)
17. Hart, S., Mas-Colell, A.: A simple adaptive procedure leading to correlated equilibrium. Econometrica **68**(5), 1127–1150 (2000). https://doi.org/10.1111/1468-0262.00153, https://onlinelibrary.wiley.com/doi/abs/10.1111/1468-0262.00153
18. Hu, J., Wellman, M.P.: Nash Q-learning for general-sum stochastic games. J. Mach. Learn. Res. **4**, 1039–1069 (2003)
19. Jin, C., Liu, Q., Wang, Y., Yu, T.: V-learning—a simple, efficient, decentralized algorithm for multiagent reinforcement learning. Mathematics of Operations Research (2023). https://doi.org/10.1287/moor.2021.0317, https://doi.org/10.1287/moor.2021.0317
20. Juditsky, A., Nemirovski, A., Tauvel, C.: Solving variational inequalities with stochastic mirror-prox algorithm. Stoch. Syst. **1**(1), 17–58 (2011). https://doi.org/10.1287/10-SSY011, https://doi.org/10.1287/10-SSY011
21. Mertikopoulos, P., Hsieh, Y.P., Cevher, V.: A unified stochastic approximation framework for learning in games. Math. Program. **203**(1), 559–609 (2024). https://doi.org/10.1007/s10107-023-02001-y, https://doi.org/10.1007/s10107-023-02001-y
22. Mertikopoulos, P., Zhou, Z.: Learning in games with continuous action sets and unknown payoff functions. Math. Program. **173**(1), 465–507 (2019). https://doi.org/10.1007/s10107-018-1254-8, https://doi.org/10.1007/s10107-018-1254-8
23. Robbins, H., Monro, S.: A Stochastic approximation method. Ann. Math. Stat. **22**(3), 400 – 407 (1951). https://doi.org/10.1214/aoms/1177729586, https://doi.org/10.1214/aoms/1177729586
24. Rosen, J.B.: Existence and uniqueness of equilibrium points for concave n-person games. Econometrica **33**(3), 520–534 (1965). http://www.jstor.org/stable/1911749
25. Shalev-Shwartz, S.: Online learning and online convex optimization. Found. Trends Mach. Learn. **4**(2), 107–194 (2012). https://doi.org/10.1561/2200000018, https://doi.org/10.1561/2200000018
26. Shapley, L.S.: Stochastic games. Proc. National Acad. Sci. **39**(10), 1095–1100 (1953). https://doi.org/10.1073/pnas.39.10.1095, https://www.pnas.org/doi/abs/10.1073/pnas.39.10.1095
27. Solan, E., Vieille, N.: Stochastic games. Proc. National Acad. Sci. **112**(45), 13743–13746 (2015). https://doi.org/10.1073/pnas.1513508112, https://www.pnas.org/doi/abs/10.1073/pnas.1513508112
28. Song, Z., Mei, S., Bai, Y.: When can we learn general-sum Markov games with a large number of players sample-efficiently?. In: International Conference on Learning Representations (2022). https://openreview.net/forum?id=6MmiS0HUJHR
29. Spall, J.C.: A one-measurement form of simultaneous perturbation stochastic approximation. Automatica **33**(1), 109–112 (1997). https://doi.org/10.1016/S0005-1098(96)00149-5, https://doi.org/10.1016/S0005-1098(96)00149-5
30. Sutton, R.S., Barto, A.G.: Reinforcement Learning: An Introduction. A Bradford Book, Cambridge, MA, USA (2018)

31. Sutton, R.S., McAllester, D., Singh, S., Mansour, Y.: Policy gradient methods for reinforcement learning with function approximation. In: Proceedings of the 12th International Conference on Neural Information Processing Systems, pp. 1057–1063. NIPS'99, MIT Press, Cambridge, MA, USA (1999)
32. Zhang, J., Mu, Y.: A payoff-based policy gradient method in stochastic games with long-run average payoffs (2024). https://arxiv.org/abs/2405.09811
33. Zhang, K., Yang, Z., Başar, T.: Multi-agent reinforcement learning: a selective overview of theories and algorithms, pp. 321–384. Springer International Publishing, Cham (2021). https://doi.org/10.1007/978-3-030-60990-0_12, https://doi.org/10.1007/978-3-030-60990-0_12
34. Zhang, R.C., Ren, Z., Li, N.: Gradient play in stochastic games: stationary points and local geometry. IFAC-PapersOnLine **55**(30), 73–78 (2022). https://doi.org/10.1016/j.ifacol.2022.11.031, https://www.sciencedirect.com/science/article/pii/S2405896322026593, 25th International Symposium on Mathematical Theory of Networks and Systems MTNS 2022

Mechanism Design for Auctions with Externalities on Budgets

Yusen Zheng[1], Yukun Cheng[2(✉)], Chenyang Xu[3], and Xiaotie Deng[1(✉)]

[1] School of Computer Science, Peking University, Beijing 100871, China
yusen@stu.pku.edu.cn, xiaotie@pku.edu.cn
[2] School of Business, Jiangnan University, Wuxi 214122, China
ykcheng@amss.ac.cn
[3] Software Engineering Institute, East China Normal University, Shanghai, China
cyxu@sei.ecnu.edu.cn

Abstract. This paper studies mechanism design for auctions with externalities on budgets, a novel setting where the budgets that bidders commit are adjusted due to the externality of the competitors' allocation outcomes—a departure from traditional auctions with fixed budgets. This setting is motivated by real-world scenarios, for example, participants may increase their budgets in response to competitors' obtained items. We initially propose a general framework with homogeneous externalities to capture the interdependence between budget updates and allocation, formalized through a budget response function that links each bidder's effective budget to the amount of items won by others.

The main contribution of this paper is to propose a truthful and individual rational auction mechanism for this novel auction setting, which achieves an approximation ratio of $1/3$ with respect to the liquid welfare. This mechanism is inspired by the uniform-price auction, in which an appropriate uniform price is selected to allocate items, ensuring the monotonicity of the allocation rule while accounting for budget adjustments. Additionally, this mechanism guarantees a constant approximation ratio by setting a purchase limit. Complementing this result, we establish an upper bound: no truthful mechanism can achieve an approximation ratio better than $1/2$. This work offers a new perspective to study the impact of externalities on auctions, providing an approach to handle budget externalities in multi-agent systems.

Keywords: Mechanism Design · Auction · Budget Constraint · Externality

---

Full version of the paper can be found at https://arxiv.org/abs/2504.14948.
This research was supported by the National Natural Science Foundation of China (NSFC) under grant numbers No. 62172012, No. 12471339 and No. 62302166, and the Key Laboratory of Interdisciplinary Research of Computation and Economics (SUFE), Ministry of Education.

1 Introduction

Auctions are a common method for resource allocation. Bidders participating in auctions typically face constraints on the amount of money they can use. Many previous studies on auctions with budget constraints model the budget as a predetermined fixed parameter [1,8,9,11,25]. Recently, some work has begun to explore scenarios where bidders do not set a hard budget parameter but instead focus on their return on investment (ROI) [4,15,16]. However, in real-world economic activities, bidders' budget decisions are often influenced by social signals, such as the behavior of others. This phenomenon is exemplified by concepts like the "herding effect" or "conspicuous consumption" commonly studied in behavioral economics. In these cases, these budget constraints are not rigid but rather flexible, meaning that bidders' internal expectations of their budget ceilings continuously adjust throughout the auction process and its outcomes. Auction mechanisms designed for fixed budget scenarios are not suitable for situations with such externalities on the budget. This paper formalizes the problem considering the specific flexible budget constraint for the first time, in which a bidder's budget is influenced by the final resource allocations obtained by others. This scenario is quite common. For example, bidders exhibiting strong herding behavior may psychologically increase their budget ceilings upon observing that other bidders receive substantial resource allocations.

We formally refer to the impact of allocation outcomes on bidders' budgets as *the allocation(-induced) externalities on budgets*. Previous works on externalities primarily focus on the effects of item allocations or the private information held by bidders on other bidders' valuations, such as [2,6,18–20]. However, our study examines the impact of item allocation on bidders' budgets, representing a new category of externality problems.

The presence of allocation externalities on budgets means that the allocation of items is no longer an isolated decision for each bidder, rather, it simultaneously affects the utilities and budgets of all bidders, including the bidder herself. This introduces new challenges in designing effective auction mechanisms. Our goal, therefore, is to develop an auction mechanism that not only provides robust incentive guarantees but also maximizes efficiency.

1.1 Our Contributions

We formalize the allocation externalities on budgets by introducing a budget impact factor, which captures these externalities. We consider a scenario with homogeneous externalities, meaning that for every bidder, the externality effect of other bidders' allocations on their budget is identical. We use liquid welfare as the metric for measuring auction efficiency, as it is a commonly used welfare measure in budget-constrained auctions [11]. Optimizing liquid welfare involves allocating items to bidders with high valuations while ensuring they possess sufficient purchasing power.

Compared to traditional welfare-maximizing auction mechanisms, the presence of allocation externalities on budgets introduces new challenges. Optimizing liquid welfare requires balancing valuation and budget, which can be in conflict. For instance, allocating more items to one bidder increases her value, but simultaneously reduces the number of items available to others, which in turn diminishes that bidder's budget due to the externalities. This conflict, combined with incentive constraints, necessitates novel technical approaches in mechanism design. Overall, our contributions are summarized as follows:

- We propose a model to capture allocation-induced externalities on budgets in auctions. In this model, a bidder's budget is represented as a function of the allocations received by other bidders, allowing us to characterize the impact of homogeneous allocation externalities on the budget.
- We provide an optimal allocation algorithm that solves the liquid welfare maximization problem (Theorem 1). The algorithm greedily allocates items in decreasing order of valuation and assigns any remaining items to the bidder whose budget is least affected by externalities (Algorithm 1).
- We design a truthful and individually rational auction mechanism (Mechanism 1) that achieves a 1/3-approximation for liquid welfare (Theorem 2). Furthermore, we prove an impossibility result showing that no truthful mechanism can achieve a better approximation ratio than 1/2 (Lemma 6).

1.2 Related Work

Mechanism Design with Budget Constraints. Budget constraints are a natural feature of economic activities, limiting the amount of money participants can spend. Many studies have focused on designing auction mechanisms that account for these constraints [1,8,9,25]. The most common approach models budget constraints as hard constraints, treating a bidder's budget as a fixed parameter, and uses budgeted quasi-linear utility functions to characterize bidders' utilities. This approach has been widely applied in revenue-maximizing auctions [1,8], welfare-maximizing auctions [10,11,21], multi-item auctions [22], and auctions in auto-bidding systems [5]. This model is reasonable in many scenarios, such as certain resource allocation settings where bidders' budgets remain fixed. However, in many real-world scenarios, bidders' budgets are not rigid but rather flexible, adjusting throughout the auction process and its outcomes. Only a few studies have explored auction mechanism design under such conditions. For instance, Goel et al. [15] model the budget as a function of the items acquired by the bidder, but do not consider the impact of other bidders' allocations on the budget. The study of auctions with externalities on budget remains a novel area of research.

Efficiency Metrics in Auctions with Budget Constraints. In their pioneering and comprehensive study of auctions with budget constraints, Dobzinski et al. [10] use *Pareto efficiency* as a measure of mechanism performance. They introduced the adaptive clinching auction mechanism and proved that it is the only mechanism that simultaneously satisfies truthfulness, individual rationality, and Pareto optimality. Subsequent works on auctions with budget constraints have also frequently relied on Pareto efficiency as a measure of efficiency [12,15,17]. However, since Pareto efficiency is a binary criterion, it does not lend itself to the development of approximation algorithms. Another commonly used metric for measuring auction efficiency, *social welfare*, has been shown to achieve no better than a $1/n$-approximation under budget constraints for any truthful and individually rational mechanism [11]. Recognizing these limitations, Dobzinski et al. [11] introduced a more quantifiable welfare metric, *liquid welfare*, which innovatively incorporates both agents' valuations of items and their purchasing power. They further leveraged this metric to design a mechanism that achieves a 1/2-approximation ratio. Building on this work, Lu and Xiao [21] improved the approximation ratio to $\frac{2}{1+\sqrt{5}}$. Many subsequent studies have used liquid welfare as a metric to evaluate the efficiency of mechanisms under budget constraints [3,13,14].

Auctions with Externalities. The study of externalities in economics has a long history. In 1920, Pigou proposed using taxes to correct market failures caused by externalities [24]. In 1960, Coase introduced the famous "Coase Theorem", emphasizing that, in the absence of transaction costs, parties can resolve externality issues through negotiation without relying on government intervention [7]. Externality problems have also been extensively studied in the field of algorithmic mechanism design, particularly in the context of auction mechanisms. In the interesting work by Jehiel et al. [19], they discuss how to design auction mechanisms to maximize the auctioneer's revenue when the allocation of items negatively impacts the utilities of other bidders. This is one of the earliest studies analyzing *allocation externalities*. The survey by Jehiel et al. [18] provides a comprehensive analysis of various novel mechanism design problems arising from the presence of externalities, including allocation externalities and *information externalities*. Krysta et al. [20] investigated externality issues in multi-item combinatorial auctions using a bidding language framework. Agarwal et al. [2] studied the impact of additive externalities in data auctions. In addition to analyzing the impact of externalities on auctions themselves, there is also work that explores how to design the structures of these externality effects. Belloni et al. [6] investigated the novel problem of how monopolistic sellers can maximize their revenue by designing network structures that capture the externality effects among buyers in the market. However, to date, no work has explored the impact of allocation-induced externalities on budgets.

1.3 Paper Organization

In Sect. 2, we introduce a model characterizing the impact of externalities on budgets. In Sect. 3, we present an allocation algorithm maximizing the liquid welfare and prove its optimality. In Sect. 4, we design a truthful and individually rational auction mechanism that achieves a 1/3-approximation for the liquid welfare. Finally, we conclude this paper in Sect. 5.

2 Preliminaries

Let us consider a sealed-bid auction for one divisible item. There are n bidders indexed by $i \in \{1, 2, \ldots, n\}$, and we assume that $n \geq 2$, as with fewer bidders (i.e., $n = 1$), there are no externalities. Each bidder i has two parameters: a private *valuation* v_i for one unit of the item, known only to bidder i; and a public *budget impact factor* $\alpha_i \in (0, +\infty)$[1], which quantifies how bidder i's budget is influenced by the allocations to other bidders. (This will be formalized later.) An *auction mechanism* $\mathcal{M}$ is an algorithm that takes the bidders' reported valuations, denoted by $\mathbf{b} = (b_1, \cdots, b_n)$, and the public budget impact factors $(\alpha_1, \cdots, \alpha_n)$ as input and outputs a tuple $(\mathbf{x}, \mathbf{p})$. To be specific, $\mathbf{x} = (x_1, \cdots, x_n)$ is the allocation of item, where $x_i \in [0, 1]$, denotes the amount of the item allocated to bidder i, satisfying $\sum_i x_i \leq 1$. Payment $\mathbf{p} = (p_1, \cdots, p_n)$, where p_i is the amount bidder i shall pay. We slightly abuse notation by using the functions $x_i(\mathbf{b})$ and $p_i(\mathbf{b})$ to denote the fraction of item allocated to bidder i and the corresponding payment, respectively, by mechanism $\mathcal{M}$ when the input bid profile is $\mathbf{b}$. For the sake of convenience, the bid profile $\mathbf{b}$ and the valuation profile $\mathbf{v}$ can be equivalently denoted as $\mathbf{b} = (b_i, \mathbf{b}_{-i})$ and $\mathbf{v} = (v_i, \mathbf{v}_{-i})$.

Unlike the traditional auction setting with hard budget constraints, we are more interested in a scenario, where bidders' budgets are subject to allocation-induced externalities. Specifically, a bidder's budget is influenced linearly by the allocations received by other bidders. Thus, given the item allocation $\mathbf{x} = (x_1, \ldots, x_n) = (x_i, \mathbf{x}_{-i})$, the *budget* of bidder i is determined as $B_i(\mathbf{x}_{-i}) = \alpha_i \cdot \sum_{j \in [n]-i} x_j$, in which, each bidder i experiences a homogeneous impact from the allocations of other bidders, as captured by the parameter α_i.

We employ a budgeted quasi-linear utility function to describe a bidder's utility. Given a reported bidding profile $\mathbf{b}$, and thus bidder i obtains her item allocation $x_i(\mathbf{b})$ with payment $p_i(\mathbf{b})$ from a mechanism. Then the utility of bidder i is defined as

$$u_i(\mathbf{b}; v_i) = \begin{cases} v_i x_i(\mathbf{b}) - p_i(\mathbf{b}), & p_i(\mathbf{b}) \leq B_i(\mathbf{x}_{-i}(\mathbf{b})) \\ -\infty, & \text{otherwise} \end{cases}.$$

A mechanism is truthful if, for each bidder reporting their true valuations is always a dominant strategy. This essential property in mechanism design ensures that a utility maximizer will always report their true valuation to the mechanism.

Definition 1 (Truthfulness). *An auction mechanism $\mathcal{M}$ is said to be* truthful, *if $u_i(v_i, \mathbf{b}_{-i}; v_i) \geq u_i(b_i, \mathbf{b}_{-i}; v_i)$ for any $i \in [n]$, v_i, b_i and $\mathbf{b}_{-i}$.*

A mechanism is individually rational (IR) if, for each bidder, bidding truthfully results in non-negative utility, regardless of how other bidders bid.

---

[1] As we will explain shortly, if $\alpha_i = 0$, this implies that bidder i's budget is zero. Consequently, we can safely exclude this bidder from consideration.

Definition 2 (Individual Rationality). *An auction mechanism is individually rational if $u_i(v_i, \mathbf{b}_{-i}; v_i) \geq 0$, for any $i \in [n]$, v_i and $\mathbf{b}_{-i}$.*

Myerson's Lemma [23] characterizes the allocation and payment functions of auction mechanisms that satisfy both truthfulness and individual rationality. For completeness, we provide a formal statement of Myerson's Lemma below.

Lemma 1 (Myerson's Lemma [23]). *An auction mechanism is truthful and individually rational if and only if the allocation and payment functions of the mechanism satisfy the following conditions:*

(1) The allocation function $x_i(v_i, \mathbf{v}_{-i})$ is non-decreasing in v_i for all $i \in [n]$ and any $\mathbf{v}_{-i}$.

(2) The payment function is given by $p_i(v_i, \mathbf{v}_{-i}) = v_i x_i(v_i, \mathbf{v}_{-i}) - \int_0^{v_i} x_i(z, \mathbf{v}_{-i}) \, dz$ for all $i \in [n]$.

Inspired by [11], we employ *liquid welfare* as the welfare measure, to evaluate the efficiency of our designed auction mechanism.

Definition 3 (Liquid Welfare). *In the auction with externalities on budgets, the liquid welfare, associated an allocation $\mathbf{x}$, is defined as*

$$LW(\mathbf{x}) = \sum_{i \in [n]} \min\{v_i x_i, B_i(\mathbf{x}_{-i})\}.$$

Optimizing liquid welfare reflects the idea of allocating the item to bidders who value it highly and have sufficient purchasing power.

For any instance $I = \{v_i, \alpha_i\}_{i \in [n]}$, let $\mathsf{OPT}(I)$ denote the optimal liquid welfare achievable by any optimal allocation. We say that a truthful auction mechanism is a ρ-*approximation* with respect to the liquid welfare if, for any instance $I = \{v_i, \alpha_i\}_{i \in [n]}$, the liquid welfare achieved by the mechanism is at least $\rho \cdot \mathsf{OPT}(I)$, where $\rho \in (0, 1]$ is the *approximation ratio*.

Our goal is to design a truthful and individually rational auction mechanism that achieves a good approximation ratio with respect to the liquid welfare.

3 Optimal Allocation for Maximizing Liquid Welfare

This section focuses on the design of an allocation algorithm that takes into account the externalities on budgets in order to maximize liquid welfare, while temporarily disregarding the incentive constraints of truthfulness and individual rationality. We begin with presenting a greedy algorithm (Algorithm 1) and subsequently prove that it achieves the optimal liquid welfare (Theorem 1).

In brief, Algorithm 1 first allocates to each bidder i a fraction of the item equal to $\min\left\{\frac{\alpha_i}{v_i + \alpha_i}, s\right\}$ in descending order of valuations, where s denotes the remaining fraction of the item. After these operations, if there is a fraction of item left, then the residual fraction is assigned to bidder ℓ, whose budget is least affected by externalities, that is $\alpha_\ell = \min_{i \in [n]} \alpha_i$. Note that Algorithm 1

Algorithm 1: Optimal Allocation

Input: Number of bidders n, valuations $\{v_i\}_{i \in [n]}$ and budget impact factors $\{\alpha_i\}_{i \in [n]}$; // Assume $v_1 \geq v_2 \geq \cdots \geq v_n$.
Output: Allocation $\mathbf{x}^* = (x_1^*, x_2^*, \cdots, x_n^*)$;

1 $s \leftarrow 1$; // s represents the remaining quantity of the item.
2 **for** $i = 1..n$ **do**
3   **if** $s \geq \frac{\alpha_i}{v_i + \alpha_i}$ **then**
4        $x_i^* \leftarrow \frac{\alpha_i}{v_i + \alpha_i}$, $s \leftarrow s - x_i^*$;
5   **else if** $s \geq 0$ **then**
6        $x_i^* \leftarrow s$, $s \leftarrow 0$;
7 **if** $s > 0$ **then**
8     $\ell \leftarrow \arg\min_i \alpha_i$, $x_\ell^* \leftarrow x_\ell^* + s$, $s \leftarrow 0$;
9 **return** $(x_1^*, x_2^*, \cdots, x_n^*)$;

distributes the entire item. Specially, if one bidder i receives $\frac{\alpha_i}{v_i + \alpha_i}$ of the item, it implies that her budget, given by $B_i(\mathbf{x}_{-i}) = \alpha_i \cdot \sum_{j \in [n]-i} x_j = \alpha_i \cdot (1 - x_i)$, is exactly equal to the value $v_i x_i$. There are two scenarios for the outcome of Algorithm 1:

- Case 1: $\sum_{i \in [n]} \frac{\alpha_i}{v_i + \alpha_i} \geq 1$. Let $r := \max\left\{r \in [n] \mid \sum_{i=1}^{r} \frac{\alpha_i}{v_i + \alpha_i} \leq 1\right\}$. Then $x_i^* = \frac{\alpha_i}{v_i + \alpha_i}$ for any $i < r$, $x_{r+1}^* = 1 - \sum_{i=1}^{r} x_i^*$ and $x_i^* = 0$ for any $i > r + 1$.
- Case 2: $\sum_{i \in [n]} \frac{\alpha_i}{v_i + \alpha_i} < 1$. Let $\ell := \arg\min_i \alpha_i$. Then $x_i^* = \frac{\alpha_i}{v_i + \alpha_i}$ for any $i \neq \ell$, and $x_\ell^* = 1 - \sum_{i \neq \ell} x_i^*$.

Theorem 1. *The allocation returned by Algorithm 1 is an optimal solution for maximizing the liquid welfare.*

The detailed proof of Theorem 1 can be found in the full version of the paper.

4 A 1/3-Approximation Truthful Mechanism

Algorithm 1 provides an optimal allocation for maximizing liquid welfare from the perspective of algorithm design. However, this allocation rule lacks monotonicity, since the allocation for the bidder with the highest valuation decreases as her valuation increases. This violates truthfulness by Myerson's Lemma (Lemma 1).

Fortunately, we can maintain the monotonicity of the allocation rule by modifying the allocation rule, while only incurring a constant loss in the approximation ratio for liquid welfare. Our mechanism design is inspired by the uniform-price auction in the work of [11]. However, the external impact of allocation on budgets makes our problem setting more complex. To achieve an ideal approximation ratio while ensuring the desirable properties of truthfulness and individual rationality, we need to explore new mechanism design techniques. Our main result is presented in Theorem 2.

Theorem 2. *For the auction with externalities on budgets, there exists a mechanism that ensures truthfulness, individual rationality, and achieves an approximation ratio of 1/3 with respect to liquid welfare.*

4.1 Mechanism Design

We present the auction mechanism (Mechanism 1) and design technique in detail.

Mechanism 1 (Uniform Price Auction with Purchase Limit). *Consider n bidders with valuations $v_1 \geq v_2 \geq \cdots \geq v_n$ and budget impact factors $\{\alpha_1, \cdots, \alpha_n\}$. Mechanism 1 proceeds as follows: First, add a dummy bidder $n+1$ with valuation $v_{n+1} = 0$ and budget impact factor $\alpha_{n+1} > 0$. Next, define*

$$k := \max\left\{\ell \in [n] \mid \sum_{i=1}^{\ell} \min\left\{\frac{\alpha_i}{v_\ell + \alpha_i}, \frac{1}{2}\right\} \leq 1\right\},$$

referred to as the division point, *and let q denote the smallest[2] non-negative root of the equation $\sum_{i=1}^{k} \min\left\{\frac{\alpha_i}{q+\alpha_i}, \frac{1}{2}\right\} = 1$, referred to as the* unifrom price. *The allocation is determined based on the following two cases:*

- *If $q > v_{k+1}$, allocate $x_i = \min\left\{\frac{\alpha_i}{q+\alpha_i}, \frac{1}{2}\right\}$ to each bidder $i \leq k$, and allocate 0 to the remaining bidders.*
- *If $q \leq v_{k+1}$, allocate $x_i = \min\left\{\frac{\alpha_i}{v_{k+1}+\alpha_i}, \frac{1}{2}\right\}$ to each bidder $i \leq k$, allocate $x_{k+1} = 1 - \sum_{j \in [k]} x_j$ to bidder $k+1$, and allocate 0 to the remaining bidders.*

Payment p_i for each bidder i is calculated according to Myerson's Lemma (Lemma 1):

$$p_i = v_i x_i(v_i, \mathbf{v}_{-i}) - \int_0^{v_i} x_i(z, \mathbf{v}_{-i})\,\mathrm{d}z,$$

where $x_i(\mathbf{v})$ denotes the fraction of item allocated to bidder i by Mechanism 1, if the reported information is $\mathbf{v}$.

The dummy bidder $n+1$ added in Mechanism 1 can help us handle the case where $k = n$. As shown in Lemma 2, adding this bidder does not affect the allocation for the "actual" bidders $\{1, 2, \ldots, n\}$, nor does it impact the liquid welfare objective. This is because the item assigned to the dummy bidder is $x_{n+1} = 0$. Consequently, her contribution to liquid welfare is zero, and she also cannot influence the contributions of other bidders through externalities. Due to space limitations, we provide the proof of Lemma 2 in the full version of the paper.

---

[2] There may be multiple roots, and choosing any of them is acceptable, as the choice does not affect the value of $\min\left\{\frac{\alpha_i}{q+\alpha_i}, \frac{1}{2}\right\}$. For simplicity, we select the smallest one.

Lemma 2. *In Mechanism 1, the allocation to dummy bidder is zero, i.e., $x_{n+1} = 0$, and the division point k satisfies $k \leq n$.*

Due to Lemma 2, it is clear that Mechanism 1 allocates the entire item to bidder $\{1, 2, \cdots, n\}$. Therefore, $B_i(\mathbf{x}_{-i}) = \alpha_i \cdot (1 - x_i)$ for all $i \in [n]$. For the case of $q \leq v_{k+1}$, bidder $k+1$ is allocated $x_{k+1} = 1 - \sum_{i \in [k]} \min\left\{\frac{\alpha_i}{v_{k+1}+\alpha_i}, \frac{1}{2}\right\} \geq 1 - \sum_{i \in [k]} \min\left\{\frac{\alpha_i}{q+\alpha_i}, \frac{1}{2}\right\} = 0$. It implies that the allocation for bidder $k+1$ is non-negative. Moreover, by the definition of k, we have $\sum_{i \in [k+1]} \min\left\{\frac{\alpha_i}{v_{k+1}+\alpha_i}, \frac{1}{2}\right\} > 1$. Therefore,

$$0 \leq x_{k+1} < \min\left\{\frac{\alpha_{k+1}}{v_{k+1}+\alpha_{k+1}}, \frac{1}{2}\right\} \quad (1)$$

The Role of the Purchase Limit. Mechanism 1 ensures that each bidder receives no more than $1/2$ of the item, which can be viewed as a form of *purchase limit*. The purchase limit is essential for the mechanism to achieve a good approximation ratio. The high-level intuition is as follows: Consider a bidder with a budget impact factor α and a valuation $v = \alpha^2$, where α is sufficiently large for this bidder to rank first in the valuation order. The uniform price is q. Without the purchase limit, this bidder would receive $x = \frac{\alpha}{q+\alpha}$, and her contribution to liquid welfare would be $\Gamma := \alpha \cdot (1-x) = \frac{q\alpha}{q+\alpha}$. However, in the optimal allocation (Algorithm 1), the bidder receives $x^* = \frac{\alpha}{v+\alpha} = \frac{1}{\alpha+1}$. Her contribution to optimal liquid welfare is $\Gamma^* := \alpha \cdot (1 - x^*) = \frac{\alpha^2}{\alpha+1}$. As $\alpha \to \infty$, $\Gamma \to q$, but $\Gamma^* \to \infty$, leading to an unbounded welfare gap. We will prove later that by imposing a purchase limit of $1/2$, the welfare gap can be bounded by a constant factor (see the proof of Lemma 5).

4.2 Theoretical Guarantees of the Mechanism

To prove that Mechanism 1 satisfies the desired attributes in Theorem 2, we first demonstrate that the allocation rule satisfies the *monotonicity*, which ensures the truthfulness of Mechanism 1 by Myerson's Lemma, and establish the property of *budget feasibility* to ensure that the payments returned by Mechanism 1 are all subject to the budgets. Due to space limits, the proofs of Lemma 3 and Lemma 4 are provided in the full version of the paper.

Lemma 3 (Monotonicity). *The allocation function of Mechanism 1 is non-decreasing in the valuation of each bidder. Specifically, for any bidder j, $x_j(v_j, \mathbf{v}_{-j})$ is non-decreasing in v_j for any $\mathbf{v}_{-j}$.*

Lemma 4 (Budget Feasibility). *Mechanism 1 is budget-feasible, that is $p_i \leq B_i$, for any $i \in [n]$.*

Despite the properties of monotonicity and budget feasibility, we also evaluate the efficiency of Mechanism 1 by proving its constant approximation ratio.

Lemma 5 (1/3-Approximation). *Mechanism 1 achieves an approximation ratio of 1/3 with respect to the liquid welfare.*

Proof. For an instance $I = \{v_i, \alpha_i\}_{i \in [n]}$, let ALG denote the liquid welfare achieved by Mechanism 1, and let OPT denote its optimal liquid welfare (for instance, provided by Algorithm 1). The corresponding allocations are denoted by $(x_1, x_2, \cdots, x_n)$ and $(x_1^*, x_2^*, \cdots, x_n^*)$, respectively. We discuss following two cases.

Case 1: $q > v_{k+1}$. Define two sets $A := \left\{ i \in [k] \mid \frac{\alpha_i}{q+\alpha_i} \leq \frac{1}{2} \right\}$ and $B := [k] - A$. Thus, $q \geq \alpha_i$ for $i \in A$ and $q < \alpha_i$ for $i \in B$. Moreover, $x_i = \frac{\alpha_i}{q+\alpha_i}$ for $i \in A$ and $x_i = 1/2$ for $i \in B$. Therefore we have

$$1 = \sum_{i \in [n]} x_i = \sum_{i \in [k]} \min\left\{ \frac{\alpha_i}{q+\alpha_i}, \frac{1}{2} \right\} = \sum_{i \in A} \frac{\alpha_i}{q+\alpha_i} + \sum_{i \in B} \frac{1}{2}.$$

Simplifying, we get

$$\sum_{i \in A} \frac{\alpha_i}{q+\alpha_i} = 1 - \frac{|B|}{2}. \tag{2}$$

Since $q \leq v_k \leq v_i$, we have $v_i x_i \geq \alpha_i(1-x_i)$ for $i \in A$. Thus,

$$\begin{aligned}
\mathsf{ALG} &= \sum_{i \in [n]} \min\{v_i x_i, \alpha_i(1-x_i)\} = \sum_{i \in A} \alpha_i(1-x_i) + \sum_{i \in B} \min\left\{ \frac{v_i}{2}, \frac{\alpha_i}{2} \right\} \\
&= q \cdot \sum_{i \in A} \frac{\alpha_i}{q+\alpha_i} + \frac{1}{2} \cdot \sum_{i \in B} \min\{v_i, \alpha_i\} \\
&= q \cdot \left(1 - \frac{|B|}{2} \right) + \frac{1}{2} \cdot \sum_{i \in B} \min\{v_i, \alpha_i\} \\
&= q + \frac{1}{2} \cdot \sum_{i \in B} (\min\{v_i, \alpha_i\} - q) \geq q > v_{k+1}.
\end{aligned} \tag{3}$$

The penultimate equality holds because of (2). The penultimate inequality holds because $q < \alpha_i$ for $i \in B$. The last inequality holds because $q > v_{k+1}$, as the condition in Case 1.

For all $i \in A$, since $q \geq \alpha_i$, we have

$$\frac{\left(\frac{v_i \alpha_i}{v_i + \alpha_i} \right)}{\left(\frac{q \alpha_i}{q + \alpha_i} \right)} = \frac{\left(\frac{v_i}{v_i + \alpha_i} \right)}{\left(\frac{q}{q + \alpha_i} \right)} \leq \frac{1}{\left(\frac{q}{q+q} \right)} = 2.$$

Therefore,

$$\sum_{i \in A} \frac{v_i \alpha_i}{v_i + \alpha_i} \leq \sum_{i \in A} \left(2 \cdot \frac{q \alpha_i}{q + \alpha_i} \right) = 2 \cdot q \cdot \left(1 - \frac{|B|}{2} \right). \tag{4}$$

The last equality holds because of (2).

Now we consider OPT:

$$\text{OPT} = \sum_{i \in [n]} \min\{v_i x_i^*, \alpha_i(1-x_i^*)\} \leq \sum_{i \in [k]} \frac{v_i \alpha_i}{v_i + \alpha_i} + \sum_{i=k+1}^{n} (v_i x_i^*)$$

$$\leq \sum_{i \in [k]} \frac{v_i \alpha_i}{v_i + \alpha_i} + v_{k+1} = \sum_{i \in A} \frac{v_i \alpha_i}{v_i + \alpha_i} + \sum_{i \in B} \frac{v_i \alpha_i}{v_i + \alpha_i} + v_{k+1}$$

$$\leq 2 \cdot \left(q \cdot \left(1 - \frac{|B|}{2}\right) + \frac{1}{2} \cdot \sum_{i \in B} \min\{v_i, \alpha_i\} \right) + v_{k+1}$$

$$\leq 2 \cdot \text{ALG} + \text{ALG} = 3 \cdot \text{ALG}.$$

The third inequality holds because of (4) and the fact that $\frac{v_i \alpha_i}{v_i + \alpha_i} \leq \min\{v_i, \alpha_i\}$ for all $i \in [n]$. The last equality holds because of the penultimate equality and the final inequality in (3).

Case 2: $q \leq v_{k+1}$. For simplicity, we reuse the symbols and redefine the sets A and B as follows: $A := \left\{ i \in [k] \mid \frac{\alpha_i}{v_{k+1} + \alpha_i} \leq \frac{1}{2} \right\}$, and $B := [k] - A$. Thus, $v_{k+1} \geq \alpha_i$ for $i \in A$ and $v_{k+1} < \alpha_i$ for $i \in B$. Moreover, $x_i = \frac{\alpha_i}{v_{k+1} + \alpha_i}$ for $i \in A$ and $x_i = 1/2$ for $i \in B$.

According to (1), $x_{k+1} \leq \frac{1}{2}$ and $x_{k+1} \leq \frac{\alpha_{k+1}}{v_{k+1} + \alpha_{k+1}}$. Thus $v_{k+1} x_{k+1} \leq \alpha_{k+1}(1 - x_{k+1})$.

Since $v_{k+1} \leq v_i$, we have $v_i x_i \geq \alpha_i(1 - x_i)$ for $i \in A$. So,

$$\text{ALG} = \sum_{i \in A} \alpha_i(1 - x_i) + \sum_{i \in B} \min\left\{\frac{v_i}{2}, \frac{\alpha_i}{2}\right\} + v_{k+1} x_{k+1}$$

$$= \sum_{i \in A} \alpha_i \cdot \left(1 - \frac{\alpha_i}{v_{k+1} + \alpha_i}\right) + \frac{1}{2} \cdot \sum_{i \in B} \min\{v_i, \alpha_i\}$$

$$+ v_{k+1} \cdot \left(1 - \sum_{i \in A} \frac{\alpha_i}{v_{k+1} + \alpha_i} - \sum_{i \in B} \frac{1}{2}\right) \quad (5)$$

$$= v_{k+1} + \frac{1}{2} \cdot \sum_{i \in B} (\min\{v_i, \alpha_i\} - v_{k+1}) \geq v_{k+1}.$$

The last inequality holds because $v_{k+1} < \alpha_i$ for $i \in B$. Furthermore, from the second equality in (5) and the non-negativity of x_{k+1} as shown in (1), we have

$$\text{ALG} \geq \sum_{i \in A} \alpha_i \cdot \left(1 - \frac{\alpha_i}{v_{k+1} + \alpha_i}\right) + \frac{1}{2} \cdot \sum_{i \in B} \min\{v_i, \alpha_i\}. \quad (6)$$

Similarly to Case 1, for all $i \in A$, since $v_{k+1} \geq \alpha_i$, we have

$$\frac{\left(\frac{v_i \alpha_i}{v_i + \alpha_i}\right)}{\alpha_i \cdot \left(1 - \frac{\alpha_i}{v_{k+1} + \alpha_i}\right)} = \frac{\left(\frac{v_i}{v_i + \alpha_i}\right)}{\left(\frac{v_{k+1}}{v_{k+1} + \alpha_i}\right)} \leq \frac{1}{\left(\frac{v_{k+1}}{v_{k+1} + v_{k+1}}\right)} = 2. \quad (7)$$

Now we consider OPT:

$$\text{OPT} \leq \sum_{i \in [k]} \frac{v_i \alpha_i}{v_i + \alpha_i} + v_{k+1} = \sum_{i \in A} \frac{v_i \alpha_i}{v_i + \alpha_i} + \sum_{i \in B} \frac{v_i \alpha_i}{v_i + \alpha_i} + v_{k+1}$$

$$\leq 2 \cdot \left(\sum_{i \in A} \alpha_i \cdot \left(1 - \frac{\alpha_i}{v_{k+1} + \alpha_i}\right) + \frac{1}{2} \cdot \sum_{i \in B} \min\{v_i, \alpha_i\} \right) + v_{k+1}$$

$$\leq 2 \cdot \text{ALG} + \text{ALG} = 3 \cdot \text{ALG}.$$

The first inequality comes from the same argument as in Case 1. The second inequality holds because of (7) and the fact that $\frac{v_i \alpha_i}{v_i + \alpha_i} \leq \min\{v_i, \alpha_i\}$ for all $i \in [n]$. The last inequality holds because of (5) and (6).

Combining the two cases, we have $\text{ALG} \geq 1/3 \cdot \text{OPT}$. □

Combining Lemmas 2 to 5, along with Myerson's Lemma (Lemma 1), directly proves Theorem 2.

4.3 Upper Bound on the Approximation Ratio

We now show that there is an upper bound on the approximation ratio with respect to liquid welfare that any truthful mechanism can achieve.

Lemma 6 (1/2-Approximation Upper Bound). *It is impossible for any truthful mechanism to achieve an approximation ratio better than 1/2 for liquid welfare.*

Proof. Consider any truthful mechanism that achieves an approximation ratio of ρ for liquid welfare. Suppose there are two bidders with valuations v_1 and v_2, and budget impact factors α_1 and α_2, where $v_2 = \alpha_2 = 1$ and $\alpha_1 > 1$. Let $x_1(v_1, v_2)$ and $x_2(v_1, v_2)$ denote the allocations for bidders 1 and 2, respectively.

Now consider the scenario in which bidder 1's valuation is $v_1 = \alpha_1^2 > v_2$. From Algorithm 1, the optimal allocation assigns $x_1^* = \frac{1}{\alpha_1+1} < \frac{1}{2}$ to bidder 1 and $x_2^* = 1 - \frac{1}{\alpha_1+1} > x_1^*$ to bidder 2. The optimal liquid welfare is

$$\text{OPT} = \min\{\alpha_1^2 x_1^*, \alpha_1 x_2^*\} + \min\{x_2^*, x_1^*\} = \frac{\alpha_1^2 + 1}{\alpha_1 + 1}.$$

Because the mechanism is ρ-approximation, we have

$$\min\{\alpha_1^2 x_1(\alpha_1^2, 1), \alpha_1 x_2(\alpha_1^2, 1)\} + \min\{x_2(\alpha_1^2, 1), x_1(\alpha_1^2, 1)\} \geq \rho \cdot \text{OPT}.$$

Additionally,

$$\min\{\alpha_1^2 x_1(\alpha_1^2, 1), \alpha_1 x_2(\alpha_1^2, 1)\} + \min\{x_2(\alpha_1^2, 1), x_1(\alpha_1^2, 1)\} \leq (\alpha_1 + 1) \cdot x_2(\alpha_1^2, 1).$$

Therefore, we have $x_2(\alpha_1^2, 1) \geq \rho \cdot \frac{\alpha_1^2+1}{(\alpha_1+1)^2}$. As $x_1(\alpha_1^2, 1) + x_2(\alpha_1^2, 1) \leq 1$, we have $x_1(\alpha_1^2, 1) \leq 1 - \rho \cdot \frac{\alpha_1^2+1}{(\alpha_1+1)^2}$.

Consider another scenario in which bidder 1's valuation is $v_1 = \sqrt{\alpha_1} > v_2$. The optimal allocation assigns $x_1'^* = \frac{\sqrt{\alpha_1}}{\sqrt{\alpha_1}+1} > \frac{1}{2}$ to bidder 1 and $x_2'^* = 1 - \frac{\sqrt{\alpha_1}}{\sqrt{\alpha_1}+1} < x_1'^*$ to bidder 2. The optimal liquid welfare is $\text{OPT}' = \min\{\sqrt{\alpha_1}x_1'^*, \alpha_1 x_2'^*\} + \min\{x_2'^*, x_1'^*\} = \frac{\alpha_1+1}{\sqrt{\alpha_1}+1}$. By the same argument, we have

$$(\sqrt{\alpha_1}+1) \cdot x_1(\sqrt{\alpha_1}, 1)$$
$$\geq \min\{\sqrt{\alpha_1}x_1(\sqrt{\alpha_1}, 1), \alpha_1 x_2(\sqrt{\alpha_1}, 1)\} + \min\{x_2(\sqrt{\alpha_1}, 1) + x_1(\sqrt{\alpha_1}, 1)\}$$
$$\geq \rho \cdot \text{OPT}' = \rho \cdot \frac{\alpha_1+1}{\sqrt{\alpha_1}+1}.$$

Therefore, we have $x_1(\sqrt{\alpha_1}, 1) \geq \rho \cdot \frac{\alpha_1+1}{(\sqrt{\alpha_1}+1)^2}$. Since the mechanism is truthful and $\alpha_1^2 > \sqrt{\alpha_1}$, it follows from Myerson's Lemma (Lemma 1) that $x_1(\alpha_1^2, 1) \geq x_1(\sqrt{\alpha_1}, 1)$. Therefore, we have

$$1 - \rho \cdot \frac{\alpha_1^2+1}{(\alpha_1+1)^2} \geq \rho \cdot \frac{\alpha_1+1}{(\sqrt{\alpha_1}+1)^2}.$$

By simple algebra, we have

$$\rho \leq 1 \Big/ \left(\frac{\alpha_1^2+1}{(\alpha_1+1)^2} + \frac{\alpha_1+1}{(\sqrt{\alpha_1}+1)^2} \right).$$

Let $\alpha_1 \to \infty$, we have $\rho \leq 1/2$. □

5 Conclusion

This paper studies a novel problem of the allocation-induced externality on budgets. This issue can be regarded as a practical extension of budget constraints in auctions and also represents a new category of allocation externality problems. For this problem, we first propose an allocation algorithm designed to maximize liquid welfare, a natural objective in budgeted settings, from the perspective of algorithm design. While this algorithm achieves optimal welfare, it lacks monotonicity, which is a challenge for us to obtain a truthful mechanism. Therefore, we are motivated by the mechanism design for the uniform-price auction, and adaptively assign items to ensure the monotonicity of the allocation rule while accounting for budget adjustments. This approach enables us to design a truthful and individually rational auction mechanism that provides a 1/3-approximation for the liquid welfare. Moreover, we derive an upper bound of the approximation ratio by proving that no truthful mechanism can achieve an approximation ratio better than 1/2. Closing this approximation gap is an interesting direction for future work. Additionally, there are many practically significant extensions of allocation externalities on budgets that merit further investigation, such as heterogeneous externality impacts and considering externalities on both valuations and budgets simultaneously. Addressing these problems requires new mechanism design and analytical techniques.

References

1. Abrams, Z.: Revenue maximization when bidders have budgets. In: Proceedings of the Seventeenth Annual ACM-SIAM Symposium on Discrete Algorithm, pp. 1074–1082. Citeseer (2006)
2. Agarwal, A., Dahleh, M., Horel, T., Rui, M.: Towards data auctions with externalities. Games Econ. Behav. **148**, 323–356 (2024)
3. Azar, Y., Feldman, M., Gravin, N., Roytman, A.: Liquid price of anarchy. In: International Symposium on Algorithmic Game Theory, pp. 3–15. Springer (2017)
4. Balseiro, S., Deng, Y., Mao, J., Mirrokni, V., Zuo, S.: Robust auction design in the auto-bidding world. Adv. Neural. Inf. Process. Syst. **34**, 17777–17788 (2021)
5. Balseiro, S., Deng, Y., Mao, J., Mirrokni, V., Zuo, S.: Optimal mechanisms for value maximizers with budget constraints via target clipping. Columbia Business School Research Paper (4081457) (2022)
6. Belloni, A., Deng, C., Pekeč, S.: Mechanism and network design with private negative externalities. Oper. Res. **65**(3), 577–594 (2017)
7. Coase, R.H.: The problem of social cost. J. Law Econ. **56**(4), 837–877 (2013)
8. Daskalakis, C., Devanur, N.R., Weinberg, S.M.: Revenue maximization and ex-post budget constraints. ACM Trans. Econ. Comput. (TEAC) **6**(3–4), 1–19 (2018)
9. Devanur, N.R., Ha, B.Q., Hartline, J.D.: Prior-free auctions for budgeted agents. In: Proceedings of the Fourteenth ACM Conference on Electronic Commerce, pp. 287–304 (2013)
10. Dobzinski, S., Lavi, R., Nisan, N.: Multi-unit auctions with budget limits. Games Econom. Behav. **74**(2), 486–503 (2012)
11. Dobzinski, S., Leme, R.P.: Efficiency guarantees in auctions with budgets. In: International Colloquium on Automata, Languages, and Programming, pp. 392–404. Springer (2014)
12. Fiat, A., Leonardi, S., Saia, J., Sankowski, P.: Single valued combinatorial auctions with budgets. In: Proceedings of the 12th ACM Conference on Electronic Commerce, pp. 223–232 (2011)
13. Fikioris, G., Tardos, É.: Liquid welfare guarantees for no-regret learning in sequential budgeted auctions. In: Proceedings of the 24th ACM Conference on Economics and Computation, pp. 678–698 (2023)
14. Fotakis, D., Lotidis, K., Podimata, C.: A bridge between liquid and social welfare in combinatorial auctions with submodular bidders. In: Proceedings of the AAAI Conference on Artificial Intelligence, vol. 33, pp. 1949–1956 (2019)
15. Goel, G., Mirrokni, V., Paes Leme, R.: Clinching auctions beyond hard budget constraints. In: Proceedings of the fifteenth ACM Conference on Economics and Computation, pp. 167–184 (2014)
16. Golrezaei, N., Lobel, I., Paes Leme, R.: Auction design for ROI-constrained buyers. In: Proceedings of the Web Conference 2021, pp. 3941–3952 (2021)
17. Hafalir, I.E., Ravi, R., Sayedi, A.: A near pareto optimal auction with budget constraints. Games Econom. Behav. **74**(2), 699–708 (2012)
18. Jehiel, P., Moldovanu, B.: Allocative and informational externalities in auctions and related mechanisms (2005)
19. Jehiel, P., Moldovanu, B., Stacchetti, E.: How (not) to sell nuclear weapons. the Am. Econ. Rev. **86**, 814–829 (1996)
20. Krysta, P., Michalak, T., Sandholm, T., Wooldridge, M.: Combinatorial auctions with externalities: basic properties and bidding languages. In: Proceedings of the 9th International Conference on Autonomous Agents and Multiagent Systems (2010)

21. Lu, P., Xiao, T.: Improved efficiency guarantees in auctions with budgets. In: Proceedings of the Sixteenth ACM Conference on Economics and Computation, pp. 397–413 (2015)
22. Lu, P., Xiao, T.: Liquid welfare maximization in auctions with multiple items. In: Bilò, V., Flammini, M. (eds.) SAGT 2017. LNCS, vol. 10504, pp. 41–52. Springer, Cham (2017). https://doi.org/10.1007/978-3-319-66700-3_4
23. Myerson, R.B.: Optimal auction design. Math. Oper. Res. **6**(1), 58–73 (1981)
24. Pigou, A.: The Economics of Welfare. Routledge (2017)
25. Richter, M.: Mechanism design with budget constraints and a population of agents. Games Econom. Behav. **115**, 30–47 (2019)

Author Index

A
Abu-Khzam, Faisal N. 1
Adamson, Duncan 16

B
Bai, Rufan 238
Berg, Magnus 33, 49
Bian, Yide 356
Boyar, Joan 49
Buld, Felix 64

C
Chen, Hongyin 222
Chen, Jing 207
Chen, Xujin 78
Chen, Zhou 252
Cheng, Yukun 400
Cohen, Johanne 279
Cordeiro, Daniel 279

D
Dardilhac, Valentin 279
Deng, Xiaotie 222, 356, 400
Deng, Xiyuan 78
Deng, Zhengyan 264
Dong, Xiaoqi 222
Duan, Zhijian 356

F
Favrholdt, Lene M. 49
Feng, Qilong 118
Fleischmann, Pamela 16
Furquim, Felipe V. 279

G
Gao, Yitian 293
Gong, Zifan 307
Guo, Xinxiang 321

H
Hu, Xiaodong 78
Huch, Annika 16

I
Isenmann, Lucas 1

J
Jansson, Jesper 92
Jena, Sangram K. 103

K
Koß, Tore 16
Kowaluk, Mirosław 92

L
Lam, Alexander 307
Larsen, Kim S. 49
Li, Bo 335
Li, Jiaxue 293
Li, Jichen 222
Li, Tiantian 181
Li, Weidong 142, 193
Lingas, Andrzej 92
Liu, Hongtao 343
Liu, Jingyi 118
Liu, Shuilian 128
Liu, Xiaofei 142
Liu, Zhonghao 142
Luo, Junjie 293
Luo, Kelin 152

M
Ma, Yunxuan 356
Manea, Florin 16
Miao, Huahua 238
Mrkaic, Momcilo 307
Mu, Yifen 321, 385

N
Navarra, Alfredo 166

P
Persson, Mia 92
Piselli, Francesco 166

Q
Qi, Qi 252
Qin, Shaowen 264
Qin, Ying 372

R
Ren, Zeyu 372

S
Schulz, Andreas S. 64
Shen, Weiran 343
Shen, Yiheng 343
Sheng, Chenliang 264
Shi, Feng 118
Sifat, Md Habibur Rahman 335
Subramani, K. 103
Sun, Ankang 335
Sun, Hao 252

W
Wang, Changjun 78
Wang, Feng 356
Wang, Jianxin 118
Wang, Lusheng 181
Wang, Zihe 372
Wu, Xiaowei 238

Wu, Zhonghai 222

X
Xiao, Bin 222
Xiao, Hanyin 193
Xiao, Man 142
Xin, Wu 222
Xu, Chenyang 400
Xu, Hao 356
Xu, Yicheng 128

Y
Yan, Yachao 307
Yang, Chenran 152
Yang, Weitao 356
Yang, Zonghan 152

Z
Zhang, Cong 238
Zhang, Jiaming 193
Zhang, Jie 372
Zhang, Junyue 385
Zhang, Yiheng 293
Zhang, Yong 128
Zhang, Yuhao 152
Zhang, Zhikang 193
Zhao, Jingshu 356
Zhao, Muyang 252
Zhao, Yingchao 307
Zheng, Yusen 264, 400
Zhou, Shengwei 238
Zhou, Wentao 207
Zhu, Daming 181

Made in the USA
Monee, IL
03 May 2026

49438524R00240